Manfred Brack

Agents Transmissible from Simians to Man

With 78 Figures and 41 Tables

Springer-Verlag Berlin Heidelberg New York
London Paris Tokyo

Manfred Brack, M. V.
German Primate Center
Department of Pathology and Veterinary Medicine
Kellnerweg 4, D-3400 Göttingen
Federal Republic of Germany

ISBN-13:978-3-642-71913-4 e-ISBN-13:978-3-642-71911-0
DOI: 10.1007/978-3-642-71911-0

Library of Congress Cataloging in Publication Data.
Brack, Manfred. Agents transmissible from Simians to man. Bibliography: p. Includes index. 1. Primates as carriers of disease. I. Title. RA641.P7B73 1987 614.4'3 87-12743
ISBN-13:978-3-642-71913-4 (U.S.)

Softcover reprint of the hardcover 1st edition 1987

2127/3145-543210

Preface

The contacts between man and nonhuman primates enable the transmission of microorganisms from one species to the other. Such contact may occur at quite different levels: man and nonhuman primates may share the same ecosystem including the presence of vectors in the countries of origins of monkeys and apes; the animals are captured to be sold or used for food; field researchers have to stay near the animals in the wild; an uncontrolled human population gets close enough to almost touch the animals in zoological gardens around the world; pet owners establish bodily contact and finally researchers doing surgery or necropsies are exposed to an increased number of pathogens liberated from the organs and body fluids.

Usually monkeys and apes are more threatened with catching the microorganisms indigenous to man than *vice versa,* but nevertheless outbreaks of true zoonoses with nonhuman primates as the source of infection have occurred. Also the retransmission of originally human pathogens via nonhuman primates to man may pose a considerable risk to human health.

Unfortunately the information on the different agents transmissible between man and his relatives is too disseminated for practical use, as it involves quite different scientific disciplines such as virology, bacteriology, parasitology, primatology, laboratory animal science etc. It seemed therefore necessary to compile the current knowledge concerning this topic in a single publication.

Human infections of simian origin may be caused by several viruses, bacteria, fungi or endoparasites. Ectoparasites, in comparison, are of little importance.

Göttingen, Spring 1987 MANFRED BRACK

Table of Contents

1 Viruses

1.1 Herpesviridae

Nonhuman primates are primary hosts to quite a number of herpesviruses (Kalter et al., 1980). Some nonhuman primate species are also highly susceptible to herpesviruses of human origin with at least the theoretical possibility of reinfection occurring in susceptible people.

Only one of the simian herpesviruses is known to induce a true zoonotic infection:

1.1.1 Herpesvirus simiae (B Virus)

Virus

H. simiae is a α-herpesvirus related to *H. hominis.* Its mature virions vary in size from 100 to 180 nm in diameter (Ressig and Melnick, 1954). Like other herpesviruses it is relatively temperature-labile, being inactivated at 36 °C within 9 days, or by heating to 56° within 30 min (Hull and Nash, 1960; Fleckenstein, 1979). It is also inactivated by extraction with ether and other liquid solvents (Fleckenstein, 1979).

H. simiae shows sequentially decreasing cross reactions with *H. hominis,* SA 8, and *H. suis* in neutralization, immunofluorescence-, Ouchterlony-, plaque-reduction and CF tests as well as in enzyme immunoassays (Brack, 1977; Eichberg et al., 1980; Hutt et al., 1981). Due to the dominance of *H. simiae,* sera of monkey origin usually react equally well with both *H. simiae* and *H. hominis* antigens, whereas sera of *H. hominis*-infected human beings usually neutralize *H. simiae* only at titers significantly below those observed with the homologous *H. hominis,* (Sabin, 1934; Burnet et al., 1939; Melnick and Banker, 1954; Hull et al., 1962; Schneweis, 1962; Cabasso et al., 1967; Watson et al., 1967; Kalter and Heberling, 1971 a, b; Martin et al., 1972).

H. simiae is a naturally occurring virus of Asian macaques, and particularly of the rhesus monkey *(Macaca mulatta).* It has been less commonly observed in cynomolgus monkeys *(M. fascicularis)* (Hull and Nash, 1960; Hartley, 1964; Gralla et al., 1966) and bonnet monkeys *(M. radiata)* (España, 1973). *H. simiae*-like viruses have been repeatedly isolated from African primate species such as patas monkeys *(Erythrocebus patas),* guerezzas *(Colobus guerezza)* (Loomis et al., 1981), and African green monkeys *(Cercopithecus aethiops)* (Sumner-Smith, 1966). Their true nature still needs to be defined.

Epidemiology

H. simiae, like many other herpesviruses, usually causes only transient, minor discomfort in its homologous host species (Macaques), thus it is easily overlooked. Rhesus monkeys usually are already infected in their natural habitat (Shah and Southwick, 1965; Pryor et al., 1970; Boulter and Grant, 1977). In one study in India approximately 73% of all adults expressed anti-herpesvirus antibodies (Orcutt et al., 1976). Transmission by bites or infected saliva has been regarded as the principal route of transmission between animals (Hackett, 1977), but recent investigations suggest that venereal transmission is the most important route (Tribe, 1982; Zwartouw and Boulter, 1984), although the virus does not seem to cross the placental barrier (Hackett, 1977). Consequently, neonates are free of B virus, and therefore any antibodies present up to 3 months of age are of maternal origin (Zwartouw et al., 1984). Spontaneous aerogenous infections are of a lesser importance, but do occur (Valerio, 1971; España, 1973) being similar to the experimental aerogenous infection of rabbits (Chapell, 1960; Benda and Polomik, 1969; Benda et al., 1969). Immature animals may act as a host to the virus without producing antibodies until they reach adulthood, the social and physiological stresses of which then trigger both virus shedding and antibody formation (Tribe, 1980). The virus has been isolated from healthy, seropositive rhesus monkeys soon after arrival (Boulter, 1975; Boulter and Grant, 1977) and up to 6 months after primary infection (Vizoso, 1975c). Transportation stress can reactivate the latent infection with increased virus shedding after arrival at the animal's destination (Hull and Nash, 1960; McCarthy, 1969; Tauraso, 1973). In monkey colonies kept in gang-cages or in semi-free conditions *H. simiae* infections may persist in the latent stage for generations (Zwartouw et al., 1984). In contrast colony-born monkeys raised in single cages usually are devoid of *H. simiae* in later generations (DiGiacomo and Shah, 1972).

Human infections usually result from either monkey bites or scratches; or from aerogenous infections, or from laboratory infections via broken glassware (Hummeler et al., 1959; Hull, 1973).

Clinical Symptoms

a) Nonhuman primates: The infection in monkeys usually runs a silent course, ie., without any recognizable signs in most cases. Occasionally transient lingual, oral or labial vesicles or ulcers develop, especially following transportation of the animals (Keeble et al., 1958; Keeble, 1960; Hartley, 1964; 1966; 1968; Cole et al., 1968; Hunt and Melendez, 1969; Hull, 1969; 1973). The lesions in the skin and mucous membranes of monkeys usually heal within 8–14 days without scarformation (but with virus excretion lasting for a longer period). In some of the very rare cases of fatal disease in monkeys a generalized necrosis of the skin, mucous membranes and the brain occurred (McClure et al., 1973; Daniel et al., 1975). Respiratory infections were also seen, accompanied by respiratory distress (España, 1973).

b) Man: 3 to 4 days following an infected monkey bite (scratches etc.), the bite wound becomes hyperemic, painful and vesiculated. This is followed by regional lymphadenopathy and a relatively minor increase in body temperature. Subsequently neurological symptoms develop, e.g. muscular pain, vomitus, intestinal spasm, nuchal stiffness, difficulty in swallowing, photophobia, hyperesthesia, paresthesia, loss of reflexes, urinary retention, ascending paralysis and finally death from respiratory paralysis (Sabin and Wright, 1934; Sabin, 1949; Willcox, 1958; Hummeler et al., 1959; Love and Jungherr, 1962; Mattingly, 1966; Sumner-Smith, 1966; Fierer et al., 1973; Hull, 1973). In aerogenous infections cough, sore throat and respiratory disease prevail (Hull, 1973). Out of 20 to 25 persons with clinical *H. simiae*-disease only five have survived (Wolf, 1974; Tribe, 1980); three of which with severe, permanent neurologic sequelae (Davidson and Hummeler, 1960), and two with disabling, but less severe neurologic disorders (Roth and Purcell, 1977; Breen et al., 1958; Bryan et al., 1975). In the fatal cases death usually occurred between the fifth day and four weeks after onset of the clinical disease.

Pathology

a) Nonhuman primates: Early lesions of lingual and oral stratified epithelia consisted in acanthosis, ballooning degeneration of the prickle cells and Malpighian cells, accompanied by giant cell - and intranuclear inclusion body formation. The resulting epithelial crevices are filled by serous or leukocytic exudates. The ensuing necrosis and ulceration may extend into the underlaying muscular tissues. Occasionally similar lesions can be observed within the esophageal tissues or the gastric pars cardiaca (Cole et al., 1968; Keeble, 1960; Valerio, 1971). In the generalized infections seen in fatal cases, necrosis of the upper digestive tract was prominent (Daniel et al., 1975) and which, in one animal, was combined with focal necrosis of the basal ganglia and surrounding white matter (Daniel et al., 1975).

In respiratory infections focal haemorrhagic pneumonia has been observed (España, 1973).

Occasionally focal glial nodule encephalitis (occurring predominantly in the *medulla oblongata* and *pons*) has been described (Melnick and Banker, 1954; Keeble et al., 1958; Keeble, 1960; Gralla et al., 1966).

b) Man: The main pathological lesions are those of a transverse necrotizing myelitis, ganglioneuritis of spinal ganglia and a necrotizing haemorrhagic encephalitis, which less commonly affects the pons, hippocampus, basal ganglia and cortical areas. [In contrast, the lesions of human *H. hominis* encephalitis are usually asymmetrical, being distributed predominantly throughout the frontobasal and temporomedial cerebral cortex (Spaar, 1976)]. Glial nodules are commonly observed, whereas inclusion bodies are rarely seen in either neurons or glial cells (Sabin and Wright, 1934; Sabin, 1949; Pierce et al., 1958; Thomas and Henschel, 1960; Davidson and Hummeler, 1960; Krücke, 1961; Hull, 1973). In the respiratory form, a haemorrhagic pneumonia with focal necrosis has been described (Hull, 1973).

Virus Isolation
H. simiae can be easily multiplied in rhesus monkey kidney cells (Krech and Lewis, 1954), VERO-, primary marmoset kidney, OMK, OMC, BS-C-1, LLCMK4-cells (Daniel et al., 1975), LLCMK2-cells (Hull et al., 1972), Hela-cells (Hummeler et al., 1959), HEK-, HEL-, WHE-, and primary rabbit kidney cells (Daniels et al., 1975). The typical herpesvirus-type CPE occurs 1 to 3 days after inoculation. Likewise, *H. simiae* can be cultivated on the CAM of chicken embryos (Krech and Lewis, 1954; Siegert, 1961).

Plaque formation can be observed in agar-overlayed VERO-cells within 21 days, progressing to round areas of 1.5 to 3 mm diameter at 5 days after inoculation (Daniel et al., 1975).

If the material has to be stored before inoculation onto cell cultures, storage in glycerin is recommended before freezing (Thomas and Henschel, 1960).

Experimental Animals
The experimental animals of choice are rabbits, which succumb to the typical local lesions, myeloencephalitis and finally death within 7 to 21 days after s.c. or i.m. inoculation; or within 3 to 8 days after intracerebral or corneal infection (Sabin and Hurst, 1935; Siegert, 1961).

Treatment
Animals with obvious clinical lesions or otherwise demonstrated virus shedding should be treated only for experimental purposes, otherwise they should be destroyed.

In man the prevention of monkey bites etc. is the most important part of any protection. Animal handlers and laboratory workers should always wear appropriate gloves, and aerosol formation should be avoided. *All* bite wounds from monkeys are *always* to be desinfected with an iodine solution. The animal in question should be observed for at least 2 weeks (Baker et al., 1982). The applications of γ-globulins or of formalin vaccines are of doubtful value. Topical treatment or i.v. injection of Acyclovir (Acycloguanisine-Burroughs Welcome Co.) has been recently recommended (Boulter et al., 1980; Juel-Jensen, 1982; Balfour, 1984). For i.v. treatment, dosages of between 10 mg/kg for 8 hrs over a 14 day period (Baker et al., 1982) and 200 mg/kg/day for a 10 day period have been recommended.

1.1.2 H. hominis

Spontaneous *H. hominis* infections have been documented in chimpanzees - *Pan troglodytes*; bonobos - *Pan paniscus* (McClure et al., 1980); gorillas - *Gorilla gorilla gorilla* (Heldstab et al., 1981); gibbons - *Hylobates lar* (Smith et al., 1969; Emmons and Lennette, 1970; Ramsey et al., 1982); owl monkeys - *Aotus trivirgatus* (Melendez et al., 1969; McClure and Keeling, 1971; McClure, 1974; Wolf, 1974); tamarins - *Saguinus spp.* (Hull, 1973) and tree shrews - *Tupaia glis* (McClure et al., 1972; McClure, 1974). So far no retransmission to man has occurred, but the possibility has to be kept in mind.

Epidemiology
The route of transmission in spontaneous infections of nonhuman primates is unknown. Contact with persons suffering from either acute or recurrent oral or genital herpes seems probable.

Clinical Disease (Nonhuman Primates Only)
In gorillas, gibbons and owl monkeys, localized cutaneous or oral vesicles or ulcerations prevail, whereas in tree shrews and sometimes in owl monkeys a generalized disease occurs. Also marmosets have been shown to be experimentally susceptible to *H. hominis*. They develop a systemic disease analogous to the disease in owl monkeys (Brack et al., 1972; Felsburg et al., 1973; Hunt et al., 1978). Type 2-infection of chimpanzees and bonobos were characterized by pustulovesicular lesions, which were noticed on the midshaft of the penis or inner labia, the gingiva and the tongue (McClure et al., 1980). *H. hominis* infection of nonhuman primates commonly leads to conjunctivitis. In gibbons, owl monkeys and tree shrews, fatal encephalitis is a frequent cause of death.

Pathology (Nonhuman Primates Only)
The cutaneous, oral and genital vesicles and ulcers are due to acantholysis, paraceratosis, coagulation necrosis and polykarocytosis of the squamous epithelia. Intranuclear inclusion bodies are present particularly in the periphery of the vesicles and ulcers (Melendez et al., 1969; McClure and Keeling, 1971; Hunt and Melendez, 1972). The generalized disease of tree shrews and owl monkeys is accompanied by focal necrosis and haemorrhages in all organs, usually with intranuclear inclusion body formation.

Meningoencephalitis in nonhuman primates due to *H. hominis* is characterized by lymphocytic perivascular cuffing, CNS-necrosis, and focal gliosis located in the cerebral cortex, perivascular white matter and pontine areas (Smith et al., 1969; Ramsey et al., 1982).

1.1.3 Varicella-Zoster Virus

On several occasions a chicken pox-like disease has been observed in chimpanzees, gorillas, and orang utans with a varicella-zoster virus of presumably human origin either being isolated or that its specific antibodies could be demonstrated (Heuschele, 1960; Klein et al., 1971; McClure and Keeling, 1971; White et al., 1972; Marennikova et al., 1973/74). The question remains, whether or not these were caused either by a true human virus or by a simian subtype, especially since at on at least one occasion chimpanzees suffering from the disease did not develop anti-varicella-zoster antibodies (Gold, 1966). So far, no retransmission to human beings has occurred, but this is theoretically possible, if the virus in question is one of human origin.

Clinical Disease (Nonhuman Primates Only)
Affected apes suffer apparently from pruritus, apathy, benign generalized pustular-vesicular rash (with the exception of palms and soles!), conjunctivitis and

slightly elevated temperature. Recovery occurs within 14 days (Klein et al., 1971; Marennikova et al., 1973/74).

References

1. Baker, C.C.; H.P. Lambert, H.P., and M.H. Wansbrough-Jones: Postexposure immunoprophylaxis against B-virus infection. Brit. Med. J. *285* (6351), 1350-1351 (1982)
2. Balfour, H.H. jr.: Acyclovir and other chemotherapy for herpes group viral infections. Ann. Rev. Med. *35*, 279-291 (1984)
3. Benda, R.; M.Fuchsova and V.Hronovsky: Course of air-borne infection caused by B-virus (Herpesvirus simiae). II. Distribution of B-virus in the organism of rabbits and histopathological changes in the organs. J.Hyg. *13*, 214-226 (1969)
4. Benda, R. and F.Polomik: Course of air-borne infections caused by B-virus (Herpesvirus simiae). I. Lethal inhalation dose of B-virus for rabbits. J.Hyg. *13*, 24-30 (1969a)
5. Boulter, E.A.: The isolation of monkey B-virus *(Herpesvirus simiae)* from the trigeminal ganglia of a healthy seropositive rhesus monkey. J.biol. Standard. *3*, 279-280 (1975)
6. Boulter, E.A. and D.P. Grant: Latent infection of monkeys with B virus and prophylactic studies in a rabbit model of this disease. J.antimicrob. Chemother. *3*, 107-113 (1977)
7. Boulter, E.A.; B.Thornton; D.J. Bauer and A.Bye: Successful treatment of experimental B virus (Herpesvirus simiae) infection with acyclovir. Brit. Med. J. *280*, 681-688 (1980)
8. Brack, M.: Morphological and Epidemiological Aspects of Simian Herpesvirus Infections; Schriftenr. Versuchstierkd. Bd.5, Paul Parey, Berlin (1977)
9. Brack, M.; P.J. Felsburg; R.L. Heberling and S.S. Kalter: Histopathological observations in marmosets *(Saguinus oedipus* and *Saguinus fuscicollis)* after *Herpesvirus hominis* vaginal infection. Bact. Proc.1972, V 286 (1972)
10. Breen, G.E.; S.F. Lamb and A.T. Otaki: Monkey bite encephalitis, report of a case with recovery. Brit. Med. J. *2*, 22-23 (1958)
11. Bryan, B.L.; C.D. España; R.W. Emmons; N.Vijayan and P.D. Hoeprich: Recovery from encephalomyelitis caused by Herpesvirus simiae. Report of a case. Arch. intern. Med. *135*, 868-870 (1975)
12. Burnet, F.M.; D.Lush and A.V. Jackson: The relationship of herpes and B viruses: immunological and epidemiological consideration. Austral. J.Exp. Biol. Med. Sci. *17*, 41-51 (1939)
13. Cabasso, V.J.; W.A. Chappell; J.E. Avampato and J.L. Bittle: Correlation of B-virus and herpes simplex virus antibodies in human sera. J.Lab. clin. Med. *70*, 170-178 (1967)
14. Chapell, W.A.: Animal infectivity of aerosols of monkey B virus. Ann. N.Y. Acad. Sci. *85*, 931-934 (1960)
15. Cole, W.C.; R.E. Bostrom and R.A. Whitney: Diagnosis and handling of B-virus in a rhesus monkey (Macaca mulatta). J.Amer. vet. med. Ass. *153*, 894-898 (1968)
16. Daniel, M.D.; F.G. Garcia; L.V. Melendez; R.D. Hunt; J.O'Connor and D.Silva: Multiple *Herpesvirus simiae* isolation from a rhesus monkey which died of cerebral infarction. Lab. Anim. Sci. *25*, 303-308 (1975)
17. Davidson, W.L. and K.Hummeler: B virus infection in man. Ann. N.Y. Acad. Sci. *85*, 970-979 (1960)
18. DiGiacomo, R.F. and K.V. Shah: Virtual absence of infection with Herpesvirus simiae in colony-reared rhesus monkeys *(Macaca mulatta)*, with a literature review of antibody prevalence in natural and laboratory rhesus populations. Lab. Anim. Sci. *22*, 61-67 (1972)
19. Eichberg, J.W.; R.L. Heberling; J.E. Guajardo and S.S. Kalter: Detection of primate herpesvirus antibodies including Herpesvirus simiae by enzyme immunoassay. Develop. biol. Standard. *45*, 61-66 (1980)
20. Emmons, R.W. and E.H. Lennette: Natural Herpesvirus hominis infection of a gibbon (Hylobates lar). Arch. ges. Virusforsch. *31*, 215-218 (1970)
21. España, C.: *Herpesvirus simiae* infection in *Macaca radiata*. Amer. J.phys. Anthrop. *38*, 447-454 (1973)
22. Felsburg, P.J.; R.L. Heberling; M.Brack, and S.S. Kalter: Experimental genital infection of the marmoset. J.med. Primatol. *2*, 50-60 (1973)

23. Fierer, J.; P. Bazeley and A.I. Braude: Herpesvirus B virus encephalomyelitis presenting as opthalmic zoster. Ann. intern. Med. *79*, 225-228 (1973)
24. Fleckenstein, B.: Oncogenic herpesviruses of non-human primates. Biochim. biophys. Acta *560*, 301-342 (1979)
25. Gold, E.: Serologic and virus-isolation studies of patients with varicella or herpes-zoster infection. New Engl. J. Med. *274*, 181-185 (1966)
26. Gralla, E.J.; S.J. Ciecura and C.S. Delahunt: Extended B-virus antibody determination in a closed monkey colony. Lab. Anim. Care *16*, 510-514 (1966)
27. Hackett, R.E.: The B-virus bogy. Primate Supply *2* (2), 26-27 (1977)
28. Hartley, E.G.: Naturally-occurring B-virus infection in cynomolgus monkeys. Vet. Rec. *76*, 555-557 (1964)
29. Heldstab, A.; D. Ruedi; W. Sonnabend and F. Deinhardt: Spontaneous generalized *Herpesvirus hominis* infection of a lowland gorilla *(Gorilla gorilla gorilla)*. J. med. Primatol. *10*, 129-135 (1981)
30. Heuschele, W.P.: Varicella (chicken pox) in three young anthropoid apes. J. Amer. vet. med. Ass. *136*, 256-257 (1960)
31. Hull, R.N.: B virus vaccine. Lab. Anim. Sci. *21*, 1068-1071 (1971)
32. Hull, R.N.: The Simian herpesviruses , pp. 389-426 in "The Herpesviruses" (A.S. Kaplan, ed.). Academic Press, New York (1973)
33. Hull, R.N.; A. Dwyer; A.W. Holmes; E. Nowakowski, F. Deinhardt; E.H. Lennette and R.W. Emmons: Recovery and characterization of a new simian herpes virus from a fatally infected spider monkey. J. nat. Cancer Inst. *49*, 225-231 (1972)
34. Hull, R.N. and J.C. Nash: Immunization against B virus infection. 1. Preparation of an experimental vaccine. Amer. J. Hyg. *71*, 15-28 (1960)
35. Hull, R.N.; F.B. Peck jr.; T.G. Ward and J.C. Nash: Immunization against B virus infection. II. Further laboratory and clinical studies with an experimental vaccine. Amer. J. Hyg. *76*, 239-251 (1962)
36. Hummeler, K.; W.L. Davidson; W. Henle; A.C. LaBocetta and H.G. Ruch: Encephalomyelitis due to infection with herpesvirus simiae (herpes B virus): A report of two fatal laboratory acquired cases. New Engl. J. Med. *261*, 64-68 (1959)
37. Hunt, R.D.; M.P. Anderson and L.V. Chalifoux: Spontaneous infectious diseases of marmosets. Primates Med. *10*, 239-253 (1978)
38. Hunt, R.D. and L.V. Melendez: Herpes virus infections in nonhuman primates: a review. Lab. Anim. Care *19*, 221-234 (1969)
39. Hunt, R.D. and L.V. Melendez: Clinical, epidemiologic, and pathologic features of cytocidal and oncogenic herpesviruses in South American monkeys. J. Nat. Cancer Inst. *49*, 261-271 (1972)
40. Hutt, R.; J.E. Guajardo and S.S. Kalter: Detection of antibodies to *Herpesvirus simiae* and *Herpesvirus hominis* in nonhuman primates. Lab. Anim. Sci. *31*, 184-189 (1981)
41. Juel-Jensen, B.E.: Postexposure immunoprophylaxis against B virus infection. Brit. med. J. *284*, 113-114 (1982)
42. Kalter, S.S. and R.L. Heberling: Serum neutralizing antibody titers for Herpesvirus in nonhuman primates. Lab. Primate Newsl. *10*, 25 and 28 (1971a)
43. Kalter, S.S. and R.L. Heberling: Comparative virology of primates. Bact. Rev. *35*, 310-364 (1971b)
44. Kalter, S.S.; D. Ablashi; C. España; R.L. Heberling; R.N. Hull; E.H. Lennette; H.H. Malherbe; S. McConnell and D.S. Yohn: Simian virus nomenclature. Intervirology *13*, 317-330 (1980)
45. Keeble, S.A.: B-virus infection in monkeys. Ann. N.Y. Acad. Sci. *85*, 960-969 (1960)
46. Keeble, S.A.; G.J. Christofinis and W. Wood: Natural virus B infection in rhesus monkeys. J. Path. Bact. *76*, 189-199 (1958)
47. Klein, M.; C. Milhaud and R. Nicolle: Note on an isolated case of viral eruption observed in a colony of chimpanzees. Lab. Primate Newsl. *10*, 20-23 (1971)
48. Krech, U. and J.L. Lewis: Propagation of B-virus in tissue cultures. Proc. Soc. exp. Biol. *87*, 174-178 (1954)
49. Krücke, W.: Über Virus-Enzephalitiden mit Kerneinschlußkörperchen beim Menschen und

die Neuropathologie der experimentellen B-Virus-Infektion. Wiener Z. Nervenheilk. *18*, 127-158 (1961)
50. Loomis, M.R.; T.O'Neill; M.Bush and R.J. Montali: Fatal Herpesvirus infection in patas monkeys and a black and white colobus monkey. J.Amer. vet. med. Ass. *179*, 1236-1239 (1981)
51. Love, F.M. and E.Jungherr: Occupational infection with virus-B of monkeys. J.Amer. med. Ass. *179*, 804-806 (1962)
52. Marennikova, S.S.; N.Maltseva; E.M. Shelukhina; L.S. Shenkman and V.I. Korneeva: A generalized herpetic infection simulating smallpox in a gorilla. Intervirology *2*, 280-286 (1973/74)
53. Martin, M.L.; E.L. Palmer and R.E. Kissling: Complement-fixing antigens of herpes simplex virus types 1 and 2: reactivity of capsid, envelope, and soluble antigens. Infect. Immun. *5*, 248-254 (1972)
54. Mattingly, S.F.: Major zoonoses of primates. J.Amer. vet. med. Ass. *149*, 1677-1680 (1966)
55. McCarthy, K.: Herpesvirus diseases. Lab. Anim. Handb. *4*, 121-124 (1969)
56. McClure, H.M.: Spontaneous *Herpesvirus hominis* infection in tree shrews *(Tupaia glis)*. Primate Zoon. Surv. Rep. *4*, 11 (1971)
57. McClure, H.M.: Observations in a necropsy series of 1512 nonhuman primates. Amer. Ass. Zoo. vet. Ann. Proc.1974, 143-149 (1974)
58. McClure, H.M. and M.E. Keeling: Viral diseases noted in the Yerkes Primate Center colony. Lab. Anim. Sci. *21*, 1002-1010 (1971)
59. McClure, H.M.; B.Olberding and L.M. Strozier: Disseminated Herpesvirus infection in a rhesus monkey *(Macaca mulatta)*. J.Med. Primatol. *2*, 190-194 (1973)
60. McClure, H.M.; R.B. Swenson; S.S. Kalter and T.L. Lester: Natural genital *Herpesvirus hominis* infection in chimpanzees *(Pan troglodytes* and *Pan paniscus)*. Lab. Anim. Sci. *30*, 895-901 (1980)
61. McLeod, D.R.E.; F.T. Shimada and M.J. Walcroft: Experimental immunization against B virus. Ann. N.Y. Acad. Sci. *85*, 980-989 (1960)
62. Melendez, L.V.; C.España; R.D. Hunt; M.D. Daniel and F.G. Garcia: Natural *herpes simplex* infection in owl monkey *(Aotes trivirgatus)*. Lab. Anim. Care *19*, 38-45 (1969)
63. Melnick, J.L. and D.D. Banker: Isolation of B virus (herpes group) from the central nervous system of a rhesus monkey. J.exp. Med. *100*, 181-194 (1954)
64. Orcutt, R.P.; G.J. Pucak; H.L. Foster; J.T. Kilcourse and T.Ferrell: Multiple testing for the detection of B virus antibody in specially handled rhesus monkeys after capture from virgin trapping grounds. Lab. Anim. Sci. *26*, 70-74 (1976)
65. Pierce, E.C.; J.D. Peirce and R.N. Hull: B-virus: its current significance. Description and diagnosis of a fatal human infection. Amer. J.Hyg. *68*, 242-250 (1958)
66. Pryor, W.H. jr.; H.S. Chiang; G.L. Raulston and L.V. Melendez: Incidence of neutralizing antibodies to herpes B virus in the Taiwan monkey *(Macaca cyclopis)*. Primates *11*, 297-301 (1970)
67. Ramsey, E.; E.L. Stair; A.E. Castro and M.I. Marks: Fatal *Herpesvirus hominis* encephalitis in a white-handed gibbon. J.Amer. vet. med. Ass. *181*, 1429-1430 (1982)
68. Ressig, M.and J.L. Melnick: The cellular changes produced in tissue cultures by herpes B virus correlated with the concurrent multiplication of virus. J.Exp. Med. *101*, 341-351 (1954)
69. Roth, A.M. and T.W. Purcell: Ocular findings associated with encephalomyelitis caused by *Herpesvirus simiae*. Amer. J.Ophthalm. *84*, 345-348 (1971)
70. Sabin, A.B.: Fatal B virus encephalomyelitis in a physician working with monkeys. J.clin. Invest. *28*, 808 (1949)
71. Sabin, A.B. and E.W. Hurst: Studies on the B virus. IV. Histopathology of the experimental disease in rhesus monkeys and rabbits. Brit. J.Exp. Path. *16*, 133-148 (1935)
72. Sabin, A.B. and A.M. Wright: Acute ascending myelitis following a monkey bite, with isolation of a virus capable of reproducing the disease. J.exp. Med. *59*, 115-135 (1934)
73. Schneweis, K.E.: Die serologischen Beziehungen der Typen 1 und 2 von Herpesvirus hominis zu Herpesvirus simiae. Z.Immunitätsforsch. *124*, 337-341 (1962)
74. Shah, K.V. and C.H. Southwick: Prevalence of antibodies to certain viruses in sera of free-living and captive monkeys. Indian J.Med. Res. *53*, 488-500 (1965)

75. Siegert, R.: Elektronenoptische Untersuchungen über die Kernveränderungen herpes-infizierter Zellen. Wien Z. Nervenheilk. *18*, 159-178 (1961)
76. Spaar, F. W.: Die menschliche Herpes-simplex-Encephalitis und -Meningitis. Gustav Fischer, Stuttgart, 1976
77. Sumner-Smith, G.: B virus in association with a monkey colony at a department of psychology. Lab. Primate Newsletter *5* (1), 1-4 (1966)
78. Tauraso, N. M.: Review of recent epizootics in nonhuman primate colonies and their relation to man. Lab. Anim. Sci. *23*, 201-210 (1973)
79. Thomas, E. and E. Henschel: Über die Herpes-B-Virus-Myelitis beim Menschen. Dtsch. Z. Nervenheilk. *181*, 494-516 (1960)
80. Tribe, G. W.: The significance of antibody to Herpesvirus simiae (B-virus). Primate Supply *5* (2), 40-42 (1980)
81. Tribe, G. W.: The pathogenesis and epidemiology of B-virus. Primate Supply *7* (1), 18-24 (1982)
82. Valerio, D. A.: Colony management as applied to disease control with mention of some viral diseases. Lab. Anim. Sci. *21*, 1011-1014 (1971)
83. Vizoso, A. D.: Simian viruses. Lancet (1), 1037-1041 (1975)
84. Vizoso, A. D.: Latency of Herpes simiae (B Virus) in rabbits. Brit. J. exp. Path. *56*, 489-493 (1975)
85. Vizoso, A. D.: Recovery of Herpesvirus simiae (B-Virus) from both primary and latent infections in rhesus monkeys. Brit. J. exp. Path. *56*, 485-488 (1975 c)
86. Watson, D. H.; P. Wildy; B. A. M. Harvey and W. I. H. Shedden: Serological relationships among viruses of the herpes group. J. Gen. Virol. *1*, 139-141 (1967)
87. White, R. J.; L. Simmons and R. B. Wilson: Chickenpox in young anthropoid apes: Clinical and laboratory findings. J. Amer. vet. med. Ass. *161*, 690-692 (1972)
88. Willcox, P. H.: Encephalomyelitis due to virus B. Proc. Roy. Soc. Med. *51*, 600 (1958)
89. Wolf, R. H.: Viral diseases of nonhuman primates. Amer. Ass. Zoo Vet. Ann. Proc. 1974, 95-108 (1974)
90. Wood, W. and F. T. Shimada: Isolation of virus B from tissue cultures of cynomolgus and rhesus kidneys. Canad. J. publ. Hlth. *45*, 509-518 (1954)
91. Zwartouw, H. T. and E. A. Boulter: Excretion of B virus in monkeys and evidence of genital infection. Lab. Anim. *18*, 65-70 (1984)
92. Zwartouw, H. T.; J. A. MacArthur; E. A. Boulter; J. H. Seamer; J. H. Marston and A. S. Chamove: Transmission of B virus infection between monkeys especially in relation to breeding colonies. Lab. Anim. *18*, 125-130 (1984)

1.2 Poxviruses

Poxviruses have only a limited host specificity. Most simian poxviruses are transferable to man, and vice versa ie. human molluscum contagiosum virus causes spontaneous infections in nonhuman primates. The question of whether spontaneous infection of monkeys and apes by the smallpox viruses occurs as suggested by some earlier reports (Gispen, 1949, 1975; Raghavan and Khan, 1968) remains unsettled. Certainly, monkeys are susceptible to the variola virus by all routes of infection, which causes a usually mild disease of two weeks duration (Hahon, 1961). The poxviruses infecting nonhuman primates under non-experimental conditions are summarized in Table 1.

Table 1. Zoonotic poxviruses in simian primates

Genus: Orthopoxvirus (Vaccinia-Variola group):
Monkeypox-virus
(White-pox viruses)
Genus not allocated: Molluscum contagiosum-group:
Yaba-virus
Tanapox (Yaba-like-, OrTeCa-)virus
Molluscum contagiosum virus

Table 2. Nonhuman Primate Species Spontaneously Affected by Monkeypox

Old World Species	New World Species
Pan troglodytes	Saimiri sciureus
Gorilla gorilla	Callithrix jacchus
Pongo pygmaeus	Saguinus oedipus
Hylobates lar	
Presbytis entellus	
Macaca fascicularis	
Macaca mulatta	
Cercopithecus hamlyni	
Cercopithecus aethiops	

1.2.1 Monkeypoxvirus

Monkeypox is a pox-disease of nonhuman primates similar to Variola in man (Cho and Wenner, 1973). Old- and New World monkeys and anthropoid apes can be affected (Gough et al., 1982), as summarized in Table 2. Monkeypoxvirus was also isolated from a wild squirrel *(Funisciurus anerythrus)* in Zaire (Khodakevich et al., 1986).

The first known outbreak was reported at the Statens Seruminstitut, Copenhagen/Denmark (von Magnus et al., 1959). Since then, 9 additional outbreaks in captive primates have occurred, with the last one being in Paris (Arita and Henderson, 1976).

Virus

Monkeypox is a rectangular virus of typical poxvirus structure and of 200 to 250 nm size (Prier and Sauer, 1960; Cho and Wenner, 1973). It is resistant to ether and relatively resistant to cold (Cho and Wenner, 1973). The typical monkeypox virus, isolated from disease outbreaks, grows well within 48 to 72 hs on the chicken CAM, in rhesus monkey kidney cells, rabbit kidney cells and patas monkey cells. It forms large plaques in VERO cells even without agarose-overlay (Prier et al., 1960; Rouhandeh et al., 1967; Marennikova et al., 1972). The CPE in the cell cultures is characterized by a rounding of the affected cells, with subsequent granulation, condensation, vacuolation, degeneration (Rondle and Sayeed, 1972) and formation of cytoplasmic inclusion bodies.

Table 3. Comparison of Monkeypox-, Variola- and Vacciniavirus-Properties

Characteristics	Variola	Vaccinia	Monkeypox
Lesions on the CAM (48 h)	small	large	small
Rabbit Skin Passage	–	+	+
Lesions in Rabbit Skin	non haemorrhagic	non haemorrhagic	haemorrhagic
Keratitis in Rabbits	+	–	+
Pathogenicity in Juvenile Mice	–	fatal	fatal
Plaque Formation in Chicken Embryo Cultures	–	+	+

The monkeypox virus produces haemagglutinating antigen in chicken CAM in higher titers than variola virus, but lower than either vaccinia- or cowpoxviruuses (Rondle and Sayeed, 1972).

Some of the characteristics of monkeypox virus, variola virus and vaccinia virus according to Arita and Henderson (1968), Cho et al. (1972), Gispen and Brand-Saathof (1972) are summarized in Table 3.

The lesions on the chicken CAM are characteristic of this virus. Typical monkeypox virus produces minute (0,5 mm diameter) and non-haemorrhagic primary pocks after 48 hs. After 96 hs secondary pocks with typical central haemorrhages develop (Rondle and Sayeed, 1972). Even more typical are the lesions in the rabbit skin. Monkeypox virus can be experimentally transmitted to rabbits by the intradermal route, with the rabbits developing haemorrhagic necrosis at the injection site.

A puzzling finding is the existence of certain monkeypox virus-strains with different *in vivo* and *in vitro* characteristics.

Several "monkeypox" virus strains isolated from monkey kidney cultures in the absence of pox like disease, did not produce any haemorrhagic lesions on the CAM or in rabbit skin; they were thus named "white" poxviruses. Additionally the original monkeypox viruses sometimes contain variable portions of the more prominent and larger white mutants besides the classical haemorrhagic strains (Gispen and Brand-Saathof, 1972, 1974). "White" poxvirus strains are virtually indistinguishable from variola virus using laboratory methods, but wild white mutants differ from variola virus by the initiation of focal necrobioses with central cytolysis in coverslip cultures of RK 13 cells (Gispen and Brand-Saathof, 1972). DNA mappings of variola and monkeypox viruses have indicated that spontaneous production of "whitepox" from monkeypox virus was genetically impossible and that the "whitepox" viruses recovered from monkeypox virus stocks had an exogenous origin (Esposito et al., 1985). Dumbell and Kapsenberg (1982) considered that the Bilthoven-strains of "whitepox"viruses were due to laboratory contaminations by Variola strains, handled at the same time in that laboratory.

The plaques formed by monkeypox virus in chicken embryo fibroblast cultures are smaller than those produced by vaccinia-, and rabbit pox viruses. Such plaques are not induced by variola mayor nor alastrim viruses (Mika and Pirsch, 1960; McConnell et al., 1964b).

In experimental infections of *M.fascicularis*, specific CF-antibodies were demonstrable approximately 1 week p.i., and the titres of which declined with

Table 4. Antigen Contents of Several Poxviridae

	Vaccinia-Ag	Variola-Ag	Monkeypox-Ag
Variola	−	+	−
White pox	−	+	−
Monkeypox	−	−	+
White mutants MP	−	−	+
Vaccinia	+	+	−
Rabbitpox	+	+	−
Buffalopox	+	+	−
Camelpox	+	+	−

regression of the lesions. CFI-antibodies appeared approximately 12 weeks p.i. (Hall et al., 1973). The cross-specificity of the monkeypox virus antibodies depends on the methods used. Rondle and Sayeed (1972) distinguished monkeypox- from vaccinia- and cowpoxvirus but not from the variola virus by HI and neutralization tests. CF tests are less specific than Ouchterlony tests using absorbed sera (Esposito et al., 1977). In Immunoprecipitation using double diffusion in agar gel, three specific antigens were found in members of the poxvirus group (Gispen and Brand-Saathof, 1974; Gispen, 1975); these are summarized in Table 4.

Epidemiology

Monkeypox is endemic in the African tropical rain forests, particularly in the Congo, Zaire, West- and Centralafrica (Breman et al., 1980; Mutombo et al., 1983). Anti-monkeypox-antibodies have been demonstrated in sera of wild living *Cercopithecus ascanius, C. nigroviridis, C. petaurista* and *Colobus badius* (Breman et al., 1980; Arita et al., 1985). The virus has repeatedly caused human infections in those areas. The first isolation of human monkeypox was from a suspected Smallpox case in Bakansuku/Zaire (Marennikova et al., 1972), and similarly most of the later human infections were usually initially suspected as being Smallpox (Ladnyi et al., 1972; Lourie et al., 1972; Arita and Henderson, 1976). Until 1984 altogether 165 cases of human monkeypox virus infections have been reported in West and Central Africa, usually in people living in small villages in the tropical rain forests (Mutombo et al., 1983; Jezek et al., 1983; Anon, 1984; Arita et al., 1985). A substantial increase in incidence was observed after 1982 (Arita et al., 1985), two years after discontinuation of smallpox vaccination in 1980. Seasonal influences were observed in northern Zaire with a peak of human monkeypox being recorded during July and August, the period of highest rainfall (Arita et al., 1985).

Person to person transmission has occurred, but only nonvaccinated persons were infected (Henderson and Arita, 1973), with spread among vaccinated contacts being rarely observed (Gispen, 1975). Among family contacts the secondary attack rate was 10% (Jezek et al., 1983). Although most African cases have involved black patients also white people can be affected.

Aerogenous transmission is considered to be the main route of transmission between nonhuman primates (Cho and Wenner, 1973) and probably also to other species like man or as in one case anteaters *(Myrmecophaga tridactyla)* (Peters,

1966). Other routes of transmission, such as biting or contact are also considered possible (Ladnyi et al., 1972; Mutombo et al., 1983).

Geographical Distribution

Human monkeypox have been reported from Zaire (148 cases), Sierra Leone (1 case), Nigeria (3 cases), Liberia (4 cases), Ivory Coast (2 cases), Cameroon (2 cases) and Central African Republic (16 cases) (Anon, 1984; Khodakevich et al., 1985).

Pathogenesis

The initial multiplication of the monkeypox virus occurs in local cellular components, most probably in either fixed or wandering connective tissue cells (Cho and Wenner, 1973). The different steps in the pathogenesis of monkeypox virus infections are summarized in Fig. 1.

In experimentally infected *M. fascicularis* a constant viremia appeared between the 3rd and the 14th day p.i. (Cho and Wenner, 1973).

Clinical Symptoms

a) Nonhuman primates: Differences exist in the susceptibility of the different host species. Anthropoid apes are usually more severely affected than monkeys, while cynomolgus monkeys suffer more than rhesus monkeys (Wenner et al., 1968).

After an incubation period of usually 3 to 4 days a sharp temperature rise heralds the onset of disease. The animals become anxious, with older ones occasionally becoming aggressive. Anorexia follows on the following day, with behavioral abnormalities such as sucking on the fingers or lips also becoming apparent.

As to the development of the pocks, two types of lesions can be distinguished:

1. Acute, marked facial edemas, with ulceration in mucous membranes and papule formation being visible, generalized lymphadenopathy, respiratory distress and death from asphyxia. Fatal cases usually belong into this group. Anthropoid apes are especially prone to such severe infection.
2. More commonly the infection occurs as a benign cutaneous eruption. 7 to 8 days after experimental infection itching and vesicular exanthema become obvious. Occasional coughing and mucopurulent nasal discharge indicate the presence of early lesions in the respiratory tract, which are mostly associated with the primary infection and virus multiplication. On the next day the first typical pocks appear as papules of 1 to 4 mm diameter, which then develop into pustules containing thick, purulent material. The vesicles become umbilicated and covered by crusts. After desquamation of the scabs or crusts within 7 to 10 days small scars remain (Prier and Sauer, 1960). The most common sites of pock formation in the monkeys are the buttocks, hands, and feet, but also the mucous membranes of the tongue, oral cavity, pharynx, larynx and trachea are commonly involved (Sauer et al., 1960); as well as the spleen, tonsils, lymph nodes, testes and ovaries (Wenner et al., 1969).

b) Man: In human monkeypox an incubation period of 7 to 21 days is assumed (Arita et al., 1985). The disease starts with a fever lasting for 2 to 4 days, which is followed by the eruption of pocks of approximately 5 mm diameter. All pocks

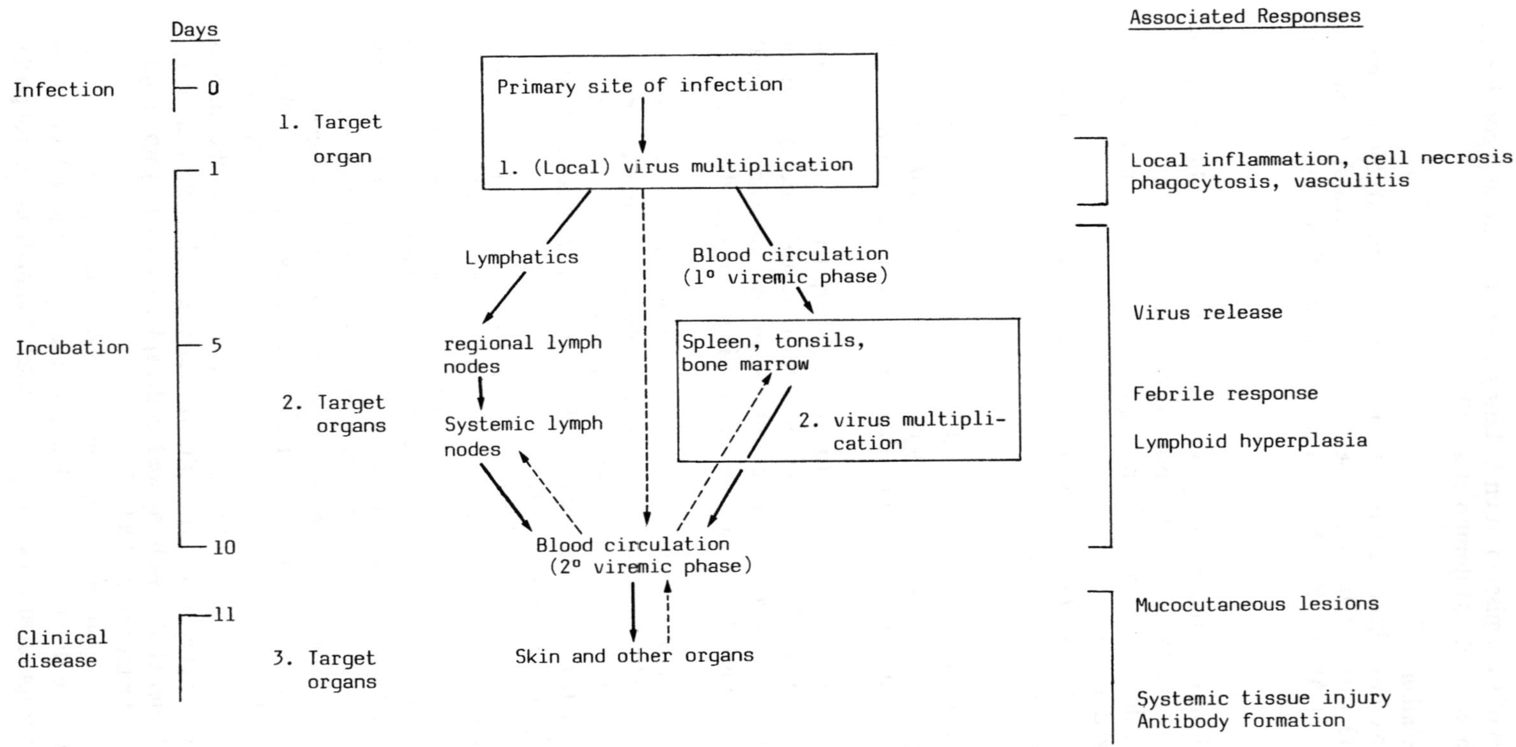

Fig. 1. Model for the pathogenesis of monkeypox. Model is based on data derived from MPV from cynomolgus monkeys infected intramuscularly

develop simultaneously from the papular stage through vesicles, umbilication and desquamation, and look very similar to smallpox lesions (Fig. 2). After shedding of the pocks small scars with initial hypopigmentation and later hyperpigmentation remain.

In minor cases the lesions are limited to the extremities, whereas in more serious infections generalization all over the entire body can be seen. One of the secondary landmarks of human monkeypox which differentiates human monkeypox from smallpox is the pronounced lymphadenopathy, expecially in the neck and inguinal regions, which either develops simultaneously with the fever or follows a day later (Jezek et al., 1983).

Human monkeypox can be fatal, with its fatality rate in man having been reported as being 14% (Jezek et al., 1983). So far 8 patients have died, all were children between 7 months and 7 years of age, and all were not vaccinated against Smallpox (Mutombo et al., 1983).

Pathology

a) Nonhuman primates: A reddened papule covered by a small crust appears 4 days p.i. at the injection site in experimental infection of monkeys. The histo-

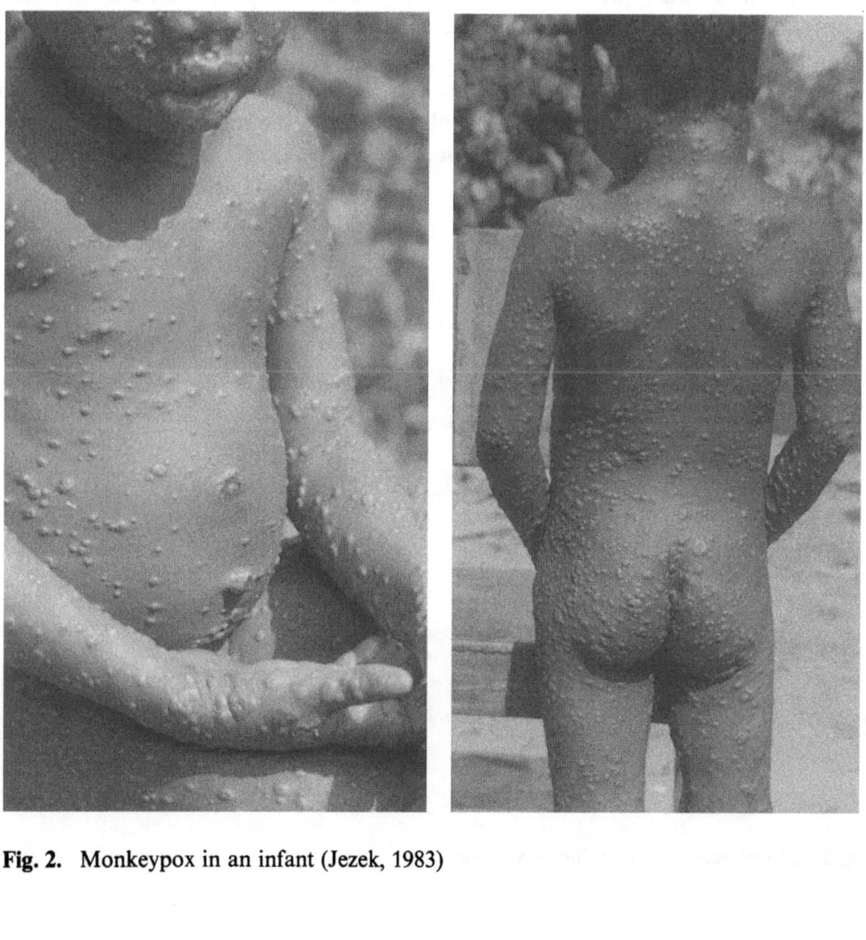

Fig. 2. Monkeypox in an infant (Jezek, 1983)

pathological characteristics of these papules consist of acanthosis, often accompanied by the formation of broad and elongated rete ridges, all of which result in a localized epidermal thickening. Guarnieri bodies have been seen within the cytoplasm of granular cells, and in a few cases intranuclear inclusion bodies were also observed (Sauer et al., 1960). The papule then develops into a typical pock lesion, covered by a central crust and surrounded by peripheral hyperaemia. The cells disintegrate and are separated by serous exudates, which result in small, multifocal liquid-filled cavities within the granular layer (Fig. 3) (Wenner et al., 1969). The cellular damages may extend down into the corium, but usually spare the basal germinal layer.

Extension into deeper tissues quite often leads to hyalinization and heavy inflammation of the skeletal muscles (Wenner et al., 1968). The lesions extend in a peripheral direction, becoming flattened and umbilicated, and due to the emigration of granulocytes the exudate becomes purulent. The final involution of the pocks is accompanied by enlargement of the crusts, decreasing hyperaemia and resolution of the edema. Reepithelization begins from the pustular borders as well as from the hair follicles. Some days later the overlaying crusts are shed. During generalization the exanthema passes through the same steps, but over a much shorter period of time - usually within 4 days.

b) Man: The monkeypox-lesions in a two year old boy who died after 5 days illness, consisted of a disseminated rash indistinguishable from smallpox, with similar lesions being present in the boy's mouth. Marked lymphadenitis was noted on the right side of the neck. A post-mortem skin biopsy of a papule revealed acanthosis with enormous elongation of the rete ridges, but no paraceratosis. The rete ridges and dermal papillae in the center of the papule were nec-

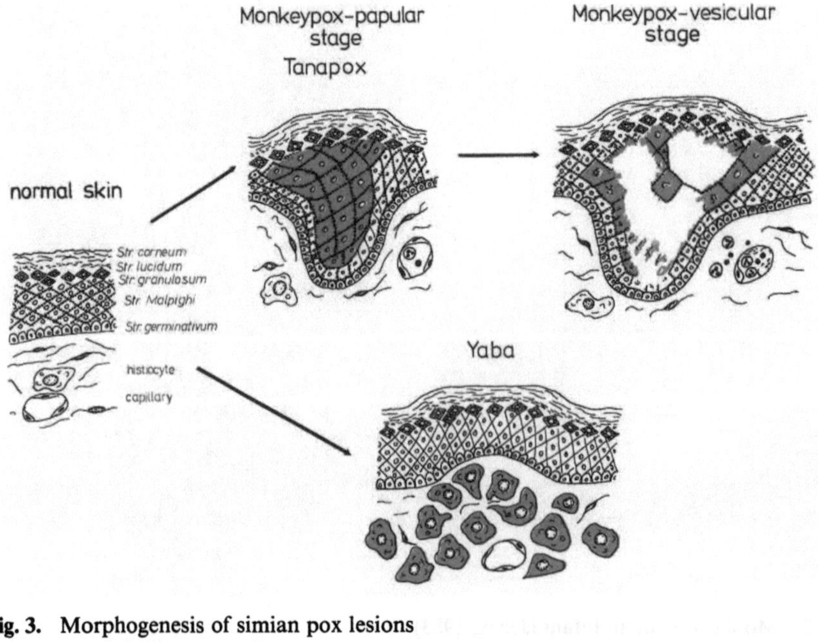

Fig. 3. Morphogenesis of simian pox lesions

rotic. The stratum spinosum above the central necrotic area was edematous. The formation of minute vesicles and the presence of multinucleated giant cells also occurred in the stratum spinosum. Guarnieri type inclusion bodies were present in lining cells of the tortuous sweat gland ducts (Stagles et al., 1985).

Treatment

The prevention of secondary infections by using antibiotics is the most important step in both man and nonhuman primates.

Endangered persons and animals can be protected by variola-vaccination, as it is effective against monkeypox as well (McConnell et al., 1964; Peters and Verlinde, 1966).

Methisazone (1-methyl-isatin-3-thiosemicarbazone) which has been used in smallpox-prophylaxis, failed to protect monkeys against experimental monkeypox infections (Peters, 1966; Cho and Wenner, 1973).

1.2.2 YABA-Virus

In 1957 a series of subcutaneous "tumorlike" growths caused by a poxvirus belonging to the molluscum contagiosum group were noticed as occurring in 20 out of 35 rhesus monkeys over a period of 5 months at the West African Council for Medical Research, Yaba/Nigeria (Bearcroft and Jamieson, 1958). Additional cases have been observed in captive-born rhesus monkeys in England (Spencer, 1985); in baboons (Niven et al., 1961; Bruestle et al., 1981; Walker et al., 1985); and in *Macaca fascicularis* and *M. speciosa* (Ambrus et al., 1969). The infection could not be transmitted to *Erythrocebus patas, Cercopithecus mona, Cercocebus torquatus, Cercocebus fuliginosus, Cebus fatuellus, Saimiri sciureus,* mice , dogs, guinea pigs, rabbits, hamsters or rats (Ambrus et al., 1963; Grace and Mirand, 1963; Ambrus and Sandström, 1966).

Virus

The virus grows in primary vervet kidney cells, VERO-, LLCMK2, MA10 and BSC-1-cells and in the CAM of 10–11 day old chicken embryos (Sandström et al., 1966; Yohn et al., 1964, 1966). The resulting CPE consists of nucleolar hypertrophy at 24 h and cytoplasmic inclusion body formation at 4 days, which stain brilliantly with FITC-labelled anti-Yabavirus-antiserum. Another type of CPE is expressed by the formation of microtumors on the chicken CAM or BSC-1 cells after inoculation with low virus titers (Yohn et al., 1964; Sandström et al., 1966).

The virus can be demonstrated by immunofluorescence using anti-Yaba-antiserum derived from monkeys in the convalescent stage in the cytoplasm of tissue culture cells and in the tumor tissues, but not in the nuclei or nucleoli (DeHaven and John, 1966; Levinthal and Shein, 1964; Taylor et al., 1968). No immunological cross reactions with Vaccinia, Monkeypox or Orf-virus have been observed (Behbani et al., 1968).

Epidemiology

The natural route of transmission is unknown, although arthropod vectors have been suggested (Hunt and Jones, 1973). In one outbreak, transmission occurred through an infected tattoo needle (Casey et al., 1967).

Experimental subcutaneous or intravenous inoculation of tumor tissue filtrates or virus suspension into rhesus monkeys, baboons or African green monkeys resulted in an analogous "tumorlike" formation within 3 to 4 weeks (Niven et al., 1961; Sproul et al., 1962, 1963). After subcutaneous inoculation only subcutaneous nodules developed in the initial infection, but after i.v. inoculation or with increasing passages, metastases developed in the lung and heart or along the lymphatics (Feltz, 1961; Grace and Mirand, 1963).

Human infection has resulted from accidental puncture by a contaminated needle (Grace and Mirand, 1963, 1965).

Clinical Diseases

Nonhuman primates: The initial clinical sign is the rapid development of small, slightly reddened papules, which, after several passages, grow larger (up to several cm in diameter) and appear as subcutaneous nodules or masses. They appear particularly at the hairless parts of the dorsovolar and dorsoplantar surfaces of the extremities and on the face, but they never spread on to the trunk (Bearcroft and Jamieson, 1958; Ambrus et al., 1969; Spencer, 1985; Walker et al., 1985).

6 weeks later the nodules spontaneously regress and because of this they should be called pseudotumors rather than true neoplasms.

CF-fixing antibodies develop shortly after the occurrence of the pseudotumors, but true protective immunity does not develop as the pseudotumors can reappear several times in the same animal.

Man: 5–7 days p.i. slowly growing nodules up to 2 cm in diameter developed in volunteers. In spontaneous cases these pseudotumors were observed particularly on the hands and feet, with an associated regional lymphadenopathy and fever. The pseudotumors disappeared spontaneously within a few weeks.

Pathology

The subcutaneous (and otherwise located) pseudotumors are essentially of histiocytomatous structures with large, pleomorphic cells growing beneath the cutaneous squamous cells (Fig. 3). These cells are characterized by large nucleoli, hyperchromasia of the nuclear walls, and during proliferation, numerous mitotic figures in the enlarged, and often burst nuclei. Occasionally large numbers of eosinophils can be seen during proliferation. In some tumor cells irregular, eosinophilic cytoplasmic inclusion bodies can be demonstrated (Andrews et al., 1959; Niven et al., 1961; Sproul et al., 1963). Intracytoplasmic oval or brick-shaped virus particles can also be demonstrated using the electron microscope (Spencer, 1985). There is no collagen fiber production in Yaba-pseudotumors in contrast to the rabbit- and deer fibromas caused by poxviruses (Behbehani et al., 1968).

Regression of the pseudotumors is usually associated with multinucleated giant cell formation.

1.2.3 Tanapox-(Yaba-Like-, OrTeCa-, Benign Epidermal Monkeypox-, BEMP-)Virus

Pseudotumors which grossly resembled early pox-lesions or pseudotumors induced by Yaba virus were found simultaneously 1967 in three monkey colonies in Oregon, Texas and California and 1970 in another colony in Maryland (McNulty, 1972). Monkey species involved were *Macaca mulatta, M. nemestrina, M. fuscata* and *Cynopithecus niger* (Hall and McNulty, 1967; Crandell et al., 1969). All animals of the 1967 outbreaks were obtained from the same dealer (McNulty, 1972). These tumors were caused by a poxvirus, whose biological properties differed markedly from those of the Yaba-virus.

Cebids and callitrichids were not affected (Hall and McNulty, 1967). Experimental infections could not be initiated in either squirrel monkeys *(Saimiri sciureus)*, lorises, galagos, sheep, swine, chicks, mice, hamster or New Zealand white rabbits (Crandell et al., 1969); whereas in German checker rabbits, a proliferative, non-necrotizing epidermal lesion was produced (Crandell et al., 1969).

Another poxvirus, serologically closely related to Tanapoxvirus has been recently isolated from a spontaneous outbreak in *Callithrix jacchus* (Gough et al., 1982).

Virus: Tanapoxvirus can be propagated in HuAm-cells, AgMK, E-BSC, R-BSC and CV-1 cells with the CPE appearing 18 days p.i. including the formation of small plaques up to 1 mm in diameter. The cytoplasmic inclusion bodies in the infected tissue cultures are smaller than those produced by Yaba-virus (Nicholas, 1970a). Tanapoxvirus does not induce any CPE in either rabbit kidney cells, chicken embryo- or HeLa cells, and it also does not initiate pock-formation on the chicken CAM (Nicholas and McNulty, 1968).

Tanapox- and Yaba-virus cross react in CF-, CFI- and neutralization tests (Nicholas, 1970b; Hall et al., 1973).

Human tanapox-Virus infection was first observed 1957 and 1962 in Kenyan children (España et al., 1971; Jezek et al., 1985). From 1978 to 1981 more than 163 cases of human Tanapox infection were discovered in Zaire (Arita and Gromyko, 1982) with the majority of tanapox cases observed in Lisala and in nearby villages located along or near the Zaire river (Jezek et al., 1985). Independent of these natural outbreaks transmission from infected monkeys to their animal handlers occurred (McNulty et al., 1968).

Clinical Disease

Nonhuman primates: After a 4–5 day incubation period small, reddened papules appeared on the face, chest, anus and perineum of infected animals. They developed into circular, flat, firm elevations up to 10 mm in diameter within 14 days, becoming umbilicated and covered by small central crusts. The surrounding erythema is only of a minor degree. The papules begin to exfoliate by the 10th day, but they never develop into either a vesicular or pustular form (Hall and McNulty, 1967; McNulty et al., 1968; Kupper et al., 1970; McNulty, 1972). Healing with scar formation takes 3 to 4 weeks.

Man: In general, the clinical course of human tanapox has two stages. In the majority of cases the illness started with pre-eruptive symptoms, the most common being fever of 38 to 39 °C, which usually lasts for 2 to 4 days. The short febrile illness is sometimes accompanied by severe headache, backache, and prostration.

The eruptive stage often started with an itching sensation at the site of development of the skin lesions and with the appearance of a circular spot. This macule is slightly elevated at the centre but lacks the central abrasion usually observed in insect bites. The macule subsequently changes to a dark, slightly raised papule with a rough surface and a readily palpable underlying induration. As the focal skin lesions developed, the fever and general symptoms abated. The papule continued to grow into an elevated lesion of pock-like appearance with little or no liquid but a considerable amount of necrotic tissue. By the end of the first week the skin lesion was usually more than 10 mm in diameter with a large erythematous areola of several centimeters diameter surrounded by swollen edematous skin. This acute stage was commonly associated with local lymphangitis and regional lymphadenitis.

The skin lesions may develop into larger nodules but more frequently brake down as a result of trauma or necrosis. The so formed ulcers were round, shallow excavations with a lightly raised perimeter and a base of soft necrotic tissue, they spread peripherally as well as into the corium. The ulcers remained about the same size for several days, the acute inflammatory reaction then diminished, the ulcer's perimeter flattened and their floor now contained granulation tissue. Re-epithelization started from the periphery and the regional lymphadenopathy diminished (Jezek et al., 1985).

In African patients the lesions were located on the front or on lower parts of the body not usually covered by clothes (Jezek et al., 1985).

Accidentally infected animal handlers developed solitary lesions along the scratched parts of their hands and also their anus (McNulty, 1972).

Pathology

Nonhuman primates: The lesions of monkey tanapox closely resemble those of monkeypox arrested at the papular stage (Fig. 3). Histologically a thickened, pale epidermis in the affected areas of the skin is seen, which is characterized by a proliferation (acanthosis) and ballooning degeneration or necrosis of the prickle cells, extending down below the hairshafts and sweatglands. The nuclei of the proliferating epithelial cells may be vacuolated. Eosinophilic inclusion bodies appear mainly in the cytoplasm, but are sometimes also seen in the enlarged nuclei (Hall and McNulty, 1967; McNulty et al., 1968). The dermis can be infiltrated by varying amounts of inflammatory cells depending upon the degree of secondary bacterial infection. The covering crusts are composed of keratin, leukocytes, dead non-keratinized epidermal cells and cellular debris.

Man: Acanthosis with elongation of the rete ridges and vacuolization of epidermal cells of all layers has been described (McNulty et al., 1968). Inclusion bodies were not identified, but many nuclei showed hyperchromatosis along the nuclear membranes. Early lesions extended only into the upper corium, whereas in the late stages the lower dermis was also involved.

Immunity

Infected animals have been proven to be immune to challenge for 2 years (McNulty, 1972). Vaccination using Vacciniavirus did not produce any productive immunity (Hall and McNulty, 1967).

1.2.4 Molluscum Contagiosum

Molluscum contagiosum, a chronic skin disease of low infectivity in man, has been diagnosed in a few chimpanzees (Douglas et al., 1967; Schmidt and Butler, 1967). Small nodules appeared in the periorbital or inguinal regions of the infected animals. Typical "molluscum contagiosum bodies" could be demonstrated in the tissues, but attempts to isolate the virus were unsuccessful.

References

1. Ambrus, J.L.; E.T. Feltz; J.T. Grace, jr. and G.Owens: A virus induced tumor in primates. Nat. Cancer Inst. Monogr. *10*, 447-458 (1963)
2. Ambrus, J.L. and H.V. Sandström: Susceptibility of Old World monkeys to Yaba Virus. Nature *211*, 876 (1966)
3. Ambrus, J.L.; H.V. Sandström and W.Kawinski: "Spontaneous" occurrence of Yaba tumor in a monkey colony. Experientia *25*, 64-65 (1969)
4. Andrews, C.H.; A.C. Allison; J.A. Armstrong; G.Bearcroft; J.S. Niven and H.G. Pereira: A virus disease of monkeys causing large superficial growth. Acta un. Internat. Cancr. *15*, 160-163 (1959)
5. Anon: The current status of human monkeypox: Memorandum from a WHO meeting. Bull Wld. Hlth Org. *62*, 703-713 (1984)
6. Arita, I.and A.Gromyko: Surveillance of orthopoxvirus infections, and associated research, in the period after smallpox eradication. Bull. Wld. Hlth Org. *60*, 367-375 (1982)
7. Arita, I.and D.A. Henderson: Smallpox and monkeypox in nonhuman primates. Bull. Wld. Hlth Org. *39*, 277-283 (1968)
8. Arita, I.and D.A. Henderson: Monkeypox and whitepox viruses in West and Central Africa. Bull. Wld. Hlth Org. *53*, 347-353 (1976)
9. Arita, I.; Z.Jezek; L.Khodakevich and K.Ruti: Human monkeypox: a newly emerged orthopoxvirus zoonosis in the tropical rain forests of Africa. Amer. J.trop. Med. Hyg. *34*, 781-789 (1985)
10. Bearcroft, W.G.C. and M.F. Jamieson: An outbreak of subcutaneous tumors in rhesus monkeys. Nature *182*, 195-196 (1958)
11. Behbehani, A.M.; C.R. Bolano; P.S. Kamitsuka and H.A. Wenner: Yaba tumor virus. I.Studies on pathogenesis and immunity. Proc. Soc. exp. Biol. *129*, 556-561 (1968)
12. Breman, J.G.; Kalisa-Ruti; M.V. Steniowski; E.Zanotto; A.I. Gromyko and A.Arita: Monkeypox chez l'homme, 1970-1979. Bull. Wld. Hlth Org.849-868 (1980)
13. Bruestle, M.E.; J.G. Golden; A.Hall III and A.R. Banknieder: Naturally occurring Yaba tumor in a baboon (Papio papio). Lab. Anim. Sci. *31*, 292-294 (1981)
14. Casey, H.W.; J.M. Woodruff and W.I. Butcher: Electron microscopy of a benign epidermal pox disease of rhesus monkeys. Amer. J.Path. *51*, 431-446 (1967)
15. Cho, C.T. and H.A. Wenner: Monkeypox virus. Bact. Rev. *37*, 1-18 (1973)
16. Cho, C.T.; C.Liu and H.A. Wenner: Monkeypox virus: responses of infection in chick embryos. Proc. Soc. exp. Biol. *139*, 1206-1212 (1972)
17. Crandell, R.A.; H.W. Casey and W.B. Brumlow: Studies of a newly recognized poxvirus of monkeys. J.Infect. Dis. *119*, 80-88 (1969)

18. DeHaven, E. and D.S. John: The fine structure of the Yaba monkey tumor poxvirus. Cancer Res. *32*, 995-1008 (1966)
19. Douglas, J.D.; K.N. Tanner; J.R. Prine; D.C. van Riper and S.K. Derwelis: Molluscum contagiosum in chimpanzees. J. Amer. vet. med. Ass. *151*, 901-904 (1962)
20. Dumbell, K.R. and J.G. Kapsenberg: Laboratory investigation of two "whitepox" viruses and comparison with two variola strains from southern India. Bull. Wld. Hlth Org. *60*, 381-387 (1982)
21. España, C.; M. Brayton and B.H. Ruebner: Electron microscopy of the tana poxvirus. Exp. Molec. Path. *15*, 34-42 (1971)
22. Esposito, J.J.; F. Obijewski and J.H. Nakano: Serological relatedness of monkeypox, variola, and vaccinia viruses. J. Med. Virol. *1*, 35-47 (1977)
23. Esposito, J.J.; J.H. Nakano and J.F. Obijewski: Can variola-like viruses be derived from monkeypox virus? An investigation based on DNA mapping. Bull. Wld. Hlth Org. *63*, 695-703 (1985)
24. Feltz, E.T.: Studies on a virus induced tumor in monkeys. Proc. Amer. Ass. Cancer Res. *3*, 224 (1961)
25. Foege, W.H. and S.O. Foster: Monkeypox in humans. Lab. Primate Newsl. *11* (3), 10-13 (1972)
26. Foster, S.O.; E.W. Brink; D.L. Hutchins; J.M. Pifer; B. Lourie; C.A. Moser; E.C. Cummings; O.E. Kuteyi; R.E.A. Eke; J.B. Titus; E.A. Smith; J.W. Hicks and W.H. Foege: Human monkeypox. Bull. Wld. Hlth Org. *46*, 569-576 (1972)
27. Gispen, R.: De Herbstmetting van Indonesie met Pokken. Nederlandske Tijdschr. Geneesk. *93*, 3686-3695 (1949)
28. Gispen, R.: Relevance of some poxvirus infections in monkeys to smallpox eradication. Trans. roy. Soc. trop. Med. Hyg. *69*, 299-302 (1975)
29. Gispen, R. and B. Brand-Saathoff: "White" poxvirus strains from monkeys. Bull. Wld. Hlth Org. *46*, 585-592 (1972)
30. Gispen, R. and B. Brand-Saathoff: Three specific antigens produced in Vaccinia, Variola, and Monkeypox infections. J. infect. Dis. *129*, 289-295 (1974)
31. Gough, A.W.; N.J. Barsoum; S.I. Gracon; L. Mitchell and J.M. Sturgess: Poxvirus infection in a colony of common marmosets *(Callithrix jacchus)*. Lab. Anim. Sci. *32*, 87-90 (1982)
32. Grace, J.T. and E.A. Mirand: Human susceptibility to a simian tumor virus. Ann. N.Y. Acad. Sci. *108*, 1123-1128 (1963)
33. Grace, J.T. and E.A. Mirand: Yaba virus infections in humans. Exp. Med. Surg. *23*, 213-216 (1965)
34. Hahon, N.: Smallpox and related poxvirus infections in the simian host. Bact. Rev. *25*, 459-475 (1961)
35. Hall, A.S. and W.P. McNulty: A contagious pox disease in monkeys. J. Amer. vet. med. Ass. *151*, 833-838 (1967)
36. Hall, R.D.; R.G. Olsen; S.P. Pakes and D.S. Yohn: Differences in clinical and convalescent phase antibodies of rhesus monkeys infected with Monkeypox, Tanapox, and Yaba poxviruses. Infect. and Immun. *7*, 539-546 (1973)
37. Henderson, D.A. and I. Arita: Monkeypox and its relevance to smallpox eradication. Wld. Hlth Org. Chron. *27*, 145-148 (1973)
38. Hunt, R.D. and T.C. Jones: Nonhuman primates in viral oncology. Advanc. vet. Sci. (Camp-Med.) *17*, 361-394 (1973)
39. Jezek, Z.; I. Arita; M. Szczeniowski; K.M. Paluku; K. Ruti, and J.H. Nakano: Human tanapox in Zaire: clinical and epidemiological observations on cases confirmed by laboratory studies. Bull Wld. Hlth Org. *63*, 1027-1035 (1985)
40. Jezek, Z.; A.I. Gromyko and M.V. Szceniowski: Human monkeypox. J. Hyg. Epidem., Microbiol. Immunol. *27*, 13-28 (1983)
41. Khodakevich, L.; Z. Jezek, and K. Kinzanzka: Isolation of monkeypox virus from wild squirrel infected in nature. Lancet *8472*, 98-99 (1986)
42. Khodakevich, L.; R. Widy-Wirski; I. Arita; S.S. Marennikova; J. Nakano et D. Meunier: Orthopoxvirose simienne de l'homme en République Centrafricaine. Bull. soc. Path. Ex. *78*, 311-320 (1985)

43. Kupper, J.L.; H.W. Casey and D.K. Johnson: Experimental Yaba and benign epidermal monkey pox in rhesus monkeys. Lab. Anim. Care *20*, 979-988 (1970)
44. Ladnyi, D.; P. Ziegler and E. Kima: A human infection caused by monkeypox virus in Basankusu Territory, Democratic Republic of the Congo. Bull. Wld. Hlth Org. *46*, 593-597 (1972)
45. Levinthal, J.M. and H.M. Shein: Propagation of a simian tumor agent (Yaba virus) in cultures of human and simian renal cells as detected by immunofluorence. Virology *23*, 268-270 (1964)
46. Lourie, B.; P.G. Bingham; H.H. Evans; S.O. Foster; J.H. Nakano and K.L. Herrmann: Human infection with monkeypox virus: Laboratory investigation of six cases in West Africa. Bull. Wld. Hlth Org. *46*, 633-639 (1972)
47. Marennikova, S.S.; E.M. Selukhina; N.N. Malceva; K.L. Cimiskjan and G.R. Macevic: Isolation and properties of the causal agent of a new variola-like disease (monkeypox) in man. Bull. Wld. Hlth Org. *46*, 599-611 (1972)
48. McConnell, S.; Y.F. Herman; D.E. Mattson; D.L. Huxsoll; C.M. Lang and R.H. Yager: Protection of rhesus monkeys against monkeypox by vaccinia virus immunization. Amer. J. vet. Res. *25*, 192-195 (1964a)
49. McConnell, S.J.; R.O. Spertzel; D.L. Huxsoll; L.H. Elliott and R.H. Yager: Plague morphology of monkeypox virus as an aid to strain identification. J. Bact. *87*, 238-239 (1964b)
50. McNulty, W.P.; W.C. Lobitz jr.; H. Fu; C.A. Maruffo and A.S. Hall: A pox disease in monkeys transmitted to man. Arch. Derm. *97*, 286-293 (1968)
51. Mika, L.A. and J.B. Pirsch: Differentiation of variola from other members of the poxvirus group by the plague technique. J. Bact. *80*, 861-863 (1960)
52. Mutombo, W.M.; I. Arita and Z. Jezek: Human monkeypox transmitted by a chimpanzee in a tropical rain-forest area of Zaire. Lancet *8327* (1), 735-737 (1983)
53. Nicholas, A.H.: A poxvirus of primates. I. Growth of the virus *in vitro* and comparison with other poxviruses. J. nat. Cancer Inst. *45*, 897-905 (1970a)
54. Nicholas, A.H.: A poxvirus of primates. II. Immunology. J. nat. Cancer Inst. *45*, 907-914 (1970b)
55. Nicholas, A.H. and W.P. McNulty jr.: *In vitro* characteristics of a poxvirus isolated from rhesus monkeys. Nature *217*, 745-746 (1968)
56. Niven, J.S.; J.A. Armstrong; E.H. Andrews; H.G. Pereira and R.C. Valentine: Subcutaneous "growth" in monkeys produced by a poxvirus. J. Path. Bact. *81*, 1-14 (1961)
57. Peters, J.C.: Eine "Monkey-Pox" Enzootie im Affenhaus des Tiergartens "Blijdorp". Kleintierpraxis *11*, 65-71 (1966)
58. Peters, J.C. and J.D. Verlinde: Die Vakzination von Affen gegen "Monkey-Pox". VIII. Internat. Symp.- Erkr. Zootiere, Leipzig, 41-47 (1966)
59. Prier, J.E. and R.M. Sauer: A pox disease of monkeys. Ann. N.Y. Acad. Sci. *85*, 951-959 (1960)
60. Prier, J.E.; R.M. Sauer; R.G. Malsberger and J.H. Sillaman: Studies on a pox disease of monkeys: II. Isolation of the etiologic agent. Amer. J. vet. Res. *21*, 381-384 (1960)
61. Raghavan, R.S. and G.A. Khan: An outbreak of pox in chimpanzees. Indian Vet. J. *45*, 75 (1968)
62. Rondle, C.J.M. and K.A.R. Sayeed: Studies on monkeypox virus. Bull. Wld. Hlth Org. *46*, 577-583 (1972)
63. Rouhandeh, H.; R. Engler; M. Taher; A. Fouad and L.L. Sells: Properties of monkey pox virus. Arch. ges. Virusforsch. *20*, 363 - 373 (1967)
64. Sandström, H.V.; J.L. Ambrus and G. Owens: Propagation of Yaba virus in embryonated hen eggs. Virology *28*, 479-481 (1966)
65. Sauer, R.M.; J.E. Prier; R.S. Buchanan; A.A. Creamer and H.C. Fegley: Studies on a pox disease of monkeys. I. Pathology. Amer. J. vet. Res. *21*, 377-380 (1960)
66. Schmidt, R.E. and T.M. Butler: Case Reports: Molluscum contagiosum in a colony born chimpanzee. Lab. Primate Newsletter *10* (4), 17 (1970)
67. Spencer, A.J.: Diagnostic exercise: Subcutaneous nodules in rhesus monkey. Lab. Anim. Sci. *35*, 79-80 (1985)
68. Sproul, E.E.; R.S. Metzgar and J.T. Grace jr.: The pathogenesis of the Yaba virus tumor in the rhesus monkey. Proc. Amer. Ass. Cancer Res. *3*, 363 (1962)
69. Sproul, E.E.; R.S. Metzgar and J.T. Grace: The pathogenesis of Yaba virus-induced histiocytomas in primates. Cancer Res. *23*, 671-675 (1963)

70. Stagles, M.J.; A.A. Watson; J.F. Boyd; I.A.R. More and D. McSeveney: The histopathology and electron microscopy of a human monkeypox lesion. Trans. roy. Soc. trop. Med. Hyg. *79*, 192-202 (1985)
71. Taylor, D.O.; M.R. Klauber; E.H. Lennette and A. Wiener: An immunofluorescent method for Yaba virus assay. J. nat. Cancer Inst. *40*, 147-155 (1968)
72. von Magnus, P.; E.K. Anderson; K.B. Peterson and A. Birch-Anderson: A poxlike disease in cynomolgus monkeys. Arch. Path. Microbiol. Scand. *46*, 156-176 (1959)
73. Walker, D.H.; F.A. Voelker; A.E. McKee and J.H. Nakano: Diagnostic exercise: Tumors in a baboon. Lab. Anim. Sci. *35*, 627-628 (1985)
74. Wenner, H.A.; C.R. Bolano; C.T. Cho and P.S. Kamitsuka: Studies on the pathogenesis of monkey pox. III. Histopathologic lesions and sites of immunofluorescence. Arch. ges. Virusforsch. *27*, 179-197 (1969)
75. Yohn, D.S.; J.T. Grace jr. and V.A. Haendiges: A quantative cell culture assay for Yaba tumour virus. Nature *202*, 881-883 (1964)
76. Yohn, D.S.; V.A. Haendiges and J.T. Grace jr.: Yaba tumor poxvirus synthesis in vitro. I. Cytopathological, histochemical, and immunofluorescent studies. J. Bact. *91*, 1977-1985 (1966)

1.3 RNA-Viruses

1.3.1 Arboviruses

Arboviruses of zoonotic properties involving nonhuman primates belong to several serogroups - Table 5

1.3.2 Chikungunya-Virus

Chikungunya virus derived its name from the main clinical symptoms in man. Chikungunya in Swahili means "painfully bent, stiff joints". It is sometimes referred to as "denguelike" because of the clinical similarities of the two arbovirus infections (McIntosh and Gear, 1981).

Table 5. Zoonotic arboviruses of nonhuman primates

Family	Genus	Species
Togaviridae	Alphavirus	Chikungunya-virus Mayaro-virus (Getah-virus) Bebaru-virus)
	Flavivirus	Yellow fever-virus Kyasanur Forest disease-virus Dengue-viruses 1-4 (Zika-virus)
Reoviridae	Orbivirus	Orungo-virus
Bunyaviridae	Bunyavirus	Oropouche-virus

The disease probably goes back as far as the 18th century, when in 1779 in Cairo, Alexandria and in Batavia (Jakarta) outbreaks of a disease were described, which closely resemble those of Chikungunya-virus infection. Other outbreaks have probably occurred in the Americas as far north as Philadelphia (Halstead, 1981), but retrospectively dengue-outbreaks can not be completely disregarded. The first ascertained outbreak of Chikungunya was observed 1952 in the Newala province of southern Tanzania (Robinson, 1955; Lumsden, 1955; Ross, 1956). Subsequent further isolations have been made in South Africa, Senegal, Uganda, Zaire, Congo, Angola, Nigeria and Zimbabwe (Gear and Read, 1957; McIntosh et al., 1963; McIntosh and Gear, 1975; Robin, 1981). In Senegal 55% of the human population showed antibodies against the virus (Robin, 1981). Human epidemics in Africa have been reported in southern Transvaal in 1957 (Gear and Reid, 1957; Gear, 1975); in Zimbabwe in 1964 (McIntosh et al., 1964); in Senegal in 1966 (Robin, 1981); in the Entebbe-area of Uganda in 1968 (Kirya and Okia, 1977); in Nigeria in 1969 and 1974 (Robin, 1981); and in Angola in 1970 ( ). Other epidemics were observed in India, Thailand, Cambodia, Vietnam, and Burma (Myers et al., 1964; Carey et al., 1969; Halstead, 1966). Anti Chik-antibodies were demonstrated also in a large of the human population of Indonesia, the frequency increasing with age (Kanamitsu et al., 1979).

In Africa *Papio ursinus*, *Cercopithecus ascanius*, *C. aethiops* and occasionally *Galago senegalensis* have been shown to harbor the virus with the development of high titered viremias (McIntosh, 1970; McIntosh et al., 1977; Kaschula et al., 1978; Robin, 1981). They are thus most probably the reservoir hosts, but isolations have also been made from a number of non-primates. The reservoir hosts in Asian countries are unknown, *M. mulatta* may be suspected because of the only mild disease associated with high titered viremias after experimental infection.

Chikungunya-antibodies have also been found in Africa born, less than 2yr old gorillas, chimpanzees and an orangutan (Harrison et al., 1967).

Virus: Chikungunya virions are spherical, approximately 42 nm in diameter with a hexagonal inner core of 25 to 30 nm diameter. It produces CPE in primary hamster kidney cells, BHK-21-, BSC-1-, VERO-, Fl-, HeLa- and rhesus monkey kidney cell lines. It replicates also in continuous *Aedes*-spp. cell lines.
Virus isolation can also be made via intracerebral inoculation of suspicious materials into 1 to 2 day old suckling mice, where it produces an acute encephalitis, and if fresh isolates are used, a haemorrhagic enteritis (Halstead, 1981).

Serology: Antisera prepared to Chikungunya-virus cross-react in CF- and neutralization tests with O'nyong nyong-, Mayaro- and Semliki Forest viruses; whereas vice versa only weak cross reactions occur (Halstead, 1981).

Epidemiology: As in yellow fever, two transmission cycles can be distinguished in Chikungunya-fever. One is monkey - mosquito - monkey, maintained with extension to forest living people via canopy-dwelling mosquitoes descending to the ground level. The other one depends on interhuman transmission by *Aedes aegypti* (Robin, 1981).

Main vectors besides *A. aegypti* are the forest dwelling *A. africanus*, *A. luteocephalus*, *A. furcifer*, *A. opok*, *A. dalzieli*, and occasionally *Culex pipiens fatigans* or *Mansonia fuscopenatta* (Halstead, 1981; Robin, 1981).

Clinical Disease

Man: After an incubation period of 3 to 12 days (in 2 accidental laboratory infections 22 and 80 hrs respectively), sudden fever with no prodromal symptoms develops. Other symptoms include headache, lymphadenopathy, joint pain, severe asthenia and anorexia, nausea and vomiting. 2 to 3 days later an irritating maculopapular or scarlatiniform rash develops over the trunk and limbs, whilst simultaneously the fever declines. The rush disappears after a few days, but the joint pain usually declines much more slowly, and, in approximately 50% of the patients apyrexial joint pain recurs intermittingly during the subsequent period, which may last up to 6 months.

In Asian patients mild haemorrhagic signs have been observed ie. petechiae, purpura and epistaxis (McIntosh and Gear, 1975).

Nonhuman primates: Information on clinical symptoms in monkeys depends on experimental infection. Rhesus monkeys develop a benign, transient, febrile illness, whilst vervet monkeys and baboons remain asymptomatic despite viremia (Robin, 1981).

Pathology

Man: The rare fatal human Chikungunya virus-infections are not well reported. The symptoms resemble those of a shock syndrome.

Nonhuman primates: No morphological lesions are reported.

Treatment

Antipyretics should be given to control the fever, but salicylates are to be avoided. Analgetics or mild sedatives may be sometimes be required to control pain for extended periods of time. Excessive losses of fluids and electrolytes due to vomiting should be treated accordingly to prevent severe dehydration.

1.3.3 Mayaro-Fever

The virus was isolated in Mayaro-county/Trinidad in 1954 (Anderson et al., 1957). 2 outbreaks of the disease in man were recognized in the Amazonian region of Brazil and one in the Bolivian rain forest (Causey and Maroya, 1957; Schaeffer et al., 1959; LeDuc et al., 1981; Pinheiro et al., 1981). Other virus isolations were made in Surinam, Panama and Colombia (Groot et al., 1961; Galindo et al., 1966; Metselaar, 1966). Serological evidences have also indicated the occurrence of human infections in Peru and Costa Rica (Downs and Anderson, 1958).

The Mayaro virus life cycle involves mammals, birds, and arthropods. Marmosets *(Callithrix argentata)* and cebids *(Alouatta belzebul, A. villosa)* have been implicated as natural reservoirs because of the high prevalence of antibodies in these primates and the occasional isolations of the virus (Anderson et al., 1957; Woodall, 1967; Hoch et al., 1981; Seymour et al., 1983).

Virus: The morphological characteristics of Mayaro-virus are essentially similar to those of other alphaviridae. It replicates and produces CPE in BHK-21, VERO, and HeLa-cell-lines; primary rhesus monkey kidney cells and in chicken- or mouse embryo fibroblasts (Pinheiro, 1981).

Mayaro virus and Semliki forest virus cross react to some degree in HI- and CF-tests, but still can be distinguished from one another (Casals and Whitman, 1957).

Epidemiology

Mayaro virus infections are widespread among forest living populations in certain neotropical areas (Pinheiro, 1981). The virus infects mammals (esp. nonhuman primates); caprimulgiform and columbiform birds and arthropods in their natural habitat.

Arthropod vectors of Mayaro virus are mosquitoes (*Haemagogus janthinomys*, *Haemagogus* sp., occasionally *Coquillettidia venezuelensis*, *Sabathes* sp., *Culex* sp., *Mansonia* sp., *Psorophora* sp., *Gigantolaelaps* sp., *Limatus flavisetosus* and *Wyeomyia aporonema*). Most of the available evidences point to a monkey - Haemagogus - monkey cycle (Groot et al., 1961; Berge, 1975; Hoch et al., 1981; Pinheiro, 1981).

Clinical Symptoms

Man: Mayaro fever is generally an acute, transient, benign, nonfatal, febrile illness accompanied by headache, epigastric pain, backache, arthralgia, generalized pain, chills, nausea and photophobia. A maculopapular or micropapular rash appears on the 5th day of illness, particularly on the trunk and the extremities, and lasts for about 3 days. Occasionally light jaundice or inguinal lymphadenopathy can also be observed (Pinheiro, 1981).

Nonhuman primates: No disease has been reported so far.

1.3.4 Getah-Virus

Antibodies to Getah virus without any disease are widespread throughout the Malaysian human population. Also neutralizing antibodies have been reported in wild macaques and *Presbytis* spp. in West Malaysia. Although the natural transmission cycle is unknown, the virus has been isolated from *Culex gelidus*, *C. tritaeniorrhynchus*, *Anopheles amictus*, and *Aedes vexans* in Malaysia, Australia and Japan. Even though these mosquitoes feed mainly on domestic animals and fowl, a jungle cycle can not be excluded based on the serological findings (Marchette, 1981).

1.3.5 Bebaru Virus

Antibodies to Bebaru Virus, known only in Malaysia, have been found in approximately 1/3 of the human sera sampled. The principal hosts are unknown, with the role of monkeys being vague. The virus has been isolated from *Culex* spp. and *Aedes butleri* (Chamberlain, 1980).

1.3.6 Kyasanur Forest Disease (KFD)

In 1956 large numbers of bonnet monkeys *(Macaca radiata)* and langurs *(Presbytis entellus)* were found dead in the Kyasanur Forest in the Shimoga-district, Mysore-State/India. At the same time a disease with 10% mortality was observed in human beings approximately one week after working in the forest (Work and Trapido, 1957; Fiennes, 1967; Rajagopalan and Anderson, 1971; Goverdhan et al., 1974). The etiological agent proved to be a flavivirus closely related serologically to Omsk haemorrhagic fever-, louping ill- and Powassan viruses (Seymour and Yuill, 1981).

Virus

The virus grows well in HeLa cells, human embryonic intestine, chicken embryos and other cell lines. Suckling mice and hamsters can easily be infected intracerebrally (Felsenfeld, 1972; Goverdhan et al., 1974).

The virus, although neutralized by anti Russian-Spring-Summer encephalitis-antisera does not show such cross reactions in Agar gel diffusion tests (Felsenfeld, 1972).

Epidemiology

Since KFD infections are highly fatal in the nonhuman primates monkeys are probably not only heterologous hosts to the virus, but may act as amplifyers, thus helping in the spread of the disease. The virus has also been isolated from forest rats *(Rattus rattus wroughtoni* and *R. blanfordi)*; shrews *(Suncus murinus)* (Boshell-M et al., 1968; Boshell-M and Rajagopalan, 1968; Rajagopalan et al., 1969), and four times from insectivorous bats *(Rhinolophus rouxi)* (Rajagopalan et al., 1969). Antibodies were demonstrated in the above mentioned species and in palm squirrels *(Funambulus tristriatus)*, gerbils *(Tatera indica hardwickei)* and forest mice *(Mus budooga)* (Berge, 1975 - cit. from Seymour and Yuill, 1981), indicating the probability of a rodent reservoir of the virus.

The KFD-virus is transmitted by several ixodid ticks, with *Haemaphysalis spinigera* and *H. turturis* being the principal hosts (Boshell-M et al., 1968; Trapido et al., 1964›). KFD-virus has also been occasionally isolated from *Ixodes petauristae*, and also from bat ticks (*Ornithodorus* spp.) collected during an epidemic (Seymour and Yuill, 1981).

The virus is transovarially transmitted in the three-host *Haemaphysalis* spp.. It also passes through all larval and nymphal stages in Ornithodorus (Bhat and Goverdhan, 1973; Shope, 1980).

Even though the virus appears also in the milk of experimentally infected monkeys 2 to 8 days p.i., it is not readily transmitted to nursing offspring (Shah, 1965).

The disease exhibits distinct seasonality in monkeys - most fatalities are observed during the post-monsoon dry season of the first half of the year, with peak incidents in February and March (Work et al., 1959; Goverdhan et al., 1974). This peak in monkey mortality corresponds well with the period of highest concentration of *Haemaphysalis* nymphs on the forest floor and their maximum activity (Rajagopalan et al., 1968; Rajagopalan and Anderson, 1971; Sreenivasan et al., 1983).

Kyasanur Forest disease has spread since its first discovery into the neighbouring southwestern and southeastern Shimoga and to North-Kanara districts (Shope, 1980). Outside India it has occurred in laboratory personel (Tauraso, 1973). Antibodies have also been demonstrated in recently imported rhesus monkeys (Meister, 1971).

Clinical Symptoms

Nonhuman primates: After experimental infection a biphasic illness ensues with viremia up to 14 days p.i. Erythrophagocytosis, erythro-, leuko- and thrombocytopenia occur 4 to 7 days p.i., followed by an encephalitic phase and finally, 24 hs prior to death, a shock-like fall of the blood pressure, bradycardia, epistaxis, diarrhea (Webb and Chatterjea, 1962; Webb, 1969) and elevated SGOT- and SGPT-values (Banerjee, 1978).

Man: In human cases myalgia, headache, nuchal rigidity, vomitus and tremor were observed and the patients had elevated SGOT- and SGPT-levels (Banerjee, 1978). In fatal cases profuse epistaxis and gastro-intestinal haemorrhage preceeded the final outcome.

Pathology

Nonhuman primates: In langurs and bonnet monkeys which died during epizootics, anal haemorrhage, pallor of the adrenal cortex, focal liver necrosis with cytoplasmic inclusion bodies were described (Iyer et al., 1960). Haemorrhages were seen in lungs, kidneys, brain and adrenals. In the brain a nonpurulent encephalitis marked by focal microgliosis and perivascular cuffing after 15 days of illness was reported (Webb and Burston, 1966).

Man: In three fatal human cases the pathology was similar to those described in monkeys (Iyer et al., 1959).

Treatment

Vaccination of endangered human beings using inactivated anti-Russian-spring-summer encephalitis vaccine was unsuccessful (Shah et al., 1962).

1.3.7 Yellow Fever

Yellow fever, one of the most devastating tropical and subtropical diseases of man has a long and interesting history. The virus probably originated in West Africa as a primary mosquito-parasite, but only after its introduction into the New World did it reach its full dimensions as a cause of dreadful epidemics in man and nonhuman primates.

Virus

Yellow fever virus is the prototype flavivirus. It is an enveloped, spherical virus of 38 nm virionic diameter, and contains a single-stranded RNA genome.

Yellow fever virus is sensitive to lipid solvents, chemical desinfectants, acids and heat (Bergold and Weibel, 1962; Robin and Beran, 1981).

YFV can be isolated from the blood at the beginning of the disease by inoculation into suckling mice or mosquito cell lines (Berge, 1975, - cit. from Seymour and Yuill, 1981; Varma et al., 1976).

For serological investigations CF-tests are recommended (Monath et al., 1980), but all serotests have to be evaluated with caution, since marked cross reactions with the other flaviviruses, especially Zika-viruses occur (Kalter and Heberling, 1971; Henderson et al., 1970).

Epidemiology

Africa: The yellow fever ecology in *Africa* follows two patterns:

a) Sylvatic cycle: Yellow fever virus is basically a commensal of West-African forest or bush living *Aedes* spp. (*A. africanus, A. vittalus, A. neoafricanus, A. furcifer-taylori, A. luteocephalus*- Felsenfeld, 1972; Monath, 1980, 1982). These transmit the infection during blood meals to monkeys living in the canopy (*Colobus badius, Cercopithecus aethiops, C. ascanius schmidti, Erythrocebus patas* - Simpson et al., 1965; Kirya and Okia, 1977; Monath et al., 1980), or as in the case of bush-dwelling mosquito species also to primate species living on the ground. The wild hosts of yellow fever virus in addition to those mentioned above are diana monkeys, mona monkeys, chimpanzees (*Pan* spp.), mangabeys (*Cercocebus* spp.), baboons (*Papio* spp.), bush babies (*Galago* spp.), and possibly hedgehogs (*Erinaceus* spp.) (Robin and Beran, 1981). Old World monkeys show only minor, transient symptoms after infection with yellow fever virus, and the ensuing viremia is of only short duration due to elimination of the virus by antibody-development (Woodall, 1968; Woodall et al., 1968; Kirya and Okia, 1977). After a yellow fever epizootic near Entebbe approximately 40% of the free living cercopithecids examined had anti-yellow fever virus antibodies (Kirya and Okia, 1977).

b) Urban cycle: Viremic monkeys - mostly *Cercopithecus* spp. roaming fields and plantations are also bitten by the peridomestic mosquito species *Aedes simpsoni* and *A. aegypti* living in villages, plantations etc. (Gear, 1977; Cordellier and Akoliba, 1981). *A. aegypti* breeding sites may be water cisterns as shown in certain villages in Ivory Coast (Lhuillier et al., 1985). Once these house-dwelling mosquitoes are infected they can subsequently transmit the infection to the villagers, and, as they are carried by the wind over large distances, can spread yellow fever epizootics rapidly. In *A. aegypti* the yellow fever virus is passed transovarially through the mosquito-generations (Beatty et al., 1980; Anon, 1982), which accounts for the survival of the virus during the dry season in emergence zones. The risk of an urban infection is much greater than that of acquiring the infection via the sylvatic cycle (Cordellier and Akoliba, 1985).

America: A. aegypti, after being shipped across the Atlantic ocean (Tribe, 1977) particularly during the period of slave-trading adapted quickly to the habitats of the Western hemisphere. In pre-Columbian times the Americas were probably free of yellow fever, at least no hints are given to any epizootic in the very detailed recording system of the pre-Columbian American cultures or in reports by the earliest Spanish or Portuguese conquerers. Once infected *A. aegypti* were intro-

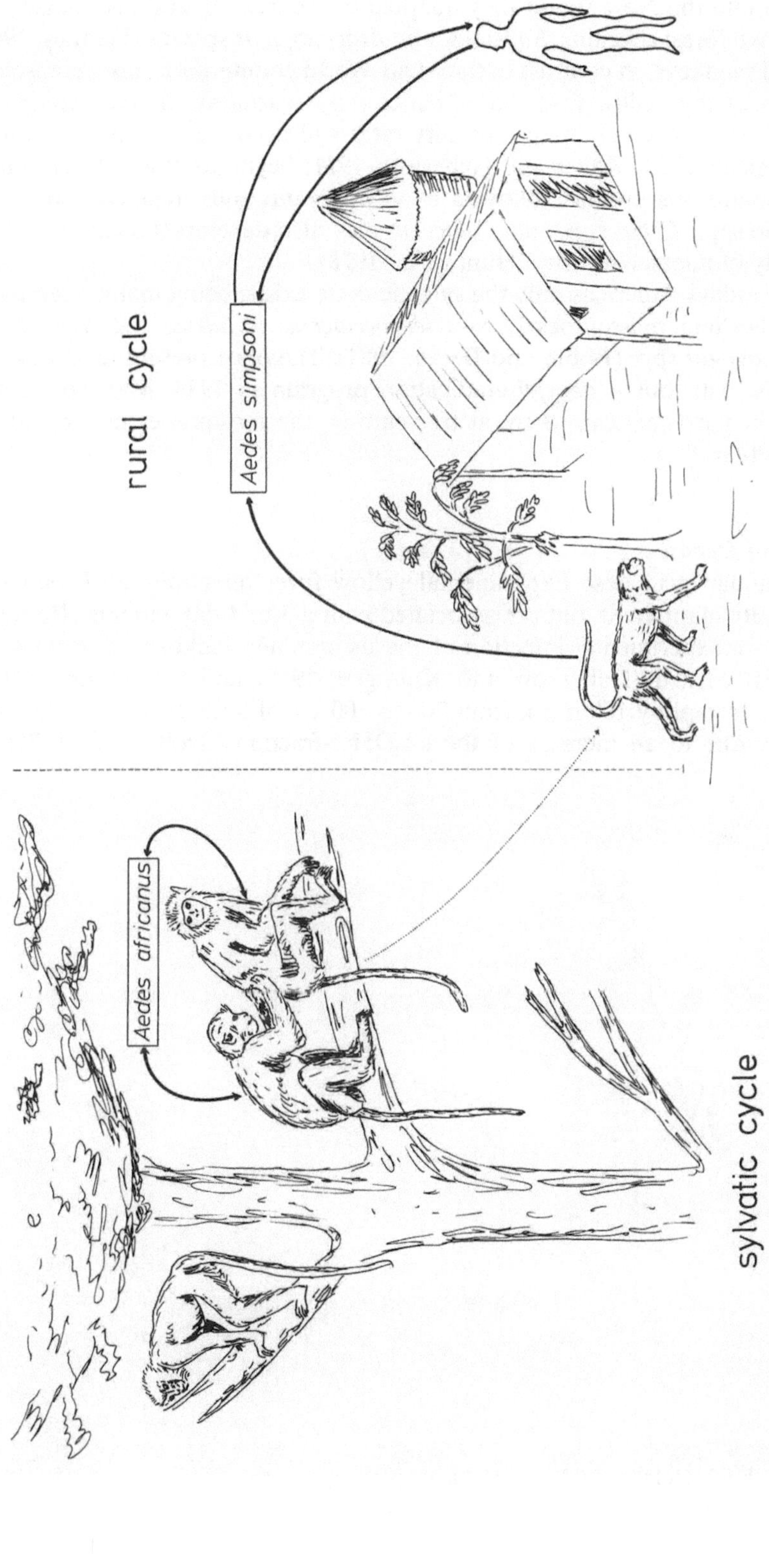

Fig. 4. Epidemiology of yellow fever in Africa

duced into the New World they adapted to the tree life and also passed the virus to native forest dwelling *Sabethes*- and *Haemagogus*-species (Fiennes, 1967). New World monkeys, in contrast to their Old World counterparts, are defenseless to the effects of the yellow fever virus, hence the devastating monkey epizootics associated with extremely high mortality rates still occur in neotropical monkeys (de Rodaniche, 1953; Kalter and Heberling, 1971; Seymour et al., 1983). *Alouatta* are particularly susceptible, followed by *A. trivirgatus* and *Saguinus* spp. *Ateles* spp., *Saimiri* spp., *Cebus* spp. and *Callicebus* may also develop clinical disease, but it is usually of a lesser severity (Hunt et al., 1978).

In todays Americas only the sylvatic cycle exists, being maintained by the canopy dwelling mosquitoes (*Aedes leucocelaenus, A. fulvus, Sabethes chloropterus, Haemagogus* spp. (Robin and Beran, 1981). The once present urban cycle ceased to exist with the *A. aegypti* eradication program in 1934. Man contracts yellow fever in the Americas when, at tree-cutting, the mosquitoes are forced down to ground-level.

Clinical Disease

Nonhuman primates: Experimental yellow fever infections of *Lemur fulvus* are clinically inapparent, but are associated with a 3 to 4 day viremia (Rodhain et al., 1985). In experimental infection of rhesus monkeys leukopenia (primarily granulocytes) ensued (Daberkow and Knüttgen, 1966) and the Serum LDH-activity increased rapidly 100 h p.i. from 200 to 300 U/ml to levels of 4000 to 5000 U/ml mostly due to an increase of the s-LDH-5-fraction (Scott et al., 1976). In fatal

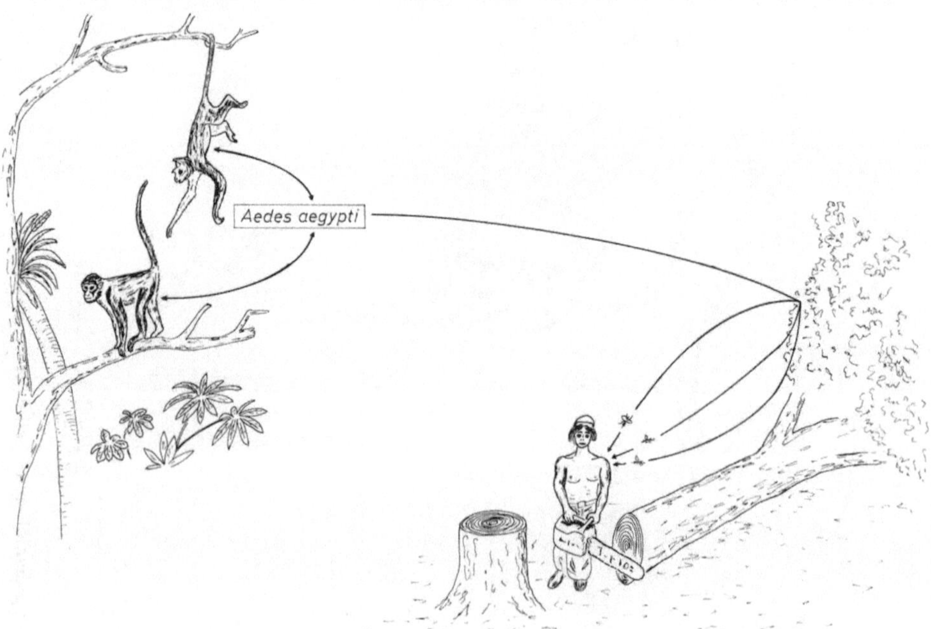

Fig. 5. Epidemiology of yellow fever neotropics

infections of nonhuman primates after a 60 h incubation period, fever and leukopenia are apparent, with death occurring at 100 h p.i.

Man: After a 3 to 10 day incubation period (Hunt et al., 1978) the disease begins with fever, headache, backache, muscle ache, nausea, epistaxis, and a lack of correlation between pulse (bradycardia) and body temperature (39–40 °C) (Faget's sign). After 2 to 3 days either recovery starts or a biphasic course of fever etc. may occur. In serious forms icterus develops at the same time. Haemorrhage and bloody vomitus are characteristics of the most serious, often fatal form of the disease (Robin and Beran, 1981). Renal failure due to lower nephron nephrosis is indicated by a significant albuminuria and anuria (Johnson, 1975).

Pathology

Nonhuman primates: At the necropsy of monkeys wich died from yellow fever, haemorrhagic diathesis and fatty liver degeneration or extensive hepatocellular necrosis were the most common lesions (Smetana, 1962). These are eventually accompanied by the occurrence of "Councilman bodies" (David-West and Smith, 1971). In experimentally infected rhesus monkeys lipids (mainly triglycerides) accumulate in the liver (Liu and Griffin, 1982).

Man: Icterus with haemorrhages in various organs are common. The gastric and intestinal contents may contain partially digested blood. The hepatic lesions are characterized by midzonal necrosis, trabecular disorganization, acidophilic degeneration and fatty metamorphosis of hepatocytes, and the formation of hyaline cytoplasmic masses, the so-called "Councilman bodies". In the kidneys haemoglobinuric lower nephron nephrosis is seen. The myocardium may show degenerative changes (Robin and Beran, 1981).

Treatment

No specific treatment is available, supportative treatment of the dehydration is indicated.

1.3.8 Dengueviruses

Currently four antigenically distinct serotypes of dengue virus, types 1 through 4, are known. Types 1 and 2 are prevalent in West Africa, 1 to 3 in South and Central America and types 1 to 4 in Asia (Shope, 1980). Dengue outbreaks have been described in the past in the latitudes of mild to tropical climates on all continents. The first isolations were accomplished in 1943/44 (Halstead, 1981) although outbreaks of disease resembling dengue fever have been described since the 18th century, particularly in tropical Asia and Australia.

Studies by Monath et al. (1974) indicated dengue virus-infections in both human beings and monkeys in the Nupeko Forest/Nigeria, which were later confirmed in serological studies (Fagbami et al., 1977). In the latter neutralizing antibodies against type 2 dengue virus were demonstrated in 74% of "monkey" sera from the gallery forest at Nupeko and in 38% of "monkey" sera from the lowland rainforest. Meister (1971) demonstrated anti-dengue-antibodies in 7% of rhesus monkeys during quarantine.

Virus

Dengue virions are spherical, approximately 50 nm in diameter and enveloped. The envelopes carry knob-like projections.

Dengue viruses grow in primary rhesus monkey kidney-, LLC-MK2-, BSC-1-, Vero- or other cell lines. Plaques are formed under agar- or methylcellulose-overlay. Explant techniques are more successful in recovery of dengue virus than others, most efficient is the co-cultivation of explants with LLC-MK2-cells (Marchette et al., 1972).

Epidemiology

Studies by Rudnick (cit. from Halstead, 1981) revealed a jungle cycle of dengue virus in Malaysia, involving forest dwelling monkeys and *Aedes niveus*, which feeds both on monkeys and man. Whether or not similar conditions exist in Africa and the tropical Americas is unknown.

In the more important urban cycle *A. aegypti* is the principal vector. The virus has also been recovered from naturally infected *A. albopictus* involved in outbreaks in the Pacific area.

Dengue viruses can be transmitted transovarially in the mosquitoes, but larvae are not infected by development in contaminated water.

Clinical Disease

Nonhuman primates: clinical symptoms in spontaneously infected monkeys are not known. After experimental infections of *Macaca* sp., *Cercopithecus* sp., *Cercocebus* sp. and *Papio* sp. an asymptomatic viremia 1 to 7 days p.i. results (Halstead, 1981).

Man: In human dengue virus-infections two types of disease can be distinguished:

One, "Dengue fever", is characterized by a biphasic fever 2 to 7 days p.i.. A sudden developing 39.5 to 41 °C fever is accompanied by frontal or retro-orbital headache. During the first two days of fever a transient, macular, generalized rash and myalgia or bone pain occur. From days 2 through to 6 of the fever, nausea and vomiting are frequent and a generalized lymphadenopathy, cutaneous hyperaesthesia, taste aberration and anorexia may develop. A second generalized maculopapular rash and fever follow 1 to 2 days after defervescence of the first symptoms, disappearing 1 to 5 days later and occasionally followed by desquamation of the epithelium. Epistaxis, petechiae and purpuric lesions are sometimes noticed.

The other form of dengue virus-infection is "Dengue haemorrhagic fever" or "Dengue-shock". This was first described in the Philippine Islands, occurring predominantly in children. An initial phase of sudden fever, malaise, vomiting, headache and cough is followed after 2 to 5 days by rapid deterioration and collapse. Petechiae on the forehead and extremities are frequent. The mortality ranges from 10 to 40%. In the surviving patients convalescence begins after a 24 to 36 hrs crisis.

Pathology

Nonhuman primates: No pathological lesions in dengue virus infected monkeys have been published.

Man: Minimal to moderate haemorrhages are usually found in several organs, and in most patients yellow to bloody effusions in the serous cavities are found. Disturbances in the normal haematopoiesis are indicated by arrested megakaryocytosis in the bone marrow with peripheral megakaryocytosis in the capillaries of the lungs, kidney, liver, and spleen. In the liver varying degrees of fatty metamorphosis to focal midzonal necrosis are seen, as well as sinusoidal non-nucleated cells with vacuolated acidophilic cytoplasm, resembling Councilman bodies.

1.3.9 Zika Virus

Zika virus was first isolated in 1947 from a sentinel rhesus in Zika forest/Uganda (Dick et al., 1952). Several isolations have been made from naturally infected human beings in Uganda, Nigeria, and Senegal, and from mosquitoes in the Central African Republic, Ivory Coast, and Malaysia (Boorman and Porterfield, 1956; Weinbren and Williams, 1958; Hadow et al., 1964; Simpson, 1964; Marchette et al., 1969).

Virus

Zika virions are of typical flavivirus-shape. They are spherical, enveloped, 18 to 45 mm in diameter, and contain single-stranded RNA. The virus is inactivated by ether, sodium deoxycholate and chloroform (Robin, 1981).

Epidemiology

The virus reservoir is unknown. West African monkeys might be included in the cycle as suggested by serological evidence (Chamberlain, 1980; Robin, 1981; McCrae and Kirya, 1982). It has been isolated from *Aedes aegypti*, *A. africanus*, *A. opok*, *A. luteocephalus*, *A. vivittatus*, *A. dalzieli*, *A. furcifer-taylori*-group, and *Anopheles gambiae*. The epidemiological pattern may be similar to that of yellow fever with *A. africanus* transmitting the virus among canopy living monkeys.

Clinical Symptoms

Nonhuman primates: No clinical symptoms have been reported in the sentinel monkeys.

Man: The few human illnesses, including one laboratory infection, were characterized as a benign febrile illness with malaise, headache, backache and possibly maculopapular rash (McIntosh et al., 1975).

1.3.10 Orungovirus

Orungovirus was first isolated in 1969 in Central Africa (Anon, 1969) and several epidemics have been reported in Nigeria (Fabiyi, 1981). The highest number of seropositive people were found in the Guinea savanna (Fabiyi, 1981). Orungovirus was also isolated during a yellow fever epidemic in the Gambia (Monath et al.,

1980 - cit. from Cordellier et al., 1982). Besides man, 24% of *Cercopithecus mona* and *aethiops* (and 50 % of sheep) in Nigeria were seropositive (Fabiyi, 1981) and the virus was also recovered from mosquitoes feeding on a *C. aethiops*.

Life cycle: Aedes dentatus, A. aegypti, A. gambiae, A. africanus, A. opok, Culex perfuscus and *Anopheles spp.* are regarded as vectors of Orungovirus (Fabiyi, 1981; Cordellier et al., 1982). Orungovirus, like other arboviruses, can be transmitted transovarially in the mosquito (Cordellier et al., 1982).

Clinical Symptoms

In nonhuman primates the disease is unknown.

Man: Orungo-fever is characterized by nausea, vomiting, myalgia, headache, conjunctivitis, leucopenia and fever lasting up to 7 days with fine papular rashes involving the face, chest, and abdomen (Fabiyi, 1981). Death is a possible sequela (Fabiyi et al., 1975).

1.3.11 Oropouche-Virus

Oropouche virus was first isolated from a febrile patient in Trinidad in 1955. It caused several epidemics in the Amazonian region of northern Brazil from 1967 to 1975 (Parkin, 1975; Pinheiro, 1981). The virus was isolated from a sloth *(Bradypus tridactylus)* in Brazil, while monkeys also seem to be involved in the epidemiology based on evidence from serological studies (Hubbert et al., 1975).

Virus

Oropouche virus is one of at least 16 simbuviruses of the bunyaviridae-family. Its enveloped virions are of 90–100 nm in diameter. The virus multiplies in primary chicken embryo-, Vero-, and BHK-21-cells. It also forms plaques in BHK-21-, MA III- and LLC-MK2 cell lines (Pinheiro, 1981). All the bunyaviridae cross react serologically to some extent, but they can be differentiated by CF-techniques. Oropouche virus contains a haemagglutinin.

Epidemiology

Serological studies have indicated that the Oropouche virus is maintained in monkeys, sloths, birds, and possibly other animals. In Trinidad many *Alouatta-* and *Cebus* spp. were seropositive and antibodies to the Oropouche virus were found in most monkeys and wild birds (particularly formicariidae) surveyed in the Amazon basin (Pinheiro, 1981).

The vectors in nature are still unknown, although the virus was isolated from *Coquilletidia venezuelensis, Aedes serratus, Culex quinquefasciatus* and *Culicoides paraensis* (Pinheiro, 1981).

Clinical Symptoms

Nonhuman primates: So far, no disease has been reported in spontaneously or experimentally infected monkeys.

Man: Oropouche fever is characterized by a fever of abrupt onset, with headache, myalgia, arthralgia, chills, dizziness, and photophobia. Nausea, vomiting, diarrhea and epigastric pains may also sometimes be present. The illness usually lasts 2 to 7 days, recurrences do occur.

Pathology

No lesions have been reported in either man or nonhuman primates.

References

1. Anderson, C.R.; W.G. Downs; G.H. Wattley; N.W. Akin and A.A. Reese: Mayaro virus: a new human disease agent. II. Isolations from blood of patients in Trinidad B.W.I. Amer. J.trop. Med. Hyg. *6*, 1012-1016 (1957)
2. Anon: Rapport annuel de l'Institut Pasteur de Bangui (1969)
3. Anon: Yellow fever surveillance. Morbidity and Mortality Weekly Report *33*, 452-453 (1982)
4. Banerjee, K.: Serum transaminases in Kyasanur Forest disease: A preliminary report. Indian J.med. Res. *67*, 1-4 (1978)
5. Beatty, B.J.; R.B. Tesh and T.H.G. Aitken: Transovarial transmission of yellow fever virus in *Stegomyia* mosquitoes. Amer. J. trop. Med. Hyg. *29*, 125-132 (1980)
6. Berge, T.O.: International Catalog of Arboviruses, 2nd ed. U.S. Dept. HEW, Publ. Hlth Educ. and Welf., Publ. Hlth Serv. (1975)
7. Bergold, G.H. and J.Weibel: Demonstration of yellow fever virus with the electron microscope. Virology *17*, 554-562 (1962)
8. Bhat, U.K.M. and M.K. Goverdhan: Transmission of Kyasanur Forest disease virus by the soft tick, Ornithodorus crossi. Acta virol. *17*, 337-342 (1973)
9. Boorman, J.P.T. and J.S. Porterfield: A simple technique for infection of mosquitoes with viruses. Transmission of Zika virus. Trans. roy. Soc. trop. Med. Hyg. *50*, 238-242 (1956)
10. Boshell-M, J.and P.K. Rajagopalan: Observation on the experimental exposure of monkeys, rodents, and shrews to infestations of ticks in forest in Kyasanur Forest disease area. Indian J.med. Res. *56*, 573-588 (1968)
11. Boshell-M, J.; P.K. Rajagopalan; A.P. Patil and K.M. Pavri: Isolation of Kyasanur Forest disease virus from ixodid ticks: 1961-1964. Indian J.med. Res. *56* (Suppl.), 541-568 (1968)
12. Carey, D.E.; R.M. Myers; C.M. de Ranitz; M.Jadhav and R.Reuben: The 1964 chikungunya epidemic at Vellore, South India, including observations on concurrent dengue. Trans. roy. Soc. trop. Med. Hyg. *63*, 434-445 (1969)
13. Casals, J.and L.Whitman: Mayaro virus: a new human disease agent. I. Relationship to other arboviruses. Amer. J.trop. Med. Hyg. *6*, 1004-1011 (1957)
14. Causey, O.R. and O.M. Maroja: Mayaro virus: A new human disease agent. III. Investigation of an epidemic of acute febrile illness on the River Guana in Para, Brazil, and isolation of Mayaro virus as causative agent. Amer. J.trop. Med. Hyg. *6*, 1017-1023 (1957)
15. Chamberlain, R.W.: Epidemiology of arthropod-borne togaviruses: The role of arthropods as hosts and vectors and of vertebrate hosts in natural transmission cycles. Pp.175-227 in "The Togaviruses" (Schlesinger, R.W., ed.). Academic Press, New York, 1980
16. Cordellier, R.et P.Akoliby: Les moustiques de la forêt sempervirente du sud-ouest ivoirien. I. Etude du contact entre l'homme et les vecteurs potentiels de fièvre jaune au niveau du sol. Cah. O.R.S.T.O.M., ser. Ent. med. et Parasitol. *XIX*, 297-301 (1981)
17. Cordellier, R.; A.Chippaux; N.Monteny ; G.Heme; B.Courtois; M.Germain and J.-P. Digoutte: Isolements du virus Orungo à partir de femelles et de males d'*Aedes* selvatiques capturé en Côte d'Ivoire. Cah. O.R.S.T.O.M., ser. Ent. med. et Parasitol. *XX* (4), 265-267 (1982)
18. Daberkow, K.und H.J. Knüttgen: Das periphere Blutbild von Rhesusaffen nach Infektion mit Gelbfieberstämmen unterschiedlicher Pathogenität. Z.Tropenmed. *17*, 176-194 (1966)
19. David-West, T.S. and J.A. Smith: Yellow fever virus infection. A correlation of complement-fixing antigen with histopathology. Brit. J.exp. Path. *52*, 114-121 (1971)
20. DeRodaniche, E.: Survey of primates captured in Panama. R.P. during the years 1952-1956 for protective antibodies against yellow fever. Amer. J.trop. Med. Hyg. *6*, 835-839 (1957)

21. Dick, G.W.A.; S.F. Kitchen and A.J. Haddow: Zika virus - (I) Isolations and serological specificity. Trans. roy. Soc. trop. Med. Hyg. *46*, 509-520 (1952)
22. Downs, W.G. and C.R. Anderson: Distribution of immunity to Mayaro virus infection in the West Indies. Indian med. J. *7*, 190-195 (1958)
23. Fabiyi, A.: Other arboviral zoonoses in Africa. Pp. 259-262 in "CRC Handbook Series in Zoonoses". (Steele, J.H., Series Editor), Section B: Viral Zoonoses, Vol. I (Beran, G.W., Section Editor). CRC Press, Boca Raton (1981)
24. Fabiyi, A.; O. Tomori and M.S.M. El-Bayoumi: Epidemics of a febrile illness associated with UgMP 359 virus in Nigeria. W. Afr. med. J. *23*, 9-11 (1975)
25. Fagbami, A.H.; T.P. Monath and A. Fabiyi: Dengue virus infections in Nigeria: a survey for antibodies in monkeys and humans. Trans. roy. Soc. trop. Med. Hyg. *71*, 60-65 (1977)
26. Felsenfeld, A.D.: The arboviruses, pp. 523 - 536 in "Pathology of Simian Primates", Part II (Fiennes, R.N.T-W., ed.), Karger, Basel (1972)
27. Fiennes, R.N.T-W.: Zoonoses of Primates. Cornell Univ. Press, Ithaca (1967)
28. Galindo, P.; S. Srihongse; E. de Rodaniche and M.A. Grayson: An ecological survey of arboviruses in Almirante, Panama, 1959-1962. Amer. J. trop. Med. Hyg. *15*, 385-400 (1966)
29. Gear, J.H.S.: Medical aspects of some zoonoses. J.S. Afr. vet. Ass. *46*,. 221-225 (1975)
30. Gear, J.H.: Hemorrhagic fevers of Africa: An account of two recent outbreaks. J.S. Afr. vet. Ass. *48*, 5-8 (1977)
31. Gear, J. and F.P. Reid: The occurrence of a dengue-like fever in the northeastern Transvaal. S. Afr. med. J. *31*, 253-257 (1957)
32. Goverdhan, M.K.; P.K. Rajagopalan; D.P. Narasimha Murthy; S. Upadhyaya; M.J. Boshell; H. Trapido and T. Ramachandra Rao: Epizootiology of Kyasanur Forest disease in wild monkeys of Shimoga district, Mysore State (1957-1964). Indian J. med. Res. *62*, 497-510 (1974)
33. Groot, H.; A. Morales and H. Vidales: Virus isolation from forest mosquitoes in San Vincente de Chucuri, Colombia. Amer. J. trop. Med. Hyg. *10*, 397-402 (1961)
34. Haddow, A.J.; M.C. Williams; J.P. Woodall; D.I.H. Simpson and L.K.H. Goma: Twelve isolations of Zika virus from *Aedes (Stegomyia) africanus* (Theobald) taken in and above a Uganda Forest. Bull. Wld. Hlth Org. *31*, 57-69 (1964)
35. Halstead, S.B.: Mosquito-borne haemorrhagic fever of South and South-East Asia. Bull. Wld. Hlth Org. *35*, 3-15 (1966)
36. Halstead, S.B.: Dengue and dengue hemorrhagic fever. Pp. 421 - 435 in "CRC Handbook Series in Zoonoses". (Steele, J.H., Series ed.), Section B: Viral Zoonoses, Vol. I (Beran, G.W., Section ed.). CRC Press, Boca Raton (1981)
37. Halstead, S.B.: Chikungunya fever. Pp. 437-447 in "CRC Handbook Series in Zoonoses" (Steele, J.H. - Series Editor), Section B: Viral Zoonoses, Vol. I (Beran, G.W. - Section Editor). CRC Press, Inc. Boca Raton (1981)
38. Harrison, V.R.; J.D. Marshall and N.B. Guilloud: The presence of antibody to chikungunya and other serologically related viruses in the sera of sub-human primate imports to the United States. J. Immunol. *28*, 979-981 (1967)
39. Henderson, B.E.; P.P. Ceshire; G.B. Kirya and M. Lule: Immunologic studies with yellow fever and selected African group B arboviruses in rhesus and vervet monkeys. Amer. J. trop. Med. Hyg. *19*, 110-118 (1970)
40. Hoch, A.L.; N.E. Peterson; J.W. LeDuc and F.P. Pinheiro: An outbreak of Mayaro virus disease in Belterra, Brazil. Amer. J. trop. Med. Hyg. *30*, 689-698 (1981)
41. Hubbert, W.T.; W.F. McCulloch and P.R. Schnurrenberger: Tabular summary of diseases, hosts, and transmission. Pp. 1117 - 1138 in "Diseases Transmitted from Animals to Man." C.C. Thomas, Springfield (1975)
42. Hunt, R.D.; M.P. Anderson and L.V. Chalifoux: Spontaneous infectious diseases of marmosets. Primates Med. *10*, 239-253 (1978)
43. Iyer, C.G.S.; R.L. Rao, T.H. Work and D.P.N. Murthy: Kyasanur Forest disease: IV. Pathological findings in three fatal human cases of Kyasanur Forest disease. Indian J. med. Sci. *13*, 1011-1022 (1959)
44. Iyer, C.G.S.; T.H. Work; D.P.N. Murthy; H. Trapido and P.K. Rajagopalan: Kyasanur Forest disease: VII. Pathological findings in monkeys, *Presbytis entellus* and *Macaca radiata* found dead in the forest. Indian J. med. Res. *48*, 276-286 (1960)

45. Johnson, K.M.: Yellow fever. pp. 929-938 in "Diseases Transmitted from Animals to Man" (Hubbert, W.T.; W.F. McCulloch and P.R. Schnurrenberger, eds.), C.C. Thomas Publ., Springfield (1975)
46. Kalter, S.S. and R.L. Heberling: Yellow fever serology in nonhuman primates. Primate Zoonoses Surveil. Rep. *5*, 13-14 (1971)
47. Kanamitsu, M.; K. Taniguchi; S. Urasawa; T. Ogata; Y. Wada and S. Saroso: Geographic distribution of arbovirus antibodies in indigenous human populations in the Indo-Australian archipelago. Amer. J. trop. Med. Hyg. *28*, 351-363 (1979)
48. Kaschula, V.R.; K.F. van Dellen and V. de Vos: Some infectious diseases of wild vervet monkeys *(Cercopithecus aethiops pygerythrus)* in South Africa. J.S.Afr. vet. Ass. *49*, 223-227 (1978)
49. Kirya, B.G. and N.O. Okia: A yellow fever epizootic in Zika Forest, Uganda, during 1972: II. Monkey serology. Trans. roy. Soc. trop. Med. Hyg. *71*, 300-303 (1977)
50. LeDuc, J.W.; F.P. Pinheiro and A.P.A. Travassos da Rosa: An outbreak of Mayaro virus disease in Belterra, Brazil. II. Epidemiology. Amer. J. trop. Med. Hyg. *30*, 682-688 (1981)
51. Lhuillier, M.; J.L. Sarthou; R. Cordellier; G.M. Gershydamet; N. Monteny; B. Bouchite and P. Calen: Epidémie rurale de fièvre jaune avec transmission interhumaine en Côte d'Ivoire de 1982. Bull. Wld. Hlth Org. *63*, 527-536 (1985)
52. Liu, C.T. and M.J. Griffin: Changes in body fluid compartments, tissue water and electrolyte distribution, and lipid concentrations in rhesus macaques with yellow fever. Amer. J. vet. Res. *43*, 2013-2018 (1982)
53. Lumsden, W.H.R.: An epidemic of virus disease in southern province, Tanganyika territory, in 1952 - 53. II. General description and epidemiology. Trans. roy. Soc. trop. Med. Hyg. *49*, 33-57 (1955)
54. Marchette, N.J.: Arboviruses of Southeast Asia. pp. 457-466 in "CRC Handbook Series in Zoonoses" (Steele, J.H. - Series ed.), Section B: Viral Zoonoses, Vol. I (Beran, G.W. - Section ed.), CRC Press, Boca Raton (1981)
55. Marchette, N.J.; R. Garcia and A. Rudnick: Isolation of Zika virus from *Aedes aegypti* mosquitoes in Malaysia. Amer. J. trop. Med. Hyg. *18*, 411-415 (1969)
56. Marchette, N.J.; S.B. Halstead; D.R. Nash and A.C. Stenhouse: Recovery of dengue viruses from tissues of experimentally infected rhesus monkeys. Appl. Microbiol. *24*, 328-333 (1972)
57. McCrae, A.W.R. and B.G. Kirya: Yellow fever and Zika virus epizootics and enzootics in Uganda. Trans. roy. Soc. trop. Med. Hyg. *76*, 552-562 (1982)
58. McIntosh, B.M.: Antibody against chikungunya virus in wild primates in southern Africa. S. Afr. J. med. Sci. *35*, 65-74 (1970)
59. McIntosh, B.M. and J.H.S. Gear: Mosquito-borne arboviruses, primarily in the eastern hemisphere. pp. 939-967 in "Diseases Transmitted from Animals to Man." (Hubbert, W.T.; W.F. McCulloch and P.R. Schnurrenberger, eds.), 6th ed., C.C. Thomas, Springfield (1975)
60. McIntosh, B.M. and J.H.S. Gear: Arboviral zoonoses in Africa. Arboviral zoonoses in Southern Africa. Chikungunya fever. pp. 217-220 in "CRC Handbook Series in Zoonoses." (Steele, J.H., Series ed.), Section B: Viral Zoonoses, Vol. I (Beran, G.W., Section ed.). CRC Press, Boca Raton (1981)
61. McIntosh, B.M.; R.M. Harwin; H.E. Paterson and M.L. Westwater: An epidemic of chikungunya in southeastern southern Rhodesia. Cent. Afr. J. Med. *9*, 351-359 (1963)
62. McIntosh, B.M.; P.G. Jupp and I. Santos: Rural epidemic of chikungunya in South Africa with involvement of *Aedes (Diceromyia) furcifer* (Edwards) and baboons. S. Afr. J. med. Sci. *73*, 267-269 (1977)
63. McIntosh, B.M.; H.E. Paterson; G.M. Gillivray and J. DeSousa: Further studies on the chikungunya outbreak in southern Rhodesia in 1962. I. Mosquitos, wild primates and birds in relation to the epidemic. Ann. trop. Med. Parasit. *58*, 45-51 (1964)
64. Meister, G.: Erfahrung bei der Quarantänisierung von Affen. Z. Versuchstierk. *13*, 236-237 (1971)
65. Metselaar, D.: Isolation of arboviruses of group A and group C in Surinam. Trop. geogr. Med. *18*, 137-142 (1966)
66. Monath, T.P.; R.B. Craven; A. Adjukievicz; M. Germain; D.B. Francy; L. Ferrara; E.M. Samba; H. N'Jie; K. Cham; S.A. Fitzgerald; P.H. Crippen; D.I.H. Simpson; E.T.W. Bowen; A. Fabiyi and J.-J. Salaun: Yellow fever in the Gambia, 1978 - 1979: Epidemiologic

aspects with observations on the occurrence of Orungo virus infection. Amer. J.trop. Med. Hyg. *29*, 912-928 (1980)
67. Monath, T.P.; V.H. Lee; A.Fagbami; O.Tomori and D.Wilson: Arbovirus studies in Nupeko forest, a possible natural focus of yellow fever virus in Nigeria. I. Description of the area and a serological survey. Trans. roy. Soc. trop. Med. Hyg. *68*, 30-38 (1974)
68. Myers, R.M.; D.E. Carey; R.Reuben; E.S. Jesudass; C.de Ranitz and M.Jadhav: The 1964 epidemic of denguelike fever in South India: Isolation of chikungunya virus from human sera and from mosquitoes. Indian J.med. Res. *53*, 694-701 (1965)
69. Nimmannitya, S.; S.B. Halstead; S.N. Cohen and M.R. Margiota: Dengue and chikungunya virus infections in man in Thailand, 1962 - 1964. I.Observations on hospitalized patients with hemorrhagic fever. Amer. J.trop. Med. Hyg. *18*, 954-971 (1969)
70. Parkin, W.E.: Mosquito-borne arboviruses other than group A primarily in the western hemisphere. Pp.984-993 in "Diseases Transmitted from Animals to Man" (Hubbert, W.T.; W.F. McCulloch, and P.R. Schnurrenberger, eds.), C.C. Thomas, Springfield (1975)
71. Pinheiro, F.P.: Oropouche fever. Pp.177-181 in "CRC Handbook Series in Zoonoses" (Steele, J.H., Series ed.), Sect. B: Viral Zoonoses, Vol.1 (Beran, G.W., Section ed.), CRC Press, Boca Raton (1981)
72. Pinheiro, F.P.: Arboviral zoonoses in South America - Mayaro fever. Pp.159-164 in "CRC Handbook Series in Zoonoses" (Steele, J.H., Series ed.), Sect. B: Viral Zoonoses, Vol.1 (Beran, G.W., Sect. ed.), CRC Press, Boca Raton (1981)
73. Pinheiro, F.P.; R.B. Freitas; J.F.S. Travassos da Rosa; Y.B. Gabbay; W.A. Mello, and J.E. LeDuc: An outbreak of Mayaro virus disease in Belterra, Brazil. I.Clinical and virological findings. Amer. J.trop. Med. Hyg. *30*, 674-681 (1981)
74. Rajagopalan, P.K. and C.R. Anderson: Further studies on ticks of wild monkeys of Kyasanur Forest disease area Shimoga District. Indian J.med. Res. *59*, 847-860 (1971)
75. Rajagopalan, P.K.; A.V. Patil and M.J. Boshell: Studies on ixodid tick populations on the forest floor in the KFD area. J.med. Res. *56* (Suppl.), 497-509 (1968)
76. Rajagopalan, P.K.; S.D. Paul and M.A. Sreenivasan: Isolation of Kyasanur Forest disease virus from the insectivorous bat *Rhinolophus rouxi* and from *Ornithodorus* ticks. Indian J.med. Res. *57*, 805-808 (1969)
77. Robin, Y.: Arboviral Zoonoses in East, Central and West Africa. Chikungunya fever. Pp.235-239 in "CRC Handbook Series in Zoonoses" (Steele, J.H., Series ed.), Sect. B: Viral Zoonoses, Vol.1 (Beran, G.W., Sect. ed.), CRC Press, Boca Raton (1981)
78. Robin, Y.: ZIKA Fever. Pp.248-250 in "CRC Handbook Series in Zoonoses" (Steele, J.H., Series ed.), Sect. B: Viral Zoonoses, Vol.1 (Beran, G.W., Sect. ed.), CRC Press, Boca Raton (1981)
79. Robin, Y.and G.W. Beran: Yellow fever. Pp.85-96 in "CRC Handbook Series in Zoonoses" (Steele, J.J., Series ed.), Sect. B: Viral Zoonoses, Vol.1 (Beran, G.W., Sect. ed.), CRC Press, Boca Raton (1981)
80. Robinson, M.C.: An epidemic of virus disease in southern province, Tanganyika territory, in 1952-53.I.Clinical features. Trans. roy. Soc. trop. Med. Hyg. *49*, 28-52 (1955)
81. Rodhain, F.; J.-J. Petter; R.Albignac; P.Coulanges, and C.Hannoun: Arboviruses and lemurs in Madagascar: Experimental infection of *Lemur fulvus* with yellow fever and West Nile viruses. Amer. J.trop. Med. Hyg. *34*, 816-822 (1985)
82. Ross, R.W.: The Newala epidemic. III. The virus: Isolation, pathogenic properties and relationship to the epidemic. J.Hyg. *54*, 177-191 (1956)
83. Schaeffer, M.; D.C. Gajdusek; A.B. Lema, and H.Eicheswald: Epidemic jungle fever among Okinawan colonists in the Bolivian rain forest. I.Epidemiology. Amer. J.trop. Med. Hyg. *8*, 372-396 (1959)
84. Scott, S.K.; P.C. Kosch, and D.E. Hilmas: Serum lactate dehydrogenase of normal, stressed and yellow fever virus infected rhesus monkeys. Lab. Anim. Sci. *26*, 436-442 (1976)
85. Seymour, C.; P.H. Peralta, and G.G. Montgomery: Serologic evidence of natural Togavirus infections in Panamanian slothes and other vertebrates. Amer. J.trop. Med. Hyg. *32*, 854-861 (1983)
86. Seymour, S.and T.M. Yuill: Arboviruses. Pp.54-86 in "Infectious Diseases of Wild Mammals", 2nd ed. (Davis, J.W.; L.H. Karstad and D.O. Trainer, eds.), Iowa State Univ. Press, Amer. (1981)

87. Shah, K.V.: Experimental infection of lactating monkeys with Kyasanur Forest disease virus. Acta Virol. *9*, 72-75 (1965)
88. Shah, K.V.; S.P. Aniker; D.P.N. Murphy; F.M. Rodrigues; M.S. Jayadeviah, and H.A. Prasanna: Evaluation of the field experience with formalin-inactivated mouse brain vaccine of Russian spring-summer encephalitis against Kyasanur Forest disease. Indian J.med. Res. *50*, 162-174 (1962)
89. Shope, R.E.: Medical significance of togaviruses: An overview of diseases caused by togaviruses in man and in domestic and wild vertebrate animals. Pp.47-82 in "The Togaviruses. Biology, Structure, Duplication" (Schlesinger, R.W., ed.), Academic Press, New York (1980)
90. Simpson, D.I.H.: Zika virus infection in man. Trans. roy. Sco. trop. Med. Hyg. *58*, 335-338 (1964)
91. Simpson, D.I.H.; A.H. Haddow; M.C. Williams and J.P. Woodall: Yellow fever in Central Uganda, 1964.4. Investigations on blood sucking diptera and monkeys. Trans. roy. Soc. trop. Med. Hyg. *59*, 449-458 (1965)
92. Smetana, H.F.: The histopathology of experimental yellow fever. Virchows Arch. Path. Anat. *335*, 411-427 (1962)
93. Sreenivasan, M.A.; P.K. Rajagopalan and H.R. Bhat: Spatial distribution of infected *Haemaphysalis* nymphs in the epizootic localities of Kyasanur forest disease. Indian J.Med. Res. *78*, 531-536 (1983)
94. Tauraso, N.M.: Review of recent epizootics in nonhuman primate colonies and their relations to man. Lab. Anim. Sci. *23*, 201-210 (1973)
95. Trapido, H.; M.K. Goverdhan; P.K. Rajagopalan, and M.Rebello: Ticks ectoparasitic on monkeys in the Kyasanur Forest disease area of Shimoga District, Mysore State, India. Amer. J.trop. Med. Hyg. *13*, 763-822 (1964)
96. Trapido, H.; P.K. Rajagopalan; T.H. Work, and M.G.R. Varma: Kyasanur Forest disease. 8. Isolation of Kyasanur Forest virus from naturally infected ticks of the genus Haemaphysalis. Indian J. med. Res. *47*, 133-138 (1959)
97. Tribe, G.W.: Simian haemorrhagic fever in perspective. Primate Supply *2* (3). 40-41 (1977)
98. Varma, M.G.R.; M.Pudney; C.J. Leake, and P.H. Peralta: Isolation in a mosquito (*Aedes pseudoscutellaris*) cell line (Mos. 61) of yellow fever virus strains from original field material. Intervirology *6*, 50-56 (1976)
99. Warren, A.J.: Landmarks in the conquest of yellow fever. Pp. 1-37 in "Yellow Fever" (Strode, G.K., ed.). McGraw-Hill, New York (1951)
100. Webb, H.E.: Kyasanur Forest disease virus infection in monkeys. Lab. Anim. Handb. *4*, 131-134 (1969)
101. Webb, H.E. and J.Burston: Clinical and pathological observations with special reference to the nervous system in *Macaca radiata* infected with Kyasanur Forest disease virus. Trans. roy. Soc. trop. Med. Hyg. *60*, 325-331 (1966)
102. Webb, H.E. and J.B. Chatterjea: Clinico-pathological observations on monkeys infected with Kyasanur Forest disease virus, with special reference to the hemopoietic system. Brit. J. Haemat. *8*, 401-413 (1962)
103. Weinbren, M.P. and M.C. Williams: Zika virus: further isolation in the Zika area, and some studies on the strains isolated. Trans. roy. Soc. trop. Med. Hyg. *52*, 263-268 (1958)
104. Woodall, J.P.: Virus research in Amazonia. Atlas do Simposio sobre a Biota Amazonica *6*, pp.31-63 (Lent, H., ed.) Conselho Nacional de Pesquisas, Rio de Janeiro (1967)
105. Woodall, J.P.: The reaction of a mangabey monkey (*Cercocebus galeritus agilis* Milne-Edwards) to inoculation with yellow fever virus. Ann. trop. Med. Parasit. *62*, 522-527 (1968)
106. Woodall, J.P.; J.R.W. Dykes, and M.C. Williams: The reaction of a species of colobus monkey to inoculation with yellow fever virus. Ann. trop. Med. Parasit. *62*, 528-535 (1968)
107. Work, T.H.; F.R. Rodriguez, and P.N. Bhatt: Virological epidemiology of the 1958 epidemic of Kyasanur Forest disease. Amer. J.publ. Hlth *49*, 869-874 (1959)
108. Work, T.H. and H.Trapido: Summary of preliminary report of investigations of the Virus Research Center on epidemic disease affecting forest villages and wild monkeys of Shimoga District, Mypore. Indian J.med. Sci. *11*, 340-341 (1957)

1.4 Rhabdoviruses

1.4.1 Rabies-Virus

Today rabies is distributed worldwide, and is still extending its geographical boundaries. The infection, although primarily carried in carnivorous animals (mainly canids), is transmissible to other mammalian species including man, and consequently, has been reported in a number of recently imported nonhuman primates of different geographical and taxonomic origin (Schmitter, 1914; Boulger, 1966; Kaplan, 1966; Richardson, 1971a, b; Richardson and Humphrey, 1971; Fiennes, 1972; Miot and Sikes 1973; Röttcher and Sawchuk, 1978). At least one human fatality has been ascribed to bites from a rabid monkey (Prakash, 1970), and thousands of human rabies-vaccinations have been required after monkey bites (Richardson, 1971).

Virus
The virological characteristics of the rabies virus are generally well known, making a lengthy discussion within the context of the present chapter unnecessary. For references see Beran, 1981.

Epidemiology
The reservoir hosts of the rabies virus are primarily carnivorous animals - in Africa: mainly dogs, jackals, hyenas, *Viveridae*, and possibly also wild felids and foxes; in subtropical and tropical Asia: dogs, foxes, wolves, jackals and mongooses; in South- and Central America: vampire bats and dogs (Beran, 1981). Occasionally monkeys are also considered to be vectors (Ressang and Umboh, 1963; Fiennes, 1971; Beran, 1981).

The routes of spontaneous infection in freeliving monkeys and apes are largely unknown. The application of live vaccine has been suggested as being the source of infection in several monkeys affected by rabies (Richardson and Humphrey, 1971; Aaron et al., 1975). Man, usually keepers, can be infected by rabid animals via biting, licking or mucosal contamination.

Pathogenesis
Following percutaneous infection the rabies virus can be recovered from the site for a brief period; after approximately 24 hrs the virus starts replication apparently usually in the myocytes. The virus remains at the site for a variable and, in many cases, for extended periods of time. The primary route by which the virus reaches the CNS is via nerve pathways, the mechanism of which being incompletely understood (Beran, 1981).

After reaching the spinal or dorsal root ganglia, the rabies virus spreads rapidly throughout the CNS. From here it spreads via the peripheral nerves throughout the body, particularly to richly innervated tissues such as the salivary glands. The appearance of rabies virus in saliva usually preceeds the onset of clinical disease; the exact period depending on the animal species involved. In Arctic foxes, Vampire bats and birds, however, salivary shedding occurs for undetermined subclinical periods (Beran, 1981).

Only fragmentary information exists on the incubation period in nonhuman primates. In spontaneous infection of monkeys it has been calculated as lasting several weeks to months (Fiennes, 1972). Experimentally infected rhesus monkeys developed furious rabies in 15 to 35 days or, in other cases, the silent variety after a 105 day incubation period (Richardson, 1971).

Clinical Symptoms (Nonhuman Primates Only)

The silent form of the disease is reported to occur in naturally infected monkeys or apes (Abdussalam, 1966; Fiennes, 1971, 1972), although a naturally infected chimpanzee developed furious rabies (Anon, 1966 b). In comparison experimentally infected rhesus monkeys developed either furious or silent rabies (Richardson, 1971 a).

The most obvious clinical signs of silent monkey rabies are salivation, automutilation and paralysis or sudden death (Fiennes, 1972). Signs of aggression towards animal caretakers including biting, which may be due to furious rabies, are difficult to evaluate because they are also part of the natural behavior pattern of all monkeys and apes.

Pathology (Nonhuman Primates Only)

As in human rabies, little is seen macroscopically other than the lesions due to automutilation. The histopathological lesions are characterized by microglial foci, perivascular lymphocytic cuffing and neuronal degeneration. Negri bodies are usually found mainly in the "cornu ammonis" and in the neighbouring cortex (Richardson, 1971 a; Fiennes, 1972), but they were missing in one rabid chimpanzee (Miot and Sikes, 1973). Virus antigen can also be demonstrated in nonhuman primate brains by FA-techniques (Richardson, 1971 a).

Treatment

Nonhuman primates living in semi-free conditions or in outdoor corrals in endemic areas should be vaccinated using inactivated vaccines. Baer et al. (1979) recommended a combination of poly IC complexed to poly L-lysine and carboxymethylcellulose together with human diploid rabies vaccine.

Rabid nonhuman primates are to be destroyed. Bite lesions inflicted by apparently normal or sick nonhuman primates must always be treated properly, not only because of possible rabies but also because of other serious infections such as tetanus or *Herpesvirus simiae.*

1. If possible, as the first step the wounds should be bled.
2. Extensive flushing with water and soap further reduces the amount of infectious agents present.
3. Antiseptic treatment using 40–70% alcohol, iodine tincture, 0,1% quaternary ammonium compounds - in the latter all traces of soap have to be removed before using these compounds.
4. Identification and observation of the monkey in question for at least 3 weeks. Brains of such animals after death or sacrifice have to be especially examined (i. e. demonstration of Negri-bodies, FA-tests, infant mouse inoculation).
5. Vaccination programs in human beings after being bitten or licked by an animal proven to be rabid.

1.4.2 Marburg-Virus

In 1967 several laboratory workers in Marburg and Frankfurt/Germany, and in Belgrad/Yugoslavia, who were involved in handling tissues or blood of African green monkeys *(Cercopithecus aethiops)* that were recently imported via London from Uganda, suffered from a hitherto unknown infectious haemorrhagic disease. The infection also proved to be transmissible to human contacts (physicians, nurses, wives). This initial outbreak involved 31 people, 7 of which died in the course of the disease.

Owing to its initial place of appearance the virus was named "Marburg Virus" (it is sometimes also referred to as "Green Monkey" or "Vervet" - Virus). Its morphological characteristics are similar to those of other rhabdoviruses except for its distinctively larger size, which lead Kiley et al. (1982) to place it into a new family: Filoviridae.

The infection was not observed again until 1975, when two hitchhikers traveling through South Africa and Zimbabwe (formerly Rhodesia) for several months, contracted the infection (Gear et al., 1975; Conrad et al., 1978). Those two travelers did not have any known contact with monkeys. One patient died, and again, the infection was passed on to a contact person - an attending nurse.

The disease surfaced again in 1980 in Kenya (Johnson et al., 1981) and Zimbabwe/Rhodesia (Slenczka, 1981). Two patients were involved in each incidence; one man died. Whether or not nonhuman primates participate regularly in the infectious cycle remains an unsolved and sometimes controversial question. Experimentally infected nonhuman primates *(C. aethiops, Macaca mulatta, Saimiri sciureus)* all died within a few days (Haas et al., 1968; Simpson et al., 1968a, b; Simpson, 1969a, b; Maass et al., 1969). All sera collected from wildliving monkeys during or shortly after the 1967 outbreaks were free of anti-Marburg virus antibodies (Johnson et al., 1981), indicating that the monkeys were incidental hosts to the virus. On the other hand, shortly before the European outbreaks occurred, numerous sick vervet monkeys were found in the Greek River area of Uganda, and at the same time an acute haemorrhagic disease of unknown mortality rate was observed in the local people of the same district (Smith, 1982). Since, according to Smith (1982), the collecting and shipping practices at that time at the single Ugandan dealers' place were well below normal hygiene standards, the possibility of the transference of any infectious disease by subclinically infected monkeys can not be discounted.

Virus

The organism found in the cytoplasm and extracellular spaces of infected organisms and tissue cultures had been classified as a rhabdovirus, as its chemical composition and its morphological appearance is similar to those of rabies and vesicular stomatitis viruses and is morphologically indistinguishable from the Ebolavirus (Johnson et al., 1981). Whether or not the slight differences in size between rabies- and vesicular stomatitis virus on the one side and Marburg- and Ebolavirus on the other justify the latter being regrouped as Filoviridae, can not be decided within the scope of the present publication; for historical reasons they will still be regarded as rhabdoviruses.

Although somewhat variable, the basic architecture of Marburg-virions consists of elongated, cylindrical particles with a diameter of 70 to 80 nm diameter and lengths up to 3000 nm. Some virions are coiled up as circle-, spiral, hook- or six-shaped forms, whilst virions originating from tissue cultures and animal tissues other than blood tend to appear as bulbous forms of 220 to 750 nm diameter (Siegert et al., 1967; Zlotnik et al., 1968; Kissling et al., 1968, 1970; Simpson, 1970; Peters et al., 1971). The complex internal structures of the virions as revealed by contrast media, consist of a dark central nucleocapsid of approximately 20 nm diameter with a pitch of 78 Å; a surrounding tubular structure of approximately 50 nm diameter with cross striation of 45 Å periodicity; an intermediate layer, and finally an approximately 100 Å thick envelope studded with spikes of about 70 A size, and arranged in roughly 100 Å distance from one another (Siegert et al., 1967; Almeida et al., 1971).

The virus is heat inactivated at 56 °C in 60 min and by gamma radiation. It is sensitive to lipid solvents, glutaraldehyde, detergents and phenol (Kiley et al., 1982; Mitchell and McCormick, 1984). It replicates in GMK-AH1-, Vero-, LLC-MK2-, BHK2-, and other cell lines, usually without producing CPE. Only in embryonic human lung fibroblasts (Slenczka, 1969; Hofmann und Kunz, 1971) or in primary vervet kidney cells (Haas et al., 1968) can a distinctive CPE be observed as a spindling and separation of the cells.

For serological studies the demonstration of the CF antibodies which persist for approximately 2 years is most useful, although this can be sometimes of a limited specificity depending on the antigen used. Neutralization tests are technically difficult and often unsatisfactory (Henderson et al., 1971; Siegert and Slenczka, 1971). The most specific test is probably the fluorescent antibody test (Bonin, 1969). Marburg-virus does not cross-react with any of the several arboviruses tested (Casals, 1971) nor with the other rhabdoviruses including the Ebola-virus (Slenczka, 1981).

Epidemiology

Although the initial 1967 outbreaks in Germany and Yugoslavia could be doubtlessly traced back to a shipment of vervet monkeys received from the Entebbe area / Lake Victoria of Uganda, this does not necessarily imply that these monkeys are the reservoir hosts of the virus. It seems more likely, that these monkeys were incidental hosts to the virus after acquiring the infection shortly before or during the shipment to their European points of destination. They might have picked up the agent from other sources in Uganda or, as suggested by Hennessen (1969), during a stopover at London airport. Experimentally infected vervet monkeys (as well as rhesus- and squirrel monkeys) uniformly succumbed to the disease within 7 to 10 days, indicating their role as heterologous hosts.

Seroepidemiological surveys of monkeys have in the past sometimes been misleading due to technical problems. Sera taken from vervet monkeys in Uganda after the initial outbreak were positive according to CF-tests using crude guinea pig liver antigen (Henderson et al., 1971), but this was later proven to be a nonspecific response (Johnson et al., 1982). Johnson et al. (1982) using more specific immunofluorescence methods after the 1980 Kenyan outbreak found 4/136 *C. aethiops* and 1/184 *Papio cynocephalus* of Kenyan origin to possess anti-Marburg virus-antibodies at titers of 1:64 to 1:128. All antibody-positive monkeys remained healthy, again questioning the specificity of the test. In Gabon no anti-Marburg virus-antibodies were found in either human or monkey sera (Ivanoff et al., 1982), whereas 1/499 human sera in the southeastern part of the Republic of Central Africa gave positive reactions in indirect fluorescence tests (Saluzzo et al., 1981).

The transmission of the Marburg virus to man in the initial cases depended on direct contact with the blood or tissues of the infected animals. In the 1975 outbreak in Zimbabwe, arthropod transmission from an unknown reservoir host has been considered, because the patient had been stung by an unidentified arthropod 6 days before clinical onset of the disease (Conrad et al., 1978). This assumption is supported by the successful propagation of Marburg virus in *Aedes aegypti* (but not in *Anopheles maculipennis* and *Ixodes ricinus* (Hofmann et al., 1969)).

In man - man transmission inadvertent percutaneous infection via small cutaneous wounds or accidental inoculation (contaminated needles or necropsy instruments) seem to be the most important route, although in one case sexual transmission 12 weeks after clinical recovery of the initial male patient is believed to have taken place (Slenczka, 1981). The agent is excreted in the urine (Siegert et al., 1968), and aerogenous infection is also considered to be possible (Simpson, 1969b).

Clinical Disease

Nonhuman primates: In experimentally infected vervet-, rhesus- and squirrel monkeys a febrile illness resulted regardless of dosage and the route of infection. The average incubation period ranged from 2 to 6 (up to 10) days, the resulting illness in all three species was almost identical and always fatal.

During the early febrile stage no overt signs other than occasional anorexia and slight weight loss were observed. The haematocrit values were diminished by 5-6% during the first 3 days (Simpson et al., 1968b). One to two days before death and only in the rhesus monkeys, a petechial skin rash appeared - principally on the flexor surfaces of the arms and thighs; on the chest, face and neck. The rash, however, was not noticed by Maas et al. (1969). At this stage, the animals did not eat or drink, sat huddled in their cages; seemed to be apathetic and did not respond to provocation (Slenczka, 1981). Immediately before death a sudden drop in body temperature occurred (Simpson, 1969 b, 1970). The animals became dyspnoic and, in some cases, suffered from diarrhea. In some individuals blood oozed from the rectum or vagina and blood samples often failed to clot. The lymphocyte numbers were reduced by as much as 90%, and the thrombocytes decreased by 50 to 80% (Slenczka, 1981).

In the survivors the rash disappeared around the 12th day, followed by desquamation at days 14 to 16 (Andrijich, 1981).

Laboratory findings are leukopenia, thrombocytopenia, elevated activated partial thromboplastin time, decreased prothrombin index, hypokalaemia, hypoproteinaemia, raised serum-urea and creatinine levels; elevated SGOT and SGPT activities (Stille and Böhle, 1971; Andrijich, 1981).

Pathology

Nonhuman primates: The specific lesions are haemorrhages in the great parenchymas of lungs, liver, and spleen; with hepato- and splenomegaly and fatty metamorphosis of hepatocytes. Histopathologically, focal necrosis could be demonstrated in virtually all organs, particularly in the liver and the reticuloendothelial system (Slenczka, 1981). In hepatocytes basophilic, Feulgen-positive inclusion bodies were described. Inflammatory responses were limited to early lesions. In the spleen and lymph nodes follicular necrosis and perifollicular haemorrhages were most noticeable. The splenic red pulp revealed a marked reduction in cell numbers. In the lungs edema and interstitial pneumonia prevailed, whilst the renal lesions consisted of glomerular microthrombosis and tubular damage (Slenczka, 1981).

Guinea pig: In the guinea pig essentially similar lesions as those described in monkeys occur, with the severity and extension of the liver damage increasing with passage number. 3 days p.i. disseminated acidophilic single cellular necrosis of the hepatocytes appeared, which further developed into regional necroses of different sizes during the next few days (Bechtelsheimer et al., 1970).

A variable glial nodule encephalitis with occasional perivascular infiltration and polioencephalitis occurs, affecting particularly the brain stem (Slenczka, 1981).

Man: At necropsy, signs of severe haemorrhagic disease in the skin, mucous membranes and large parenchymas predominate. The gut is filled with blood from diffuse haemorrhages without any ulcers or erosion. These hemorrhages in the Marburg-virus disease are probably due to intravascular coagulation (Gagal et al., 1970). Histopathologically, again focal necroses in almost all organs except the bones and skeletal muscles are a constant feature (Gedigk et al., 1971). The necrosis is most pronounced in liver, lymphatic organs and gonads with only a minimal inflammatory response occurring. Peculiar basophilic bodies of 1 to 4 &m appeared in the necrotic foci in the liver; signs of mitotic regeneration were marked (Bechtelsheimer et al., 1970, 1972). All patients which died from the infection showed glial nodule type panencephalitis with only discrete perivascular lymphocytic cuffing and haemorrhagic lesions varying from petechial meningeal haemorrhages to massive cerebral haemorrhages (Jacob, 1971; Slenczka, 1981).

Treatment

There is no treatment for Marburg-virus infected animals; they have to be destroyed.

In man convalescent serum can be administered, accompanied by symptomatic and supportive treatment.

Shock is to be treated accordingly. Vitamin K, platelet infusions, fibrinogen and epsilonaminocaproic acid should be given to counteract the bleeding.

Oliguric patients are given mannitol, anuric patients are to be placed on a dialysis program.

References

1. Aaron, E.; I. Kamei; E.V. Bayer; R.W. Emmons, and J. Chin: Probable vaccine-induced rabies in a pet marmoset. CDC Morbidity and Mortality Weekly Report, March 15, p.99 (1975)
2. Abdussalam, M.: Natural rabies in a laboratory monkey. Lab. Primate Newsl. *5* (3), 13 (1966)
3. Almeida, J.D.; A.P. Waterson, and D.I.H. Simpson: Morphology and morphogenesis of the Marburg agent. Pp.84-97, in "Marburg Virus Disease" (Martini G.A. and R. Siegert, eds.), Springer, Berlin (1971)
4. Anon: Rabies in a chimpanzee - Georgia. CDC Vet. Publ. Hlth Notes, Sept. 1972
5. Andrijich, V.B.: Marburg virus disease. The diagnosis and management of suspected cases. South Afr. med. J. *6*, 751-753 (1981)
6. Baer, G.M.; S.A. Moore; J.H. Shaddock, and H.B. Levy: An effective rabies treatment in exposed monkeys: a single dose of interferon inducer and vaccine. Bull. Wld. Hlth Org. *57*, 807-813 (1979)
7. Bechtelsheimer, H.; G. Korb und P. Gedigk: Die "Marburg-Virus"-Hepatitis. Untersuchungen bei Menschen und Meerschweinchen. Virchows Arch. Abt. A. Path. Anat. *351*, 273-290 (1970)
8. Bechtelsheimer, H.; G. Korb und P. Gedigk: The morphology and pathogenesis of "Marburg Virus" hepatitis. Hum. Path. *3*, 255-264 (1972)
9. Beran, G.W.: Rabies and infections by rabies-related viruses. Pp.57-135, in "CRC Handbook Series in Zoonoses" (Steele, D.H., Series ed.), Section B: Viral Zoonoses, Vol. II (Beran, G.W., Section ed.), CRC Press Boca Raton (1981)
10. Bonin, O.: The cercopithecus monkey disease in Marburg and Frankfurt (Main), 1967. Acta zool. path. antverp. *48*, 319-331 (1969)
11. Boulger, L.R.: Natural rabies in a laboratory monkey. Lancet *1*, 941-943 (1966)
12. Casals, J.: Absence of serological relationship between the Marburg virus and some arboviruses, pp. 98-104 in "Marburg Virus Disease" (Martini, G.A. and R. Siegert, eds.), Springer, Berlin (1971)
13. Conrad, J.L.; M. Isaacson; E.B. Smith; H. Wulf; M. Crees; P. Geldenhuys, and J. Johnston: Epidemiologic investigation of Marburg virus disease, Southern Africa, 1975. Amer. J. trop. Med. Hyg. *27*, 1210-1215 (1978)
14. Fiennes, R.N.T-W.: Rabies in primates. Pp. 683-685 in "Medical Primatology 1970" (E.I. Goldsmith and J. Moor-Jankowski, eds.), Karger, Basel (1971)
15. Fiennes, R.N.T-W.: Rabies. Pp.646-662, in "The Pathology of Simian Primates". Part II (Fiennes, R.N.T-W., ed.), Karger, Basel (1972)
16. Gagel, C.; M. Lindner; G. Müller-Berghaus, and H.G. Lasch: Virus infection and blood coagulation. Thrombos. Diathes. haemorrh. *23*, 1-11 (1970)
17. Gear, J.H.S.; I. Spence; M. Kirsh; J. Ryan; T. Bothwell; J. Gear; G. Cassell, and J. Davies: Marburg virus disease - South Africa. CDC Morbidity and Mortality Weekly Rep. *24* (10), 89-90 (1975)
18. Gedigk, P.; H. Bechtelsheimer, and G. Korb: Pathologic anatomy of the Marburg virus disease. Pp. 50-53 in "Marburg Virus Disease" (Martini, G.A. and R. Siegert, eds.), Springer, Berlin (1971)
19. Haas, R.; G. Maass; J. Müller und W. Oehlert: Experimentelle Infektionen von Cercopithecus aethiops mit dem Erreger des Frankfurt-Marburg-Syndroms (FMS). Zschr. Med. Mikrobiol. Immunol. *154*, 210-220 (1968)
20. Henderson, B.E.; R.E. Kissling; M.C. Williams; G.W. Kafuko, and M. Martin: Epidemiological studies in Uganda relating to the "Marburg"-Agent, pp. 166 - 176 in "Marburg-Virus Disease" (Martini, G.A. and R. Siegert, eds.), Springer, Berlin (1971)
21. Hennessen, W.: Epidemiology of Marburg virus disease. Lab. Anim. Handb. *4*, 137-142 (1969)
22. Hofmann, H. and C. Kunz: Cultivation of the Marburg virus (Rhabdovirus simiae) in cell cultures, pp.112-116 in "Marburg Virus Disease" (Martini, G.A. and R. Siegert, eds.), Springer, Berlin (1971)
23. Hofmann, H.; C. Kunz; H. Aspöck und A. Radda: Zur Ökologie des sogenannten "Marburg Virus" (Rhabdovirus simiae). Zbl. Bakt., I. Abt. Orig. *212*, 168-173 (1969)
24. Ivanoff, B.; P. Du Quesnoy; G. Languillat; J.F. Saluzzo; A. Georges; J.P. Gonzalez, and J. McCormick: Haemorrhagic fever in Gabon. I. Incidence of Lassa, Ebola and Marburg Viruses in Haut-Ogooue. Trans. roy. Soc. trop. Med. Hyg. *76*, 719-720 (1982)

25. Jacob, H.: The neuropathology of the Marburg disease in man. Pp. 54-61 in "Marburg Virus Disease" (Martini, G.A. and R.Siegert, eds.), Springer, Berlin (1971)
26. Johnson, B.K.; L.G. Gitau; A.Gichogo; P.M. Tukei; J.G. Else; M.A. Suleman, and R.Kimani: Marburg and Ebola virus antibodies in Kenyan primates. Lancet *1*, 1420-1421 (1981)
27. Johnson, B.K.; L.G. Gitau; A.Gichogo; P.M. Tukei; J.G. Else; M.A. Suleman; R.Kimani, and P.D. Sayer: Marburg, Ebola and Rift Valley fever virus antibodies in East African primates. Trans. roy. Soc. trop. Med. Hyg. *76*, 307-310 (1982)
28. Kaplan, C.: Rabies in non-human primates. Lab. Anim. Handb. *4*, 117-118 (1969)
29. Kiley, M.P.; E.T.W. Bowen; G.A. Eddy; M.Isaacson; K.M. Johnson; J.B. McCormick; F.A. Murphy; S.R. Pattyn; D.Peters; O.W. Prozesky; R.L. Regnery; D.I.H. Simpson; W: Slenczka; P.Sureau; G.van der Groen; P.A. Webb, and H.Wulff: Filoviridae: A taxonomic name for Marburg and Ebola viruses? Intervirology *18*, 24-32 (1982)
30. Kissling, R.E.; F.A. Murphy, and B.E. Henderson: Marburg virus. Ann. N.Y. Acad. Sci. *174*, 932-945 (1970)
31. Kissling, R.E.; R.Q. Robinson; F.A Murphy, and S.G. Whitfield: Agent of disease contracted from Green monkeys. Science *160*, 888-890 (1968)
32. Maass, G.; R.Haas and W.Oehlert: Experimental infections of monkeys with the causative agent of the Frankfurt-Marburg-Syndrome (FMS). Lab. Anim. Handb. *4*, 155-165 (1969)
33. Miot, M.R. and R.K. Sikes: Rabies in a chimpanzee. Lab. Primate Newsletter *12* (1), 6 (1973) - from CDC Vet. Publ. Hlth Notes, Sept. 1972
34. Mitchell, S.W. and J.B. McCormick: Physicochemical inactivation of Lassa, Ebola, and Marburg viruses and effect on clinical laboratory analyses. J.Clin. Microbiol. *20*, 486-489 (1984)
35. Peters, D.; G.Müller, and W.Slenczka: Morphology, development, and classification of the Marburg virus; pp.68-83 in "Marburg-Virus Disease" (Martini, G.A. and R.Siegert, eds.), Springer, Berlin (1971)
36. Ressang, A.A. and T.F.H. Umboh: The concept of veterinary public health in Indonesia. Communication Veterinariae *6*, 1-11 (1962)
37. Richardson, J.H.: Rabies in nonhuman primates. Primate Zoonoses Surveill. Rep. *4*, 18-20 (1971a)
38. Richardson, J.H.: Rabies in nonhuman primates. Primate Zoonoses Surveill. Rep. *5*, 14 (1971b)
39. Richardson, J.H. and G.L. Humphrey: Rabies in imported nonhuman primates. Lab. Anim. Sci. *21*, 1082-1083 (1971)
40. Röttcher, D.and A.M. Sawchuk: Wildlife rabies in Zambia. J.Wildlife Dis. *14*, 513-517 (1978)
41. Saluzzo, J.-F.; J.P. Gonzalez; A.-J. Georges, and K.M. Johnson: Mise en évidence d'anticorps vis-à-vis du virus Marburg parmi les populations humaines du sud-est de la République central-africaine. Compt. Rend. des Séances de l'academie des sciences. Serie III: Sciences de la vie. *292*, 29-31 (1981)
42. Schmitter, F.: Hydrophobia in a young Philippine monkey. J.Amer. Med. Res. *62*, 598 (1914)
43. Siegert, R.: Marburg-, Lassa- und Ebola-Virus als Erreger hämorrhagischen Fiebers. Dtsch. med. Wschr. *103*, 1176-1181 (1978)
44. Siegert, R.; Hsin-Lu Shu; W.Slenczka; D.Peters und G.Müller: Zur Ätiologie einer unbekannten, von Affen ausgegangenen, menschlichen Infektionskrankheit. Dtsch. Med. Wschr. *92*, 2341-2342 (1967)
45. Siegert, R.; Hsin-Lu Shu, and W.Slenczka: Nachweis des "Marburg Virus" beim Patienten. Dtsch. Med. Wschr. *93*, 616-619 (1968)
46. Siegert, R.and W.Slenczka: Laboratory diagnosis and pathogenesis. Pp.155-160 in "Marburg Virus Disease" (Martini, G.A. and R.Siegert, eds.), Springer, Berlin (1971)
47. Simpson, D.I.H.: Marburg agent disease in monkeys. Trans. roy. Soc. trop. Med. Hyg. *63*, 303-309 (1969a)
48. Simpson, D.I.H.: Marburg virus disease: Experimental infection of monkeys. Lab. Anim. Handb. *4*, 149-154 (1969b)
49. Simpson, D.I.H.: Marburg virus. A review of laboratory studies. Pp. 39-44 in "Infection and Immunosuppression in Subhuman Primates." (Balner, H.and W.I.B. Beveridge, eds.), Munksgaard, Copenhagen (1970)
50. Simpson, D.I.H.; E.T.W. Bowen, and W.F. Bright: Vervet monkey disease: Experimental

infection of monkeys with the causative agent, and antibody studies in wild-caught monkeys. Lab. Anim. Handb. *2*, 75-81 (1968a)
51. Simpson, D.I.H.; P.Zlotnik, and D.A. Rutter: Vervet monkey disease. Experimental infection of guinea pigs and monkeys with the causative agent. Brit. J.exp. Path. *49*, 458-464 (1968b)
52. Slenczka, W.: Growth of Marburg virus in vero cells. Lab. Anim. Handb. *4*, 143-147 (1969)
53. Slenczka, W.: Marburg and Ebola virus disease. Marburg virus disease. Pp. 41-55 in "CRC Handbook Series in Zoonoses" (J.H. Steele, Senior Editor) Section B: Viral Zoonoses. Vol. II, (Beran, G.W., Section Editor). CRC Press, Boca Raton (1981)
54. Smith, M.W.: Field aspects of the Marburg virus outbreak: 1967. Primate Supply *7*(1), 11-15 (1982)
55. Stille, W. and E. Böhle: Clinical course and prognosis of Marburg Virus ("green-monkey") disease. Pp. 10-18 in "Marburg Virus Disease" (Martini, G.A. and R. Siegert, eds.), Springer, Berlin (1971)
56. Zlotnik, I.; D.I.H. Simpson, and D.M.R. Howard: Structure of the vervet monkey disease agent. Lancet (1), 26-28 (1968)

1.5 Picornaviruses

1.5.1 Encephalomyocarditis - Viruses (EMC-Virus)

The EMC-viruses constitute a group of several antigenically related viruses of the genus *Enterovirus*, which share as a major feature the ability to induce myocarditis. The most commonly reported ones are the encephalomyocarditis-, Columbia -SK-, M.M. -, and Mengo-Viruses. They have been recovered worldwide from a variety of animals including man. Swine are the most commonly and seriously affected species.

EMC-viruses have also been isolated from nonhuman primates, e.g. in 1945 from a chimpanzee in Florida (Helwig and Schmidt, 1945); a rhesus monkey in Uganda (Dick et al., 1948; Dick, 1949); another chimpanzee in 1948 (Schmidt, 1948); a baboon and a "monkey" in Florida (Kissling et al., 1956); a baboon and two chimpanzees at the Tarpon Spring/Florida-Zoo (Gainer, 1967); two adult rhesus monkeys (Dshikidze et al., 1974) and a *Hapalemur griseus* in Madagascar (Coulanges et al., 1976).

Virus

The isometric EMC-virions have a diameter of approximately 30 nm. They are ether resistant, acid stable and withstand freezing temperatures down to −70° C. They are inactivated by 60 °C for 30 min. EMC-viruses pass through Berkefeld- and Seitz-filters or gradocol membranes with 30 nm pore size. They replicate in embryonating hens'eggs and produce CPE in embryonic and other cell cultures of chicken, mouse, monkey, hamster, swine and cattle origin (Murnane, 1981). Plaques of different sizes are formed after agar-overlay, depending on the individual virus strains. The virus is pathogenic in newborn and adult mice, young guinea pigs and newborn rabbits (Dshikidze et al., 1974).

Epidemiology

Rats, other wild rodents, and swine are suspected to be the reservoir hosts of EMC-viruses (Gainer, 1974; Murnane, 1981) with the pig being the most susceptible animal species of all. Infected rats excrete the virus in their feces. EMC-virus has also been recovered from swamp water in endemic areas. Pigs excrete the virus only for a short period of time after infection.

The infection of susceptible animals or man is probably acquired through food or water contaminated by infectious animal droppings or by consumption of viscera or meat of EMC infected animals. Although so far no monkey - man transmission of EMC-virus has been known, one has to be aware of the possibility at least, particularly if working with animals imported from the wild.

Clinical Disease

Nonhuman primates: In all EMC-virus infections of nonhuman primates the infected animals were either found dead or moribund, with the development of paralysis shortly before death (Helwig and Schmidt, 1945; Kissling et al., 1956; Roca-Garcia and Sanmartin-Barberi, 1957; Gainer, 1967; Coulanges et al., 1976). Experimentally infected *C. aethiops* showed sluggishness, anorexia, pale and edematous skin, tachycardia and, in the final stages, convulsions. ECG-changes were characterized by an increased PR interval and RS-T segment shifts, but these were not always present (Dshikidze et al., 1976).

Man: The best description is probably that of an accidental laboratory Mengo virus infection. The worker suffered from intermittent delirium, headache, nuchal rigidity, emesis and fever over a 4 day period. He developed a transient nerve deafness, but otherwise recovered uneventfully. In other human EMC-patients sudden fever of 2 to 3 days duration, severe headaches, pharyngitis and nuchal rigidity were reported, with a few patients being comatose on admission to the hospital. All patients recovered promptly with no residual complications (Murnane, 1981).

Pathology

Nonhuman primates: Interstitial myocarditis, often with massive myocardial necrosis, pulmonary edema and minor encephalitis have been reported in dead monkeys and apes (Gainer, 1967; Dshikidze et al., 1976; Murnane, 1981). In a *Hapalemur griseus* cerebral haemorrhages were the most conspicious features (Coulanges et al., 1978).

Man: So far no fatal infections have been reported in man, and thus the lesions in man induced by EMC-viruses are unknown.

1.5.2 Polioviruses

Human poliomyelitis used to be a worldwide threat as a crippling disease before vaccine-development. All three human polio-virus-strains are experimentally transmissible to macaques and apes. Spontaneous infections have occurred particularly in the pre-vaccine era in chimpanzees (Lindau, 1960; Douglas et al., 1970); gorillas (Froeschle and Allmond, 1965; Guilloud et al., 1969), and orangutans (Froeschle and Allmond, 1965; Guilloud et al., 1969). Recently an outbreak caused by type 1 was described in 3 captive *Colobus abyssinicus kikuyuensis* at the Institute of Primate Resarch in Kenya (Suleman et al., 1984).

Although no poliovirus transmission from nonhuman primates to man has been proven so far, the possibility at least exists theoretically.

Virus

The virions have a diameter of 20 to 30 nm; an icosahedral symmetry and lack an envelope. They are stable at pH 3 and are ether resistant, but are inactivated by temperatures of 50 to 60° in 30 min.. Synthesis of the virions occurs primarily in the cytoplasm of intestinal cells (Beran, 1981). Human poliovirus-strains are differentiated into 3 serotypes (Types 1-3) and so are the identical polioviruses isolated from nonhuman primates.

The polioviruses replicate best of all in HEK cells or in primary rhesus monkey cells and to a lesser extent in WI-38 cells (Melnick, 1976).

Specific identification of the serotype depends mostly on serum neutralization tests, although HI-CF-, Immunofluorescence-, or precipitation-tests may also be useful.

Epidemiology

All enteroviruses are excreted via the feces, with transmission occurring either via infected sewage or water. Direct fecal - oral transmission is also possible and has to be avoided especially when handling animals. Chimpanzees, e.g. can be symptomless carriers (Fiennes, 1967).

Clinical Symptoms

Nonhuman primates: Poliovirus-infections of nonhuman primates, as in man, do not necessarily induce clinical illness. The symptoms observed so far consisted of anorexia, weakness, paresis, hemiplegia or paraplegia. The disease may end fatally, as observed in a colobus monkey (Suleman et al., 1984).

Man: Human poliovirus-infections can remain subclinical, in other cases paralytic disease with fever, headache, gastrointestinal disease, nuchal rigidity and flaccid paralysis may develop. This latter may be life threatening if it extends to the respiratory muscles.

Pathology

Nonhuman primates: The poliomyelitic lesions of nonhuman primates are identical to those of human beings (Godglück, 1960). They consist of a nonpurulent myelo-encephalitis with extensive loss of ganglial cells and an associated glial-cell proliferation particularly in the spinal cord, *formatio reticularis*, cerebrallar nuclei, diencephalon and the cerebral cortex except for the temporal lobes, and the frontal- and occipital poles. The ultrastructural lesion in experimentally infected cynomolgus monkeys were characterized by membrane-bound vesicles in the cytoplasm of degenerating motoneurons. No virus particles were found (Hashimoto et al., 1984).

Man: The morphological lesions of human poliomyelitis are essentially similar to those of nonhuman primates, for a more detailed description see Ule, 1974.

1.5.3 Coxsackieviruses

Coxsackieviruses are differentiated from the polioviruses by their unique virulence in suckling mice. Human strains are readily transferable to a number of domestic animals, they have also been reported in wild animals, usually after human contact. Captive monkeys and apes have been found to be infected, occasionally even already upon arrival at the laboratory (Kalter and Heberling, 1971). Rhesus-, cynomolgus- and bonnet monkeys and chimpanzees have been successfully experimentally infected by serotypes A 2, 3, 4, 9, 12 and 16 and B 1, 3, 4, 5 and 6 (Kraft and Melnick, 1953; Melnick and Kaplan, 1953; Lou and Wenner, 1962; Schmidt et al., 1965; Tobe et al., 1967; Chandy et al., 1980). As virusexcretion has been demonstrated in naturally infected nonhuman primates, the possibility of re-transmission to man although so far not reported, has to be kept in mind; therefore the infection will be discussed briefly.

Virus

The morphological and cultural characteristics of coxsackieviruses are those of enteroviruses in general (for description see Kalter, 1982). Human coxsackieviruses are divided into serogroups A and B, and are subgrouped by numbers. Their serological typing depends primarily on neutralization and to a lesser extend on HI- and CF-tests. Most of the reported natural coxsackievirus infections of nonhuman primates have been identified by such serological evidence (Kalter and Heberling, 1971; Yamane, 1974), whilst in a few cases the agents have been actually recovered (Heberling, 1972; Kelly et al., 1978).

Clinical Disease

Nonhuman primates: Coxsackievirus infections of monkeys are symptomless in most cases (Heberling, 1972). Only in an infant chimpanzee, which died at 81 days of age due to a coxsackie B-5 infection (and upper respiratory infection by ß-hemolytic streptococci), were an associated myocarditis and interstitial pneumonia reported (Kelly et al., 1978).

Man: Although most human infections are probably subclinical, others, particularly if occurring in children, may be followed by a variety of syndromes. These range from symptomless fever to severe disease, including herpangina, meningoencephalitis, paralysis, exanthema, upper respiratory disease with or without pneumonia, pleurodynia, hepatitis, myo- and pericarditis (Gear and Measrock, 1973; Dery et al., 1974; Beran, 1981).

References

1. Beran, G.W.: Animal infections by human enteroviruses; pp.154-161 in "CRC Handbook Series in Zoonoses" (Steele, J.H., Series editor), Section B: Viral Zoonoses, Vol. II (Beran, G.W., Section editor). CRC Press, Boca Raton, 1981
2. Chandy, K.G.; P.Mukundan, and T.J. John: Evidence for natural infection with group B coxsackieviruses in bonnet monkeys. J.Med. Primatol. 9, 323-324 (1980)

3. Coulanges, P.; P. Sureau, and Y. Robin: Premier isolement du virus Mengo à Madagascar au cours d'une affection mortelle du lemurien *Happalemur griseus.* Pouvoir pathogéne expérimental pour le *Lemur fulvus.* Arch. Inst. Pasteur Madagascar *45,* 115-126 (1976)
4. Dery, P.; M. I. Marks, and R. Shaper: Clinical manifestations of coxsackievirus infection in children. Amer. J. Dis. Child. *128,* 464-468 (1974)
5. Dick, G. W. A.: The relationship of Mengo encephalomyelitis, encephalomyocarditis, Columbia-SK, and M. M. viruses. J. Immunol. *62,* 375-386 (1949)
6. Dick, G. W. A.; K. C. Smithburn, and A. J. Haddon: Mengo encephalomyelitis virus: Isolation and immunological properties. Brit. J. exp. Path. *29,* 547-558 (1948)
7. Dshikidze, E. K.; Z. V. Shevtsova; R. I. Krylova; E. Y. Balaeva; N. A. Voskanjan und V. I. Uvarova: Zur Ätiologie der Enzephalomyokarditis von Rhesusaffen. Z. Versuchstierk. *16,* 142-149 (1974)
8. Dshikidze, E. K.; Z. V. Shevtsova; R. L. Krylova; N. A. Voskanyan; G. M. Cherkovic; V. I. Uvarova, and I. G. Zodelava: Experimental encephalomyocarditis in some monkey species. Exp. Path. *12,* 242-249 (1976)
9. Fiennes, R. N. T-W.: Zoonoses of Primates. The Epidemiology and Ecology of Simian Diseases in Relation to Man. Weidenfels and Nicolson, London, 1967
10. Froeschle, J. and B. Allmond: Polio outbreak among primates at Yerkes Primate Center. Lab. Primate Newsletter *4* (2), 6 (1965)
11. Gainer, J. H.: Encephalomyocarditis virus infections in Florida, 1960-1966. J. Amer. vet. med. Ass. *151,* 421-425 (1967)
12. Gainer, J. H.: Viral myocarditis in animals. Advanc. Cardiol. *13,* 94-105 (1974)
13. Gear, J. H. S. and V. Measrock: Coxsackievirus infections of the newborn. Progr. med. Virol. *15,* 42-62 (1973)
14. Godglück, G.: Poliomyelitis und Virus-Enteritis bei Affen. Proc. 2. Internat. Symp. on Zoo Anim. Path., Warsaw 135-145 (1960)
15. Guilloud, N. B.; B. W. Allmond; D. E. Froeschle, and F. L. Fitzgerald: Paralytic poliomyelitis in laboratory primates. J. Amer. vet. med. Ass. *155,* 1190-1193 (1969)
16. Hashimoto, I.; A. Hagiwara, and T. Komatsu: Ultrastructural studies on the pathogenesis of poliomyelitis in monkeys infected with poliovirus. Acta Neuropathol. *64,* 53-60 (1984)
17. Heberling, R. L.: The simian picornaviruses; pp. 497-515 in "Pathology of Simian Primates", Part II (Fiennes, R. N. T-W., ed.) Karger, Basel (1972)
18. Helwig, F. C. and E. C. H. Schmidt: Filter-passing agent producing interstitial myocarditis in anthropoid apes and small animals. Science *102,* 31-33 (1945)
19. Kalter, S. S.: Enteric viruses of nonhuman primates. Vet. Path. *19* (Suppl. 7), 33-43 (1982)
20. Kalter, S. S. and R. L. Heberling: Comparative virology of primates. Bact. Rev. *35,* 310-364 (1971)
21. Kelly, M. E.; K. Soike; K. Ahmed, and M. J. Iatropoulos: Coxsackievirus in an infant chimpanzee. J. med. Prim. *7,* 119-121 (1978)
22. Kissling, R. E.; J. M. Vanella, and M. Schaffer: Recent isolation of encephalomyocarditis virus. Proc. Soc. exp. Biol. *91,* 148-152 (1956)
23. Kraft, L. M. and J. L. Melnick: Quantitative studies of the virus host relationship in chimpanzees after inapparent infection with coxsackie-viruses. II. The development of complement-fixing antibodies. J. exp. Med. *97,* 401-414 (1953)
24. Lindau, K. H.: Poliomyelitis bei Menschenaffen. Proc. 2. Internat. Symp. on Zoo Anim. Path. Warsaw, 213-238 (1960)
25. Lou, T. Y. and H. A. Wenner: Experimental infection with enteroviruses. V. Studies on virulence and pathogenesis in cynomolgus monkeys. Arch. Virusforsch. *12,* 302-322 (1962)
26. Melnick, J. L.: Enteroviruses, p. 163-207 in "Viral Infections of Humans. Epidemiology and Control" (Evans, A. S. ed.) Wiley & Sons, London (1976)
27. Melnick, J. L. and A. S. Kaplan: Quantitative aspects of the virus - host relationship in chimpanzees after inapparent infection with coxsackie viruses. I. The virus carrier state and the development of neutralizing antibodies. J. exp. Med. *97,* 367-400 (1953)
28. Murnane, T. G.: Encephalomyocarditis. Pp. 137-147 in "CRC Handbook Series in Zoonoses" (Steele, J. H., Series Editor) Section B: Viral Zoonosis, Vol. II (Beran, G. W., Section Editor). CRC Press, Boca Raton, 1981
29. Roca-Garcia, M. and C. Sanmartin-Barberi: The isolation of encephalomyocarditis virus from *Aotus* monkeys. Amer. J. trop. Med. Hyg. *6,* 840-852 (1957)

30. Schmidt, E.C.H.: Virus myocarditis - Pathologic and experimental studies. Amer. J. Path. *24*, 97-118 (1948)
31. Schmidt, N.J.; J. Dennis; E.H. Lennette, and T.T. Shinomoto: Antibody responses of rhesus (Macaca mulatta) monkeys experimentally infected with coxsackievirus of group B and group A, type 9. J. Immunol. *95*, 54-69 (1965)
32. Suleman, M.A.; B.J. Johnson; R. Tarara; P.D. Sayer; M.D. Ochieng; J.M. Muli; E. Mbete; P.M. Tukei; D. Ndirangu; S. Kago, and J.G. Else: An outbreak of poliomyelitis caused by poliovirus type 1 in captive black and white colobus monkeys (*Colobus abyssinicus kikuyuensis*) in Kenya. Trans. roy. Soc. trop. Med. Hyg. *78*, 665-669 (1984)
33. Tobe, T.; Y. Horikoshi; C. Hamada, and Y. Hamashima: Virus infection as a trigger of appendicitis; experimental investigation of coxsackie B 5 virus infection in monkey intestine. Surgery *62*, 927-934 (1967)
34. Ule, G.: Nervensystem, pp. 9-1-9-92 in "Organpathologie" (Doerr, W., ed.), Stuttgart (1974)
35. Yamane, Y.: Natural virus infection in green and cynomolgus monkeys. Kitasato Arch. of exp. Med. *47*, 149-200 (1974)

1.6 Reoviridae

1.6.1 Orthoreoviruses

Reoviruses are a group of double-stranded RNA-viruses which occur in human beings, and, according to serological studies, in a number of animal species (Rosen, 1968).

The mammalian reoviruses are divided into serotypes 1, 2, and 3, natural infections have been reported in several species of nonhuman primates. Types 1 (= SV12) and 2 (= SV 59) were described in macaques (Hull et al., 1958) and chimpanzees (Sabin, 1959; Rogers et al., 1967; Soike et al., 1969); type 1 (= SA3) in African green monkeys (Malherbe et al., 1963) and type 3 in marmosets (Deinhardt et al., 1967). All three serotypes were described in gorillas, chimpanzees, orangutans, gibbons, baboons, geladas, African green monkeys, patas monkeys, rhesus monkeys, cynomolgus monkeys, bonnet monkeys, stumptailed macaques, Japanese macaques and langurs (Bhatt et al., 1966; Kalter and Heberling, 1971). Type 3 has been shown to be the most common type in nonhuman primates (Kalter and Heberling, 1971). Experimental infections of nonhuman primates are easily achieved. Interspecies transmission from man to animals and vice versa are known to occur (Thein and Scheid, 1981). Human infections generally occur early in childhood (Scharpe and Fields, 1985), certain reoviruses such as the rotaviruses are major pathogens.

Virus

The reoviruses possess segmented double stranded RNA. They are isometric, nonenveloped, and 50 to 130 nm in diameter (Atoynatan and Hsiung, 1964; Soike et al., 1967). They have a double capsid shell with the outer, indistinct, capsid layer consisting of 92 capsomeres, and the inner capsid being of icosahedral symmetry and of 45 nm in diameter.

Simian reoviruses grow well on primary monkey kidney cells (Heberling, 1972). Like other reoviruses they are assembled in the cytoplasm, forming large perinuclear cytoplasmic inclusion bodies (Easterbrook and Rozee, 1971) Mammalian reoviruses are distinguished into types 1, 2 and 3. They all haemagglutinate human type 0 erythrocytes, with types 1 and 2 also haemagglutinating squirrel monkey erythrocytes and type 3 less extensively cattle erythrocytes. The simian reoviruses SV 12 (Hull et al., 1956) and SA 3 (Malherbe and Harwin, 1957) are serologically identical to type 1, whilst SV 59 (Hull et al., 1958) corresponds to type 2 (Kalter, 1982).

Reoviruses are resistant to ether and acid treatment. They are quite stable at −20 °C or lower.

Epidemiology

Judging from serological studies (Kalter and Heberling, 1971) reovirus infections are common in both wild and captive monkeys. Most reovirus infections are inapparent or cause only mild symptoms in man or animals and consequently go frequently unrecognized. The viruses are excreted via the respiratory or intestinal tract and are horizontally transmitted.

Pathogenesis

Reoviruses are thought to enter the host primarily via the M-cells of the gastrointestinal tract. They then multiply in the lymphoid tissues of the Peyer's patches (type 1) and within 4 to 6 days spread to more distant organs in increasing concentrations. Another port of entry is the respiratory system, with multiplation of the viruses occurring in the peribronchial lymphoid tissues.

Reoviruses produce a lytic infection *in vivo*, but can also produce persistent cellular infections *in vitro*. In the lytic infection the host cell's metabolism and cytoskeletal organization are altered, whereas in the persistent infections the virus and the host cells adapt to each other, with the host cell being able to support viral replication without lysis (Sharpe and Fields, 1985).

Both type 1 and type 3 reoviruses can infect the CNS of experimentally infected newborn mice, with type 1 initiating a nonlethal infection of primarily ependymal cells with hydrocephalus as a possible result. Type 3 infects the neuronal tissues thus causing an acute and fatal encephalitis (Sharpe and Fields, 1985).

Clinical Symptoms

Nonhuman primates: Natural infections with type 1 have been reported in *Macaca mulatta, M. fascicularis* (Hull et al., 1956, 1958; Hull and Minner, 1957; Hull, 1968) and *Cercopithecus aethiops* (Malherbe and Harwin, 1957); those with type 2 in *M. mulatta* (Hull et al., 1958) and chimpanzees (Sabin, 1959); whilst type 3 has been isolated solely from marmosets (Deinhardt et al., 1967) and type 5 strains from fecal specimens of newly imported chimpanzees (Soike et al., 1969).

In the only outbreak associated with clinical disease, respiratory symptoms ranging from rhinitis to interstitial pneumonia predominated (Sabin, 1959).

In experimental type 2 infections of chimpanzees an upper respiratory tract illness was produced (Sabin, 1959).

Man: Most human infections are asymptomatic, but sometimes the viruses can be isolated in a variety of diseases.

Reovirus types 1, 2, and 3 have been recovered from patients with hepatitis and encephalitis, but their etiological relationship remains open to question (Thein and Schein, 1981). Types 2 and 3 cause febrile exanthema, diarrhea, and steatorrhea in children. In newborn babies reoviruses may cause congenital pneumonia, pneumopathy and diarrhea (Thein and Schein, 1981).

Pathology

Nonhuman primates: In natural reoviral infections in macaques interstitial pneumonia has been observed (Hull et al., 1956). Experimental infection resulted in hepatitis, meningitis and necrosis of the chorioid plexus (Hull et al., 1956, 1958; Thein and Scheid, 1981).

Man: Diffuse interstitial pneumonia with thickening of the alveolar septae and metaplasia of the pneumocytes, hyaline membranes and alveolar necrosis have been found in an infant infected with type 3 virus, and in a type 1 infection. In the latter focal myocardial degeneration, focal liver degeneration and centrilobular necrosis of hepatocytes were also observed (Thein and Scheid, 1981).

References

1. Atoynatan, T. and G. D. Hsiung: Ultrafiltration of simian viruses. Proc. Soc. Exp. Biol. Med. *116*, 852-856 (1964)
2. Bhatt, P. N.; C. D. Brandt; R. A. Weiss; J. P. Fox, and M. F. Shaffer: Viral infections of monkeys in their natural habitat in Southern India. II. Serological evidence of viral infection. Amer. J. trop. Med. Hyg. *15*, 561-566 (1966)
3. Deinhardt, F.; A. W. Holmes; J. Devine, and J. Deinhardt: Marmosets as laboratory animals. IV. The microbiology of laboratory kept marmosets. Lab. Anim. Care *17*, 48-70 (1967)
4. Easterbrook, K. B. and K. R. Rozee: The ultrastructure of the reovirus inclusion: A freeze etching study. J. Ultrastruct. Res. *34*, 303-315 (1971)
5. Heberling, R. L.: The simian reoviruses; pp. 516-522 in "Pathology of Simian Primates", Part II (Fiennes, R. N. T-W., ed.) Karger, Basel (1972)
6. Hull, R. N.: The simian viruses. Virol. Monogr. *2*, 1-66, Springer, New York (1968)
7. Hull, R. N. and J. R. Minner: New viral agents recovered from tissue cultures of monkey kidneys. II. Problems of isolation and recovery. Ann. N. Y. Acad. Sci. *67*, 413-423 (1957)
8. Hull, R. N.; J. R. Minner, and C. C. Mascoli: New viral agents recovered from tissue cultures of monkey kidneys. III. Recovery of additional agents both from cultures of monkey tissues and directly from tissues and excreta. Amer. J. Hyg. *68*, 31-44 (1958)
9. Hull, R. N.; J. R. Minner, and J. W. Smith: New viral agents recovered from tissue cultures of monkey kidneys. I. Origin and properties of cytopathologic agents SV 1, SV 2, SV 4, SV 5, SV 6, SV 11, SV 12, and SV 15. Amer. J. Hyg. *63*, 204-215 (1956)
10. Kalter, S. S.: Enteric viruses of nonhuman primates. Vet. Pathol. *19* (Suppl. 7), 33-43 (1982)
11. Kalter, S. S. and R. L. Heberling: Comparative virology of primates. Bact. Rev. *35*, 310-364 (1971)
12. Malherbe, H. and R. Harwin: Several viruses isolated from the vervet monkey. Brit. J. exp. Path. *38*, 539-541 (1957)
13. Malherbe, H.; R. Harwin, and M. Ulrich: The cytopathic effects of vervet monkey viruses. S. Afr. med. J. *37*, 407-411 (1963)
14. Rogers, N. G.; M. Basnight; C. J. Gibbs, jr., and D. G. Gajdusek: Latent viruses in chimpanzees with experimental Kuru. Nature *216*, 446-449 (1967)

15. Rosen, L.: Reoviruses. Pp.71-107 in "Virology Monographs", Vol. I, Springer, New York (1968)
16. Sabin, A. B.: Reoviruses. A new group of respiratory and enteric viruses formerly classified as ECHO type 10 is described. Science *130*, 1387-1389 (1959)
17. Sharpe, A. H. and B. N. Fields: Pathogenesis of viral infections. Basic concepts derived from the reovirus model. New Engl. J. Med. *312*, 486-497 (1985)
18. Soike, K. F.; F. Coulston; P. Day; R. Deibel, and H. Plager: Viruses of the alimentary tract of chimpanzees. Exp. mol. Path. *7*, 259-303 (1967)
19. Soike, K. F.; J. D. Douglas; H. Plager, and F. Coulston: Further studies of the viruses of the chimpanzee. II. Viral infection in recently imported chimpanzees captured in Africa. Exp. Path. *11*, 333-339 (1969)
20. Thein, P. and R. Scheid: Reoviral infections. Pp.191-216 in "CRC Handbook Series in Zoonoses". (Steele, J. H., Series ed.), Section B: Viral Zoonoses, Vol. II (Beran, G. W., Sect. ed.), CRC Press, Boca Raton (1981)

1.7 Myxoviruses

1.7.1 Orthomyxoviruses - Influenza Viruses

Influenza viruses are of only limited host specificity, ie. porcine influenza viruses also cause disease in man and human influenza has been shown to infect a variety of avian species. Natural infections of nonhuman primates by human strains have been repeatedly reported - mostly by the demonstration of specific antibodies (Bhatt et al., 1966; Kalter et al., 1967, 1969; Nigi and Tanaka, 1973; O'Brien and Tauraso, 1973; Yamane, 1974; Romvary et al., 1976; Kalter and Heberling, 1978). They have also been sometimes associated with clinical disease (Johnsen et al., 1971; Romvary et al., 1976). Clinical symptoms and death resulted in some experimental infections of *M. mulatta* and *M. fascicularis* with type A influenza virus (Saslaw et al., 1946; Saslaw and Carlisle, 1965). Outbreaks of fatal respiratory disease related to human influenza epidemics without virological confirmation have been reported in *Pan sp.* and *Cebus apella* (Mouquet, 1926; Ratcliffe, 1942). These, however, could have been caused by other respiratory agents such as parainfluenza or respiratory syncytial virus.

Virus

Orthomyxoviruses are slightly stranded RNA viruses of 80 to 120 nm particle size, consisting of a nucleoprotein helix surrounded by a lipoprotein envelope. The envelope is studded with spike-like processes - the haemagglutinins. Orthomyxoviruses possess a mucinolytic enzyme (neuraminidase) capable of removing neuraminic acid side chains from mucoproteins.

They are sensitive to lipid solvents (ether) and heat above 37 °C. They are also destroyed by low pH, ultraviolet irradiation and formaldehyde.

All influenza viruses grow well in the allantois of embryonated chicken eggs and in a variety of tissue cultures, thereby producing polykaryotic giant cells. All myxoviruses react through their neuraminidases with the mucoproteins of erythrocytes, thus inducing haemagglutination (Hsiung and Swack, 1972). Influenza viruses are serogrouped into A, B, or C-strains depending on their internal ribonucleoprotein and matrix protein (Dowdle and Coleman, 1974). A-viruses include human and animal influenza viruses, all of which are antigenically related (Easterday, 1981) B- and C-strains are restricted to man.

Epidemiology

Influenza viruses are aerogenically transmitted under nonexperimental circumstances. They can be excreted from naturally or experimentally infected nonhuman primates (Kalter et al., 1969; Johnsen et al., 1971). Hull et al. (1956) isolated Asian influenza virus from uninoculated monkey kidney cell cultures; infected animals might therefore be a source of human infection.

Clinical Symptoms

Nonhuman primates: Most incidental infections occur without any clinical symptoms, only gibbons have been reported to develop illness and sometimes death (Johnsen et al., 1971; Romvary et al., 1976). The animals became febrile with temperatures of up to 105 °F. They developed a serous to purulent rhinitis and suffered from coughing, anorexia, depression, weight loss and gastrointestinal disturbances. The disease lasted for 3 to 6 days, 2 naturally (and 2 experimentally) infected animals out of a total of 36 died after A2/Hong Kong/68-infection (Johnsen et al., 1971). One gibbon at the Budapest Zoo died of Victoria/75 induced respiratory disease (Romvary et al., 1976). Experimentally infected rhesus monkeys developed leukopenia at 3 to 7 days p.i. (Saslaw et al., 1946).

Man: Influenza is an acute respiratory disease of epidemic dimensions. The incubation period ranges from 1 to 3 days, followed by general malaise, chills, fever, headache, myalgias and respiratory disease shown usually as a light nonproductive cough.

Pathology

Nonhuman primates: The few fatally infected gibbons all died of acute pneumonia, which, in the naturally infected animals was of a purulent character, whereas in the experimental infections haemorrhagic-fibrinous bronchopneumonias prevailed. Some of the desquamated bronchiolar epithelium resembled multi-nucleated giant cells (Johnsen et al., 1971). PAS-positive plasma exudates or hyaline membranes were lacking in the lesions.

Man: The viral neuraminidase also destroys the mucoproteins of respiratory cellular walls, resulting in cellular death. Uncomplicated influenza is usually characterized by a focal fibrinous tracheitis. Peracute, fatal cases also show signs of circulatory collapse, with pulmonary haemorrhage and edema (Giese, 1974).

1.7.2 Paramyxoviridae

1.7.2.1 SV5 (Simian Haemadsorbing Virus - SHV)

SV5 has initially been recovered from spontaneously infected rhesus monkey kidney cell cultures by Hull et al. (1956). It has frequently been isolated from latently infected kidney cell cultures of Old World monkeys since 1960 (Emery and York, 1960; Chanock et al., 1961; Yoshida et al., 1965; Tribe, 1966; Rhim and Schell; Hsiung and Atoynatan, 1969). The isolation rate varied widely between 0 (Atoynatan and Hsiung, 1969) up to 70% (Emery and York, 1960) of the tissue cultures tested. SV5 is also classified as paramyxovirus 5. Other para-5-strains closely related to, or identical with, SV5 have been isolated from human sources and designed as SA-virus (Schultz and Habel, 1959); DA-virus (Hsiung, 1959; Hsiung et al., 1962; van Euler, 1963), clone-I-virus (Krim et al., 1961); WB (Liebhaber et al., 1965) or AT-7 (Behbehani et al., 1965). Anti-SV5-antibodies have been demonstrated in nonhuman primates and man as well (Hsiung et al., 1963; Aulisio et al., 1964; Kalter et al., 1967).

Virus
The morphological and cultured characteristics of paramyxoviruses are very similar to those of orthomyxoviruses. They are, however, slightly larger with 120 to 460 nm diameter (Choppin and Stoeckenius, 1964), and they do not replicate in chicken eggs. SV5 replicates in primary monkey kidney cells. The ensuing CPE in primary rhesus kidney cells is poor, but it is much more intensive in BHK-21-cells with the formation of large syncytia and intracytoplasmic inclusion bodies within 12 to 18 hrs (Holmes and Choppin, 1960). A highly sensitive method to recognize paramyxovirus infections is the ability of the paramyxovirus infected cells to haemadsorb erythrocytes (Hsiung, 1981). Other approved methods are neutralization tests and CP-tests (Atoynatan and Hsiung, 1969). Antigenic relationships among the different paramyxoviruses are very close, and the resulting cross reactions make serological diagnosis difficult (Hsiung, 1981).

Epidemiology
Despite the frequent isolation of SV5 from laboratory monkey tissues, sera taken either from monkeys in the field or on their arrival in captivity have only insignificant antibody-levels (Hsiung et al., 1962; Atoynatan and Hsiung, 1969; Nishikawa et al., 1974). Sera taken from indoor bred cynomolgus monkeys lacked anti-SV5-antibodies altogether (Sasagawa et al., 1981). During 30 to 90 days of quarantine the conversion rate to SV5 averaged 50% (Atoynatan and Hsiung, 1969; Nishikawa et al., 1974).

In the human populations tested, 50% of the samples taken at 2 to 5 yrs of age were seropositive with the frequency climbing to 80% at the age of 10 or greater (Hsiung et al., 1963; Kalter et al., 1967; Hsiung and Swack, 1972). For these reasons the status of SV5 as a monkey or human agent is unclear (Hsiung, 1981), as the infection of monkeys might occur during transit or soon after the first human contact (Hsiung and Swack, 1972).

The virus is easily spread by airborne transmission or by contact among monkeys (Larin et al., 1967). SV5 can also experimentally infect mice, hamsters, dogs and cats (Hsiung, 1981).

After infection of nonhuman primates at least, the development of circulating antibodies soon inhibits the propagation of the virus in cell cultures and possibly also limits the infectivity of the animals.

Clinical Symptoms

Nonhuman primates: The great majority of either natural or experimental SV5 infections of nonhuman primates are asymptomatic. Only very rarely were mild respiratory signs observed after experimental (intranasal) infections, or by the in-contact animals as a transient rhinorrhea and pharyngitis (Heath et al., 1966; Larin et al., 1967).

Man: So far, no clinical symptoms have been associated with the (serologically) described human infections.

Pathology

No lesions have been described in either nonhuman primates or man.

1.7.2.2 Parainfluenzavirus 3 (Haemadsorption Virus Type-1 (= HA 1)

The human parainfluenzavirus 3 is also able to infect and sometimes produce clinical illness in nonhuman primates (Churchill, 1963; Shah and Southwick, 1965; Kalter et al., 1967; Deinhardt et al., 1967; Nishikawa et al., 1974; Martin and Kaye, 1983; Jones et al., 1984). Although no retransmission to man has been reported, the theoretical possibility exists.

Virus

The morphological criteria of para-3-virus are similar to those of SV5. For cultivation rhesus monkey kidney cell monolayers are used, but the resulting CPE is indistinctive in most cases. Detection of viral growth is usually accomplished by haemadsorption with guinea-pig erythrocytes (Fulginiti and Stahl, 1974). Para-3-virus can be distinguished into human and bovine (shipping fever) strains. Those viruses isolated from outbreaks in nonhuman primates were related to the human variety (Churchill, 1963). In serotests antigenic cross-reactions have been reported with para-1, -2, and -5 and with the mumps-viruses (Hsiung and Swack, 1982).

Epidemiology

Similar to the situation in SV5 infections most imported monkeys are free of detectable antibodies upon arrival, but within a few weeks the frequencies and titers of antibodies start to climb (Atoynatan and Hsiung, 1969; Kalter and Heberling, 1971; Nishikawa et al., 1974), indicating the occurrence of infection either at or shortly after capture.

Clinical Symptoms

Nonhuman primates: Most PI-3 infections are asymptomatic, but outbreaks of clinical disease and even fatalities have been reported in *Erythrocebus patas* (Churchill, 1963); "marmosets" (Deinhardt et al., 1967), *Hylobates lar* (Martin and Kaye, 1983) and chimpanzees (Jones et al., 1984). The symptoms observed ranged from mild upper respiratory disease, associated with a serous rhinorrhea, coughing, anorexia and lethargy (Martin and Kaye, 1983), to a rapidly fatal pneumonia (Churchill, 1963; Deinhardt et al., 1967). Experimentally infected *E. patas* showed increasing temperature 5 to 6 days p.i., followed by a slight nasal discharge in 2/4 animals 10 or 12 days p.i. (Churchill, 1963).

Man: Parainfluenza viruses belong to the most important viral pathogens for infants and children. They can affect the entire respiratory tract, initiating common colds, pharyngitis, laryngotracheobronchitis, bronchitis, bronchopneumonia or bronchiolitis. Occasionally infection of liver, central nervous system or salivary glands can occur (Fulginiti and Stahl, 1974).

Pathology

Nonhuman primates: The primary types of lesions were those of an acute bronchopneumonia, which, in less advanced cases affected the tips of the cardiac lobes most commonly, and in fatal cases extended throughout the entire lobes (Churchill, 1963; Deinhardt et al., 1967). Other lesions were pleurisy, pericarditis and peritonitis. Some of the advanced lesions were probably of bacterial origin, because respiratory pathogens, e.g. pneumococci, were invariably also isolated from such lungs.

1.7.2.3 Paramyxovirus-I (Sendaivirus-Type-D-Influenza)

Sendaivirus, which primarily infects rodents, is frequently isolated from infants with croup (Hsiung, 1981). The infectivity for nonhuman primates seems to be low, according to the failure to detect anti-para I-antibodies in most serological surveys (Kalter, 1967; Atoynatan and Hsiung, 1969; Nishikawa et al., 1974). In contrast to the serological studies, one outbreak of respiratory disease due to Sendaivirus has been reported in *Callithrix jacchus* (Flecknell et al., 1983).

Epidemiology

Like other myxoviruses, Sendaivirus is transmitted by dropletinfection. The way of introduction of the virus into the indoorkept marmoset-colony is unknown, but rodents should not have access to the animal-houses. There exists, however, the possibility of infection occurring via 6 young male marmosets brought into a colony 2 months prior to the outbreak of disease.

Clinical Symptoms

Nonhuman primates: In the afore mentioned outbreak animals of all ages were affected, with the most serious cases occurring in the young ones. The initial signs were inappetance, depression, tachypnoea and dyspnoea. Other symptoms were persistent sneezing, ocular and nasal discharge of a serous or purulent nature.

Man: Para I-virus outbreaks caused principally laryngotracheobronchitis (croup) in children between 4 months and 5 years of age (Glezen et al., 1976).

Pathology
Nonhuman primates: 2 infants died due to a purulent pneumonia of all lobes of the lungs, accompanied by extensive pulmonary oedema (Flecknell et al., 1983).

Man: In Sendaivirus infections of newborns the principal lesions are focal, haemorrhagic pneumonia with peribronchial or alveolar oedema and a remarkable thickening of the alveolar septae due to cellular infiltration. The bronchial epithelia, however, show only minor lesions (Siegert, 1967).

1.7.2.4 Pneumoviruses: Respiratory Syncytial Virus (RSV) Chimpanzee Coryza Agent

Respiratory syncytial virus was first isolated in 1955 from chimpanzees affected with coughing, sneezing and rhinorrhea (Morris et al., 1956). Subsequently the virus was demonstrated as a major cause of respiratory infection in man (Chanock and Finberg, 1957; Chanock et al., 1957). Respiratory viruses have also been isolated from cattle with acute respiratory disease (Philipps et al., 1981).

Virus
The virions contain single-stranded RNA. They are enveloped, spherical and have a size of approximately 90 to 130 nm. The surface is studded by regularly shaped projections, but the virions do not possess haemagglutinin or haemadsorption activity.

The virus grows well in tissue cultures of human, simian, bovine, porcine, and hamster origin (human strains best in man-derived cells and bovine strains best in cell cultures of bovine origin).
CPE appears 1 to 14 days p.i., and is characterized by the formation of syncytial giant cells (Morris et al., 1956; Chanock et al., 1957). RS-viruses are very labile, being inactivated by heating above 37 °C or reversely by freezing at −15 to −25 °C for several days, and by ether or low pH.

Epidemiology
RSV infections are worldwide distributed in animals and man. In nonhuman primates the virus has been isolated exclusively from chimpanzees (Morris et al., 1956; Dick and Dick, 1974), although antibodies against RSV have been found additionally in captive orangutans *(Pongo pygmaeus)*, gorillas *(Gorilla gorilla)*, gibbons *(Hylobates lar)*, cynomolgus *(Macaca fuscicularis)*, rhesus *(Macaca mulatta)*, vervet *(Cercopithecus aethiops)*, squirrel *(Saimiri sciureus)* and spider monkeys *(Cebus apella, Cebus albifrons)* (Kalter and Heberling, 1971; Richardson-Wyatt et al., 1981).

A high prevalence of RSV-infections has been shown to occur in domestic animals and man. Nonhuman primates are therefore most probably infected from man or domestic animals following capture. The virus is highly contagious, with asymptomatic or only mildly ill children being considered to be the reservoir, particularly during the winter season.

Transmission of the virus is via an aerogenous route either by direct contact or droplets.

Clinical Symptoms

Nonhuman primates: Coughing, sneezing and mucopurulent nasal discharge have been observed in chimpanzees (Morris et al., 1956). Experimentally infected *S. fuscicollis* failed to develop any clinical signs (Coates and Chanock, 1962).

Man: RSV- infections vary in severity from the asymptomatic to a fatal disease. Severe disease is more often seen in infants. The actual signs of RSV-infections are low grade fever, runny nose, cough and respiratory disease. In fatalities the disease rapidly progressed to respiratory failure, coma and death (Philipps et al., 1981).

Pathology

Nonhuman primates: No fatal infections have been observed so far.

Man: In infants which died of RSV-infections necrosis of bronchocytes and interstitial pneumonia were found. Cytoplasmic inclusion bodies have been seen in pulmonary parenchymal cells (Philipps et al., 1981).

References

1. Atoynatan, T. and G. D. Hsiung: Epidemiologic studies of latent virus infections in captive monkeys and baboons. II. Serologic evidence of myxovirus infections with special reference to SV 5. Amer. J. Epidem. *89*, 472-479 (1969)
2. Aulisio, C. G.; D. C. Wong, and J. A. Morris: Neutralizing antibodies against simian viruses SV5 and SV20 in human sera. Proc. Soc. exp. Biol. *117*, 6-11 (1964)
3. Behbehani, A. M.; J. L. Melnick, and M. E. de Bakey: A paramyxovirus isolated from human atheromatous lesion. Exp. molec. Path. *4*, 606-619 (1965)
4. Bhatt, P. N.; C. D. Brandt; R. A. Weiss; J. P. Fox, and M. F. Shaffer: Viral infections of monkeys in their natural habitat in Southern India. II. Serological evidence of viral infection. Amer. J. trop. Med. Hyg. *15*, 561-566 (1966)
5. Chanock, R. and L. Finberg: Recovery from infants with respiratory illness of virus related to chimpanzee coryza agent (CCA). II. Epidemiologic aspects of infection in infants and young children. Amer. J. Hyg. *66*, 291-300 (1957)
6. Chanock, R. M.; K. M. Johnson; M. K. Cook; D. C. Wong, and A. Vargosko: The hemadsorption technique with special reference to the problem of naturally occurring simian parainfluenza virus. Amer. Rev. resp. Dis. *83/2*, 125-129 (1961)
7. Chanock, R.; B. Roizman, and R. Meyers: Recovery from infants with respiratory illness of virus related to chimpanzee coryza agent (CCA). Isolation, properties, and characterization. Amer. J. Hyg. *66*, 281-290 (1957)
8. Choppin, P. W. and W. Stoeckenius: The morphology of SV5 virus. Virology *23*, 195-202 (1964)
9. Churchill, A. E.: The isolation of parainfluenza 3 virus from fatal cases of pneumonia in *Erythrocebus patas*. Brit. J. Exp. Path. *45*, 529-537 (1963)
10. Coates, H. V. and P. M. Chanock: Experimental infection with respiratory syncytial virus in several species of animals. Amer. J. Hyg. *76*, 302-311 (1962)
11. Deinhardt, F.; A. W. Holmes; J. Devine, and J. Deinhardt: Marmosets as laboratory animals. IV. The microbiology of laboratory-kept marmosets. Lab. Anim. Care *17*, 48-70 (1967)
12. Dick, E. C. and C. R. Dick: Natural and experimental infection of nonhuman primates with respiratory viruses. Lab. Anim. Sci. *24*, 177-181 (1974)
13. Dowdle, W. R. and M. T. Coleman: Influenza virus, pp. 678-685 in "Manual of Clinical Microbiology", 2nd ed. (Lennette, E. H., E. H. Spaulding, and J. P. Truant, eds.). Amer. Soc. Microbiol., Washington D. C. (1974)

14. Easterday, B.C.: Influenza. Swine Influenzas; pp.225-231 in "CRC Handbook Series in Zoonoses" (Steele, J.H., Series ed.), Section B: Viral Zoonoses. Vol. II (Beran, G.W., Section ed.), CRC Press, Boca Raton (1981)
15. Emery, J.B. and C.J. York: Occurrence of a hemadsorption virus in normal monkey tissue culture. Virology *11*, 313-315 (1960)
16. Flecknell, P.A.; R.Parry; J.R. Needham; R.M. Ridley; H.F. Baker, and P.Bowes: Respiratory disease associated with parainfluenza type I (Sendai) virus in a colony of marmosets (*Callithrix jacchus*). Lab. Anim. *17*, 111-113 (1983)
17. Fulginiti, V.A. and M.Stahl: Parainfluenza and respiratoy syncytial viruses; pp. 686-694 in "Manual of Clinical Microbiology", 2nd ed. (Lennette, E.H., E.H. Spaulding, and J.P. Truant, eds.), Amer. Soc. Microbiol., Washington, D.C. (1974)
18. Giese, W.: Atemwege und Lungen. Pp. 3/1-3/108 in "Organpathologie" (Doerr, W., Series ed.), Bd. I: Herz und Gefäße, Blut und blutbereitende Organe, Atemwege und Lungen (Doerr, W., W.Giese, L.D. Leder und W.Remmele, Section eds.). Thieme, Stuttgart (1974)
19. Glezen, W.P.; F.A. Loda, and F.W. Denny: The parainfluenza viruses; pp. 337-349 in "Viral Infections of Humans. Epidemiology and Control" (Evans, A.S., ed.), Wiley and Sons, London (1976)
20. Heath, R.B.; I.E. Falk; J.E. Stark; R.H. Herbst-Laier, and N.M. Laurin: Induced upper respiratory infection in the vervet monkey: A model of acute viral respiratoy disease in man. Brit. J. exp. Path. *47*, 93-98 (1966)
21. Holmes, K.V. and P.W. Choppin: On the role of the response of the cell membrane in determinating virus virulence. Contrasting effects of the parainfluenza virus SV5 in two cell types. J.exp. Med. *124*, 501-520 (1966)
22. Hsiung, G.D.: The use of agar overlay cultures for detection of new virus isolates. Virology *9*, 717-719 (1959)
23. Hsiung, G.D.: Parainfluenza virus infection; pp.247-259 in "CRC Handbook Series in Zoonoses". (Steele, J.H., Series ed.), Section B: Viral Zoonoses. Vol. II (Beran, G.W., Section ed.), CRC Press, Boca Raton (1981)
24. Hsiung, G.D. and T.Atoynatan: Incidence of naturally acquired virus infections of captive monkeys. Amer. J.Epidem. *83*, 38-47 (1966)
25. Hsiung, G.D.; P.Isacson, and R.W. McCollum: Studies of a myxovirus isolated from human blood. I.Isolation and properties. J. Immunol. *88*, 284-290 (1962)
26. Hsiung, G.D.; P.Isacson, and G.Tucker: Studies of parainfluenza viruses. II. Serologic interrelationships in humans. Yale J.Biol. Med. *35*, 534-544 (1963)
27. Hsiung, G.D. and N.S. Swack: Myxoviruses and pseudomyxovirus groups; pp.537-571 in "Pathology of Simian Primates", Part II (Fiennes, R.N.T-W., ed.), Karger, Basel (1972)
28. Hull, R.N.; D.R. Minner, and J.W. Smith: New viral agents recovered from tissue cultures of monkey kidney cells. I.Origin and properties of cytopathogenic agents. SV1, SV2, SV4, SV5, SV6, SV11, SV12, and SV15. Amer. J.Hyg. *63*, 204-215 (1956)
29. Johnsen, D.O.; W.L. Wooding; P.Tanticharoenyos, and C.Karnjanaprakorn: An epizootic of A2/Hong Kong/68 influenza in gibbons. J.infect. Dis. *123*, 365-370 (1971)
30. Jones, E.E.; P.L. Alford; A.L. Reingold; H.Russell; M.E. Keeling, and C.V. Broome: Predisposition to invasive pneumococcal illness following parainfluenza type 3 virus infection in chimpanzees. J.Amer. vet. Med. Ass. *185*, 1351-1353 (1984)
31. Kalter, S.S. and R.L. Heberling: Comparative virology of primates. Bact. Rev. *35*, 310-364 (1971)
32. Kalter, S.S. and R.L. Heberling: Serologic response of primates to influenza viruses. Proc. Soc. exp. Biol. *159*, 414-417 (1978)
33. Kalter, S.S.; R.L. Heberling; T.E. Vice; F.S. Lief, and A.R. Rodriguez: Influenza (A2/Hong Kong/68) in the baboon (*Papio* sp.). Proc. Soc. exp. Biol. *132*, 357-361 (1969)
34. Kalter, S.S.; J.Ratner; G.V. Kalter; A.R. Rodriguez, and C.S. Kim: A survey of primate sera for antibodies to viruses of human and simian origin. Amer. J.Epidem. *86*, 552-568 (1967)
35. Krim, M.; S.C. Wong, and E.D. Kilbourne: Isolation of a virus from a clone of a human cell line in continous cultivation. Fed. Proc. *20*, 442 (1961)
36. Larin, N.M.; R.H. Herbst-Laier; M.P. Copping, and R.B.M. Wenham: Controlled investigation in baboon *(Papio cynocephalus)* on transmission of SV5 virus by contact. Nature *213*, 827-828 (1967)

37. Liebhaber, H.; S. Krugman; D. McGregor; J. P. Giles: Studies of a myxovirus recovered from patients with infectious hepatitis. I. Isolation and characterization. J. exp. Med. *122*, 1135-1150 (1965)
38. Martin, D. P. and H. S. Kaye: Epizootic of parainfluenza-3 virus infections in gibbons. J. Amer. vet. med. Ass. *183*, 1185-1187 (1983)
39. Morris, J. A.; R. E. Blount, jr. and R. E. Savage: Recovery of cytopathogenic agent from chimpanzees with coryza. Proc. Soc. exp. Biol. *92*, 544-549 (1956)
40. Mouquet, A. D.: Maladie à allure grippale chez des chimpanzés. Soc. Cent. Med. Vet. *1926*, 46-50 (1926)
41. Nigi, H. and T. Tanaka: Clinical examinations on the Japanese monkey (Macaca fuscata). Exp. Anim. *22*, 461-470 (1973)
42. Nishikawa, F.; T. Sugiyama; M. Takasaka, and S. Honjo: Epidemiology of myxo-paramyxovirus infection among cynomolgus monkeys imported for laboratory use. Jap. J. med. Sci. Biol. *27*, 249-262 (1974)
43. O'Brian, T. C. and N. M. Tauraso: Antibodies to type A influenza viruses in sera from nonhuman primates. Arch. ges. Virusforsch. *40*, 359-365 (1973)
44. Phillips, C. F.; B. R. Forsyth, and C. A. Phillips: Respiratory syncytial virus infection; pp. 293-297 in "CRC Handbook Series in Zoonoses." (Steele, J. H., Series ed.), Section B: Viral Zoonoses, Vol. II (Beran, G. W., Section ed.), CRC Press, Boca Raton (1981)
45. Ratcliffe, H. L.: Deaths and important diseases. Rep. Penrose Res. Lab. 1942, 11-25 (1942)
46. Rhim, J. S. and K. Schell: Cytopathic effects of the parainfluenza virus SV5 in vero cells. Nature *216*, 271-272 (1967)
47. Richardson-Wyatt, L. S.; R. B. Belshe; W. T. London; D. L. Sly; E. Camargo, and R. M. Chanock: Respiratory syncytial virus antibodies in nonhuman primates and domestic animals. Lab. Anim. Sci. *31*, 413-415 (1981)
48. Romvary, J.; J. Meszaros; J. Tanyi; J. Rosza, and L. Fabian: Spreading of virus infection among wild birds and monkeys during the influenza epidemic caused by the Victoria (3) 75 variant of A (H3H2) virus. Acta vet. Acad. Sci. Hung. *26*, 369-376 (1976)
49. Sasagawa, A.; T. Inayoshi; F. Nakamura; M. Suzuki, and S. Honjo: Absence of antibodies against SV5, measles and herpes simplex in indoor colony-bred cynomolgus monkeys. Jap. J. med. Sci. Biol. *34*, 381-384 (1981)
50. Saslaw, S. and H. N. Carlisle: Aerosol exposure of monkeys to influenza virus. Proc. Soc. exp. Biol. *119*, 838-843 (1965)
51. Saslaw, S.; H. E. Wilson; L. A. Doan; O. C. Woolpert, and J. L. Schwab: Reaction of monkeys to experimentally induced influenza virus A infection. An analysis of the relative roles of humoral and cellular immunity under conditions of deficient nutrition. J. exp. Med. *84*, 113-125 (1946)
52. Schultz, E. W. and K. Habel: SA virus - A new member of the myxovirus group. J. Immunol. *82*, 274-278 (1959)
53. Shah, K. V. and C. H. Southwick: Prevalence of antibodies to certain viruses in sera of free-living rhesus and of captive monkeys. Indian J. med. Res. *53*, 488-500 (1965)
54. Siegert, R.: Parainfluenzavirus-Infektionen; pp. 403-418 in "Infektionskrankheiten" Vol. I, Part I (Gsell, O. and W. Mohr, eds.), Springer, Berlin (1967)
55. Tribe, G. W.: An investigation on the incidence, epidemiology and control of simian virus 5. Brit. J. exp. Path. *47*, 472-479 (1966)
56. Van Eueler, L.; F. S. Kantor, and G. D. Hsiung: Studies of parainfluenza viruses. I. Clinical, pathological and virological observations. Yale J. Biol. Med. *35*, 523-533 (1963)
57. Yoshida, E.; H. Yamamoto, and H. Shimajo: Contamination of cynomolgus monkey kidney cell cultures by hemagglutinating simian virus (SV5). Jap. J. Med. exp. *18*, 151-156 (1965)

1.8 Pseudomyxoviruses

1.8.1 Measles Virus (Monkey Intra Nuclear Inclusion Agent = MINIA)

Measles virus quite commonly infects nonhuman primates of many species under natural conditions.

Affected species comprise *M. mulatta* (Habermann and Williams, 1957; Ruckle, 1956; Potkay et al., 1966; Hall et al., 1971; Renne et al., 1973; Remfry and Hovell, 1976; Remfry, 1976; Saha et al., 1979); *M. fascicularis* (Shishido, 1966; Yamanouchi et al., 1973); *M. radiata* (Saha et al., 1979), *Presbytis cristatus* (Montrey et al., 1980); *Colobus guereza* (Hime et al., 1975; Scott and Keymer, 1975; Hime, 1976); *Pan troglodytes* (MacArthur et al., 1979); *Callithrix jacchus, Saguinus oedipus, S. fuscicollis* (Levy and Mirkovic, 1971; Tribe, 1978); *Aotus trivirgatus* (O'Brien et al., 1981) and *Saimiri sciureus* (Kalter and Heberling, 1972; MacArthur et al., 1979).

Virus

The pseudomyxoviridae differ from the paramyxoviridae by their lack of neuraminidase. Within the pseudomyxoviridae the measles virus can be distinguished from other members by its ability to haemagglutinate primate erythrocytes, which is not prevented by receptor destroying enzymes (Swack, 1981). The human and simian measles virus strains - the monkey virus formerly designed as MINIA - are identical (Ruckle-Enders, 1962; Meyer et al., 1962).

Measles viruses are spherical to pleomorphic with 120 to 270 nm diameter, with occasionally filamentous forms occurring. The core consists of a single stranded RNA. The entire nucleocapsid is helically wound and the outer lipoprotein membrane contains numerous projections (Swack, 1981). It is heat-, acid-, ether- and radiation-labile.

In vitro isolation of measles virus is time consuming. The best results are obtained with primary human kidney cells. Primary cynomolgus monkey kidney cells and primary rhesus monkey kidney cells although supporting the replication of measles virus satisfactory, might be contaminated with either measles virus or foamy virus. Diploid cell lines of human origin, BS-C-I cells and VERO-cells are acceptable alternatives (Black, 1974). The CPE in measles virus infected tissue cultures is dominated by syncytial formation.

Antigenic relationships have been demonstrated between measles, canine distemper and rinderpest virus; whereas no cross reaction occurs between measles and foamy viruses (Ruckle, 1956).

Serological diagnosis depends on haemagglutination-inhibition-, CF-, haemolysis inhibition-, γ-M-antibody- or neutralization tests (Black, 1974).

Epidemiology

Nonhuman primates living in their natural habitat without human contacts are evidently free of measles virus infections (Meyer et al., 1962; Bhatt et al., 1966;

Shishido, 1966; Kalter et al., 1967; Kalter and Heberling, 1972). They obviously acquire the infection from man soon after capture. The measles infection rate in developing countries is quite high (Saha et al., 1979). 40% of children have been known to die in village epidemics in tropical areas (Remfry, 1976). The chance therefore of transmitting the infection to newly exposed and susceptible monkeys is accordingly high. As monkeys are highly susceptible to measles virus infections, a high percentage of urban rhesus monkeys are already infected and may iniatiate a monkey-monkey cycle in their urban habitat (Saha et al., 1979). Colony born cynomolgus monkeys, on the other hand, have been shown in one study to be free of anti-measles virus antibodies until the age of 3 months (Sasagawa et al., 1981), whereas in another series (Fujimoto et al., 1983) the newborn cynomolgus had appreciable HI-titers at birth which subsequently lowered linearly with a half-life period of 3,4 ± 0,4 weeks. This might simulate the human situation. Human maternal anti-measles-antibodies do cross the placenta, but are destroyed during the first week of life (MacArthur et al., 1979) and must, thereafter be actively produced.

Regardless of the immediate source of infection (man or monkey), once the infection is introduced into a monkey population it spreads very rapidly. In the survey by Meyer et al. (1962) none of 108 rhesus monkeys trapped in the jungles had measles antibodies, whereas all 108 monkeys expressed anti-measles virus antibody activity 8 weeks later. As a result, the infection rate of Old World monkeys transported to laboratories in the Northern hemisphere increases rapidly after capture or after arrival at the laboratory, and, consequently, most outbreaks of clinical monkey measles occur in recently imported animals (Ruckle, 1956; Potkay et al., 1966; Shishido, 1966; Valerio, 1971; Hime et al., 1975; Scott and Keymer, 1975; Remfry and Hovell, 1976).

As in man, the route of transmission to monkeys is probably by droplet infection from children or monkeys during their infectious period before appearance of the rash.

Clinical Symptoms

Nonhuman primates: In nonhuman primate, species or genus dependent differences in the course of the infection seem to exist. The disease in macaques is normally mild with clinical signs very often being missed entirely (Hall et al., 1971). In contrast, the disease in a marmoset colony resulted in 326 deaths (Levy and Mirkovic, 1971) and in colobines a fatality rate of up to 100% has occurred (Hime, 1975; Scott and Keymer, 1975).

In macaques the incubation period is from 6 to 10 days (Blake and Trask, 1921a, b; Taniguchi et al., 1954), and in colobus monkeys from 10 to 14 days (Hime et al., 1975). In both species the main symptoms of experimental infections were fever, leukopenia, conjunctivitis, and maculo-papular exanthema (Taniguchi et al., 1954). Respiratory signs such as dry coughs and nasal discharges (Scott and Keymer, 1975; Remfry 1976) are sometimes the leading symptoms. These are occasionally accompanied by conjunctivitis and a maculopapular rash, which particularly involves the ventral body and always spares the palms and soles (Potkay et al., 1966; Montrey et al., 1980).

Koplik's spots (small red spots often with a white center appearing in the buccal mucous membranes ahead of the cutaneous rashes) are usually missed (Potkay et al., 1966), possibly because of technical reasons. Other clinical signs of measles in monkeys are facial edema (particularly as periorbital edema), erythema, progressive lethargy, anorexia and severe diarrhea (Levy and Mirkovic, 1971; Valerio, 1971; Hime et al., 1975; Scott and Keymer, 1975; Hime, 1976).

The non pruritic rashes are followed by a powdery to scaly desquamation of the exanthemas lasting several days (Potkay et al., 1966; Montrey et al., 1980). The scales may reach a diameter of up to 10 mm (seen in langurs), and the desquamation can continue sometimes for approximately 2 weeks after disappearance of the initial rashes.

Man: After a 12 to 14 day incubation period the early symptoms of chills, sneezing, conjunctivitis, cough, fever and characteristic Koplik's spots appear. 3 to 5 days later, during which time both fever and cough worsen, a maculopapular rash appears, beginning on the forehead and behind the ears. This rash spreads within the next few days over the face, neck, trunk and limbs, and once it is fully developed, the body temperature drops to normal levels (Black, 1974).

Pathology

Nonhuman primates: The predominating lesion in fatal cases is a proliferative, giant cellular, focal to extensive interstitial pneumonia (Valerio, 1971; Renne et al., 1973; Remfry, 1975). The alveolar walls are thickened by an infiltration of mononuclear cells or sometimes more fibrotic tissue (Levy and Mirkovic, 1971; Scott and Keymer, 1975). Similar cells emigrate into the alveolar spaces, which, on the other hand, remain virtually free of fibrin exudates (Scott and Keymer, 1975). Syncytial giant cells which appear in the alveolar or bronchiolar/bronchial epithelia are sometimes referred to as "Warthin-Finkeldey" cells (Scott and Keymer, 1975). This term, however, might be incorrect as by their definition, Warthin-Finkeldey cells are derived from lymphoid tissues. True Warthin-Finkeldey cells are found in monkeys, often in germinal centers of lymphfollicles (Nii et al. 1964; Tajima and Kudow, 1976). From their cellular origin the giant cells in lymphoid tissues can be distinguished into

a) lymphoid - the true Warthin-Finkeldey cells with somewhat irregular nuclei and often a common nuclear membrane,
b) reticular giant cells with a wide rim of cytoplasm with long cytoplasmic processes,
c) plasmacytic giant cells and
d) phagocytic giant cells (Nii et al., 1964).

The typical Warthin-Finkeldey giant cells rarely contain intranuclear or intracytoplasmic inclusions, whereas the syncytial giant cells in the bronchial/bronchiolar/alveolar epithelia in measles virus infections usually contain such intranuclear or cytoplasmic inclusion bodies.

Warthin-Finkeldey cells (and probably also other measles induced giant cells) have only a short life-span of 5 to 12 days subsequent to inoculation. The rapid necrosis is due to lethal cytopathic effects of the measles virus (Tajima and Kudow, 1976). Typical Warthin-Finkeldey cells appear also in the enlarged lymphnodes, spleen and lymphoid follicles of the intestinal wall (Levy and Mirkovic, 1971; Renne et al., 1973).

Epithelial syncytial cells similar to those appearing in the respiratory epithelia are also sometimes formed in the thyroid, intestinal epithelia, urinary bladder, renal pelvis, liver or even the brain (Hall et al., 1971).

The skin lesions (biopsies taken during the rash) are most pronounced in the epidermis and the hair follicles. They consist of acanthosis, hyperkeratosis and parakeratosis of the skin with necrosis of hair follicles. Syncytia of epithelial cells containing up to 20 nuclei were found in the epithelium and hair follicles (Hall et al., 1971).

Man: Human measles is characterized by formation of multinucleated Warthin-Finkeldey cells in the tonsils, lymphnodes, spleen, Peyer's patches, and lymphoid tissue of the appendix. They *do not* develop intranuclear or intracytoplasmic inclusion bodies.

In the lungs giant cellular macrocytic interstitial pneumonia is prevalent. Syncytia in the respiratory tract generally contain intranuclear and cytoplasmic inclusion bodies, the bronchial and bronchiolar epithelia show squamous metaplasia (Archibald et al., 1971).

The skin lesions are due to necrosis and vacuolization of the epidermal cells. They are also accompanied by dermal inflammation.

Subacute sclerosing panencephalitis and multiple necrosis have been linked to measles infections (Swack, 1981).

References

1. Archibald, R.W.R.; R.O. Weller, and S.R. Meadow: Measles pneumonia and the nature of the inclusion-bearing giant cells: A light- and electron-microscope study. J. Path. *103*, 27-34 (1971)
2. Bhatt, P.N.; C.D. Brandt; R.A. Weiss; J.P. Fox, and M.F. Shaffer: Viral infections of monkeys in their natural habitat in southern India. II. Serological evidence of viral infection. Amer. J. trop. Med. Hyg. *15*, 561-566 (1966)
3. Black, F.L.: Measles virus; pp. 709-715 in "Manual of Clinical Microbiology". 2nd ed. (Lennette, E.H.; E.H. Spaulding and J.P. Truant, eds.) Amer. Soc. Microbiol., Washington D.C. (1974)
4. Blake, F.G. and J.D. Trask: Studies on measles. I. Susceptibility of monkeys to the virus of measles. J. exp. Med. *33*, 385-412 (1921a)
5. Blake, F.G. and J.D. Trask: Studies on measles. II. Symptomatology and pathology in monkeys experimentally infected. J. exp. Med. *33*, 413-422 (1921b)
6. Fujimoto, K.; K. Terao; F. Cho, and S. Honjo: Immunoglobulins and measles antibody in sera of newborn cynomolgus monkeys and in milk of their mothers. Jap. J. Med. Sci. Biol. *36*, 209-214 (1983)
7. Habermann, R.T. and F.P. Williams: Diseases seen at necropsy of 708 *Macaca mulatta* (rhesus monkey) and *Macaca philippinensis* (cynomolgus monkey). Amer. J. vet. Res. *18*, 419-426 (1957)
8. Hall, W.C.; R.M. Kovatch; P.H. Herman, and J.G. Fox: Pathology of measles in rhesus monkeys. Vet. Path. *8*, 307-319 (1971)
9. Hime, J.M.: Report of the veterinary officer. Regent 's Park: 1974-1975. J. Zool. *178*, 447-455 (1976)
10. Hime, J.M.; I.F. Keymer, and C.J. Baxter: Measles in recently imported colobus monkeys *(Colobus guerezza)*. Vet. Rec. *97*, 392 (1975)
11. Kalter, S.S. and R.L. Heberling: Measles and rubella antibody in nonhuman primates. Primate Zoonoses Surveill. Rep. *7*, 14-17 (1972)

12. Kalter, S.S.; J. Ratner; G.V. Kalter; A.R. Rodriguez, and C.S. Kim: A survey of primate sera for antibodies to viruses of human and simian origin. Amer. J. Epidem. *86*, 552-568 (1967)
13. Levy, B.M. and R.R. Mirkovic: An epizootic of measles in a marmoset colony. Lab. Anim. Sci. *21*, 33-39 (1971)
14. MacArthur, J.A.; P.G. Mann; V.Oreffo, and G.B.D. Scott: Measles in monkeys: an epidemiological study. J. Hyg. *83*, 207-212 (1979)
15. Meyer, H.M., jr.; B.E. Brooks; R.D. Douglas, and N.G. Douglas: Ecology of measles in monkeys. Amer. J. Dis. Child. *103*, 307-313 (1962)
16. Montrey, R.D.; D.L. Huxsoll; P.K. Hildebrandt; B.W. Booth, and S. Arimbalam: An epizootic of measles in captive silvered leaf-monkeys *(Presbytis cristatus)* in Malaysia. Lab. Anim. Sci. *30*, 694-697 (1980)
17. Nigi, H. and T. Tanaka: Clinical examination of the Japanese monkey *(Macaca fuscata)*. Exp. Anim. *22*, 461-470 (1973)
18. Nii, S.; J. Kamahora; Y. Mori; M. Takahashi, and Y. Okuno: Experimental pathology of measles in monkeys. Biken's J. *6*, 271-297 (1964)
19. O'Brien, A.W.; L.N. Binn, R.H. Hall; R.J. Beattle, and R.H. Marchwicki: Measles virus antibodies in a laboratory colony of owl monkeys *(Aotus trivirgatus)*. Lab. Anim. *15*, 343-345 (1981)
20. Potkay, S.; J.R. Ganaway; N. Rogers, and R. Kinard: An epizootic of measles in a colony of rhesus monkeys *(Macaca mulatta)*. Amer. J. vet. Res. *27*, 331-334 (1966)
21. Remfry, J.: A measles epizootic with 5 deaths in newly-imported rhesus monkeys (*Macaca mulatta*). Lab. Anim. *10*, 49-57 (1976)
22. Remfry, J. and G.J.R. Hovell: Measles in monkeys. Vet. Rec. *98*, 38 (1976)
23. Renne, R.A.; R. McLaughlin, and A.B. Jenson: Measles virus-associated endometritis, cervicitis, and abortion in a rhesus monkey. J. Amer. vet. med. Ass. *163*, 639-641 (1973)
24. Ruckle, G.: Measles in humans and in monkeys, report of isolation from cynomolgus monkeys of an agent immunologically related to human measles virus. Fed. Proc. *15*, 610 (1956)
25. Ruckle-Enders, G.: Comparative studies of monkey and human measles virus strains. Amer. J. Dis. Child. *103*, 297-307 (1962)
26. Saha, S.M.; S.B. Sharma, and M. Balasubrahmanyan: Antibodies against measles antigen in the laboratory housed rhesus and bonnet monkeys. Indian J. med. Res. *69*, 213-216 (1979)
27. Sasagawa, A.; T. Inayoshi; F. Nakamura; M. Suzuki, and S. Honjo: Absence of antibodies against SV5, measles and herpes simplex in indoor colony-bred cynomolgus monkeys. Jap. J. med. Sci. Biol. *34*, 381-384 (1981)
28. Scott, G.B.D. and I.F. Keymer: The pathology of measles in Abyssinian colobus monkeys *(Colobus Guereza)*: A description of an outbreak. J. Path. *117*, 229-233 (1975)
29. Shishido, A.: Natural infection of measles virus in laboratory monkeys. Jap. J. med. Sci. Biol. *19*, 215-227 (1966)
30. Swack, N.S.: Pseudomyxoviral infections. Measles, canine distemper, and rinderpest; pp. 275-282 in "CRC Handbook Series in Zoonoses" (Steele, J.H., Series ed.) Section B: Viral Zoonoses, Vol. II (Beran, G.W. Section CRC Press, Boca Raton (1981)
31. Tajima, M. and S. Kudow: Morphology of the Warthin-Finkeldey giant cells in monkeys with experimentally induced measles. Acta path. jap. *26*, 367-380 (1976)
32. Taniguchi, T.; J. Kamahora; S. Kato, and K. Hagiwara: Pathology in monkeys experimentally infected with measles virus. Med. J. Osaka Univ. *5*, 367-396 (1954)
33. Taniguchi, T.; Y. Okuno; J. Kamahora; A. Aoyama, and S. Kato: Animal experiments of measles on monkeys. Med. J. Osaka Univ. *4*, 381-399 (1954)
34. Valerio, D.A.: Colony management as applied to disease control with mention of some viral diseases. Lab. Anim. Sci. *21*, 1011-1014 (1971)
35. Yamanouchi, K.; F. Chino; F. Kobune; H. Kodama, and T. Tsuruhara: Growth of measles virus in the lymphoid tissues of monkeys. J. infect. Dis. *128*, 795-799 (1973)
36. Yamanouchi, K.; A. Shishido, and S. Honjo: Natural infection of cynomolgus monkey with measles virus. Exp. Anim. *122*, 389-393 (1973)

1.9 Retroviruses

1.9.1 Spumavirinae = Foamy Viruses

Foamy viruses are probably among the most common latent viruses occurring in primary monkey kidney cell cultures. They have been of little concern as zoonotic agents in the past and their biological properties and potential still lie very much in the dark. Possibly they have to be reevaluated because simian foamy viruses have been isolated from human tumor materials (Achong et al., 1971; Young et al., 1973; Brown et al., 1978).

Foamy viruses are most often found in aged tissue cultures through their characteristic induction of multinucleated giant cells with large intracellular vacuoles. This characteristic appearance resembles the CPE induced by measles virus except for the absence of eosinophilic intranuclear inclusions in foamy virus infected cells (Hsiung and Swack, 1971). Foamy viruses are not restricted to primates, but have been isolated from cattle, cats and hamsters as well (Barahona et al., 1976).

Agents

Foamy viruses contain single-stranded RNA and an antigenically specific reverse transcriptase. The nucleocapsid has a diameter of 725 Å, surrounded by an envelope with 125 Å projections (Clarke and Attridge, 1968). The enveloped virions finally are of approximately 100 nm diameter (Kalter et al., 1980). Foamy viruses *in vitro* typically bud into intracisternal vesicles leading to dilation of the endoplasmic reticulum (Peries and Todaro, 1977).

Foamy viruses possess a RNA-dependent DNA-polymerase like the other retroviridae (Spiegelman et al., 1970; Parks et al., 1971).

So far, 12 simian foamy viruses (Table 6) can be distinguished by their biological properties, e.g. the ability to produce syncytia on primary rabbit kidney cells (Kalter et al., 1980). They are usually sensitive to chloroform and low pH (Johnston, 1971; Hooks and Gibbs, 1975) and they are inactivated by 60 °C for 30 min. The commonly known lack of haemagglutinin might not hold true since at least simian foamy virus I does exhibit haemagglutinin activity when guinea pig erythrocytes are used (Peries and Todaro, 1977).

Serological surveys depend on neutralization, immunofluorescence and CF-tests, but they are unreliable because foamy viruses are often isolated from seronegative animals (Hsiung and Swack, 1972). Circulating antibodies, on the other hand, do not inhibit the maintenance and growth of the foamy viruses *in vivo.*

Epidemiology

Most isolations of simian foamy viruses in nonhuman primates were achieved from primary kidney tissue cultures, particularly from aging ones. Others were isolated from lymphocytes or lymphoblastoid lines (Feldman et al., 1975; Barahona et al., 1976; Agrba et al., 1978; Rhodes-Feuilette et al., 1979; Neumann-Haefelin et al., 1983); brain (Hooks et al., 1972, 1973); throat swabs (Johnston, 1961; Swack et al., 1970) and in one case from ectopic endometrium (DiGiacomo et al., 1977).

Table 6. Simian Foamy Viruses

Virustype	Host species	References
1 = SAI	M. mulatta, M. fascicularis	Rustigan et al., 1955;
	M. cyclopis	Brown, 1957; Malherbe and Harwin, 1957; Johnston, 1961
	M. nemestrina, C. aethiops	Stiles et al., 1964; Swack et al., 1970; Digiacomo et al., 1977
2	M. cyclopis	Johnston, 1961
	M. mulatta	Hsiung et al., 1969
	P. hamadryas	Agrba et al., 1978
3	C. aethiops	Stiles et al., 1964
	M. mulatta	Stiles, 1968
4	S. sciureus	Johnston, 1971
5	S. sciureus	Johnston, 1971
	G. crassicaudatus	
6	P. troglodytes	Rogers et al., 1976
		Hooks et al., 1972
7	P. troglodytes	Rogers et al., 1967
		Hooks et al., 1972
8	Lagothrix sp.	Hooks et al., 1973
ULAFV	Cacajao rubicundus	Barahona et al., 1976
LK-3	C. aethiops	Neumann-Haefelin et al., 1983
FXV	C. jacchus	Marczynska et al., 1981
FB 10	P. cynocephalus	Rhodes-Feuilette et al., 1979

Foamy viruses are ubiquitous in wild and laboratory primates (Feldman et al., 1975; Kaschula et al., 1978), rhesus monkeys e.g. are heavily infected with type 1 (Stiles et al., 1964; Stiles, 1968). 50 out of 117 newly arrived, wild caught *M. fascicularis* were seropositive (Osterhaus and van Steenis, 1981).

Transmission of the foamy viruses within a primate population probably occurs via the respiratory route, and once the infection has become established in an animal, it appears to persist for a long time, probably for life (Swack and Hsiung, 1975). Additionally Agrba et al. (1978) and Kaschula et al. (1978) have presented evidence of an intrauterine transmission in a baboon.

Human infection may take the same route - another possible way is by contaminated vaccines. The frequency of human infection by simian foamy viruses is seemingly comparatively low. Johnston (1961) found no human infection with type 1 and 3 in Taiwan, Neumann-Haefelin (1983) reported two LK-3 infections in staff members working with infected monkeys. Brown et al. (1978) identified type 6 as the isolate from a human nasopharyngeal cancer.

Pathogenesis

Spumavirinae, like oncovirinae, possess a RNA-dependent DNA-polymerase (reverse transcriptase) which enables the virus to include a step into its life cycle in which the viral RNA is copied into the host cell DNA. This is the reverse of the direction in which cellular genetic information commonly flows and its potential still has to be evaluated. There are associations of foamy viruses with malignancies in both homologous and heterologous hosts, but the role of the virus still has to be clarified.

Clinical Symptoms

Nonhuman primates: So far, no clinical symptoms have been ascribed to simian foamy viruses in their homologous hosts, although they have been isolated from baboons suffering from haemoblastomas (Agrba et al., 1978) and the ectopic uterine tissue of a pigtailed macaque (DiGiacomo et al., 1971).

Man: The isolation of simian foamy virus from a nasopharyngeal cancer (Achong et al., 1971; Brown et al., 1978) or from a case of leukemia (Young et al., 1973) might have been incidental.

Table 7. Oncovirinae of Nonhuman Primates

Type	Host	Tissue	References
B-type			
–	Papio cynocephalus	prostate gland	Kalter et al., 1975
C-type			
SSV-1	Lagothrix lagotricha	fibrosarcoma	Theilen et al., 1971
GALV	Hylobates lar	malignant lymphoma	Gallo et al., 1978
GALV-1 GALV-SEATO GBr-1 GBr-3	Hylobates sp.	lymphosarcoma leukemia	Kawakami et al., 1972, 1976 Hehlmann, 1976
MAC-1	Macaca arctoides	continuous splenic cell line	Todaro et al., 1978a
MMC-1	Macaca mulatta	esophageal carcinoma	Rabin et al., 1979
	Macaca mulatta	placenta	Mayer et al., 1974
M 7	Papio cynocephalus	ovum, placenta, fetus	Kalter et al., 1973, 1975
M 28	Papio sp.		Todaro et al., 1976
BAB-8-K	Papio sp.		Todaro et al., 1976
PP-I-Lu	Papio papio		Todaro et al., 1976
BILN	Papio hamadryas	malignant lymphoma	Voevodin and Lapin, 1976
CPC-1	Colobus polykomos	kidney fibroblasts	Sherwin and Todaro, 1979
–	Pan troglodytes	placenta	Kalter et al., 1975
–	Saguinus fuscicollis	placenta	Kalter et al., 1975
–	Cebus sp.	urinary bladder	Kalter et al., 1974
OMC-1	Aotus trivirgatus	kidney cells	Todaro et al., 1978
TRV-1	Tupaia belangeri	placenta	Flügel et al., 1978
D-type			
Mason Pfizer Virus	Macaca mulatta	mammary carcinoma	Chopra and Mason, 1970
		lactating mammary gland	Ahmed et al., 1973
		placenta	Ahmed et al., 1974
SMRV	Saimiri sciureus	embryonic lung	Heberling et al., 1977; Colcher et al., 1977a
PO-1-La	Presbytis obscurus	lung cell cultures	Benveniste and Todaro, 1977; Todaro et al., 1978b
SAIDS	Macaca mulatta	blood (Simian AIDS)	Marx et al., 1984
SAIDS	Macaca cyclopis	lymphocytes (Simian AIDS)	Daniel et al., 1984

1.9.2 Oncoviruses

The significance of simian oncoviruses (Tab. 7) to man is still largely unknown. Sequences of simian oncovirus genoms have been demonstrated in human specimen and structural, genetic and serologic relationships between certain simian and human oncoviruses are known to exist.

Oncoviruses are capable of encoding their nucleotid sequences into the host cell genetic information (Fig. 6). Normal cells contain DNA sequences very similar to the viral *src*-gene. All retroviruses contain a sequence, called long term repeat (LTR) at the ends, and LTR causes cancer in birds when inserted near the cellular counterpart of the *myc* (Marx, 1984).

1.9.2.1 Baboon Endogenous Viruses (M7)

M7 is one of several C-type viruses related to feline RD114 which occurs in baboon placenta, fetal and prostatic tissues (Kalter et al., 1973, 1975a,c; Benveniste et al., 1974). M7-provirus was demonstrated in leukemic patients (Wong-Staal et al., 1976).

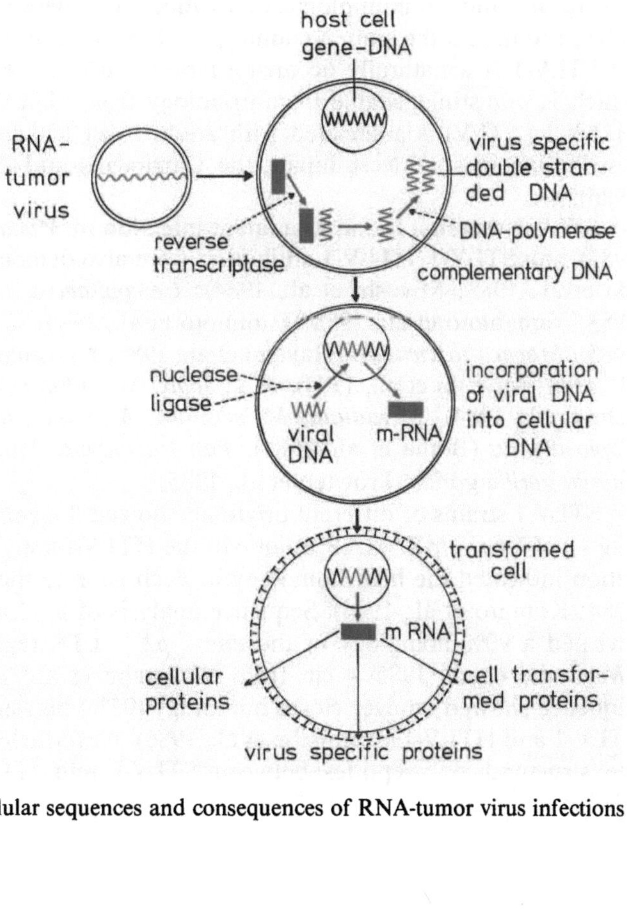

Fig. 6. Cellular sequences and consequences of RNA-tumor virus infections

1.9.2.2 SSV-1

SSV-1 is another C-type virus, which was initially isolated from a fibrosarcoma of a pet *Lagothrix lagotricha* (Theilen et al., 1971; Hunt and Jones, 1973). Its RNA can be hybridized with the RNA of human leukemic cells (Gallo et al., 1973; Sherr and Todaro, 1975), the nucleotide sequences of its transforming region *(v-sis)* show a 91% homology to those of the normal cellular locus *(c-sis)*, which includes one chain of human platelet-derived growth factor (Joseph et al., 1984). Peripheral leukocytes of a patient suffering from myelogenous leukemia produced a C-type virus (HL 23 V), which contained SSV-1 nucleotid sequences (Reitz et al., 1976). More than 50% of normal human sera have been shown to possess antibodies against SSV-1, "Simian Sarcoma Associated Virus" (SSAV) and the closely related "Gibbon Ape Leukemia Virus" (GALV)-group (Aoki et al., 1976; Kurth and Schmitt, 1977; Kurth et al., 1977; Kurth and Mikschy, 1978). A C-type virus derived from the HEL-12 strain of normal human embryonic lung fibroblasts is immunologically related to SSV-1 (and BaEv and RD114) (Prochownik and Kirsten, 1977).

1.9.2.3 STLV-1

The greatest impact on human health problems is possibly resulting from the close structural and immunological relationship between the human HTLV (HIV)-group and the simian counterpart, the STLV-group.

STLV-1 is a naturally occurring type C virus of certain nonhuman primates which is undistinguishable in morphology from HTLV-1 (Miyoshi et al., 1984). HTLV-1 (ATLV) is associated with adult T-cell leukemia of man with endemic distribution in southwest Japan, the Carribeans and Africa (s. Watanabe et al., 1986).

STLV-1 was first found as a latent infection in *Macaca fuscata* (Miyoshi et al., 1982), anti STLV-1/HTLV-1 antibodies were also demonstrated in *M. fuscata* (Ishida et al., 1983; Miyoshi et al., 1984); *Cercopithecus aethiops* (Hunsmann et al., 1983; Yamamoto et al., 1983; Tsujimoto et al., 1985), *C. pygerythrus* (Botha et al., 1985), *Macaca fascicularis* (Hayami et al., 1983; Hunsmann et al., 1983; Homma et al., 1984; Miyoshi et al., 1984), *M. cyclopis, M. mulatta* (Homma et al., 1984; Miyoshi et al., 1984), *M. radiata, M. speciosa, M. nemestrina* (Miyoshi et al., 1984), *Papio ursinus* (Botha et al., 1985), *Pan troglodytes* (Hunsmann et al., 1983) and *Gorilla gorilla gorilla* (Prowten et al., 1985).

STLV-1 strains of different origin all showed the genomic arrangement LTR - *gag* - *pol* - *env* - pX - LTR unique to the HTLV-family and nucleic acid hybridization indicated the high homology of each gene to that of HTLV-1 (Guo et al., 1984; Komuro et al., 1984). Sequence analysis of a *M. nemestrina* STLV-1 isolate revealed a 90% homology of the *env* - *pX* - LTR region with that of HTLV-1 (Watanabe et al., 1985 - cit. from Watanabe et al., 1986), the LTR nucleotid sequence showed an ever closer homology (95%) between the African subtype of STLV-1 and HTLV-1 (Watanabe et al., 1986). Nevertheless there are differences in the structural polypeptides between STLV-I and HTLV-1 (Yamamoto et al., 1984b).

Clinical and Pathological Features

The majority of STLV I-positive monkey sera were derived from apparently healthy donors. In several cases, however, these infections were associated with lymphoproliferative diseases or lymphomas (Homma et al., 1984; Fujimoto et al., 1985; Prowten et al., 1985).

HTLV-1, the human lymphotropic C-type retrovirus has been isolated from adult T-cell leukemia/lymphoma from various parts of the world, particularly in Japan, the Carribean and Africa (s. Hunsmann and Hinuma, 1984). HTLV-1 and its proviral DNA were also isolated from some homosexual AIDS and ELAS-patients (Gallo et al., 1983; Gelmann et al., 1983).

Experimental infections of cynomolgus monkeys with HTLV-1 resulted in the appearance of HTLV-1 specific antigens in peripheral lymphocytes and in the development of specific antibodies, but no animal developed any signs of leukemia (Yamamoto et al., 1984).

1.9.2.4 STLV III

STLV III is an exogenous type C retrovirus of nonhuman primates similar to HTLV III of man in its growth characteristics, T4 tropism, ultrastructural morphology and viral proteins (Kanki et al., 1985b).

STLV III strains were isolated from four macaques (STLV III $_{mac}$), three with immunodeficiency syndromes and one with transmitted lymphoma (Daniel et al., 1985; Kanki et al., 1985c) and from healthy wild caught *C. aethiops* - STLV III $_{Agm}$ (Kanki et al., 1985a). Anti STLV III antibodies were demonstrated in healthy *C. aethiops* (Kanki et al., 1985b). So far, no indications of STLV III infections were found in anthropoid apes, New World monkeys or prosimians (Hayami et al., 1985).

STLV III, like HTLV III, can be isolated preferentially from T4 lymphocytes rather than from T8 lymphocytes (Letvin et al., 1985), STLV III-infected cells revealed virus-specific proteins of 160, 120, 55 and 24 kilodaltons similar in size to the major *gag* and *env* proteins encoded by HTLV III (Kanki et al., 1985c).

Serological evidences and virus isolations suggest natural STLV III-like infections of man in Senegal (Barin et al., 1985; Kanki et al., 1986). The samples reacted strongly with all major viral antigens of STLV III $_{Agm}$, but showed only indistinct reactivity with the major viral antigens of HTLV III in special techniques (Kanki et al., 1986).

HTLV III is the first candidate as causative agent of AIDS in man (Gallo et al., 1983; Marx, 1984; Popovic et al., 1984). Other retroviruses incriminated in the etiology of human AIDS are HTLV-1 (Gallo et al., 1983) and LAV I and II / IDAV (Montagnier - cit. from Walgate, 1986), which are related to, but not identical with HTLV III (Weiss, 1984).

(The etiological agent of simian AIDS = SAIDS, although a retrovirus, is only distantly related to the HTLV-group since it is not a C-type but a D-type virus (Daniel et al., 1984; Marx et al., 1984; Desrosiers et al., 1985; Gardner, 1985)).

Clinical-Pathological Features

Although most STLV III infected animals appeared healthy, some of these infections were associated with definite disease.

One *M. mulatta* died with a lymphoma 26 months after being inoculated with materials from another spontaneous *M. mulatta* lymphoma. Inoculation of other macaques with lymphoma tissues resulted in death 47 to 511 days p.i., all from opportunistic infections such as candidiasis, cytomegalovirosis or cryptosporidiosis.

Two other *M. mulatta* of the same study died from a combination of candidiasis, cryptosporidiosis and a peculiar form of encephalitis characterized by macrophage infiltrates.

The fourth rhesus monkey developed diarrhea, facial rash, generalized lymphadenopathy, and splenomegaly (Daniel et al., 1985).

Experimentally STLV III $_{mac}$ infected rhesus monkeys developed a wasting syndrome, opportunistic infections, encephalitis and a drop in peripheral T 4 lymphocytes (Letvin et al., 1985).

1.9.2.5 Mason Pfizer Monkey Virus (MPMV)

Mason Pfizer monkey virus is the prototype of an exogenous simian tumor D-type oncovirus. MPMV was first isolated from a spontaneous mammary carcinoma of a rhesus monkey (Chopra and Mason, 1970) and subsequently also from normal, lactating mammary tissues or placenta of rhesus monkeys (Ahmed et al., 1973, 1974; Yaniv et al., 1974). Healthy rhesus monkeys from several primate centers harboured anti MPMV-antibodies (Fine et al., 1978).

MPMV-related antigen was found in human breast cancer (Yeh et al., 1975) and in leukemic spleens (Kim, 1977), the RNA's of MPMV and of human breast cancers showed remarkable sequence homologies (Colcher et al., 1974).

References

1. Achong, B.G.; P.W.A. Mansell; M.A. Epstein, and P. Clifford: An unusual virus in cultures from a human nasopharyngeal carcinoma. J. nat. Cancer Inst. *46*, 299-307 (1971)
2. Agrba, V.Z.; L.V. Kokosa; G.N. Cuvirov; D. Bierwolf; R. Widmaier; A. Graffi; B.A. Lapin; I.A. Sangulia, and M. Rudolph: Isolierung des Foamy-Virus Typ II aus den Zellen der blutbildenden Organe von gesunden und an Hämoblastosen erkrankten Mantelpavianen. Arch. Geschwulstforsch. *48*, 97-111 (1978)
3. Ahmed, M.; W. Korol; G. Schildlowsky; J. Vidrine, and S. Mayyasi: Detection of Mason-Pfizer monkey virus in normal monkey mammary tissue and embryonic cultures. Proc. Amer. Ass. Cancer Res. *14*, 34 (1973)
4. Ahmed, M.; G. Schildlovsky; W. Korol; G. Vidrine, and J.L. Cicmanec: Occurrence of Mason-Pfizer monkey virus in healthy rhesus monkeys. Cancer Res. *34*, 3504-3508 (1974)
5. Aoki, T.; M.J. Walling; G.S. Bushar; M. Liu, and K.C. Hsu: Natural antibodies in sera from healthy humans to antigens on surfaces of type C RNA viruses and cells from primates. Proc. natl. Acad. Sci. *73*, 2491-2495 (1976)
6. Barahona, H.; F.G. Garcia; L.V. Melendez; N.W. King; J.K. Ingalls, and M.D. Daniel: Isolation and characterization of a lymphocyte-associated foamy virus from a red uakari monkey. J. med. Primatol. *5*, 253-265 (1976)

7. Barin, F.; S. M'Boup; F. Denis; P. Kanki; J. S. Allan; T. H. Lee, and M. Essex: Serological evidence for virus related to simian T-lymphotropic retrovirus III in residents of West Africa. Lancet II, *8469*, 1387-1389 (1985)
8. Benveniste, R. E.; M. M. Lieber; D. M. Livingston; C. J. Sheer; G. J. Todaro, and S. S. Kalter: Infectious C-type virus isolated from a baboon placenta. Nature *248*, 17-20 (1974)
9. Benveniste, R. E. and G. J. Todaro: Evolution of primate oncornaviruses: an endogenous virus from langurs *(Presbytis sp.)* with related virogene sequences in other Old World monkeys. Proc. natl. Acad. Sci. *74*, 4557-4561 (1977)
10. Botha, M. C.; M. Jones; W. A. de Klerk, and N. Yamamoto: Spread and distribution of human T-cell leukemia virus type 1-reactive antibody among baboons and monkeys in the northern and eastern Transvaal. S. Afr. Med. J. *67*, 665-668 (1985)
11. Brown, L. V.: Pathogenicity for rabbit kidney cell cultures of certain agents derived from "normal" monkey kidney tissue. I. Isolation and propagation. Amer. J. Hyg. *65*, 189-209 (1957)
12. Brown, P.; G. Nemo, and D. C. Gajdusek: Human foamy virus: Further characterization, seroepidemiology, and relationships to chimpanzee foamy viruses. J. Infect. Dis. *137*, 421-427 (1978)
13. Chopra, H. C. and N. M. Mason: A new virus in a spontaneous mammary tumor of a rhesus monkey. Cancer Res. *30*, 2081-2086 (1970)
14. Clarke, J. K. and J. T. Attridge: The morphology of simian foamy agents. J. gen. Virol. *3*, 185-190 (1968)
15. Colcher, D.; R. L. Heberling; S. S. Kalter, and J. Schlom: Squirrel monkey retrovirus: an endogenous virus of a New World primate. J. Virol. *23*, 294-301 (1977)
16. Colcher, D.; S. S. Spiegelman, and J. Schlom: Sequence homology between the RNA of Mason-Pfizer monkey virus and the RNA of human malignant breast tumors. Proc. natl. Acad. Sci. USA *71*, 4975-4979 (1974)
17. Daniel, M. D.; N. W. King; N. L. Letvin; R. D. Hunt; P. K. Sehgal, and R. C. Desrosiers: A new type D retrovirus isolated from macaques with an immunodeficiency syndrome. Science *223*, 602-605 (1984)
18. Daniel, M. D.; N. L. Letvin; N. W. King; M. Kannagi; P. K. Sehgal; R. D. Hunt; P. J. Kanki; M. Essex, and R. C. Desrosiers: Isolation of T-cell tropic HTLV-III-like retrovirus from macaques. Science *228*, 1201-1204 (1985)
19. Desrosiers, R. C.; M. D. Daniel; C. V. Butler; D. K. Schmidt; N. L. Letvin; R. D. Hunt; N. W. King; C. S. Barker, and E. Hunter: Retrovirus D/New England and its relation to Mason-Pfizer monkey virus. J. Virol. *54*, 552-560 (1985)
20. DiGiacomo, R. F.; J. J. Hooks; M. P. Sulima; C. J. Gibbs, jr., and D. C. Gajdusek: Pelvic endometriosis and simian foamy virus infection in a pigtailed macaque. J. Amer. vet. med. Ass. *171*, 589-861 (1977)
21. Feldman, M. D.; N. R. Dunnick; D. W. Barry, and P. D. Parkman: Isolation of foamy virus from rhesus, African green and cynomolgus monkey leukocytes. J. med. Primatol. *4*, 287-295 (1975)
22. Fine, D. L.; S. G. Devare; L. O. Arthur; H. P. Charman, and J. R. Stephenson: Type D retroviruses: Occurrence of natural antibodies to Mason-Pfizer monkey viruses in rhesus monkeys. Virology *86*, 567-571 (1978)
23. Flügel, R. M.; H. Zentgraf; K. Munk, and C. Darai: Activation of an endogenous retrovirus from *Tupaia* (tree shrew). Nature *271*, 543-545 (1978)
24. Gallo, R. C.; N. R. Miller; W. C. Saxinger, and D. Gillespie: Primate RNA tumor virus-like DNA synthesized endogenously by RNA-dependent DNA polymerase in virus-like particles from fresh human acute leukemic blood cells. Proc. natl. Acad. Sci. *70*, 3219-3224 (1973)
25. Gallo, R. C.; P. S. Sarin; E. P. Gelman; M. Robert-Guroff; E. Richardson; V. S. Kalyanaraman; D. Mann; G. D. Sidhu; R. E. Stahl; S. Zolla-Pazner; J. Leibowitch, and M. Popovic: Isolation of human T-cell leukemia virus in acquired immune deficiency syndrome (AIDS). Science *220*, 865-867 (1983)
26. Gallo, R. C.; R. E. Gallagher; F. Wong-Staal; T. Aoki; P. D. Markham; H. Schetters; F. Ruscetti; M. Valerio; M. J. Walling; R. T. O'Keffe; W. J. Saxinger; R. G. Smith; D. H. Gillespie, and M. S. Reitz, jr.: Isolation and tissue distribution of type-C virus and viral components from a gibbon ape *(Hylobates lar)* with lymphocytic leukemia. Virology *84*, 359-373 (1978)

27. Gardner, M.; P. Marx; D. Maul; K. Osborn; C. Miller; N. Lerche, and R. Henrickson: Simian AIDS: Evidence for a retrovirus etiology. Front. Rad. Ther. Onc. *19*, 26-32 (1985)
28. Gelman, E. P.; M. Popovic; D. Blayney; H. Masur; G. Sidhu; R. E. Stahl, and R. C. Gallo: Proviral DNA of a retrovirus, human T-cell leukemia virus, in two patients with AIDS. Science *220*, 862-865 (1983)
29. Guo, H.-G.; F. Wong-Staal, and R. Gallo: Novel viral sequences related to human T-cell leukemia virus in T cells of a seropositive baboon. Science *223*, 1195-1197 (1984)
30. Hayami, M.; K. I. Ishikawa; A. Komuro; Y. Kawamoto; K. Nozawa; K. Yamamoto; T. Ishida; Y. Hinuma: ATLV antibody in cynomolgus monkeys in the wild. Lancet *8350* (II), 620 (1983)
31. Hayami, M.; Y. Ohta; T. Hattori; H. Nakamura; K. Takatsuki; A. Kashiwa; K. Nozawa; I. Miyoshi; T. Ishida; Y. Tanioka; T. Fujiwara, and S. Honjo: Detection of antibodies to human T-lymphotropic virus type III in various non-human primates. Japan J. Exp. Med. *55*, 251-255 (1985)
32. Heberling, R. L.; S. T. Barker; S. S. Kalter; G. C. Smith, and R. J. Helmke: Oncornavirus: Isolation from a squirrel monkey *(Saimiri sciureus)* lung culture. Science *195*, 289-292 (1977)
33. Hehlmann, R.: RNA tumor viruses and human cancer. Current topics in microbiol. Immunology *73*, 141-215 (1976)
34. Homma, T.; P.J. Kanki; N.W. King jr.; R.D. Hunt; M.J. O'Connell; N.L. Letvin; M.D. Daniel; R.C. Desrosiers; C.S. Yang, and M. Essex: Lymphoma in macaques: Association with virus of human T lymphotrophic family. Science *22*, 716-718 (1984)
35. Hooks, J. J. and C. J. Gibbs, jr.: The foamy viruses. Bact. Rev. *39*, 169-185 (1975)
36. Hooks, J. J.; C. J. Gibbs, jr.; S. Chou; R. Howk; M. Lewis, and D. C. Gajdusek: Isolation of a new simian foamy virus from a spider monkey brain culture. Infect. and Immun. *8*, 804-813 (1973)
37. Hooks, J. J.; C. J. Gibbs, jr.; E. C. Cutchins; N. G. Rogers; P. Lampert, and D. V. Gajdusek: Characterization and distribution of two new foamy viruses isolated from chimpanzees. Arch. ges. Virusforsch. *38*, 38-55 (1972)
38. Hsiung, G. D.; T. Atoynatan, and C. W. Lee: Epidemiologic studies of latent virus infection in captive monkeys and baboons. I. Overall plans and virus isolations with special reference to SV 40 and foamy viruses isolated from chimpanzees. Amer. J. Epidem. *89*, 464-471 (1969)
39. Hsiung, G. D. and N. S. Swack: Myxovirus and pseudomyxovirus groups. Pp. 537-571 in "Pathology of Simian Primates" Part II (Fiennes, R. N. T-W., ed.), Karger, Basel (1972)
40. Hunsmann, G. and Y. Hinuma: Human adult T-cell leukemia virus and its association with disease. Advanc. Viral Oncol. *5*, 147-172 (1984)
41. Hunsmann, G.; J. Schneider; J. Schmitt, and N. Yamamoto: Detection of serum antibodies to adult T-cell leukemia virus in non-human primates and in people from Africa. Int. J. Cancer *32*, 329-332 (1983)
42. Hunt, R. D. and I. C. Jones: Nonhuman primates in viral oncology. Advanc. vet. Sci. comp. Med. *17*, 361-394 (1973)
43. Ishida, T.; K. Yamamoto; R. Kaneko; E. Tokita, and Y. Hinuma: Seroepidemiological study of antibodies to adult T-cell leukemia virus-associated antigen (ATLA) in free-ranging Japanese monkeys *(Macaca fuscata)*. Microbiol. Immunol. *27*, 297-301 (1983)
44. Johnston, P. B.: A second immunologic type of simian foamy virus. Monkey throat infections and unmasking by both types. J. Infect. Dis. *109*, 1-9 (1961)
45. Johnston, P. B.: Taxonomic features of seven serotypes of simian and ape foamy viruses. Infect. Immun. *3*, 793-799 (1971)
46. Josephs, S. F.; C. Guo; L. Ratner, and F. Wong-Staal: Human protooncogene nucleotide sequences corresponding to the transforming region of simian sarcoma virus. Science *223*, 487-491 (1984)
47. Kalter, S. S.; D. Ablashi; C. Espana; R. L. Heberling; R. N. Hull; E. H. Lennette; H. H. Malherbe; S. McConnell, and D. S. Yohn: Simian virus nomenclature, 1980. Intervirology *13*, 317-330 (1980)
48. Kalter, S. S.; R. L. Heberling; A. Hellman; G. J. Todaro, and M. Panigel: C-type particles in baboon placenta. Proc. roy. Soc. Med. *68*, 135-140 (1975 a)
49. Kalter, S. S.; R. L. Heberling; R. J. Helmke; M. Panigel; G. C. Smith; D. C. Kraemer; A. Hellman; A. K. Fowler, and J. E. Strickland: A comparative study on the presence of C-type viral

particles in placentas from primates and other animals; pp. 391 - 401 in "Comparative Leukemia Research 1973, Leukemogenesis" (Ito, Y. and R.M. Dutcher, eds.), Karger, Basel (1975)
50. Kalter, S.S.; R.L. Heberling; A. Hellman; G.J. Todaro, and M. Panigel: C-type particles in baboon placenta. Proc. roy. Soc. Med. *68*, 135-140 (1975a)
51. Kalter, S.S.; R.L. Heberling; G.C. Smith, and R.J. Helmke: Brief communication: C-type viruses in chimpanzees *(Pan sp.)* placentas. J. natl. Cancer Inst. *55*, 735 (1975b)
52. Kalter, S.S.; R.J. Helmke; M. Panigel; R.L. Heberling; P.J. Felsburg, and L.R. Axelrod: Observations of apparent C-type particles in baboon *(Papio cynocephalus)* placentas. Science *179*, 1332-1333 (1973)
53. Kalter, S.S.; R.E. Kuntz; R.L. Heberling; R.J. Helmke, and G.C. Smith: C-type viral particles in a urinary bladder neoplasm induced by Schistosoma haematobium. Nature *251*, 440 (1974)
54. Kalter, S.S.; S.A. Shain; G.C. Smith; B. McCullough; R.L. Heberling, and A.J. Dalton: Brief communication: Oncorna-like viruses in baboon prostate tissue. J. nat. Cancer Inst. *55*, 1237-1241 (1975c)
55. Kanki, P.J.; J. Alroy, and M. Essex: Isolation of T-lymphotropic retrovirus related to HTLV-III/LAV from wild-caught African green monkeys. Science *230*, 951-954 (1985a)
56. Kanki, P.J.; F. Barin; S. M'Boup; J.S. Allan; J.L. Romet-Lemonne; R. Marlink; M.F. McLane; T.-H. Lee, B. Arbeille; F. Denis, and M. Essex: New human T-lymphotropic retrovirus related to simian T-lymphotropic virus type III (STLV-III_{AGM}). Science *232*, 238-243 (1986)
57. Kanki, P.J.; R. Kurth; W. Becker; G. Dreesman; M.F. McLane, and M. Essex: Antibodies to simian T-lymphotropic retrovirus type III in African green monkeys and recognition of STLV-III viral proteins by AIDS and related sera. Lancet I, 1330-1332 (1985b)
58. Kanki, P.J.; M.F. McLane; N.W. King, jr.; N.L. Letvin; R.D. Hunt; P. Sehgal; M.D. Daniel; R.C. Desrosiers, and M. Essex: Serologic identification and characterization of a macaque T-lymphotropic retrovirus closely related to HTLV-III. Science *228*, 1199-1201 (1985c)
59. Kaschula, V.R.; A.F. van Dellen, and V. de Vos: Some infectious diseases of wild vervet monkeys *(Cercopithecus aethiops pygerythrus)* in South Africa. J.S. Afr. vet. Ass. *49*, 223-227 (1978)
60. Kawakami, T.G.; S.D. Huff; P.M. Buckley; D.L. Dungsworth; S.P. Snyder, and R.V. Gilden: C-type virus associated with gibbon lymphosarcoma. Nature new Biol. *235*, 170-171 (1972)
61. Kawakami, T.G.; T.S. McDowell; D.L. Johnson; A.L. Breznock, and J.B. Harrold: Oncornavirus-host interaction in gibbons. Bibl. haemat. *43*, 184-186 (1976)
62. Kim, B.S.: Cross-reaction between antigens of human myelogenous leukemia and Mason-Pfizer monkey virus. Europ. J. Cancer *13*, 932-941 (1977)
63. Komuro, A.; T. Watanabe; I. Miyoshi; M. Hayami; H. Tsujimoto; M. Seikl, and M. Yoshida: Detection and characterization of simian retrovirus homologous to human T-cell leukemia virus type I. Virology *138*, 373-378 (1984)
64. Kurth, R. and U. Mikschy: Human antibodies reactive with purified envelope antigens of primate type-C tumor viruses. Proc. nat. Acad. Sci. *75*, 5692-5696 (1978)
65. Kurth, R. and C. Schmitt: Human humeral antibodies specific for primate C-type viral antigens. Med. Microbiol. Immunol. *164*, 167-177 (1977)
66. Kurth, R.; N.M. Teich; R. Weiss, and R.T.D. Oliver: Natural human antibodies reactive with primate type C viral antigens. Proc. nat. Acad. Sci. *74*, 1237-1241 (1977)
67. Letvin, N.L.; M.D. Daniel; P.K. Sehgal; R.C. Desrosiers; R.D. Hunt; L.M. Waldron; J.J. MacKey; D.K. Schmidt; L.V. Chalifoux, and N.W. King: Induction of AIDS-like disease in macaque monkeys with T-cell tropic retrovirus STLV-III. Science *230*, 71-73 (1985)
68. Malherbe, H. and R. Harwin: Seven viruses isolated from the vervet monkey. Brit. J. Exp. Pathol. *38*, 539-541 (1957)
69. Marczynska, B.; C.J. Jones, and L.G. Wolfe: Syncytium-forming virus of common marmosets *(Callithrix jacchus jacchus)*. Infect. Immun. *31*, 1261-1269 (1981)
70. Marx, J.L.: Strong new candidate for AIDS agent. A newly discovered member of the human T-cell leukemia virus family is very closely linked to the immunodeficiency disease. Science *224*, 475-477 (1984a)

71. Marx, J.L.: What do oncogenes do? Science *223*, 673-676 (1984)
72. Marx, P.A.; D.H. Maul; K.G. Osborn; N.W. Lerche; P.Moody; J.L. Lowenstine; R.V. Henrickson; L.O. Arthur; R.V. Gilden; M.Gravell; W.T. London; J.L. Sever; J.A. Levy; R.J. Munn, and M.B. Gardner: Simian AIDS: Isolation of a type D retrovirus and transmission of the disease. Science *223*, 1083-1086 (1984c)
73. Mayer, R.J.; R.G. Smith, and R.C. Gallo: Reverse transcriptase in normal rhesus monkey placenta. Science *185*, 864-867 (1974)
74. Miyoshi, I.; Y.Ohtsuki; M.Fujishita; S.Yoshimoto; I.Kubonishi, and M.Minezawa: Detection of type C virus particles in Japanese monkeys seropositive to adult T-cell leukemia-associated antigens. Gann *73*, 848-849 (1982)
75. Miyoshi, I.; H.Taguchi; S.Yoshimoto; M.Fujishita, and Y.Ohtsuki: Natural infection in monkeys with adult T-cell leukemia virus or a closely related agent; pp.313-321 in "Human T-Cell Leukemia/Lymphoma Virus: The Family of Human T-Lymphotropic Retroviruses: Their Role in Malignancies and Association With AIDS". (Gallo, R.C.; M.E. Essex, and L.Gross, eds.) Cold Spring Harbor Lab., Cold Spring Harbor (1984)
76. Neumann-Haefelin, D.; A.Rethwilm; G.Bauer; F.Gudat, and H.zur Hausen: Characterization of a foamy virus isolated from Cercopithecus aethiops lymphoblastoid cells. Med. Microbiol. Immunol. *172*, 75-86 (1983)
77. Parks, W.P.; G.J. Todaro; E.M. Scolnick, and S.A. Aaronson: RNA dependent DNA polymerase in primate syncytium-forming (foamy) viruses. Nature *229*, 258-260 (1971)
78. Peries, J.and G.J. Todaro: A haemagglutination assay for the primate syncytium-forming (foamy) viruses. J.gen. Virol. *34*, 195-199 (1977)
79. Popovic, M.; M.G. Sarngadharan; E.Read, and R.C. Gallo: Detection, isolation, and continous production of cytopathic retrovirus (HTLV-III) from patients with AIDS and pre-AIDS. Science *224*, 497-500 (1984)
80. Prochownik, E.V. and W.H. Kirsten: Nucleic acid sequences of primate type C viruses in normal and neoplastic human tissues. Nature *267*, 175-177 (1977)
81. Prowten, A.W.; R.V. Lee; R.M. Krishnamsetty; S.K. Satchidanand, and B.I.S. Srivastava: T-cell lymphoma associated with immunologic evidence of retrovirus infection in a lowland gorilla. J.Amer. vet. med. Ass. *187*, 1280-1282 (1985)
82. Rabin, H.; C.V. Benton; M.A. Tainsky; N.Rice, and R.V. Gilden: Isolation and characterization of an endogenous type C virus of rhesus monkeys. Science *204*, 841-842 (1979)
83. Reitz, M.S.; N.R. Miller; F.Wong-Staal; R.E. Gallagher; R.V. Gallo, and D.H. Gillespie: Primate type-C virus nucleic acid sequences (woolly monkey and baboon type) in tissues from a patient with acute myelogenous leukemia and in viruses isolated from cultured cells of the same patient. Proc. natl. Acad. Sci. *73*, 2113-2117 (1976)
84. Rhodes-Feuillette, A.; F.Saal; J.Lasneret; P.Dubouch, and J. Peries: Isolation and characterization of a new simian foamy virus serotype from lymphocytes of *Papio cynocephalus* baboon. J.med. Primatol. *8*, 308-320 (1979)
85. Rogers, N.G.; M.Basnight; C.J. Gibbs, jr., and D.C. Gajdusek: Latent viruses in chimpanzees with experimental Kuru. Nature *216*, 446-449 (1967)
86. Rustigian, R.; P.Johnston, and H.Reihart: Infection of monkey kidney tissue cultures with virus-like agents. Proc. Soc. exp. Biol. Med. *88*, 8-16 (1955)
87. Sherr, C.and G.Todaro: Primate type C virus p30 antigen in cells from humans with acute leukemia. Science *187*, 844-848 (1975)
88. Sherwin, S.A. and G.J. Todaro: A new endogenous primate type C virus isolated from the Old World monkey *Colobus polykomos*. Proc. natl. Acad. Sci. *76*, 5041-5045 (1979)
89. Spiegelman, S.; A.Burny; M.R. Das; J.Keydar; J.Schlom; M.Travnicek, and K.Watson: Characterization of the products of RNA-directed DNA polymerases in oncogenic RNA viruses. Nature *227*, 563-567 (1970)
90. Stiles, G.E.: Serologic screening of rhesus and grivet monkeys for SV 40 and the foamy viruses. Proc. Soc. exp. Biol. *127*, 225-230 (1968)
91. Stiles, G.E.; J.L. Bittle, and V.J. Cabasso: Comparison of simian foamy virus strains including a new serological type. Nature *201*, 1350-1351 (1964)
92. Swack, N.S. and G.D. Hsiung: Pathogenesis of simian foamy virus infection in natural and experimental hosts. Infect. Immun. *12*, 470-474 (1975)

93. Swack, N.S.; R.A. Schoentag, and G.D. Hsiung: Foamy virus infection of rhesus and green monkeys in captivity. Amer. J. Epidem. *92*, 79-83 (1970)
94. Theilen, G.H.; D.Gould; M.Fowler, and D.L. Dungworth: C-type virus in tumor tissue of a woolly monkey *(Lagothrix spp.)* with fibrosarcoma. J.nat. Cancer Inst. *47*, 881-889 (1971)
95. Todaro, G.J.; R.E. Benveniste; C.J. Sherr; J.Schlom; G.Schidlovsky, and J.R. Stephenson: Isolation and characterization of a new type D retrovirus from the Asian primate, *Presbytis obscurus* (Spectacled langur). Virology *84*, 189-194 (1978b)
96. Todaro, G.J.; C.J. Sherr, and R.E. Benveniste: Baboons and their close relatives are unusual among primates in their ability to release nondefective endogenous type C viruses. Virology *72*, 278-282 (1976)
97. Todaro, G.J.; C.J. Sherr; A.Sen; N.King; M.D. Daniel, and B. Fleckenstein: Endogenous New World primate type-C viruses isolated from owl monkey (*Aotus trivirgatus*) kidney cell line. Proc. natl. Acd. Sci. *75*, 1004-1008 (1978c)
98. Tsujimoto, H.; M.Seiki; H.Nakamura; T.Watanabe; I.Sakakibara; A.Sasagawa; S.Honjo; M.Hayami, and M.Yoshida: Adult T-cell leukemia-like disease in monkey naturally infected with simian retrovirus related to human T-cell leukemia virus type 1. Gann *76*, 911-914 (1985)
99. Voevodin, A.F. and B.A. Lapin: Antigenic characterization of C-type oncornavirus isolated from lymphatous baboon *(Papio hamadryas)*. Arch. Geschwulstforsch. *46*, 105-109 (1976)
100. Walgate, R.: New human retrovirus: One causes AIDS. Nature *320*, 385 (1986)
101. Watanabe, T.; M.Seiki; Y.Hirayama, and M.Yoshida: Human T-cell leukemia virus type 1 is a member of the African subtype of simian viruses (STLV). Virology *148*, 385-388 (1986)
102. Weiss, R.: Retroviruses linked with AIDS. Nature *309*, 12-13 (1984)
103. Wong-Staal, F.; D.Gillespie, and R.C. Gallo: Proviral sequences of baboon endogenous type-C RNA virus in DNA of human leukemic tissues. Nature *262*, 190-195 (1976)
104. Yamamoto, N.; M.Hayami; A.Komuro; J.Schneider; G.Hunsmann; M.Okada, and Y.Hinuma: Experimental infection of cynomolgus monkeys with a human retrovirus, adult T-cell leukemia virus. Med. Microbiol. Immunol. *173*, 57-64 (1984a)
105. Yamamoto, N.; Y.Hinuma; H.zur Hausen; J.Schneider, and G.Hunsmann: African green monkeys are infected with adult T-cell leukemia virus or a closely related agent. Lancet I, *8318*, 240-241 (1983)
106. Yamamoto, N.; M.Okada; Y.Hinuma; F.W. Hirsch; T.Chosa; J.Schneider, and G.Hunsmann: Human adult T-cell leukemia virus is distinct from a similar isolate of Japanese monkeys. J.gen. Virol. *65*, 2259-2264 (1984b)
107. Yaniv, A.; T.Ohno; D.Kacian; D.Colcher; S.Witkin; J.Schlom, and S.Spiegelman: Serological analysis of reverse transcriptase of the Mason-Pfizer monkey virus. Virology *59*, 335-338 (1974)
108. Yeh, Y.; M.Ahmed; S.Mayyasi, and A.A. Alessi: Detection of an antigen related to Mason-Pfizer virus in malignant human breast tumours. Amer. Ass. Advanc. Sci. *190*, 583-584 (1975)
109. Young, D.; J.Samuels, and J.K. Clarke: A foamy virus of possible human origin isolated in BHK-21 cells. Arch. ges. Virusforsch. *42*, 228-234 (1973)

1.10 Hepatitis Viruses

Since the first report by Hillis (1961) of viral hepatitis in veterinarians exposed to newly imported chimpanzees, more than 200 cases of nonhuman primate derived human cases of viral hepatitis have been reported (Riopelle, 1963; Anon, 1964; 1974; Smetana, 1965; 1972; Mosley et al., 1967; Altman and Marzinsky, 1968; McInerney et al., 1968; Friedman et al., 1971a, b; Hinthorn and Price, 1971; McCollum et al., 1971; Dickerson et al., 1973; Pattison et al., 1975; Rill et al., 1975). Of 173 human cases 151 were associated with chimpanzees, 9 with woolly monkeys, 5 with siamangs, 4 with gorillas, and 4 with celebes crested apes (Anon, 1973). As far as determined all simian primate derived human infections were caused by type A-hepatitis virus (Dickerson et al., 1972; Kovacs et al., 1974) evidence of chimpanzee-associated type B hepatitis in human beings is lacking (Pattison et al., 1975), although chimpanzees and cynomolgus monkeys are susceptible to type B-hepatitis virus under natural conditions as well (Maynard et al., 1971; Kornegay et al., 1985).

Viruses

At least 3 forms of viral hepatitis exist in man: A-, B-, and non A-, non B- (C) hepatitis (Krugman, 1978; Zuckermann and Howard, 1979).

A-hepatitis virus particles are roughly spherical, and nonenveloped. They have a diameter of 24 to 29 nm (Zuckermann and Howard, 1979). As they contain a single stranded RNA, they are classified within the family Picornaviridae, genus enteroviruses. The thermal stability of hepatitis A-virus (HAV) is greater than that of picornaviruses, *in vitro* replication is not blocked by guanidine, nor by other known inhibitors of poliovirus replication. HAV is antigenically distinct from other picornaviruses. The *in vitro* replication of HAV, although it is more difficult than that of other picornaviruses, is possible in marmoset liver explants, normal foetal rhesus monkey kidney cell lines or other primate cell cultures (Provost and Hillemann, 1975; Daemer et al., 1981; Binn et al., 1984). Unlike the typical picornaviruses HAV induces neither shutdown of host cell macromolecular synthesis nor a CPE but tends to induce a persistent infection (Lemon, 1985). The HAV antigen can be visualized in such cells by immunofluorescence which reveals a characteristic granular accumulation of antigen.

The hepatitis B-virus possesses a lipid envelope and the particle-size ranges around 42 nm. It contains DNA within its core similar to the Parvoviridae (Melnick, 1981) and an associated DNA polymerase. Antigenically hepatitis B core antigen (HbcAG) can be distinguished from hepatitis B-surface antigen (HBsAG). Another mark of hepatitis B virus infection is the e-antigen (HBeAG), indicating ongoing hepatic disease and a poor prognosis in man (Mayumi, 1978). As with hepatitis A-virus most attempts to cultivate hepatitis B-virus have been fruitless, only Ishida et al. (1978) achieved growth of hepatitis virus in human embryonic liver cells.

The term non-A, non-B hepatitis has been used for the classification of hepatitis of man that is not caused by the commonly recognized agents, namely hepatitis A or B-virus (Bradley, 1985). Two distinct agents are responsible for posttransfusion non-A, non-B hepatitis, both were also recovered from experimentally infected chimpanzees (Zuckermann and Howard, 1979; Bradley, 1985) with the $CHCl_3$ resistant non A:non B hepatitis virus probably visualized in liver biopsies of an experimentally infected chimpanzee as particles measuring 25 to 30 nm in diameter. The antigen was identified by double immunodiffusion assays (Zuckermann and Howard, 1979).

Epidemiology

Hepatitis viruses are infectious only to man and nonhuman primates, although they can be transmitted by other, especially water living animals.

Hepatitis A-virus is spread by the intestinal-oral route, including contaminated water (Grabow, 1976), shellfish or nonhuman primates. It has also been transmitted spontaneously between nonhuman primates such as chimpanzees (Smetana et al., 1970) or owl monkeys *(Aotus trivirgatus)* (Deluc et al., 1981). 33% of wild living cynomolgus monkeys (Burke and Heisey, 1984) and 25% of the chimpanzee population of the London Zoo and particularly all captive born chimpanzees of that zoo (and a hand reared orang-utan) had anti-HAV antibodies (Kessler et al., 1982). Similarly 36,6% of the chimpanzees, 38,2% of the baboons, 57,9% of the vervet monkeys, 40% of *Cebus albifrons,* and 50% of the common marmosets held at Southwest Foundation for Research and Education, San Antonio harboured anti-HAV antibodies (Eichberg and Kalter, 1980). Spontaneously acquired anti-HAV antibodies have also been demonstrated in *Cynopithecus niger, Ateles geoffroyi, Hylobates lar,* and *Mandrillus sphinx* (Smith et al., 1980).

Hepatitis A virus can experimentally be transmitted to chimpanzees, *S. mystax S. fuscicollis, S. oedipus,* and to lesser bushbabies *(Galago senegalensis moholi)* (Grabow, 1976), but does not cause overt clinical signs in tamarins (Deinhardt et al., 1967; Appleton, 1975a, b; Lorenz et al., 1970; Hilleman et al., 1977; Karayiannis et al., 1986).

In hepatitis B-virus infections the basic reservoir is the chronic, asymptomatic carrier of HBsAG. The principal infectious body fluid is the blood plasma (Maynard, 1978). Accordingly hepatitis B was initially classified as transfusion hepatitis in the past, as it is transmitted by contaminated blood conserves, syringes, needles etc. However, evidence suggests the occurrence of other possible routes of transmission as well, since HBsAG has been found in a variety of body fluids - amniotic fluid, saliva, menstrual and vaginal discharges, seminal fluid, colostrum and breast milk, urine, bile, faeces, sweat etc. HBsAG has also been detected in captive gorillas (Linnemann et al., 1984), and approximately 50% of the chimpanzees when tested using passive haemagglutination and radio-immuno assays, and recently an outbreak of hepatitis B has been reported at the London Zoo (Zuckermann et al., 1978). Also up to 25% of wild caught chimpanzees had antibodies to hepatitis B-virus when tested several weeks or months after capture; also 10 of 26 chimpanzees in a holding compound in Africa possessed HBsAG (cit. from Kessler et al., 1982). Finally spontaneous hepatitis B was diagnosed in two *M. fascicularis* imported from Indonesia. The animals harboured anti HB_sAg-antibodies, respectively HB_sAg (Kornegay et al., 1985).

Hepatitis B-virus is experimentally transmissible to chimpanzees (Barker et al., 1973; 1975; Drucker et al., 1979; Sly et al., 1979). It is occasionally followed by chronic carrier stages also in the apes (Shoval et al., 1980; Thung et al., 1981).

Non A: Non B hepatitis so far has been diagnosed as a post transfusion hepatitis, the agent is also transmissable to chimpanzees (for references see Zuckermann and Howard, 1979).

Both types A and B hepatitis viruses are usually transmitted by the fecal-oral route or by infected environments from man to apes or monkeys and reverse, although exceptions have occurred in the past. 5 chimpanzee-man transmissions were related to bite or scratch lesions inflicted by the animals (Altman and Marzinsky, 1968). The initial infection of chimpanzees has been blamed upon their inoculation with pooled human blood in their country of origin (Hillis, 1961).

Transmission of hepatitis viruses between nonhuman primates is certainly helped by coprophagia. Infection of human beings from nonhuman primates is favoured by poor hygienic standards such as not wearing protective clothing, kissing the animals, negligence of personal hygiene etc. (Krushak, 1970; Friedmann et al., 1971b).

Clinical Symptoms

Nonhuman primates: Most spontaneous and experimental hepatitis virus infections of nonhuman primates are asymptomatic or nonspecific (Kessler et al., 1982). Clinical signs, if at all present, occur spontaneously particularly in newly arrived animals and often represent transportation stress rather than hepatitis virus infection. The symptoms range from mild respiratory infection to nonspecific gastrointestinal illness (McCollum et al., 1971; Dickerson et al., 1973; Friedmann et al., 1971a). Sometimes anorexia and persistant diarrhea have been reported (Dickerson et al., 1973; Rill et al., 1975). The most significant signs of virus hepatitis in chimpanzees were elevations of the serum liver enzymes and of bilirubin. SGOT values rose from normally 0 to 15 IU to 85 IU, bilirubin from 0.1 to 0.5 mg% to 2.0 mg% (Rill et al., 1975).

At least 2 fulminant spontaneous type A infections have been reported in chimpanzees (Smetana, 1965; Abe and Shikata, 1982). The animals came down with clinical disease a few weeks after arrival, and after 4, resp. 6 days of an illness characterized by fever, anorexia, vomiting, diarrhea and jaundice, they died.

In experimental HAV infections of chimpanzees both SGPT- and SGOT-levels became elevated 3 to 4 weeks after inoculation of the infective material. Whereas the changes of the SGPT-levels were of great significance, those of the SGOT-levels were not very reliable, since minor elevations occurred repeatedly without association to other disease parameters (Dienstag et al., 1975).

In the hepatitis A-virus infections of tamarins and galagos serum alanine-aminotransferase and isocitric dehydrogenase levels became elevated (Grabow, 1976; Purcell and Dienstag, 1978; Karayiannis et al., 1986).

Hepatitis B-virus infections of chimpanzees were evidenced by the demonstration of antigens/antibodies only. Experimentally type B-infected chimpanzees sometimes developed anorexia, lethargy and jaundice 27 weeks after inoculation of HBsAG positive plasma. The jaundice lasted for 10 days (Sly et al., 1979). During the illness the serum alanine aminotransferase-, aspartate aminotransferase-, alkaline phosphatase and bilirubin levels increased. Spontaneous hepatitis B infections of two *M. fascicularis* were associated with anorexia, lethargy and hepatomegaly (Kornegay et al., 1985).

Spontaneous non A: non B hepatitis virus infections so far have not been observed in nonhuman primates. Experimental transmission although inducing typical histological lesions, was not always followed by any clinical or biochemical abnormalities (see Zuckermann and Howard, 1979). Valenza and Muchmore (1985) recognized significant elevations of gamma glutamyl transferase in addition to aspartate aminotransferase and alanine transferase increase in experimentally infected chimpanzees.

Man: The clinical picture of viral hepatitis in man varies from inapparent infection to mild gastro-intestinal symptoms, to acute anicteric hepatitis, to acute icteric illness and finally chronic liver disease. Extrahepatic lesions in hepatitis B include cases of arthritis, polyarteritis nodosa, glomerulonephritis, polymyalgia rheumatica or infantile papular acrodermatitis.

The incubation period, although occasionally overlapping, is 28–30 days in hepatitis A, and 40–180 days in hepatitis B (Zuckermann and Howard, 1979).

Pathology

Nonhuman primates: In the experimentally infected tamarins acute hepatitis with hepatocytic swelling and mononuclear infiltration developed near peak alanine aminotransferase levels (Karayiannis et al., 1986). The first signs of experimental type A viral hepatitis in chimpanzees occurred 3 weeks p.i. as focal accumulation of lymphocytes and macrophages at the periphery of the liver lobules with replacement of lost hepatocytes. The portal tracts were densely infiltrated by mononuclear cells, predominantly lymphocytes. Subsequently the necroses as well as the hepatocytic alterations developed further, occasionally progressing to the formation of acidophilic bodies. 6 weeks p.i. the portal inflammation was maximal, in places extending into the parenchyma, with an associated erosion of the limiting plate. Areas of focal necrosis occasionally connected inflamed portal tracts. The central lobular zone was always spared, septal or intralobular fibrosis was never observed.

6–8 weeks p.i. the lesions normally receded (Dienstag et al., 1975; 1976). In the chimpanzees which died of spontaneous hepatitis A, the macroscopial appearance of the liver was that of acute yellow atrophy. Histologically, massive hepatocyte necrosis with diffuse haemorrhages and mononuclear infiltration (Abe and Shikata, 1982) or "glandular transformation of liver cell cords" associated with varying inflammatory infiltrates (Smetana, 1965) prevailed. In hepatitis B inoculated chimpanzees, the morphologic lesions were far more variable. 12 weeks after exposure, histiocytes and lymphocytes accumulated focally throughout all parts of the lobule, especially in the centrolobular areas. Simultaneously, small areas of necrosis appeared and the sinusoidal lining cells were distinctively activated. One week later, focal necrosis and sinusoidal cell activation were more pronounced. Hepatocellular changes included sparse acidophilic bodies. The inflammatory infiltrates increased towards the climax between 15 and 22 weeks p.i. (Dienstag et al., 1976). Barker et al. (1973) described striking focal necrosis of single hepatocytes with hydropic changes in biopsies taken 25 weeks p.i.

Similar results were obtained in experimental hepatitis A infections of owl monkeys (Miller Keenan et al., 1984).

In liver biopsies taken from experimentally non A: Non B-infected chimpanzees the following lesions were described after an incubation period of 10 weeks: collections of mainly mononuclear inflammatory cells occurred in the portal tracts. Inflammatory infiltrates and focal intralobular necrosis was present. Kupffer cells were very prominent (Zuckermann and Howard, 1979).

The ultrastructural changes in experimentally non-A, non-B hepatitis infected chimpanzees were marked by peculiar convoluted membranes derived from the hepatocytic smooth endoplasmic reticulum (Bradley, 1985).

Man: The great variety of symptoms makes it difficult to describe the development of morphological lesions in the liver, but two features are constant in acute viral hepatitis of both type A and B: Necrosis of hepatocytes and histiocytic periportal inflammation. The hepatocytes are swollen and show ballooning or "ground glass" appearance of their cytoplasm. Shrunken cells give rise to acidophilic bodies. The necrosis frequently tends to be zonal, being particularly extensively in the centrolobular areas (Zuckermann and Howard, 1979).

References

1. Abe, K. and T. Shikita: Fulminant type A viral hepatitis in a chimpanzee. Acta path. jap. *32*, 143-148 (1982)
2. Altman, R. and P. Marzinsky: Nonhuman primate-associated hepatitis. New Jersey. Lab. Primate Newsl. *7* (4), 24 (1968)
3. Anon: Chimpanzee-associated hepatitis - 1963. Lab. Primate Newsl. *3* (3), 5-6 (1964)
4. Anon: Hepatitis-A in humans associated with nonhuman primates. Ohio. CDC Morb. Mortal Weekly Rep. *22* (49), 407-408 (1973)
5. Anon: Chimpanzee-associated hepatitis. Lab. Primate Newsl. *10* (3), 21-22 (1974)
6. Appleton, H.: Hepatitis A in marmosets. Developm. biol. Standard *30*, 405-407 (1975 a)
7. Appleton, H.: Hepatitis transmission in marmosets. J. infect. Dis. *132*, 500-505 (1975 b)
8. Barker, L. F.; F. V. Chisari; P. P. McGrath; D. W. Dalgard; R. L. Kirschstein; J. D. Almeida; T. S. Eddington; D. G. Sharp, and M. R. Peterson: Transmission of type B viral hepatitis to chimpanzees. J. infect. Dis. *127*, 648-662 (1973)
9. Barker, L. F.; J. E. Maynard; R. H. Purcell; J. H. Hoofnagle; K. R. Berquist; W. T. London; R. J. Gerety, and D. W. Krushak: Hepatitis B virus infection in chimpanzees. Titration of subtypes. J. infect. Dis. *132*, 451-458 (1975)
10. Binn, L. N.; S. M. Lemon; R. H. Marchwicki; R. R. Redfield; N. L. Gates, and W. H. Bancroft: Primary isolation and serial passage of hepatitis A virus strains in primate cell cultures. J. Clin. Microbiol. *20*, 28-33 (1984)
11. Bradley, D. W.: The agents of non-A, non-B viral hepatitis. J. Virol. Meth. *10*, 307-319 (1985)
12. Burke, D. S. and G. B. Heisey: Wild Malaysian cynomolgus monkeys are exposed to hepatitis A virus. Amer. J. trop. Med. Hyg. *33*, 940-944 (1984)
13. Deinhardt, F.; A. W. Holmes; R. B. Capps, and H. Popper: Studies on the transmission of human viral hepatitis to marmoset monkeys. J. exp. Med. *125*, 673-688 (1967)
14. DeLuc, J. W.; A. Escajadillo, and S. M. Lemon: Hepatitis A virus among captive Panamanian owl monkeys. Lancet *II*, 1427-1428 (1981)
15. Dickerson, M. S.; F. Soifer; A. G. Randall; R. A. MacLean; E. V. Bayer, and R. R. Roberto: Chimpanzee associated hepatitis - Texas. Lab. Primate Newsl. *12*, 7-8 (1973) from CDC Morb. Mortal. Weekly Rep. *25*, 343-348 (1972)
16. Dienstag, J. L.; S. M. Feinstone; R. H. Purcell; J. H. Hoofnagle; L. F. Barker; W. T. London; H. Popper; J. M. Peterson, and A. Z. Kapikian: Experimental infection of chimpanzees with hepatitis A virus. J. infect. Dis. *132*, 532-533 (1975)
17. Dienstag, J. L.; H. Popper, and R. H. Purcell: The pathology of viral hepatitis types A and B in chimpanzees. Amer. J. Path. *85*, 131-148 (1976)
18. Drucker, J.; E. Tabor; R. J. Gerety; D. Jackson, and L. F. Barker: Simultaneous acute infections with hepatitis A and hepatitis B viruses in a chimpanzee. J. infect. Dis. *139*, 338-342 (1979)
19. Eichberg, J. W. and S. S. Kalter: Hepatitis A and B: Serologic survey of human and nonhuman primate sera. Lab. Anim. Sci. *30*, 541-543 (1980)
20. Friedmann, C. T. H.; J. Bernstein; M. Dinnes; I. Kamei, and E. Aaron: Three recent outbreaks of chimpanzee-associated hepatitis. (Outbreak 1) Primate Zoonoses Surveill. Rep. *4*, 13 (1971 a)
21. Friedmann, C. T. H.; M. R. Dinnes; J. F. Bernstein, and G. A. Heidbreder: Chimpanzee-associated infectious hepatitis among personnel at an animal hospital. J. Amer. vet. med. Ass. *159*, 541-545 (1971 b)

22. Grabow, W.O.K.: Progress in studies on the type A (infectious) hepatitis virus in water. Water SA *2*, 20-24 (1976)
23. Hilleman, M.R.; P.J. Provost; V.M. Villarejos; E.B. Buynak; W.J. Miller; O.L. Ittensohn; B.S. Wolanski, and W.J. McAleer: Infectious hepatitis (Hepatitis A) research in nonhuman primates. Bull. Wld Hlth Org. *11*, 140-152 (1977)
24. Hillis, W.D.: An outbreak of infectious hepatitis among chimpanzee handlers at a United States Air Force base. Amer. J.Hyg. *73*, 316-328 (1961)
25. Hinthorn, D.R. and E.R. Price: Three recent outbreaks of chimpanzee-associated hepatitis (Outbreak 3). Primate Zoonoses Surveill. Rep. *4*, 15 (1971)
26. Ishida, N.; M.Watanabe, and H.Ohori: Further evidence for the growth of hepatitis B virus in oval cells originating from human embryonic liver. Pp.85-103 in "Hepatitis Viruses" (Oda, T., ed.), Univ. Park Press, Baltimore (1978)
27. Kessler, H.; K.N. Tsiquaye; H.Smith; D.M. Jones, and A.J. Zuckermann: Hepatitis A and B at the London Zoo. J.infect. Dis. *4*, 63-67 (1982)
28. Kornegay, R.W.; W.E. Giddens, jr.; G.L. van Hoosier, jr. and W.R. Morton: Subacute nonsuppurative hepatitis associated with hepatitis B virus infection in two cynomolgus monkeys. Lab. Anim. Sci. *35*, 400-401 (1985)
29. Kovacs, J.; P.Marzinsky, and R.Altman: Probable chimpanzee-associated hepatitis A - New Jersey. CDC Morb. Mortal. Weekly Rep. *23* (43), 372 (1974)
30. Krugman, S.: Perspectives in viral hepatitis infection: Past, present and future; pp. 3-10 in "Viral Hepatitis. A Contemporary Assessment of Etiology, Epidemiology, Pathogenesis and Prevention." (Vyas, G.N.; S.N. Cohen, and R.Schmid, eds.), Franklin Inst. Press, Philadelphia (1978)
31. Krushak, D.H.: Application of preventive health measures to curtail chimpanzee-associated infectious hepatitis in handlers. Lab. Anim. Care *1*, 52-56 (1970)
32. Lemon, S.M.: Type A viral hepatitis. New developments in an old disease. New Engl. J.Med. *313*, 1059-1067 (1985)
33. Linneman, C.C. jr.; L.W. Kramer, and P.A. Askey: Familial clustering of hepatitis B infections in gorillas. Amer. J.Epidem. *119*, 424-430 (1984)
34. Lorenz, D.; L.Barker; D.Stevens; M.Peterson, and R.Kirschstein: Hepatitis in the marmoset, *Saguinus mystax*. Proc. Soc. exp. Biol. *135*, 348-354 (1970)
35. Maynard, J.E.; W.V. Hartwell, and K.R. Berquist: Hepatitis-associated antigen in chimpanzees. J.infect. Dis. *126*, 660-664 (1971)
36. Mayumi, M.: e antigen/anti-e system and hepatitis B virus, pp. 63-73 in "Hepatitis Viruses" (Oda, T., ed.), Univ. Park Press, Baltimore (1978)
37. McCollum, R.; L.O. Nezevesky; H.Austrian, and J.C. Hart: Three recent outbreaks of chimpanzee-associated hepatitis. (Outbreak 2). Primate Zoonoses Surveill. Rep. *4*, 14 (1971)
38. McInerney, T.; F.Cordon; T.M. Barr, and D.B. Coohon: Nonhuman-primate-associated hepatitis - Oakland County, Michigan. Lab. Primate Newsl. *7* (4), 25-26 (1968)
39. Melnick, J.L.: The classification of vertebrate viruses. Pp. 3-25 in "CRC Handbook Series in Zoonoses" (Steele, J.H., Series ed.), Section B: Viral Zoonoses, Vol. I (Beran, G.W., Sect. ed.), CRC Press, Boca Raton (1981)
40. Miller Keenan, C.; S.M. Lemon; J.W. LeDuc; G.A. McNamee, and L.N. Binn: Pathology of hepatitis A infection in the owl monkey *(Aotus trivirgatus)*. Amer. J.Path. *115*, 1-8 (1984)
41. Mosley, J.W.; H.P. Reinhardt, and R.F. Hassler: Chimpanzee-associated hepatitis. An outbreak in Oklahoma. J.Amer. med. Ass. *199*, 695-697 (1967)
42. Pattison, C.P.; J.E. Maynard, and J.S. Byran: Subhuman primate associated hepatitis. J.infect. Dis. *132*, 478-480 (1975)
43. Provost, P.J. and M.R. Hilleman: Propagation of human hepatitis A virus in cell culture in vitro. Proc. Soc. exp. Biol. *160*, 213-221 (1979)
44. Purcell, R.H. and J.L. Dienstag: Experimental hepatitis A virus infection. Pp.3-12 in "Hepatitis Viruses" (Oda, T., ed.) Univ. Park Press, Baltimore (1978)
45. Rill, R.; W.D. Schrack, jr.: E.J. Witte; W.E. Parkin; J.Jacobs, and F.Rumberger: Nonhuman primate-associated hepatitis - Pennsylvania. CDC Morbid. Mortal. Weekly Rep. *28*, 116 (1975)
46. Riopelle, A.J.: Chimpanzee-associated hepatitis. Proc. Conf. Res. with Primates. Pp.19-26; Trektonik Found. Beaverton, Ov. (1963)

47. Shouval, D.; P.R. Chakraborty; N. Ruiz-Opazo; S. Baum; I. Spigland; E. Muchmore; M.A. Gerber; S.N. Thung; H. Popper, and D.A. Shafritz: Chronic hepatitis in chimpanzee carriers of hepatitis B virus: Morphologic, immunologic, and viral DNA studies. Proc. nat. Acad. Sci. *77*, 6147-6151 (1980)
48. Sly, J.L.; W.T. London, and R.H. Purcell: Illness in a chimpanzee inoculated with hepatitis B virus. J. Amer. vet. med. Ass. *175*, 987-988 (1979)
49. Smetana, H.F.: Experimental and spontaneous viral hepatitis in primates. Lab. Invest. *14*, 1366-1374 (1965)
50. Smetana, H.F.: Infectious hepatitis. Pp. 663-683 in "Pathology of Simian Primates", Part II (Fiennes, R.N.T-W., ed.) Karger, Basel (1972)
51. Smetana, H.F.; A.D. Felsenfeld, and A.J. Riopelle: Human viral hepatitis and chimpanzees. The Chimpanzee, Vol. 3, pp. 26-55, Karger, Basel (1970)
52. Smith, M.S.; P.J. Swanepoel, and M. Bootsma: Hepatitis A in nonhuman primates in nature. Lancet *8206* (II), 1241-1242 (1980)
53. Thung, S.N.; M.A. Gerber; R.H. Purcell; W.T. London; K.B. Mihalik, and H. Popper: Chimpanzee carriers of hepatitis B virus. Amer. J. Path. *105*, 328-332 (1981)
54. Valenza, F.P. and E. Muchmore: The clinical chemistry of chimpanzees: II. Gamma glutamyl transferase levels in hepatitis studies. J. med. Primatol. *14*, 305-315 (1985)
55. Zuckermann, A.J. and C.R. Howard: Hepatitis Viruses of Man. Academic Press, London (1979)
56. Zuckermann, A.J.; A. Thornton; C.R. Howard; K.N. Tsiquaye; D.M. Jones, and M.R. Brambell: Hepatitis B outbreak among chimpanzees at the London Zoo. Lancet *II*, 652-654 (1978)
57. Zwart, P.: Hepatitis infectiosa bei einem Schimpansen. XI. Internat. Symp. Erkr. Zootiere, Zagreb, pp. 77-80 (1969)

1.11 Papovaviruses - SV 40

SV 40 produces a natural latent infection in the renal tissues of *M. mulatta, M. fuscata,* and *M. cyclopis,* less commonly of *M. fascicularis* (Sweet and Hilleman, 1960; Ashkenazi and Melnick, 1963; Yang et al., 1967; Shah and Nathanson, 1976). It is tumorogenic in newborn hamsters (Gerber and Kirschstein, 1962; Girardi et al., 1962; Gerber, 1963; Bonin et al., 1964; Unterharnscheidt et al., 1964; Allison et al., 1967) and transforms human cells *in vitro* (Koprowski et al., 1962; Shein and Enders, 1962; Chang and Sinskey, 1968; Steinberg and Defendi, 1979; Sack, 1981). The position of SV 40 as a true zoonotic agent is at best uncertain, since live SV 40 has been isolated from contaminated polio- and adenovirus-vaccines between 1954 and 1963, and consequently had been administered parenterally to a vast number of especially young people during that time (Gerber, 1967; Shah et al., 1972; Shah and Nathanson, 1976).

Virus

SV 40, like other papovaviruses, is small in size (45 nm), its capsid being of icosahedral symmetry. The virions contain doublestranded RNA and do not possess an envelope. They multiply in the nucleus (Kalter, 1972).

SV 40 growth is propagated in primary African green monkey kidney-, LLCMK2-, BSC-1 cells and other systems such as 3T3 cells of mouse origin or in human cells. SV 40 is capable of both lytic and non-lytic infection of the cell cultures depending on the cell system used (Sweet and Hilleman, 1960; Oxman,

1962; Ashkenazi and Melnick, 1962; Rabson and Kirschstein, 1962; Shein and Enders, 1962a, b; Black and Rowe, 1963a, b; Shein et al., 1963; Carp and Gilden, 1966). Monkey cells support productive or lytic infection, whilst mouse cells become transformed without production of virus. Human cells may become transformed and still produce the virus (Sack, 1981).

CPE appears, between days 7 and 29 p.i. in lytic infections of primary vervet monkey kidney cells, as a vacuolation of the cells - a marker which gave the virus its name as a vacuolating agent - followed by granulation, condensation and rounding of the cells (Ashkenazi and Melnick, 1962; Girardi et al., 1962; Meyer et al., 1962).

Some products in SV 40 infected cells appear to be the direct result of the expression of SV 40 genetic information (Oxman, 1967). One is the SV 40-T antigen detectable by FA-staining in the nuclei of SV 40 infected and SV 40 transformed cells. The antigen is never present in uninfected cells or in cells infected with other oncogenic viruses. It is synthesized before viral DNA-replication begins and has the characteristics of an early viral protein (Oxman, 1967).

SV 40 is relatively resistant to formalin, being inactivated in a biphasic curve. 99,9% of the virus is inactivated within the first 50 hrs (Gerber et al., 1961), but the residual SV 40 has been found alive even after 30 days (Hull, 1968).

Epidemiology

SV 40 is primarily a virus of Asian macaques. In the wild nearly all adult rhesus monkeys, but only a small proportion of juveniles possess anti-SV 40 antibodies (Meyer et al., 1962; Shah and Southwick, 1965). In captive situations SV 40 is not only readily transmitted to other macaque species but also to African monkeys, which are normally free of antibodies under natural conditions (Hsiung and Gaylord, 1961; Meyer et al., 1962; Hsiung et al., 1969; Kalter and Heberling, 1971).

SV 40 like other papovaviruses, appears to infect primarily the urinary tract in a latent form, and although the exact manner of transmission is unknown, urinary excretion seems to be most likely (Shah and Nathanson, 1976). Experimentally infected vervets developed viremia between the 2nd and 16th day p.i. The infection was further transmitted to noninfected control animals without induction of any visible changes (Meyer et al., 1962). Virus excretion via urine was maximal between 14 and 16 days p.i. and continued for up to 60 days (Fabiyi et al., 1967). Another way of virus excretion has been found via nasopharyngeal secreta during the viremic phase (Meyer et al., 1962).

Human exposure may result from close contact to infected monkeys. Shah (1966) found 14 of 161 human sera in Uttar Pradesh and 10 out of 37 animal handlers, all without vaccination record, to be seropositive. A much more common way of human infection has been the administration of SV 40 contaminated adeno- or poliovirus-vaccines to a great number of children and young adults particularly in the United States. The vaccines came first into general use in 1954 and were withdrawn in 1961, but could still have been used till 1963, the expiration date of the last lot manufactured in 1961 (Shah and Nathanson, 1976). Infected human beings, like macaques, may shed the virus with nasopharyngeal secreta (Morris et al., 1961).

Clinical Symptoms

Nonhuman primates: SV 40 infections of Cercopithecinae usually are thought to be asymptomatic (Ashkenazi and Melnick, 1962; Meyer et al., 1962). Fabiyi et al. (1968), however, reported increased urinary sediments including erythrocytes in experimentally infected vervet monkeys. Douglas-Sheffield et al. (1980) isolated SV 40 from a spontaneous fatal interstitial pneumonia and renal tubular necrosis in a possibly immune deficient young rhesus monkey. Gribble et al. (1975) found papova-like virions in 7 *M. mulatta* and 1 *M. speciosa* of the California Primate Research Center, suffering from spontaneous progressive leukoencephalopathy. Two of these viruses were tested further and proved to induce SV 40-T antigen and reacted in plaque reduction assays and restriction endonuclease digestion analysis like SV 40 (Holmberg et al., 1976).

Experimental intracranial inoculation of rhesus monkey fetuses induced hydrocephalus, generalized myoclonic seizures and severe incoordination which required anticonvulsants to control at 3 years of age (London et al., 1983).

Man: The major exposure of the population in the United States by SV 40 contaminated vaccines was not followed by any detectable immediate consequences, but Heinonen et al. (1973 - cit. from Shah and Nathanson, 1976) reported a higher incidence of malignancies, especially of neural tumors in children born to mothers who had received poliovirus vaccine during pregnancy.

SV 40 has repeatedly been seen in or isolated from patients suffering from progressive multifocal leukoencephalopathy (Penney et al., 1972 - cit. from Shah and Nathanson, 1976; Weiner et al., 1972; Peters et al., 1980; Hayashi et al., 1985) which is clinically apparent as ataxia, tremor, dysdiadochokinesia, and aproxia. It was also isolated from a malignant melanoma (Soriano et al., 1974). SV 40-T antigen has been demonstrated in meningeoma cells (Weiss et al., 1975 - cit. from Shah and Nathanson, 1976; Tabuchi et al., 1978; Zang and May, 1979).

Pathology

Nonhuman primates: The pathological lesions in the one reported fatal interstitial pneumonia consisted of hypertrophy and hyperplasia of pneumocytes, with interstitial infiltrates of primarily histiocytes. The renal tubular damage in this case involved the collecting tubules of the inner cortex and medulla with hypertrophy of the epithelial cells, some of which contained basophilic cytoplasmic and intranuclear inclusions. Necrotic or degenerating tubular epithelial cells, some with inclusions, were common. Additionally marked lymphoid depletion was observed in thymus, spleen, lymphnodes and bone marrow and was considered to be a sign of immune deficiency (Sheffield et al., 1980).

In the leukoencephalopathic monkey brains foci of demyelination on some with signs of confluence, were adjacent to blood vessels in the white matter of the brain and spinal cord. Increased microglia and large astrocytes were present throughout and adjacent to the foci of demyelination.

Basophilic and eosinophilic intranuclear inclusion bodies were present in the oligodendroglia and astrocytes. Inflammatory responses consisted of perivascular cuffing (lymphocytes, monocytes, and rarely plasma cells) (Gribble et al., 1975).

Man: In biopsy specimens taken from a patient with leukoencephalopathy and SV 40 infection, the leading lesions were demyelination with marked activity of the macrophages and astrocytes; some of which contained small inclusion bodies. Marked chronic inflammatory reactions were present around blood vessels (Peters et al., 1980).

The malignant melanoma which originated from the skin of the back of a 73 year old patient had extensively metastasized to most internal organs (Soriano et al., 1974). The brain tumors were classified as meningeomas.

Hamsters: The experimental tumors of hamsters depended partially on the infected organs and thus were described as subcutaneous or muscular sarcomas (Bonin et al., 1964; Unterharnscheidt et al., 1964); or after intracerebral inoculation either as ependymomas (Gerber and Kirschstein, 1962) or fibrosarcomas (Girardi et al., 1962).

References

1. Allison, A.C.; F.C. Chesterman, and S. Baron: Induction of tumors in adult hamsters with simian virus 40. J. nat. Cancer Inst. *38*, 567-572 (1967)
2. Ashkenazi, A. and J.L. Melnick: Induced latent infection of monkeys with vacuolating SV 40 papovavirus. Virus in kidneys and urine. Proc. Soc. exp. Biol. *111*, 367-372 (1962)
3. Black, P.H. and W.P. Rowe: Transformation in hamster kidney monolayers by vacuolating virus. SV 40. Virology *19*, 107-109 (1963 a)
4. Black, P.H. and W.P. Rowe: SV 40 induced proliferation of tissue culture cells of rabbit, mouse, and porcine origin. Proc. Soc. exp. Biol. *114*, 721-727 (1963 b)
5. Bonin, O.; F. Unterharnscheidt; I. Schmidt, and K. Schmidt: Die Pathomorphologie von virusinduzierten (SV40) Neoplasmen bei neugeborenen Goldhamstern. I. Mitteilung: Polymorphzellige Sarkome in Unterhautbindegewebe und Muskulatur im Bereich der Injektionsstelle. Acta neuropath. *3*, 351-361 (1964)
6. Carp, R.I. and R.V. Gilden: A comparison of the replication cycles of simian virus 40 in human diploid and African green monkey kidney cells. Virology *28*, 150-162 (1966)
7. Chang, R.S. and T.J. Sinskey: Observations on the transformation of human amnion cell cultures by simian virus 40. J. nat. Cancer Inst. *40*, 505-512 (1968)
8. Douglas-Sheffield, W.; J.D. Strandberg; L. Braun; K. Shah, and S.S. Kalter: Simian virus 40-associated fatal interstitial pneumonia and renal tubular necrosis in a rhesus monkey. J. infect. Dis. *142*, 618-622 (1980)
9. Fabiyi, A.; P.L. Calcagno; J.L. Sever; T. Antonovych, and F. Wolman: Isolation of SV 40 virus and its effect on renal function in African green monkeys. Nature *215*, 88-89 (1967)
10. Gerber, P.: Tumors induced in hamsters by simian virus 40: Persistant subviral infection. Science *140*, 889-890 (1963)
11. Gerber, P.: Patterns of antibodies to SV 40 in children following the last booster with inactivated poliomyelitis vaccines. Proc. Soc. exp. Biol. *125*, 1284-1287 (1967)
12. Gerber, P.; G.A. Hottle, and R.E. Grubbs: Inactivation of vacuolating virus (SV40) by formaldehyde. Proc. Soc. exp. Biol. *108*, 205-209 (1961)
13. Gerber, R. and R.L. Kirschstein: SV 40 induced ependymomas in newborn hamsters. I. Virus tumor relationships. Virology *18*, 582-588 (1962)
14. Girardi, A.J.; B.H. Sweet; V.B. Slotnick, and M.R. Hilleman: Development of tumors in hamsters inoculated in the neonatal period with vacuolating virus SV 40. Proc. Soc. exp. Biol. *109*, 649-660 (1962)
15. Gribble, D.H.; C.C. Haden; L.W. Schwartz, and R.V. Henrickson: Spontaneous progressive multifocal leukoencephalopathy (PML) in macaques. Science *254*, 602-604 (1975)
16. Hayashi, T.; A. Sumiyoshi; M. Tanaka; S. Araki, and Y. Minamishima: Progressive multifocal leukoencephalopathy. A case with special reference to SV 40 etiology. Acta Pathol. Jap. *35*, 173-181 (1985)
17. Hsiung, G.D.; T. Atoynatan, and C.W. Lee: Epidemiologic studies of latent virus infections in

captive monkeys and baboons. I. Overall plans and virus isolation with special reference to SV 40 and foamy viruses. Amer. J. Epidem. *89*, 464-471 (1969)
18. Hsiung, G. D. and W. H. Gaylord: The vacuolating virus of monkeys. I. Isolation, growth characteristics and inclusion body formation. J. exp. Med. *114*, 975-986 (1961)
19. Hull, R. N.: The Simian Viruses. Virology Monographs No. 2, Springer, New York (1968)
20. Kalter, S. S.: Identification and study of viruses; pp. 382-468 in "Pathology of Simian Primates", Part II (Fiennes, R. N. T-W., ed.), Karger, Basel (1972)
21. Kalter, S. S. and R. L. Heberling: Comparative virology of primates. Bact. Rev. *35*, 310-364 (1971)
22. Koprowski, H.; J. A. Ponten; F. Jensen; R. G. Ravdin; P. Moorehead, and E. Saksola: Transformation of cultures of human tissue infected with simian virus, SV 40. J. cell. Physiol. *59*, 281-292 (1962)
23. London, W. T.; M. Gravell; B. Curfman; A. J. Martinez, and J. L. Sever: Hydrocephalus, seizures, and incoordination in rhesus monkeys infected in utero with SV-40 virus. Lab. Anim. Sci. *33*, 482-483 (1983)
24. Meyer, H. M. jr.; H. E. Hopps; N. G. Rogers; B. E. Brooks; B. C. Bernheim; W. P. Jones; A. Nisalak, and R. D. Douglas: Studies on simian virus 40. J. Immunol. *88*, 796-806 (1962)
25. Morris, J. A.; K. M. Johnson; C. G. Aulisio; R. M. Chanock, and V. Knight: Clinical and serological responses in volunteers given vacuolating virus (SV40) by respiratory route. Proc. Soc. exp. Biol. *108*, 56-59 (1961)
26. Oxman, M. N.: Some behavioral studies of simian virus 40 (SV 40). Arch. ges. Virusforsch. *22*, 171-187 (1967)
27. Peters, A. C. B.; J. Versteeg; G. T. A. M. Bots; W. Boogerd, and J. Vielvoye: Progressive multifocal leukoencephalopathy. Immunofluorescent demonstration of simian virus 40 antigen in CSF cells and response to cytarabine therapy. Arch. Neurol. *37*, 497-501 (1980)
28. Rabson, A. S. and R. L. Kirschstein: Induction of malignancy *in vitro* in newborn hamster kidney tissue infected with simian vacuolating virus (SV 40). Proc. Soc. exp. Biol. *3*, 323-328 (1962)
29. Sack, G. H. jr.: Human cell transformation by simian virus 40 - A review. In Vitro *17*, 1-19 (1981)
30. Shah, K. V: Neutralizing antibodies to simian virus 40 (SV40) in human sera from India. Proc. Soc. exp. Biol. *121*, 303-307 (1966)
31. Shah, K. V.; F. R. McCrumbs jr.; R. W. Daniel, and H. L. Ozer: Serologic evidence for a Simian-Virus-40-like infection of man. J. nat. Cancer Inst. *48*, 557-561 (1972)
32. Shah, K. and N. Nathanson: Human exposure to SV 40: Review and comment. Amer. J. Epidem. *103*, 1-12 (1976)
33. Shah, K. V. and C. H. Southwick: Prevalence of antibodies to certain viruses in sera of free-living rhesus and of captive monkeys. Indian J. med. Res. *53*, 488-500 (1965)
34. Shein, H. M. and J. F. Enders: Multiplication and cytopathogenicity of simian vacuolating virus 40 in cultures of human tissues. Proc. Soc. exp. Biol. *109*, 495-500 (1962)
35. Shein, H. M. and J. F. Enders: Transformation induced by simian virus 40 in human renal cell culture. I. Morphology and growth characteristics. Proc. nat. Acad. Sci. *48*, 1164-1172 (1962a)
36. Shein, H. M.; J. F. Enders; J. D. Levinthal, and A. E. Burket: Transformation induced by simian virus 40 in newborn Syrian hamster renal cell cultures. Proc. nat. Acad. Sci. *49*, 28-34 (1963)
37. Soriano, F.; C. E. Shelburne, and M. Gökcen: Simian virus 40 in a human cancer. Nature *249*, 421-424 (1974)
38. Steinberg, M. L. and V. Defendi: Altered patterns of growth and differentiation in human keratinocytes infected by simian virus 40. Proc. nat. Acad. Sci. *76*, 801-805 (1979)
39. Sweet, B. H. and M. R. Hilleman: The vacuolating virus SV 40. Proc. Soc. exp. Biol. *105*, 420-427 (1960)
40. Tabuchi, K.; W. M. Krisch; M. Low; D. Gaskin; J. V. Buskirk, and S. Maa: Screening of human brain tumors for SV 40-related T-antigen. Int. J. Cancer *21*, 12-17 (1978)
41. Tan, K. B.: Metabolism in simian virus 40-infected cells and incorporation into virions. Proc. nat. Acad. Sci. *74*, 2805-2809 (1977)
42. Unterharnscheidt, F.; O. Bonin; K. Schmidt, and I. Schmidt: Die Pathomorphologie von virusinduzierten (SV40) Neoplasmen bei neugeborenen Goldhamstern. II. Mitteilung: Hyperpla-

sien und Papillome des Plexus chorioideus der Hirnventrikel nach subcutaner bzw. intramuskulärer Injektion des Virus. Acta neuropath. *3*, 362-371 (1964)
43. Weiner, L. P.; R. M. Herndon; O. Narayan; R. T. Johnson; K. Shah; L. J. Rubinstein; T. J. Preziosi, and F. K. Conley: Isolation of virus related to SV 40 from patients with progressive multifocal leukoencephalopathy. N. Eng. J. Med. *286*, 385-390 (1972)
44. Yang, C.; C. Kuo, and C. Chen: A study of natural infection of simian virus 40 (SV40) in Taiwan monkeys *(Macaca cyclopis)*. Formosan Med. Ass. *66*, 143-148 (1967)
45. Zang, K. D. and G. May: Expression of SV 40-related T-antigen in cell cultures of human meningiomas. Naturwissenschaften *60*, 59 (1979)

2 Mollicutes

2.1 Mycoplasmataceae

Presently seven species of mycoplasmas are indigenous to man. Four human species of mild pathogenicity in man have also been occasionally reported in nonhuman primates.

2.1.1 Mycoplasma Hominis

M. hominis has been isolated from *Pan troglodytes* (Cole et al., 1970), *Papio anubis* and *Macaca mulatta* (Hill, 1977).

Agent
M. hominis like most other mycoplasmas requires sterol for growth. It prefers microaerophilic conditions and hydrolyses arginine, but no glucose or urea.

Clinical Symptoms
Nonhuman primates: All *M. hominis* strains were recovered from materials of "healthy" animals.

Man: *M. hominis* rarely may cause pelvic inflammatory disease and urinary tract infections (Finegold and Martin, 1982).

2.1.2 Mycoplasma Salivarium

M. salivarium was isolated from *Cercopithecus aethiops* (Del Giudice et al., 1969) and *Callithrix jacchus* (Hill et al., 1978).

Agent
M. salivarium differs from *M. hominis* by its microaerophilic growth requirements (Kenny, 1978).

Clinical Symptoms
M. salivarium was isolated predominantly from the pharynx, occasionally from the cervix or prostatic urethra of apparently healthy *C. aethiops*, and from the naso-

pharynx and throat of normal marmosets (Del Giudice et al., 1969; Hill, 1978). In man it is a common inhabitant of the gingiva, its etiological role in periodontal disease is not exactly known (Kenny, 1974).

2.1.3 Mycoplasma orale

M. orale was once isolated from the throat of an apparently healthy *C. aethiops* (Del Giudice et al., 1969), in man it is considered a normal pharyngeal organism.

2.1.4 Mycoplasma fermentans

M. fermentans was recovered from the cervix of a *C. aethiops* (Del Giudice et al., 1969).

Agent
M. fermentans is distinguished from *M. hominis* (and *M. salivarium* and *M. orale*) by its glucose fermentation, it is rarely isolated from the genitourinary tract of man (Kenny, 1974).

2.2 Ureaplasmataceae

2.2.1 Ureaplasma Sp.

Ureaplasmas, formerly called T-mycoplasmas, are morphologically similar to mycoplasmas except that the colonies growing on solid media are very small. They differ from the mycoplasmas metabolically by being ureolytic.

They were first recognized in man (Shepard, 1954) and subsequently were isolated from a variety of animals.

Ureaplasmas cause pneumonia in calves, keratoconjunctivitis and conjunctivitis in cattle (Ross and Nitzschke, 1985) and nongonococcal urethritis (Csonka et al., 1966; Ford, 1967; Shepard, 1970; McDonald et al., 1982; Taylor-Robinson, 1984) or reproductive failure in man (Kundsin et al., 1967; Gnarpe and Friberg, 1973). Experimental intraurethral infections resulted in urethritis of the volunteers (Taylor-Robinson et al., 1977). Ureaplasmas have also been recovered from *Pan troglodytes* (Brown et al., 1976), *Miopithecus talapoin, Erythrocebus patas, Macaca fascicularis* (Kundsin et al., 1975), *Saimiri sciureus* (Taylor-Robinson et al., 1971), *Callithrix jacchus* (Furr et al., 1976; 1979; 1984; Hill et al., 1978), *Callithrix geoffroyi* and *Saguinus oedipus* (own findings - unpublished). Experimental urethral infection of several monkey species did not induce any clinical signs (Moller and

Freundt, 1983) whereas in chimpanzees urethritis resulted (Taylor-Robinson, 1984).

The taxonomic relationship or identity of the ureaplasmas from different sources is still open to question.

References

1. Brown, W.J.; N.F. Jacobs jr.; E.S. Arum, and R.J. Arko: T-strain mycoplasma in the chimpanzee. Lab. Anim. Sci. *26*, 81-83 (1976)
2. Cole, B.C.; I.R. Ward; L. Golightly-Rowland, and C.E. Graham: Characterization of mycoplasmas isolated from the great apes. Canad. J. Microbiol. *16*, 1331-1339 (1970)
3. Csonka, G.W.; R.E.O. Williams, and J. Corse: T-strain mycoplasma in non-gonococcal urethritis. Lancet *1*, 1292 (1966)
4. Del Giudice, R.A.; T.R. Carski; M.F. Barile; H.M. Yamashiroya, and J.E. Verna: Recovery of human mycoplasmas from simian tissues. Nature *222*, 1088-1089 (1969)
5. Finegold, S.M. and W.J. Martin: Diagnostic Microbiology. 6th ed. Mosby, St. Louis (1982)
6. Ford, D.K.: Relationships between mycoplasma and the etiology of nongonococcal urethritis and Reiter's syndrome. Ann. N.Y. Acad. Sci. *143*, 501-504 (1967)
7. Furr, P.M.; C.M. Hetherington, and D. Taylor-Robinson: Ureaplasmas in the marmoset *(Callithrix jacchus)*: Transmission and elimination. J. med. Primatol. *8*, 321-326 (1979)
8. Furr, P.M.; D. Taylor-Robinson, and C.M. Hetherington: The occurrence of ureaplasmas in marmosets. Lab. Anim. *10*, 393-398 (1976)
9. Furr, P.M.; D. Taylor-Robinson, and C.M. Hetherington: The inhibitory effect of flurofamide on ureaplasmas and their elimination from marmosets by its use. J. Antimicrob. Chemother. *14*, 613-618 (1984)
10. Gnarpe, H. and J. Friberg: T-mycoplasmas as a possible cause for reproductive failure. Nature *242*, 120-121 (1973)
11. Hill, A.: The isolation of mycoplasmas from non-human primates. Vet. Rec. *101*, 117 (1977)
12. Hill, A.C.; J.A. Turton, and J. Bleby: Bacterial and mycoplasma flora of a laboratory colony of the common marmoset *(Callithrix jacchus)*. Vet. Rec. *103*, 824-827 (1978)
13. Kenny, G.E.: Mycoplasma; pp. 333-337 in "Manual of Clinical Microbiology", 2nd ed. (Lennette, E.H.; E.H. Spaulding, and J.P. Truant, eds.) Amer. Soc. Microbiol., Washington (1974)
14. Kundsin, R.B.; S.G. Driscoll, and P.J. Ming: Strains of mycoplasma associated with human reproductive failure. Science *157*, 1573-1574 (1967)
15. Kundsin, R.B.; T. Rowell; M.C. Shepard; A. Parreno, and C.D. Lunceford: T-strain mycoplasmas and reproductive failure in monkeys. Lab. Anim. Sci. *25*, 221-224 (1975)
16. McDonald, M.I.; M.H. Lam; D.F. Birch; A.D. D'Arcy; K.F. Fairley, and E.R.J. Pavillard: Ureaplasma urealyticum in patients with acute symptoms of urinary tract infection. J. Urol. *128*, 517-519 (1982)
17. Moller, B.R. and E.A. Freundt: Monkey animal model for study of mycoplasmal infections of the urogenital tract. Sex. Transm. Dis. *11*, 359-362 (1983)
18. Ross, R.F. und E. Nitzschke: Mollicutes; pp. 314-446 in "Handbuch der bakteriellen Infektionen bei Tieren." Bd. V (Blobel, H. und T. Schliesser, eds.). Gustav Fischer, Stuttgart (1985)
19. Shepard, M.C.: The recovery of pleuropneumonia-like organisms from Negro men with and without nongonococcal urethritis. Amer. J. Syph. Gonorrh. Ven. Dis. *38*, 113-124 (1954)
20. Shepard, M.C.: Nongonococcal urethritis associated with human strains of T mycoplasmas. J. Amer. med. Ass. *211*, 1335-1340 (1970)
21. Taylor-Robinson, D.: Ureaplasmas as a cause of disease in man and animals: fact or fancy? Isr. J. Med. Sci. *20*, 843-847 (1984)
22. Taylor-Robinson, D.; G.W. Csonka, and M.J. Prentice: Human intraurethral inoculation of ureaplasmas. Q.J. Med. *46*, 309-326 (1977)
23. Taylor-Robinson, D.; C. Martin-Bourgon; T. Watanabe, and J.P. Addey: Isolation of T-mycoplasmas from dogs and squirrel monkeys: biological and serological comparison with those isolated from man and cattle. J. Gen. Microbiol. *68*, 97-107 (1971)

3 Bacteria

3.1 Spirochaetaceae

3.1.1 Borrelia

Borrelia spp. are the causative agents of two diseases in man and sometimes in nonhuman primates. One is the complex of relapsing fevers, occurring in man, nonhuman primates, and a variety of nonprimates, particularly rodents. The other one is Vincent's disease of man and nonhuman primates.

Relapsing Fevers
Because of considerable technical obstacles in the isolation of Borreliae most reports of *Borrelia* infections are based on dark field preparations of fresh blood smears with no species identification possible. The serological differentiation of Borreliae is hampered as well and of little diagnostic value because of the many antigenic types and the antigenic variability of the *Borrelia* strains (Pinkerton, 1972).

Borrelia infections of relapsing fever type have been reported in simian primates living under feral conditions (Sellards, 1930), in particular in *Callithrix (Leontopithecus), Cebus, Ateles* and *Aotus* (Dunn and Clark, 1933), and in *Erythrocebus* and *Cercopithecus* (Thiroux and Dufougere, 1910; Garnham, 1947).

Borrelia mazzottii, the causative agent of American tick born relapsing fever has also been found in "monkeys" (Felsenfeld, 1979).

Agents
Morphologically all Borreliae look alike as helical organisms of 5 to 25 × 0,2 × 0.5 µm dimensions with 4 to 30 uneven coils. They are very difficult to cultivate especially for primary isolations. Kelly's medium gives feasible results in phenotyping studies. Borreliae are anaerobic, they require longchain fatty acids for nutrition. Dextrose is metabolized to lactic acid (Felsenfeld, 1972).

Vectors
Relapsing fever-type Borreliae rely on arthropod vectors for transmission. In classical relapsing fever caused by *B. recurrentis* it is the human body and head louse *Pediculus humanus humanus*, in other relapsing fevers involving animal hosts as well, the usual vectors are argasid ticks. *B. mazzottii* e.g. is transmitted by *Ornithodorus talaje*. As *Ornithodorus* ticks require low humidity and moderate temper-

atures they are not found in tropical rain forests but rather infest animal caves, abandoned houses and other biotopes rarely entered by man. *Ornithodorus* ticks are also usually nocturnal feeders and not all *Ornithodorus talaje* strains bite man. The danger to man is thus low, although *O. talaje* is spread all throughout the western United States, Central and South America.

Pathogenesis
After being transferred by their arthropod vectors the circulating Borreliae multiply, but subsequently are mostly destroyed by antibodies. Borreliae having survived the attack of specific antibodies retreat into the host's RES and undergo antigenic variations, thus enabling another explosive multiplication of the bacteria – again limited by newly formed antibodies.

Clinical Symptoms
Nonhuman primates: No clinical symptoms are reported in spontanous infections. Experimental relapsing fever in monkeys resembles that of man very closely. Two or three days after subcutaneous inoculation with the patient's blood a 3–4 days lasting pyrexia occurs followed by 2 to 4 relapses at 2–8 days intervals (Pinkerton, 1972).

Man: The first attack usually begins without prodromal signs as sudden chills and fever up to 41 °C particularly in the American tick born relapsing fever. Violent occipital or occipitofrontal headaches or retroorbital pain, myalgia and backache, and in some outbreaks a macular skin eruption accompany the fever. The fever remains high for 3 to 7 days, it ends with a crisis within a few hours, seldom with a lysis lasting 1 to 3 days.

Thrombocytopenia results in purpura, epistaxis, melena, haematuria and metrorrhagia. Dry cough is frequently reported, possibly caused by haemodynamic changes.

The intervals between the attacks are irregular, varying from 3 to 36 days. During the intervals the patient is usually weak, subfebrile and frequently shows an increase of mononuclear cells in the peripheral blood.

Pathology
Man and nonhuman primates: The most common lesions are splenomegaly with focal necroses or infarcts and hepatomegaly. The kidneys may be enlarged, in histological slides fibrinous exudates in Bowman's capsules may be present. The skin may be icteric and subcutaneous haemorrhages may be seen.

3.1.2 Borrelia Vincenti

Vincent's disease has been observed in chimpanzees (van Riper and Finey, 1967); all of 18 healthy captive chimpanzees investigated were found to be harbouring *Borrelia vincenti* in gum scrapings. In baboons the overall incidence of *B. vincenti* was low, but tended to be correlated with a Vincent's like syndrome (Pinkerton, 1972).

3.1.3 Leptospira

Many wild and domesticated animals including monkeys (Babudieri, 1958; Ball, 1966) have been found to host various species of *Leptospira,* with some causing spontaneous disease in nonhuman primates. The first outbreak in simian primates was described in 1927 by Wilbert and Delorme in chimpanzees in Africa - 23 of 24 animals spontaneously infected with *L. icterohaemorrhagiae* died. Fear et al. (1968) observed clinical symptoms in *L. ballum* infected captive baboons. The symptoms in the baboons included occasional diarrhea and an increase of the general abortion rate. Shive et al. (1969) described 3 fatal *L. icterohaemorrhagiae*-infections in *M. sylvana,* whilst Johnson and Porter (1969) reported 2 fatalities in woolly monkeys, again due to *L. icterohaemorrhagiae.*

Serological surveys of more than 1400 simian sera of 22 Old World and 12 New World primate species showed 107 positive sera, with all but one belonging to Old World monkeys (Minette, 1966). Only one *S. oedipus* harboured anti-Leptospira-antibodies. *L. icterohaemorrhagiae* was the most frequently encountered serotype, being predominantly found in the sera of baboons, chimpanzees, patas monkeys and the one marmoset. Pinkerton (1972) reported 25 of 89 newly imported baboons to be seropositive to *L. icterohaemorrhagiae.*

Agents

Leptospires are flexuous, helical bacteria of 6 to 20 × 0.1 to 0.2 μm dimensions. They are tightly coiled, hooked at one or both ends under in vivo conditions, whereas in culture media they may become straight. As leptospires are invisible in ordinary illumination, the dark field technique is necessary for observation. The normal staining procedure is by silver impregnation. Although leptospires are gram-positive gram staining does not give good results. Leptospires are motile by means of an axostyle, which is composed of two axial filaments inserted subterminally at both ends.

For primary isolation the inoculation of susceptible laboratory animals (3 to 4 week old hamsters) is still the best procedure. For in vitro growth 10% serum has to be added to the nutrient media. Additionally leptospires require specific fatty acids as a source of carbon and energy and there are absolute minimal requirements for vitamin B12, thiamine and various metal salts. Leptospires are strictly aerophilic, or sometimes slightly microaerophilic. They grow best in liquid or semiliquid media, in the latter a few millimeters beneath the surface, but even under optimum conditions the *in vitro* growth of leptospires is slow. Optimal incubating temperature is 28 °C, whilst the optimal pH is 7.2 to 7.4 (Torten, 1979).

Leptospires are very susceptible to dry environments and will die within minutes of exposure to dry conditions.

Classification of leptospires is still based mainly on serology (two way cross absorption - agglutination). Today the genus Leptospira is subdivided into the saprophytic *L. biflexa*-serovars, and the parasitic and pathogenic *L. interrogans* - a group comprising all medically important serovars (Table 8).

For diagnostic serology agglutination-, haemolytic- and haemagglutination-, complement-fixation-, and FA-tests are used (Torten, 1979).

Table 8. Leptospira interrogans serogroups

L. australis	L. icterohaemorrhagiae
L. autumnalis	L. javanica
L. ballum	L. Kazachstanica
L. bataviae	L. panama
L. bufonis	L. pomona
L. canicola	L. pyrogenes
L. celledoni	L. ranarum
L. cynopteri	L. saxkoebing
L. grippotyphosa	L. shermanni
L. hebdomadis	L. tarassovi
L. hyos	

Epidemiology
Leptospires can be directly transmitted and therefore they may always induce sporadic infections in either man or animals, depending upon the presence of a susceptible host in the immediate vicinity of a vector shedding live, virulent leptospires in its urine. Outbreaks of leptospirosis require the survival of the organisms in the environment outside the host for a minimum length of time. Arid lands, deserts and polar regions for this reason will not support outbreaks of leptospirosis. Fertilized wet soil, on the other hand, enables the survival of the leptospires in the environment, which is further favoured by the ammonium salts in the dungs, which can be used as the sole nitrogen source by leptospires. Dangerous areas are also man-made or natural oases or lakes within a desert area because of the concentration of man, wild and domesticated animals at such places. The introduction of a carrier animal (or man) can quickly lead to the formation of a well-defined endemic zone. In case of leptospirosis outbreaks in captive non-human primates infection from rats is usually suspected.

The leptospires enter the body through small skin lacerations or wounds or by the oral route with food and water contaminated with infected urine. The principal nesting site of leptospires within the body are the renal tubules, from where they can be shed for lifetime.

Clinical Symptoms
Nonhuman primates: Data on clinical symptoms in sporadic infections are rarely reported, as most infections are clinically inapparent (Lapin, 1962 - cit. from Pinkerton, 1972). In *L. ballum* infection of baboons, in addition to the increased abortion rate, occasional diarrhea and discoloration of serum samples were observed. Nouvel et al. (1957) found jaundice associated with a leptospirosis-outbreak in monkeys.

In experimental *L. icterohaemorrhagiae* infections of macaques, cebus monkeys, baboons, spider monkeys, and marmosets minimal symptomatology such as fever, thirst, malaise and conjunctival infections were reported. Only one patas monkey died of icterus and spirochaetal bacteriuria, whilst an experimentally infected *M. sinica* developed icterus, conjunctival infection, fever and prostration (for references see Pinkerton, 1972). In another *L. icterohaemorrhagiae*-study using adult cebus-, rhesus- and vervet monkeys the animals responded with febrile temperatures, weight loss, antibody production and bacteriuria despite the lack of any frank illness (Minette and Shaffer, 1968). Similarly infected squirrel monkeys developed clinical disease, in some cases even fatal, with leptospiruria, pronounced icterus and increased indirect bilirubin levels.

Man: After an incubation period of 3 to 20 days three stages of illness begin:

a) An initial 2 to 3 day period of influenza-like fever with chills, headache, myalgia, hyperemia, conjunctivitis, albuminuria, haematuria and intestinal disturbances.
b) Three days of icterus, liver-edema, splenomegaly and cutaneous haemorrhages.
c) 7 to 13 days p.i. beginning convalescence, jaundice, neural and cardiac symptoms.

Occasionally acute renal failure may lead to death.

Pathology

Nonhuman primates: The macroscopical findings in the three fatal cases in *M. sylvana* were icterus and haemorrhages in the skin, lymphnodes, visceral organs and heart. Histopathological lesions included focal liver necrosis, renal congestion and renal tubular necrosis (Shive et al., 1969). In the woolly monkeys marked icterus, gastroenteritis, hepatomegaly, renal and adrenal enlargement were observed.

Man: The most prominent lesions are diffuse hepato-cellular damage with little or no intrahepatic biliary destruction in the liver, acute tubular degeneration in the kidney and small or massive haemorrhages in any organ.

Treatment

In guinea pigs chloramphenicol and tetracycline are relatively effective (Hubert and Miller, 1967). In human pomona-leptospirosis best results were obtained with oxytetracyclin and chlortetracycline.

3.1.4 Treponema

Treponema spp. in man are the causative agents of syphilis *(T. pallidum)*, yaws or frambösia *(T. pertenue)* and pinta *(T. carateum)*.

The existence of similar or possibly identical treponema infections in nonhuman primates was first indicated by comparative studies of human and simian globulins (Fribourg-Blanc et al., 1963), and was confirmed by bacteriological and epidemiological results (Fribourg-Blanc et Mollaret, 1969). Simian treponematosis is closely related or identical to *T. pertenue* infection in man and they coexist with endemic yaws in certain African regions (Fribourg-Blanc, 1970). The agents originating from nonhuman primates have been provisionally named *T. Fribourg-Blanc* (Fribourg-Blanc, 1972); whilst others are believed to belong to the Bangkok-strains of *T. pertenue* (Sepetijian et al., 1969). Felsenfeld and Wolf (1972) demonstrated treponemal reagins in 13 of 125 nonhuman primate sera *(P. troglodytes, C. aethiops, E. patas, C. moloch)*, which were believed to be *T. pertenue* rather than *T. pallidum*.

Agents
Members of the Treponemataceae cannot be distinguished from each other by morphological criteria. All pathogenic treponemes are thin, filamentous, 6 to 10 μm × 0.2 μm in size. Their body is coiled by 6 to 14 spirals and they are motile.

Treponemes, like leptospires, are not stained sufficiently with ordinary stains and silver impregnation is the method of choice (Kelly, 1974).

Pathogenic treponemes can not be cultivated *in vitro,* but can be maintained for several days under anaerobic conditions in media containing serum ultrafiltrate, pyruvate, albumin and a sulfhydryl compound, e.g. the medium of Nelson and Mayer (Fribourg-Blanc, 1972). *In vivo* they can be maintained in hamsters.

It is also possible to separate *T. pallidum, T. pertenue* and *T. carateum* immunologically from each other. Anti *T. pallidum* antisera will react with the other treponemes in immobilization- and immunofluorescence tests (Fribourg-Blanc and Mollaret, 1969), a fact used in serological surveys. Treponemes, like spirochetes, are quite sensitive to drying and temperatures.

Epidemiology
T. pertenue and *T. carateum* are transmitted by nonveneral contact through skin lacerations, with flies being suspected of being the vectors (Cousins, 1972; Hallmann and Burkhardt, 1974; Kelly, 1974).

Pathogenesis
In human *T. pertenue*-infections the disease begins, after an incubation period of two to eight weeks, as strictly extragenital skin lesions. These are followed by secondary and tertiary stages which might also involve the bones. Yaws is considered to be the cause of goundou.

Clinical Symptoms
Man: The symptoms of yaws come in three chronological stages:

a) General malaise, pyrexia, and development of an extragenital primary skin nodule, the mother yaw, at the site of infection.
b) Approximately 3 months later a rash, consisting of circular blotches with papules or yaws forming the centre of each blotch, develops. These then break down and leave sores with central yellowish scabs. Because of this raspberry-like appearance the disease was named "framboise" by the French.
   Highly infectious mucoid patches sometimes develop in the oral cavity.
c) The final tertiary stages consist of deep-seated lesions extending to the bones as osteitis and osteomyelitis. These lead to deformities, with widespread soft-tissue sloughing and ulcers (Cousins, 1972).
   Goundou, a disease endemic in West Africa and characterized by inflammatory exostoses of the perinasal maxillary bones is considered to be a secondary or tertiary stage of yaws (Bortreau-Roussel – cit. from Schultz and Starck, 1977).

Nonhuman primates: Spontaneous clinical *T. pertenue* infections in nonhuman primates are difficult to assess. Lesions very similar to yaws or goundou have been described, e.g. as facial skin ulcers in 3 baboons captured in Senegal (Fribourg-Blanc, 1972), they have also been described in hominoids. Some, however, might rather be due to leprosy, which also effects nonhuman primates as has been recently learned, and also the skin lesions of rhinophycomycosis *(Entomophtora coronata)* of chimpanzees looked very similar (Roy and Cameron, 1972). The facial bone deformities referred to as goundou in nonhuman primates by some authors are considered to be the expression of nutritional hyperparathyreoidosis by others (Schultz and Starck, 1977).

Pathology

Man: The cutaneous lesions consist of granulomatous papules and macules. The papules are formed by epithelial hyperplasias, mixed cellular infiltrations, and by exudation of serum. The papillae are elongated and infiltrated, the interpapillary pegs are often thickened. This process might be followed by hyperkeratosis of the overlying epithelium, with superficial erosions leading to shallow and sharply defined ulcers. Invasion of skeletal tissues produces osteitis and periostitis leading to bone deformities, which may be extensively destructive in nasal and palatine bones (Hunter et al., 1964).

Nonhuman primates: Since the infection of nonhuman primates normally goes undetected no pathological data are present at this time.

3.1.5 Spirochetes

About 12% of clinically normal rhesus monkeys and 3% of healthy human subjects have large intestinal spirochete infections, which are indistinguishable from each other (Takeuchi and Zeller, 1972a; Takeuchi et al., 1974). Similar microorganisms can be found in the gastric mucosa of rhesus monkeys, other animals and man (Sato and Takeuchi, 1982). The significance of the gastrointestinal spirochetes as related to intestinal disease in man or animal is still controversial.

Agents

Intestinal spirochetes are anaerobic. They share the general characteristics of the other spirochetes. Their size averages 3 to 6 μm × 0.2 to 0.4 μm, with their body being coiled by 2 to 6 spirals (Takeuchi and Zeller, 1972b).

Pathology

Intestinal spirochetosis is characterized by massive infestation of the colonic surface epithelium by uniform, large, basophilic and argentophilic spirochetes (Takeuchi et al., 1971). The spirochetes lying in direct contact with epithelial plasmalemma and replacing the glycocalyx, cause microlesions of the brush border. The Microvilli and the terminal web disappear, with the brush border appearing broader than normal (Takeuchi et al., 1971; Takeuchi and Zeller, 1972b).

References

1. Babudieri, B.: Animal reservoirs of leptospirosis. Ann. N.Y. Acad. Sci. *70*, 393-413 (1958)
2. Ball, M.G.: Animal hosts of leptospires in Kenya and Uganda. Amer. J.trop. Med. Hyg. *15*, 523-530 (1966)

3. Clark, H.C.; L.H. Dunn, and J. Benavides: Experimental transmission to man of a relapsing fever spirochaete in a wild monkey of Panama - *Leontocebus geoffroyi* (Pucheran). Amer. J. trop. Med. Hyg. *11*, 243-257 (1931)
4. Cousins, D.: Diseases and injuries in wild and captive gorillas, *Gorilla gorilla*. Int. Zoo Yearb. *12*, 211-218 (1972)
5. Diesch, S.L. and H.C. Ellinghausen: Leptospirosis; pp. 436-462 in "Diseases Transmitted from Animals to Man" (Hubbert, W.T., W.F. McCulloch, and P.R. Schnurrenberger, eds.). C.C. Thomas, Springfield (1975)
6. Dunn, L.H. and H.C. Clark: Notes on relapsing fever in Panama with special reference to animal hosts. Amer. J. trop. Med. Hyg. *13*, 201-209 (1933)
7. Fear, F.A.; M.E. Pinkerton; J.A. Cline; F. Kriewaldt, and S.S. Kalter: A leptospirosis outbreak in a baboon (*Papio* sp.) colony. Lab. Anim. Care *18*, 22-28 (1968)
8. Felsenfeld, O.: Borreliosis, pp. 79-96 in "CRC Handbook Series in Zoonoses" (Steele, J.H., ed.), Section A: Bacterial, Rickettsial, and Mycotic Diseases. Vol. I (Stoenner, H.; W. Kaplan, and M. Torten, eds.), CRC Press, Boca Raton (1979)
9. Felsenfeld, O. and R.H. Wolf: Serological reactions with treponemal antigens in nonhuman primates and the natural history of treponematosis in Man. Folia primatol. *16*, 294-305 (1972)
10. Fribourg-Blanc, A.: La treponamatose naturelle des primates. La Revue de Medecine *22*, 1301-1304 (1970)
11. Fribourg-Blanc, A.: Treponema; pp. 255-262 in "Pathology of Simian Primates", Vol. 2 (Fiennes, R.N.T-W., ed.), Karger, Basel (1972)
12. Fribourg-Blanc, A. and H.H. Mollaret: Natural treponematosis of the African primate. Primates in Med. *3*, 113-121 (1969)
13. Fribourg-Blanc, A.; G. Niel, et H.H. Mollaret: Note sur quelques aspects immunologiques du cynocephale africain. Bull. Soc. Path. exot. *56*, 476-585 (1963)
14. Garnham, P.C.C.: A new blood spirochaete in the grivet monkey *Cercopithecus aethiops*. E. Afr. med. J. *24*, 47-51 (1947)
15. Hackett, C.J.: Natural history of framboesia. Bull. Wld Hlth Org. *29*, 7-41 (1963)
16. Hallmann, L. and F. Burkhardt: Klinische Mikrobiologie. Georg Thieme, Stuttgart (1974)
17. Hubbert, W.T. and J.N. Miller: Leptospirosis infection and development of acquired resistance in guinea pigs. Amer. J. vet. Res. *28*, 861-864 (1967)
18. Hudson, E.H.: Treponematosis and man's social evolution. Amer. Anthropologist *67*, 885-901 (1965)
19. Hunter, G.W.; W.W. Frye, and J.C. Swartzwelder: A Manual of Tropical Medicine. W.B. Saunders, Philadelphia (1964)
20. Johnson, P. jr. and R. Morter: A natural outbreak of leptospirosis in a monkey colony; p. 69-71 in: "XI. Internat. Symp. Erkr. Zootiere", Zagreb (1969)
21. Kelly, R.T.: Treponema; pp. 358-360 in "Manual of Clinical Microbiology". 2nd ed. (Lennette, E.H., E.H. Spaulding, and J.P. Truant, eds.). Amer. Soc. Microbiol., Washington D.C. (1974)
22. Minette, H.P.: Leptospirosis in primates other than man. Amer. J. trop. Med. Hyg. *15*, 190-198 (1966)
23. Nouvel, J.; P. Bullier et J. Rinjard: Rapport sur la mortalité et la natalité enregistrée au Parc zoologique pendant l'annee. Bull. Mus. Hist. nat. *29*, 297-307 (1957)
24. Pinkerton, M.: Spirochaeta, Spirillum, Leptospira; pp. 243-254 in "Pathology of Simian Primates", Part II (Fiennes, R.N.T-W., ed.) Karger, Basel (1972)
25. Roy, A.D. and H.M. Cameron: Rhinophycomycosis entomopthorae occurring in a chimpanzee in the wild in East Africa. Amer. J. trop. Med. Hyg. *21*, 234-237 (1972)
26. Sato, T. and A. Takeuchi: Infection by spirilla in the stomach of the rhesus monkey. Vet. Path. *19* (Suppl. 7), 17-25 (1982)
27. Schultz, M. and D. Starck: Neue Beobachtungen und Überlegungen zur Pathologie des Primatenschädels. Folia primatol. *28*, 81-108 (1977)
28. Sellards, A.W.: The cultivation of treponemata from the blood of normal monkeys (Macaca rhesus) and from the blood of monkeys infected with yellow fever. Proc. nat. Acad. Sci. (Wash.) *16*, 222-228 (1930)
29. Sepetjian, M.; F.T. Guerraz; D. Salussola; J. Thivolet, et J.C. Monier: Contributions à l'etude du treponeme du singe par A. Fribourg-Blanc. Bull. Wld Hlth Org. *40*, 141-151 (1969)

30. Shive, R.J.; S.S. Green; L.B. Evans, and F.M. Garner: Leptospirosis in barbary apes. J. Amer. vet. med. Ass. *155*, 1176-1178 (1969)
31. Takeuchi, A.; H. Sprinz, and A. Sohn: Intestinal spirochetosis in the monkey and man. Lab. Invest. *24*, 450 (1971)
32. Takeuchi, A. and J.A. Zeller: Scanning electron microscopic observations on the surface of the normal and spirochete-infested colonic mucosa of the rhesus monkey. J. Ultrastruct. Res. *40*, 313-324 (1972a)
33. Takeuchi, A. and J.A. Zeller: Ultrastructural identification of spirochetes and flagellated microbes at the brush border of the large intestinal epithelium of the rhesus monkey. Infect. and Immun. *6*, 1008-1018 (1972b)
34. Thiroux, A. et W. Dufougere: Persistance de l'infection des meninges chez un singe gueri sans médication d'une infection sanguine à spirilles naturelles. Bull. Soc. Path. exot. *3*, 23-24 (1910)
35. Torten, M.: Leptospirosis; pp. 363-421 in "CRC Handbook Series in Zoonoses". (Steele, J.H. Series ed.), Section A: Bacterial, Rickettsial, and Mycotic Diseases, Vol. I (Stoenner, H., W. Kaplan, and M. Torten, eds.). CRC Press, Boca Raton (1979)
36. van Riper, C.D. and J. Finey: Vincent's disease in chimpanzees. J. Amer. vet. med. Ass. *151*, 905-907 (1967)
37. Wilbert, R. et M. Delorme: Sur une spirochetose ictéro-hémorragique du chimpanzé transmissible à l'homme. Ann. Inst. Pasteur *41*, 1139-1155 (1927)

3.2 Spirillaceae

Campylobacter Spp.

Since the first report in 1946 by Levy *Campylobacter fetus/jejuni* have been demonstrated as causative agents of enteritis in man in an increasing number of cases (King, 1957; Wheeler and Borcher, 1961; Kilo et al., 1965; Butzler et al., 1973; Skirrow, 1977; Blaser et al., 1979; Bokkenheuser et al., 1979; Karmali and Fleming, 1979).

In nonhuman primates *C. fetus/jejuni* was first isolated from a rhesus monkey with chronic diarrhea (Valerio et al., 1969) and from a baboon with acute haemorrhagic-necrotic enteritis (Boncyk et al., 1972). As in man, the number of Campylobacter isolations increases every year (Tab. 9).

Table 9. Campylobacter in Nonhuman Primates

Campylobacter fetus withouth further specification	Campylobacter jejuni
Valerio et al., 1969	Tribe et al., 1979
Boncyk et al., 1972	Bauwens and DeMeurichy, 1981
	Kaplan et al., 1981
	Lauwers et al., 1981
	Luechtefeld et al., 1981
	Fox, 1982
	Kaplan et al., 1982
	Bryant et al., 1983
	Goodwin et al., 1983
	Isenbügel et al., 1983
	Morton et al., 1983
	Tribe and Fleming, 1983
	Yoshimura et al., 1984

At least one case of simultaneous campylobacteriosis in nonhuman primates and intestinal disease in the caretaker has been reported (Tribe and Frank, 1980). For a summary of human infections originating from other animals see Fox (1982) and Blaser et al. (1983).

Agents

Campylobacter fetus spp. are microaerophilic (5 to 10% CO_2), curved rods, which are motile by means of a single, polar flagellum. Only *C. jejuni* tends to grow as coccoid forms. *Campylobacter* spp. are gram negative, but usually stain only weakly. Better results are obtained using a 1% aqueous basic fuchsin solution which reveals a characteristic sea-gull-like configuration in both cultivated materials and stool specimen (Park et al., 1983). They grow neither under aerobic nor under strict anaerobic conditions on initial isolation but can be adapted to an aerobic atmosphere in later passages.

The isolation of *Campylobacter* spp. requires special selective procedures, e.g., a filtration technique which depends on the ability of *Campylobacter* to pass through 0.65 μm filter pores; or cultivation on special agarose containing a variety of antibiotics. Most useful agaroses are *Skirrow's medium,* containing vancomycin (10 mg/ml), polymyxin B sulphate B (2,5:U/ml), and trimethoprime lactate (5 μg/ml); *Butzler's medium* with added bacitracin (25 iU/ml), novobiocin (5 μg/ml), actidione (cycloheximide) (50 μg/ml), colistin (10 U/ml) and cetazolin (15 μg/ml); and *Campy BAP* with vancomycin (10 mg/l), polymyxin B (2500 U/l), trimetoprim (5.0 mg/l), amphotericin B (2.0 mg/l) and cephalothin (15.0 mg/l).

However, as most media only insufficiently inhibit the growth of Enterobacteriaceae and *Ps. aeruginosa,* a new medium was developed by Goosens et al. (1983). This consists of Butzler's medium plus 15 mg cetoperazone, 10 mg rifampicin, 10.000 i.a. colistin, and 2 mg amphotericin B per liter of distilled water. Finally, a blood free campylobacter-medium has been developed by Bolton and Coates (1983), though the results are somewhat inferior to those obtained with the other campylobacter media. The isolation rate of *C. jejuni* from fecal specimens is enhanced by overnight refrigeration in Campy-thio broth before streaking onto one of the Campy-agars (Luechtefeld et al., 1981b; Rubin and Woodard, 1983) or by placing the first sample in a selective enrichment medium (Martin et al., 1983; Barot and Bokkenhuiser, 1984). The best results are obtained with Doyle and Roman broth (Rothenberg et al., 1984).

For transport of the stool specimen to the laboratory, transport media such as Cary Blair medium, thioglycolate broth or Campy-thio broth with refrigeration temperatures are recommended during transportation (Wang et al., 1983).

The colonies growing on the surface of the agar-plates are usually smooth, round and colourless, but growth as the R-form occurs sometimes as opaque, cut-glass-like, cream-coloured, mucoid colonies. The temperature requirements for *in vitro* cultivation depends on the *Campylobacter* subspecies involved (Table 10).

C. jejuni is oxidase positive, produces catalase, reduces nitrate to nitrite and is intolerant of 3.5% NaCl in growth media. Campylobacters do not metabolize carbohydrates, but utilize the tricarbonic acid cycle. *C. jejuni* is hippurate-positive in contrast to *C. coli* (Munroe et al., 1983; Hebert et al., 1983). Most *C. coli* are resistant to pyronin, but *C. jejuni* is not (Bolton et al., 1984). The terms *C. jejuni* and

Table 10. Differential Characteristics of Campylobacter Fetus

	C. fetus subsp. *Fetus*		*C. fetus* subsp. *intestinalis*		*C. fetus* subsp. *jejuni*
Group	A-1	A sub-1	A-2	B	C
Biotype	1	Sub 1	2	2	Not designated
Heat-stable antigen	A	A	A	B	C
Sensitive H_2S[a]	−	+	+	+	+
Glycine tolerance[b]	−	−	+	+	V
Growth at 25 °C	+	+	+	+	−
Growth at 42 °C	V[c]	V[c]	V[c]	V[c]	+
Growth at 45 °C	−	−	−	−	+ (−)[d]

[a] Test is conducted by suspending paper strips satuated with lead acetate over semi-solid medium containing cysteine HCl or L-cystine.
[b] Semi-solid medium containing 1% glycine used.
[c] Valuable results.
[d] Ninety per cent or more of the isolates are positive.

C. coli come from a historical basis. *C. jejuni* have originally been isolated from the jejunum of cattle with enteritis. *C. coli*, on the other hand, was first isolated from colonic contents of pigs with dysentery.

Campylobacter contain heat stable antigens A, B, and C. Considering diagnostic aspects *Campylobacter* probably possess O, H, and K antigens (McCoy et al., 1976a, b). Typing *C. jejuni* can be done using passive haemagglutination (Penner and Hennessy, 1980; Penner et al., 1982; 1983; Lauwers, 1982), tube agglutination (Butzler and Skirrow, 1979), slide agglutination (Rogol et al., 1983; Lior and Butzler, 1986), bactericidal assays (Karmali and Fleming, 1979), indirect immunofluorescence (Karmali and Fleming, 1979) and immuno-electrophoresis (Kosunen et al., 1980). So far 34 serotypes have been recognized with types 8, 11, 12, 18 and 3 being the most common (Munroe et al., 1983; Rogol et al., 1983). *C. fetus* survives and multiplies in bile at 37 °C for 2 months; in feces, milk, water and urine it survived better at 4 °C than at 25 °C. Its maximal viability in feces or milk being 3 weeks at 4 °C, in water 4 weeks, and in urine 5 weeks (Blaser et al., 1980).

Epidemiology

C. jejuni, the only species reported in nonhuman primates so far, is normally carried by sheep. It is transmitted orally through contaminated food or water or direct contact with infected animals. Milk borne outbreaks are well-known in man. However, the route of contamination of the milk is unknown. In nonhuman primates *C. jejuni* has also been isolated from gastric contents of aborted fetuses and from placentas in abortions (Tribe and Fleming, 1981), suggesting uterine infections.

Perinatal infections also occur in children (Blaser et al., 1983b).

C. jejuni has been proven to induce carrier stages in carrion-eating birds thus enabling rapid spread over large distances. It also sometimes infects the human gallbladder (Schwartz et al., 1966) followed by prolonged shedding of the organisms in the feces of such persons. However, silent *Campylobacter* infections occur less commonly in man than they are reported in nonhuman primates (Tribe and Frank, 1980; Tribe and Fleming, 1983; Morton et al., 1983).

Experimental infections of *M. mulatta* required at least 106 live *Campylobacter jejuni* to induce clinical illness (Fox, 1982).

Pathogenesis

The pathogenesis of *Campylobacter jejuni*-enteritis is not clearly understood. Invasion of the intestinal mucosa is suggested (Blaser et al., 1983a; Butzler and Skirrow, 1979). The *Campylobacter* reaching the propria are phagocytosed mainly by mononuclear cells. However, only 20 to 30% of the patients show evidences of intestinal invasion such as blood, mucus or leukocytes in the stools. In the majority of cases the illness is associated with a secretory diarrhea which suggests another mechanism.

C. jejuni has also been shown to produce a heat-labile cholera-like toxin and a heat-stable toxin in vitro (McCardell et al., 1984; Ruiz-Palacios et al., 1983; Johnson and Lior, 1984). The heat-labile toxin induces intraluminal fluid secretion in rat ileal loops, but not in rabbit ileal loops. It induces elongation and increased intracellular AMP-levels in CHO-cells (Ruiz-Palacios et al., 1983). It is immunologically related to cholera/coli groups of heat-labile enterotoxins (Klipstein and Engert, 1984a). Strains of *C. jejuni* vary in their capacity to elaborate a heat-labile enterotoxin (Klipstein and Engert, 1984 b).

Clinical Symptoms

Nonhuman primates: Many *C. jejuni*-infections of nonhuman primates are asymptomatic (Tribe and Fleming, 1983; Morton, 1983). Clinical disease in others is characterized by passing of semisolid stools or even mucohaemorrhagic diarrhea (Boncyk et al., 1972; Tribe and Fleming, 1983; Bryant et al., 1983; Goodwin et al., 1983). This latter sometimes is of an intermittent type with repeated episodes (Luechtefeld et al., 1981). Long lasting diarrhea may lead to rectal prolapses (Bryant et al., 1983). Microhaematocrit and haemoglobin values are subnormal, leukocyte counts and blood urea nitrogen are elevated (Goodwin et al., 1983). 50% of experimentally infected young rhesus monkeys were anorectic on the 4th and 5th days. Most animals had pyrexia on days 2 to 5, with soft stools or diarrhea on days 3 and 4 (Fitzgeorge et al., 1981).

Man: Symptomatic infection due to *C. jejuni* occur as a gastrointestinal illness. Mild infections may cause symptoms similar to those of viral gastroenteritis lasting for just one day. Severe disease, on the other hand, may appear as a fulminant illness, which may mimic the acute relapses of ulcerative colitis. Illness often begins with fever, headache, myalgia, arthralgia, and backache for one to two days. This is followed by diarrhea, abdominal pain, malaise, nausea and vomiting. Most patients have occult blood in the stools, these may even be grossly bloody. Most patients recover in less than a week, but relapses are common. Extraintestinal manifestations include cholecystitis, pancreatitis, cystitis and arthritis. The disease may be fatal in debilitated or elderly patients (Blaser et al., 1983). About one-fourth of human cases displayed neurological symptoms, mostly those of meningoencephalitis (Gunderson and Sack, 1971).

About 90% of sera collected from patients recovering from *Campylobacter* enteritis contained specific IgM, IgG and IgA like those usually seen in other acute infections, whereas in patients with intermittent or prolonged enteritis the antibody response was rather poor (Kaldor et al., 1983).

Pathology

Nonhuman primates: The lesions in the acute disease consisted of serous exudation into the bowel lumen with only mild inflammatory responses, and frequently a low graduate peritonitis. Severe cases were characterized by haemorrhagic-necrotizing enteritis (Boncyk et al., 1972; Tribe and Fleming, 1981).

In *Erythrocebus patas* with chronic *Campylobacter* diarrhea the most prominent lesions were thickened, turgid terminal ileal walls, with the distal jejunum and the ileum being filled with semisolid to watery contents. The mesenteric lymphnodes of these animals were grossly enlarged. Histopathologically the villi of jejunum and ileum were short and thick and lined with low columnar to cuboidal epithelium. The lacteals were dilated, the propria was infiltrated with large, eosinophilic macrophages, lymphocytes and necrotic cells. Epithelial hyperplasia was seen in the rectum and colon, whilst crypt abscesses were present in the colonic wall (Bryant et al., 1983).

Man: In biopsies taken at laparotomy, acute inflammation of the jejunum has been seen (DuPont and Pickering, 1980). Blaser et al. (1980) found *C.jejuni* to be associated with acute colitis.

Treatment

The treatment of choice is ciprofloxacin, a new fluoroquinoline, given perorally twice a day for 5 days (dosage in marmosets: 5 mg twice daily). Ciprofloxacin prevented relapses (Goodman et al., 1986). Less effective is erythromycin (Blaser and Reller, 1981; Holländer, 1983; Shanker and Sorrell, 1983) given in doses of 250 mg twice daily for 10 days to *E.patas* (Bryant et al., 1983). In nonhuman primates despite erythromycin curing the symptoms recurrences were later observed (Tribe and Fleming, 1981; Bryant et al., 1983) with excretion of live *Campylobacter* resumed within 10 days after the end of therapy (Goodman et al., 1986).

Tylosin can also be used, but it is even less reliable than erythromycin (Tribe and Fleming, 1983). Gentamycin and furazolidone have been proven to be unsuccessful in one outbreak (Luechtefeld et al., 1981 a), whereas they were the most active agents in another study (Fox et al., 1984).

Resistance to e.g. tetracycline is mediated by plasmids (Taylor et al., 1983; Tenover et al., 1985). There was, however, no homology noted between these plasmids and the tetracycine R-factors of *Enterobacteriaceae* (Tenover et al., 1985).

References

1. Barot, M.S. and V.D. Bokkenheuser: Systematic investigation of enrichment media for wild-type *Campylobacter jejuni* strains. J.Clin. Microbiol. *20*, 77-80 (1984)
2. Bauwens, L.and W.DeMeurichy: The occurrence of thermophilic campylobacters in zoo animals. Acta zool. path. antverp. *76*, 181-189 (1981)
3. Blaser, M.J.; I.D. Berkowitz; F.M. LaForce; J.Cravens; L.B. Reller, and W.-L.L. Wang: Campylobacter enteritis: Clinical and epidemiologic features. Ann. intern. Med. *91*, 179-185 (1979)
4. Blaser, M.J.; D.J. Duncan; G.Warren, and W.-L.L. Wang: Experimental *Campylobacter jejuni* infection of adult mice. Infect. and Immun. *39*, 908-916 (1983 a)
5. Blaser, M.J.; H.L. Hardesty; B.Powers, and W.-L.L. Wang: Survival of *Campylobacter fetus* subsp. *jejuni* in biological milieus. J.clin. Microbiol. *11*, 309-313 (1980)

6. Blaser, M.J.; R.B. Parsons, and W.L. Wang: Acute colitis caused by Campylobacter fetus v.jejuni. Gastroenterology *78*, 448-453 (1980)
7. Blaser, M.J. and L.B. Reller: Campylobacter enteritis. New Engl. J.Med. *305*, 1444-1452 (1981)
8. Blaser, M.J.; D.N. Taylor, and R.A. Feldman: Epidemiology of *Campylobacter jejuni* infection. Epidemiol. Rev. *5*, 157-176 (1983b)
9. Bokkenheuser, V.D.; N.J. Richardson; J.H. Bryner; P.J. Roux; A.B. Schutte; H.J. Koornhof; J.Freiman, and E.Hartman: Detections of enteric campylobacterosis in children. J.clin. Microbiol. *9*, 227-232 (1979)
10. Bolton, F.J. and D.Coates: Development of blood-free *Campylobacter* medium; screening tests on basal media and supplements, and the ability of selected supplements to facilitate aerotolerance. J.appl. Bact. *54*, 115-125 (1983)
11. Bolton, F.J.; A.V. Holt, and D.N. Hutchinson: Campylobacter biotyping scheme of epidemiological value. J.clin. Pathol. *37*, 677-681 (1984)
12. Boncyk, L.H.; M.Brack, and S.S. Kalter: Hemorrhagic-necrotic-enteritis in a baboon *(Papio cynocephalus)* due to *Vibrio fetus*. Lab. Anim. Sci. *22*, 734-738 (1972)
13. Bruce, D.and W.Zuchowski: Campylobacter infections in cats and dogs. Vet. Rec. *107*, 200-201 (1980)
14. Butzler, J.P.; P.Dekeyser; M.Detrain, and F.Dehaen: Related vibrios in stools. J.Pediat. *82*, 493-495 (1973)
15. Butzler, J.P. and M.B. Skirrow: Campylobacter enteritis. Clin. Gastroent. *8*, 737-765 (1979)
16. Bryant, J.L.; H.F. Stills; R.H. Lentsch, and C.C. Middleton: *Campylobacter jejuni* isolated from patas monkeys with diarrhea. Lab. Anim. Sci. *33*, 303-305 (1983)
17. DuPont, H.L. and L.K. Pickering: Infection of the Gastrointestinal Tract. Microbiology, Pathophysiology, and Clinical Features. Plenum, New York (1980)
18. Firehammer, B.D.: Campylobacteriosis. Pp.295-308 in "CRC Handbook Series in Zoonoses" (Steele, J.H., Series ed.), Section A: Bacterial, Rickettsial, and Mycotic Diseases, Vol. I (Stoenner, H.; W.Kaplan, and M.Torten, Section eds.), CRC Press, Boca Raton (1979)
19. Fitzgeorge, R.B.; A.Baskerville, and K.P. Lander: Experimental infection of rhesus monkeys with a human strain of *Campylobacter jejuni*. J.Hyg. *86*, 343-351 (1981)
20. Fox, J.G.: Campylobacteriosis - A "new" disease in laboratory animals. Lab. Anim. Sci. *32*, 625-637 (1982)
21. Fox, J.G.; J.L. Dzink, and J.I. Ackermann: Antibiotic sensitivity patterns of *Campylobacter jejuni/coli* isolated from laboratory animals and pets. Lab. Anim. Sci. *34*, 264-267 (1984)
22. Goodman, L.J.; R.L. Kaplan; R.M. Petrak; R.M. Fliegelman; D.Taff; F.Walton; J.L. Penner, and G.M. Trenholme: Effects of erythromycin and ciprofloxacin on chronic fecal excretion of *Campylobacter* species in marmosets. Antimicrob. Agents Chemother. *29*, 185-187 (1986)
23. Goodwin, T.; M.R. Adams; N.D.M. Lehner, and C.P. Jerome: Campylobacter enteritis in patas monkeys. Lab. Anim. Sci. *33*, 481 (1983)
24. Goosens, H.; M.de Boeck, and J.P. Butzler: A new selective medium for the isolation of *Campylobacter jejuni* from human faeces. Europ. J.clin. Microbiol. *2*, 389-394 (1983)
25. Hebert, G.A.; D.G. Hollis; R.E. Weaver; A.G. Steigerwalt; R.M. McKinney, and D.J. Brenner: Serogroups of *Campylobacter jejuni, Campylobacter coli,* and *Campylobacter fetus* defined by direct immunofluorescence. J.clin. Microbiol. *17*, 529-538 (1983)
26. Holländer, R.: Die Aktivität von 23 Chemotherapeutika in vitro gegen aus Stuhlproben isolierte *Campylobacter jejuni/coli*-Stämme. Zbl. Bakt. Hyg. A *256*, 196-201 (1983)
27. Isenbügel, E.; R.Largo, and P.Fehlmann: *Campylobacter jejuni*-Infektionen bei jungen Primaten. Pp. 281-285 in 25.Internat. Symp. Erkr. Zootiere, Wien (1983)
28. Johnson, W.M. and H.Lior: Toxins produced by *Campylobacter jejuni* and *Campylobacter coli*. Lancet 1984, 229-230 (1984)
29. Kaldor, J.; H.Pritchard; A.Serpell, and W.Metcalf: Serum antibodies in Campylobacter enteritis. J.clin. Microbiol. *18*, 1-4 (1983)
30. Kaplan, R.L.; L.J. Goodman; J.E. Barrett, and W.Landau: Comparison of rectal swabs to stool cultures for the detection of *Campylobacter fetus* subsp. *jejuni*. Abstr. Ann. Meet. Amer. Soc. Microbiol. *81*, 298 (1981)
31. Kaplan, R.L.; L.J. Goodman, J.E. Barrett; G.M. Trenholme, and W. Landau: Comparison of

rectal swabs and stool cultures in detecting *Campylobacter fetus* subsp. *jejuni*. J.clin. Microbiol. *15*, 959-960 (1982)

32. Karmali, M.A. and P.C. Fleming: Campylobacter enteritis in children. J.Pediat. *94*, 527-533 (1979)
33. Kilo, C.; O.Hagemann, and J.Marzi: Septic arthritis and bacteremia due to *Vibrio fetus*. Amer. J.Med. *38*, 962-971 (1965)
34. King, E.O.: Human infections with *Vibrio fetus* and a closely related *Vibrio*. J.infect. Dis. *101*, 119-128 (1957)
35. Klipstein, F.A. and R.F. Engert: Purification of *Campylobacter jejuni* enterotoxin. Lancet 1984, 1123-1124 (1984 a)
36. Klipstein, F.A. and R.F. Engert: Properties of crude *Campylobacter jejuni* heat-labile enterotoxin. Infect. and Immun. *45*, 314-319 (1984b)
37. Kosunen, T.U.; D.Danielsson, and J.Kjellander: Serology of *Campylobacter fetus* ss. *jejuni* (related campylobacters). Acta path. Microbiol. scand., Sect. B *88*, 207-218 (1980)
38. Lauwers, S.: Serotyping of *C.jejuni*: A useful tool in the epidemiology of *Campylobacter diarrhoea*, pp. 96-97 in "Campylobacter: Epidemiology, pathogenesis and biochemistry (Newell, D.G., ed.) M.T.P. Press Ltd., Lancaster/England (1982)
39. Lauwers, S.; E.Berge; A.Naessens; J.P. Butzler: Monkeys as a reservoir for *Campylobacter jejuni*. Abstr. Ann. Meet. Amer. Soc. Microbiol. *81*, 298 (1981)
40. Levy, A.J.: A gastroenteritis outbreak probably due to a bovine strain of vibrio. Yale J.Biol. Med. *18*, 243-258 (1946)
41. Lior, H.and J.P. Butzler: Serotyping Campylobacter. Lancet I, 1381-1382 (1986)
42. Luechtefeld, N.W.; R.C. Cambre, and W.-L.L. Wang: Isolation of *Campylobacter fetus* subsp. *jejuni* from Zoo Animals. J.Amer. vet. med. Ass. *179*, 1119-1122 (1981)
43. Luechtefeld, N.W.; W.-L.L. Wang; M.J. Blaser, and L.B. Reller: Evaluation of transport and storage techniques for isolation of *Campylobacter fetus* subsp. *jejuni* from turkey fecal specimen. J.clin. Microbiol. *14*, 438-443 (1981)
44. Martin, W.T.; C.M. Patton; G.K. Morris; M.E. Potter, and N.D. Puhr: Selective enrichment broth medium for isolation of *Campylobacter jejuni*. J.clin. Microbiol. *17*, 853-855 (1983)
45. McCardell, B.A.; J.M. Madden, and E.C. Lee: Production of cholera-like toxin by *Campylobacter jejuni/coli*. Lancet 1984, 448-449 (1984)
46. McCoy, E.C.; H.A. Wiltberger, and A.J. Winter: Antibody mediated immobilization of *Campylobacter fetus*: Inhibition of a somatic antigen. Infect. and Immun. *13*, 1266-1272 (1976a)
47. McCoy, E.C.; H.A. Wiltberger, and A.J. Winter: Major outer membrane protein of *Campylobacter fetus*: Physical and immunological characterization. Infect. and Immun. *13*, 1258-1265 (1976b)
48. McFadyean, J.and S.G. Stockmann: Epizootic Abortion. Britain Board of Agriculture and Fish. Rept. Dept. Comm. London (1913)
49. Morton, W.R.; M.Bronsdon; G.Mickelsen; G.Knitter; S.Rosenkranz; L.R. Kuller, and D.Sajuthi: Identification of *Campylobacter jejuni* in *Macaca fascicularis* imported from Indonesia. Lab. Anim. Sci. *33*, 187-188 (1983)
50. Munroe, D.L.; J.F. Prescott, and J.L. Penner: *Campylobacter jejuni* and *Campylobacter coli* serotypes isolated from chickens, cattle and pigs. J.clin. Microbiol. *18*, 877-881 (1983)
51. Park, C.H.; D.L. Hixon; A.S. Polhemus; C.B. Ferguson; S.L. Hall; C.C. Risheim, and C.B. Cook: A rapid diagnosis of *Campylobacter* enteritis by direct smear examination. Amer. J.clin. Path. *80*, 388-390 (1983)
52. Penner, J.L. and J.N. Hennessy: Passive haemagglutination technique for serotyping *Campylobacter fetus* subsp. *jejuni* on the basis of soluble heat stable antigens. J.clin. Microbiol. *12*, 732-737 (1980)
53. Penner, J.L.; J.N. Hennessy, and M.M. Goodby: Development of a scheme for serotyping *Campylobacter jejuni*; pp. 89-91 in "Campylobacter: Epidemiology, pathogenesis and biochemistry" (Newell, D.G., ed.) M.I.P. Press Ltd., Lancaster/England (1982)
54. Penner, J.L.; J.N. Hennessy; S.D. Millis, and W.C. Bradbury: Application of serotyping and chromosomal restriction endonuclease digest analysis in investigating a laboratory-acquired case of *Campylobacter jejuni* enteritis. J.clin. Microbiol. *18*, 1427-1428 (1983)
55. Rogol, M.; I.Sechter; I.Braunstein, and C.B. Gerichter: Extended scheme for serotyping

Campylobacter jejuni: Results obtained in Israel from 1980 to 1981. J. clin. Microbiol. *18*, 282-286 (1983)
56. Rothenberg, P.J.; N.J. Stern, and D.C. Westhoff: Selected enrichment broths for recovery of *Campylobacter jejuni* from foods. Appl. Environ. Microbiol. *48*, 78-80 (1984)
57. Rubin, S.J. and M. Woodard: Enhanced isolation of *Campylobacter jejuni* by cold enrichment in campy-thio-broth. J. clin. Microbiol. *18*, 1008-1010 (1983)
58. Ruiz-Palacios, G.M.; J. Torres; N.I. Torres; E. Escamilla; B.R. Ruiz-Palacios, and J. Tamayo: Cholera-like enterotoxin produced by *Campylobacter jejuni*. Lancet 1983, 250-253 (1983)
59. Schwartz, R.; E. Hirsch; J.E. Mule, and L. Bluestone: Antral mucosal diaphragm. - First reported case of *Vibrio fetus* in human bile. Amer. J. Gastroent. *45*, 366-373 (1966)
60. Shanker, S. and T.C. Sorrell: Susceptibility of *Campylobacter jejuni* to twenty three antimicrobial agents. Pathology *15*, 61-63 (1983)
61. Skirrow, M.B.: Campylobacter enteritis - A "new" disease. Brit. med. J. *2*, 9-11 (1977)
62. Smith, T. and M.S. Taylor: Some morphological and biological characters of the spirilla (*Vibrio fetus*, N. Sp.) associated with disease of the fetal membrane in cattle. J. exp. Med. *30*, 299-311 (1919)
63. Taylor, D.E.; R.S. Garner, and B.J. Allan: Characterization of tetracycline resistance plasmids from *Campylobacter jejuni* and *Campylobacter coli*. Antimicrob. Agents Chemother. *24*, 930-935 (1983)
64. Tenover, F.C.; S. Williams; K.P. Gordon; C. Nolan, and J.C. Plorde: Survey of plasmids and resistance factors in *Campylobacter jejuni* and *Campylobacter coli*. Antimicrob. Agents Chemother *27*, 37-41 (1985)
65. Tribe, G.W. and M.P. Fleming: Biphasic enteritis in imported cynomolgus *(Macaca fascicularis)* monkeys infected with *Shigella*, *Salmonella* and *Campylobacter* species. Lab. Anim. *17*, 65-69 (1983)
66. Tribe, G.W. and A. Frank: Campylobacter in monkeys. Vet. Rec. *106*, 365-366 (1980)
67. Valerio, D.A.; R.L. Miller; J.R.M. Innes; D.A. Courtney; A.J. Palotta, and R.M. Guttmacher: *Macaca mulatta*, Management of a Laboratory Breeding Colony. Academic Press, New York (1969)
68. Wang, W.-L.L.; L. Barth-Reller; B. Smallwood; N.W. Luechtefeld, and M.J. Blaser: Evaluation of transport media for *Campylobacter jejuni* in human fecal specimen. J. clin. Microbiol. *18*, 803-807 (1983)
69. Wright, E.: Attempts to isolate *Campylobacter jejuni* from various body sites. J. clin. Path. *36*, 667-669 (1983)
70. Yoshimura, M.; M. Ebukuro; N. Kagiyama; A. Shimuzu, and Y. Saito: Isolation of *Campylobacter jejuni* from experimental dogs and monkeys in Japan. Exp. Anim. *33*, 209-212 (1984)

3.3 Pseudomonadaceae

3.3.1 Pseudomonas Aeruginosa

Ps. aeruginosa causes serious diseases in both man and animals. Although not a truly zoonotic agent, *Ps. aeruginosa* can infect the human environment - or vice versa -, it may also be possibly transmitted by direct contact.

Agent
Ps. aeruginosa is an aerobic, gram negative, straight or slightly curved rod. It is motile by means of usually one flagella at each pole. It grows in mineral base media without growth factors, which contain ammonium ions as the sole source of nitrogen, with glucose as the source of carbon and energy. On blood agar it

usually produces haemolysis. After 24 hrs of incubation at 37 °C, flat colonies of 2 to 3 mm diameter with a slightly bluish color and a characteristic sweetish odor appear.

The main biochemical characteristics are (Pulverer and Korth, 1962; Devos et al., 1966; Hugh and Gilardi, 1974):

Table 11. Biochemical Characteristics of *Pseudomonas aeruginosa*

Cytochromoxidase	+	Glucose	+ (Ø)	Raffinose	Ø
Katalase	+	Maltose	Ø	Salicin	Ø
Peroxidase	+	Dulcitol	Ø	D-Sorbitol	Ø
		Galactose	+	Citrate	+
		Inulin	Ø (+)	Indophenol	+
		Lactose	Ø	Lysine	Ø
		D-Mannose	+ (−)	Saccharose	Ø
		H_2S	+	Urea	Ø
		Arabinose	+ (Ø)	Indol	Ø

Ps. aeruginosa is the only *Pseudomonas* sp. known to produce pyocyanin, a water- and chloroform soluble, nonfluorescent, blue phenazin pigment. It may also synthesize various combinations of pyocyanin, pyoverdin (fluorescent), pyorubin and pyomelanin (Hugh and Gilardi, 1974).

Subtyping can be achieved using serotyping (Verder and Evans, 1961; Homma et al., 1970; Yabuuchi et al., 1971), bacteriophage typing (Sutter et al., 1965) and typing by pyocine production (Darrell and Wahba, 1964; Osman, 1965; Gillies and Govan, 1966); all of which have been used for epidemiological studies. There is evidence that certain types of *Ps. aeruginosa* vary in their virulence for human beings (Bobo et al., 1973).

Ps. aeruginosa survives in the environment up to 6 months independent of the relative humidity (Wahba, 1965).

Epidemiology

Ps. aeruginosa is a common ubiquitous inhabitant of soil. The normal mammalian intestine does not appear to be a major habitant, although *Ps. aeruginosa* is found in approximately 10% of normal human stools (Hugh and Gilardi, 1974). It is also found in normal appearing animals (Srivastava and Mathews, 1980), particularly in some nonhuman primate colonies (Anon, 1983). *Ps. aeruginosa* has been isolated from hospital equipment, water faucets, floors, baths and sinks. *Ps. aeruginosa* on the other hand infects wounds, sinuses, the external ear and sometimes also lungs, causing sepsis in untreated patients.

Ps. aeruginosa often flourishes after antibiotic treatment which eradicates the normal bacterial flora keeping the *Pseudomonas* in check. Therefore, for control of intestinal *Ps. aeruginosa* infections antibiotics should be discontinued to allow restoration of the normal flora (Yow, 1952).

Pathogenesis

Ps. aeruginosa if it invades tissues, leads to ulcers, necrosis and inflammation in various organs. It has also been shown to produce a heat-labile enterotoxin that causes accumulation of fluid in ligated loops of rabbit ileum (Kubota and Liu, 1971).

Clinical Symptoms

Nonhuman primates: The disease caused by *Ps. aeruginosa* in nonhuman primates varies from endocarditis (Stills et al., 1979) to air sacculitis (Strobert and Swenson, 1979), pneumonia and enteritis (Deinhardt et al., 1967; Anon, 1983; own results) to pyelonephritis (Fiennes, 1967), peritonitis (Schüppel et al., 1980) or endometritis (Schneider, 1984).

Man: The spectrum of diseases in man ranges from wound infections to otitis externa, urinary tract infections, diarrhea, acute or chronic respiratory disease and meningitis (Stanley, 1947; Hardy et al., 1954; Curtin et al., 1961). In eye infections *Ps. aeruginosa* can be rapidly destructive to the corneal tissues leading to corneal ulcers and panophthalmitis. *Ps. aeruginosa* is of great importance particularly in children, causing enteritis, otitis including mastoiditis, rhinitis, conjunctivitis and skin-eruptions (ecthyma gangraenosum) (Struwe, 1964).

Pathology

Nonhuman primates: Lung lesions were reported as an acute necrotizing-haemorrhagic bronchopneumonia with extensive vascular destruction (Deinhardt et al., 1967; Stills et al., 1979; personal experience). The cardiac lesions secondarily associated with *Ps. aeruginosa*-pneumonias, consisted of acute vegetative endocarditis or acute myocarditis with focal to extensive liquefaction necrosis of the myocardium and of the intramural coronary arteries (Deinhardt et al., 1967; Stills et al., 1979; Schüppel et al., 1980).

Man: Ear infections produce otitis externa, otitis media, mastoitis and pericarditis, all with pronounced tissue damage (Stevenson, 1964). Meningitis is usually of the purulent type (Struwe, 1964).

Children which died of *Ps. aeruginosa* suffered from ulcerative enterocolitis leading to perforation and peritonitis (Paul and Marget, 1963). Ectyma gangraenosum consists of purple, raised, confluent areas of the skin which show a tendency for ulceration and necrosis (Struwe, 1964).

Treatment

Ps. aeruginosa infections in man are known for their resistance to antibiotic treatment. Antibiotic resistance of *Ps. aeruginosa* is of an intrinsic nature, the mechanism of which is more likely to be a feature of cell wall construction in contrast to the plasmid-mediated, extrinsic antibiotic resistance of enterobacteriaceae (Rhame, 1980; Godfrey and Bryan, 1984).

Polymyxin B and its analogues (colistin) are considered most effective. 96% of *Ps. aeruginosa* strains were found to be sensitive in vitro to polymyxin B-sulphate; 88% to gentamycin; 28% to streptomycin and 14% to kanamycin (Srivastava and Mathews, 1980). *In vivo* studies in *M. mulatta* revealed good recovery rates after 2 mg/kg/day of colistin or 50 mg/kg/day of carbenicillin; lower dosages were less effective (Saslaw et al., 1973).

3.3.2 Pseudomonas Pseudomallei (Malleomyces Ps., Loefflerella Ps., Pfeifferella Ps.)

Ps. pseudomallei is the causative agent of melioidosis or pseudoglanders, a disease which has been known in morphine addicts and in people working in the rice paddies of south-east Asia since 1910 (Whitmore, 1913). When due to troop movements, during and after World War II cases of melioidosis occurred in military personnel outside Asia the disease became increasingly known to western medicine and due to the long incubation period was sometimes called the "Vietnamese medical time bomb" (Groves, 1979).

Unlike the true glanders agent *Ps. pseudomallei* in animals is not primarily associated with equines but occurs in swine, sheep, goats and other animals as well. Most cases of human or animal melioidosis have originated in south-east Asia and tropical Australia, although because of the long incubation period some became clinically apparent outside the endemic areas.

Several cases have been reported in captive nonhuman primates. So far, no animal to man transmission has occurred and only one direct man to man transmission has been documented (McCormick et al., 1975 - cit. from Groves, 1979). Melioidosis is not a true zoonosis, but infected animals may contaminate the human environment.

Agent
Ps. pseudomallei is a motile, aerobic, Gram-negative rod of 1 to 2 × 0.5 μm size, which grows well on ordinary mineral base media at 37 to 42 °C with a characteristic mouldy odor. Older colonies are distinctively wrinkled or sometimes bizarre (Hugh and Gilardi, 1974; Redfearn and Palleroni, 1975). *Ps. pseudomallei* has polar tufts of three or more flagella. It metabolizes cellobiose and lactose, and usually also maltose, inositol, D-sorbitol and dulcitol. It is gelatin- and 2-ketogluconate negative and does not produce pyocyanin or pyoverdin.

Agglutination and fluorescent-antibody reactions are helpful for the identification of *Ps. pseudomallei.* C.F. or indirect haemagglutination tests may be used in sero-epidemiological studies (Groves, 1979).

Guinea pigs are very susceptible to *Ps. pseudomallei,* as few as 10 organisms are fatal after intraabdominal inoculation.

Epidemiology
The natural reservoirs of *Ps. pseudomallei* are the soil and contaminated water (Groves, 1979). *Ps. pseudomallei* is naturally distributed throughout Burma, Malaysia, Vietnam, Singapore, Cambodia, Thailand, Java, Borneo, New Guinea and northern Australia. Sporadic cases have been reported from Central India, Sri Lanka, Niger, Chad, Madagascar, USA, Ecuador, Panama, Bahamas and in Iran even well above 30 °N latitude (Redfearn et al., 1966; Redfearn and Palleroni, 1975; Spilsbury et al., 1977). The infection in these cases was apparently contracted in endemic areas.

Ps. pseudomallei is isolated from surface water more frequently following heavy rains (Strauss et al., 1969b). It was also isolated after an outbreak in

M. mulatta from river water flowing through a zoo (Strauss et al., 1969 c). The diseases in man and animals in endemic areas are coincident with the rainy season (Laws and Hall, 1964; Johnson, 1967). Contaminated surface waters are the most likely vehicle for transmission (Fournier 1960; Strauss et al., 1969 a).

The exact mode of transmission of *Ps. pseudomallei* to man and animals has not yet been established, percutaneous or respiratory routes have been suggested (Piggott and Hochholzer, 1970). Whether or not the experimental transmission by means of rat flea *Xenopsylla cheopis* and mosquitoes *(Aedes aegypti)* plays any role under natural circumstances remains to be proved, but the organisms multiplied in the insect's intestinal tracts, rendering the insect infective for 50 days (Laws and Hall, 1964).

In clinical melioidosis the organism has been recovered from nasal discharges, urine, uterine discharges and dermal abscesses (see Groves, 1979); therefore infected human beings and animals should be regarded as potentially hazardous.

Clinical Symptoms

Nonhuman primates: Melioidosis has been observed in chimpanzees (Alexander, 1970; Kaufmann et al., 1970; Butler et al., 1971 - cit. from Groves, 1979); orangutans (Tammemagi and Johnston, 1963; Kaufmann et al., 1970; Smith and Damit, 1982), a gibbon (Lim and Mukundan, 1968), a spider monkey (Lim and Mukundan, 1968); *Macaca mulatta, M. nemestrina* and *M. arctoides* (Anon, 1969 a, b; 1970; Strauss et al., 1969 c; Alexander et al. 1970; Kaufmann, 1970) and a *Presbytis melalophos* (Mutalib et al., 1984).

Melioidosis in nonhuman primates is presented as either broncho-pneumonia or as a cutaneous disease characterized by subcutaneous abscesses, pyodermia and lymphadenitis. One case of osteomyelitis has been reported as well (Britt et al., 1981).

Man: Melioidosis in man mimics a variety of other diseases. The pulmonary form mimics pulmonary tuberculosis in many aspects. Patients usually suffer from upper-lobe pulmonary lesions, productive cough, blood tinged sputum, chest pain, fever and anemia. This form, if untreated, leads to fatal septicemic melioidosis with multiple abscesses throughout various internal organs.

Extrapulmonary or localized melioidosis is characterized by osteomyelitis, arthritis or subcutaneous abscesses. In one unusual case the signs of melioidosis were limited to meningoencephalitis (Spilsbury et al., 1977).

The most common form seems to be a subclinical or undiagnosed inapparent chronic infection. Subclinical melioidosis or chronic inapparent melioidosis may both exacerbate months or years later - cases have been reported with clinical melioidosis up to 24 years after leaving an endemic area (see Groves, 1979).

Pathology

Nonhuman primates: Yellowish coloration of the subcutaneous tissues has been reported in an orangutan and a macaque as characteristic lesions (Tammemagi and Johnston, 1963; Retnasabapathy and Joseph, 1966), otherwise cutaneous melioidosis in nonhuman primates is marked by subcutaneous abscesses filled with a milky, white material (Anon, 1969 a, b; Strauss et al., 1969 a; Douglas et al., 1971).

Pulmonary melioidosis expresses itself as bronchopneumonia affecting all lobes (Smith and Damit, 1982), as lobar pleuropneumonia or Tbc-like whitish-confluent nodules containing fine, grey, granular pus (Tammemagi and Johnston, 1963; Retnasabapathy and Joseph, 1966; Mutalib et al., 1984). At least the pulmonary melioidosis is associated with hepatomegaly. Generalization of the infection resulted in abscesses in liver, spleen, and lymphnodes (Anon, 1969b).

Man: Very acute infections are characterized by numerous small abscesses in many internal organs, including the brain. The abscesses containing polymorphonuclear leucocytes and histiocytes are associated with much fibrinous exudate and necrosis.

In older lesions the inflammatory cells are replaced by increasing numbers of histiocytes and megakaryocyte-like giant cells. Around pulmonary abscesses a considerable fibrinous edema extended into the surrounding unaffected alveolar tissues. Finally, in chronic melioidosis the abscess cavities of varying size contain greenish pus. The centers of the abscesses are filled with necrotic debris and some polymorphonuclear leucocytes, surrounded by a wall of histiocytes and occasional giant cells, plasma cells, lymphocytes, and fibrous tissue (Spencer, 1973; Spilsbury et al., 1977).

Treatment

Melioidosis is difficult to treat in man and animals.

Trimethoprim and sulphadoxine could not prevent the fatal outcome in a young orangutan (Smith and Damit, 1982). Sequentially given penicillin, cephalotin and gentamicin were equally unsuccessful in a 61 year old American tourist (Spilsbury et al., 1977).

According to an editorial note (Anon, 1977), therapy with a combination of tetracycline and chloramphenicol has been recommended.

References

1. Alexander, A.D.: A serologic survey of macaques for *Pseudomonas pseudomallei* antibodies. Primate Zoonoses Surveill. Rep. *1*, 14-20 (1970)
2. Anon: Melioidosis in an imported monkey. Lab. Primate Newsletter *8* (3), 21-22 (1969a)
3. Anon: Report on another case of melioidosis in a monkey. Lab. Primate Newsl. *8* (4), 10-11 (1969b)
4. Anon: Spontaneous melioidosis in a monkey: Report on another case. Lab. Primate Newsl. *9* (2), 3 (1970)
5. Anon: Editorial note - melioidosis - Pennsylvania. CDC Morb. Mortal. Wkly. Rep. *25* (52), 419-420 (1977)
6. Anon: Infection rate of *Pseudomonas aeruginosa* in cynomolgus monkeys and the isolated serovar. Tsukuba Primate Center News *2* (1), 5 (1983)
7. Bobo, R.A.; E.J. Newton; L.F. Jones; L.H. Farmer, and J.J. Farmer III: Nursery outbreak of *Pseudomonas aeruginosa*: Epidemiological conclusions from five different typing methods. Appl. Microbiol. *25*, 414-420 (1973)
8. Britt, J.O.; E.B. Howard; C.J. Kean, and J.Jones: Melioidotic osteomyelitis in an imported primate. J.Amer. vet. med. Ass. *179*, 1303-1306 (1981)
9. Curtin, J.A.; R.C. Petersdorf, and I.L. Bennett jr.: *Pseudomonas* bacteremia; Review of ninety-one cases. Ann. intern. Med. *54*, 1077-1107 (1961)

10. Darrell, J.H. and A.H. Wahba: Pyocine-typing of hospital strains of *Pseudomonas pyocyanea.* J.clin. Path. *17*, 236-242 (1964)
11. Deinhardt, J.B.; J.Devine; M.Passovoy; R.Pohlman, and F.Deinhardt: Marmosets in the laboratory, pathology and outline of statistical evaluation of data. Lab. Anim. Care *17*, 11-29 (1967)
12. Devos, A.; J.van Impe; N.Viaene, and L.Spanoghe: *Pseudomonas aeruginosa* infectie bij Chinchilla's. Vlaams Diergeneesk Tijdschr. *35*, 222-232 (1966)
13. Douglas, J.D.; R.J. Cronin, and A.F. Kaufmann: Spontaneous melioidoses in recently imported monkeys. Pp. 742-747 in "Medical Primatology 1970" (Goldsmith, E.I. and J.Moor-Jankowski, eds.) Karger, Basel (1971)
14. Fiennes, R.N.T-W.: Zoonoses of Primates. Cornell University Press, Ithaca (1967)
15. Fournier, J.: Une zoonose qui gagne du terrain: la meliodose. Med. Egypte *9*, 23-45 (1960)
16. Gillies, R.R. and I.R.W. Govan: Typing *Pseudomonas pyocyanea* by pyocyine production. J.path. Bacteriol. *91*, 339-345 (1966)
17. Godfrey, A.J. and L.E. Bryan: Resistance of *Pseudomonas aeruginosa* to new β-lactamase-resistant β-lactams. Antimicrob. Agents Chemother. *26*, 485-488 (1984)
18. Groves, M.G.: Melioidosis; pp.465-472 in "CRC Handbook Series in Zoonoses." (Steele, J.H., Series ed.), Section A: Bacterial, Rickettsial, and Mycotic Diseases. Vol. I (Stoenner, H., W.Kaplan, and M.Torten, Sect. Ed.), CRC Press, Boca Raton (1979)
19. Hardy, A.V.; R.B. Mitchell; M.Schreiber; W.R. Hoffert; E.Yawn, and F.Young: Bacterial studies of otitis externa. Laryngoscope *64*, 1020-1024 (1954)
20. Homma, J.Y.; K.S. Kim; H.Yamada; M.Ito; H.Shinoya, and Y.Kawabe: Serological typing of *Pseudomonas aeruginosa* and its cross-infection. Jap. J.Exp. Med. *40*, 347-359 (1970)
21. Hugh, R.and G.L. Gilardi: Pseudomonas; pp.250-269 in "Manual of Clinical Microbiology". 2nd ed. (Lennette, E.H., E.H. Spaulding, and J.P. Truant, eds.) Amer. Soc. Microbiol. Washington D.C. (1974)
22. Johnson, D.W.: Melioidosis. A report of four cases from Torres Strait. Med. J Aust. *2*, 587-588 (1967)
23. Kaufmann, A.F.; A.D. Alexander; A.M. Allen; R.D. Cronin; L.A. Dillingham; J.D. Douglas, and T.D. Moore: Melioidosis in imported nonhuman primates. J.Wildlife Dis. *6*, 211-219 (1970)
24. Kubota, Y.and P.V. Liu: An enterotoxin of *Pseudomonas aeruginosa.* J.infect. Dis. *123*, 97-98 (1971)
25. Laws, L.and W.T.K. Hall: Melioidosis in animals in North Queensland. IV. Epidemiology. Aust. vet. J. *40*, 309-314 (1964)
26. Lim, S.Y. and V.Mukundhan: Melioidosis in a spider monkey and a gibbon. Kajian Vet. *1*, 180 (1968)
27. Mutalib, A.R.; A.R. Sheikh-Omar, and M.Zamri: Melioidosis in a banded leaf-monkey *(Presbytis melalophos).* Vet. Rec. *115*, 438-439 (1984)
28. Osman, M.A.M.: Pyocine typing of *Pseudomonas aeruginosa.* J.clin. Path. *18*, 200-202 (1965)
29. Paul, R.and W.Marget: Pseudomonas aeruginosa - Infektionen (Pyocyaneus-Infektionen) des Säuglings und Kleinkindes. Dtsch. med. Wschr. *88*, 1638-1643 (1963)
30. Piggott, A.and L.Hochholzer: Human melioidosis. A histopathologic study of acute and chronic melioidosis. Arch. Path. *90*, 101-111 (1970)
31. Pinkerton, M.: Miscellaneous organisms. Pp. 283-313 in "Pathology of Simian Primates", Part II (Fiennes, R.N.T-W., ed.) Karger, Basel (1972)
32. Pulverer, G.und H.Korth: Zur Biochemie und Fermentstruktur von *Pseudomonas aeruginosa.* Zbl. Bakt., I.Abt. Orig. *186*, 36-54 (1962)
33. Redfearn, M.S. and N.J. Palleroni: Glanders and melioidosis; pp. 110-128 in "Diseases Transmitted From Animals to Man" (Hubbert, W.T., W.F. McCulloch, and P.R. Schnurrenberger, eds.) C.C. Thomas, Springfield (1975)
34. Redfearn, M.S.; N.J. Palleroni, and R.Y. Stanier: A comparative study of *Pseudomonas pseudomallei* and *Bacillus mallei.* J.gen. Microbiol. *43*, 293-313 (1966)
35. Retnasbapathy, A.and P.G. Joseph: A case of melioidosis in a macaque monkey. Vet. Rec. *79*, 72-73 (1966)
36. Rhame, F.S.: The ecology and epidemiology of *Pseudomonas aeruginosa*; pp.31-51 in "Pseu-

domonas aeruginosa. The Organism, Diseases it Causes, and Their Treatment." (Sabath, L. D., ed.) Hans Haber, Bern (1980)
37. Saslaw, S.; H. N. Carlisle, and M. Moheimani: Comparison of colistin-carbenicillin, colistin, and carbenicillin in *Pseudomonas* sepsis in monkeys. Antimicrob. Agents Chemother. *3*, 118-124 (1973)
38. Schneider, H. E.: Weitere Erkrankungen bei Orang-Utans im Zoologischen Garten Dresden. XXVI. Internat. Symp. Erkr. Zootiere, Brno, 189-193 (1984)
39. Schüppel, K.-F.; H.-J. Selbitz; S. Seifert, and D. Altmann: Weitere Befunde bei Primatensektionen. XXII. Internat. Symp. Erkr. Zootiere, Arnhem, 21-29 (1980)
40. Smith, N. R. and M. Damit: Fatal bronchopneumonia in a young orangutan caused by *Pseudomonas pseudomallei.* Vet. Rec. *110*, 251 (1982)
41. Spencer, H.: Non-intestinal bacterial infections in tropical regions; pp. 27-66 in "Tropical Pathology". (Spencer, H., A. D. Dayan, J. B. Gibson, R. G. Huntsman, M. S. R. Hutt, G. C. Jenkins, F. Köberle, B. G. Maegraith, and K. Salfelder, eds.), Springer, New York (1973)
42. Spilsbury, P.; W. B. Tyler, and W. E. Parkin: Melioidosis - Pennsylvania. CDC Morb. Mortal. Wkly. Rep. *25* (52), 419 (1977)
43. Srivastava, L. and T. Mathews: Carriage of *Pseudomonas aeruginosa* by laboratory animals. Indian J. Path. Microbiol. *21*, 181-184 (1980)
44. Stanley, M. M.: *Bacillus pyocyaneus* infections; a review, report of cases and discussion of newer therapy including streptomycin. Amer. J. Med. *2*, 253-277 (1947)
45. Stevenson, E. W.: Bacillus pyocyaneus perichondritis of the ear. An effective method of treatment. Laryngoscope *74*, 255-259 (1964)
46. Stills, H. F. jr.; M. G. Bond, and B. C. Bullock: Bacterial myocarditis in African green monkeys *(Cercopithecus aethiops).* Vet. Path. *16*, 376-380 (1979)
47. Strauss, J. M.; D. W. Ellison; E. Gan; S. Jason; J. L. Marcarelli, and G. Rapmund: Melioidosis in Malaysia. IV. Intensive ecological study of Carey Island, Selangor for *Pseudomonas pseudomallei.* Med. J. Malaysia *24*, 94-100 (1969 a)
48. Strauss, J. M.; M. G. Groves; M. Mariappan, and D. W. Ellison: Melioidosis in Malaysia. II. Distribution of *Pseudomonas pseudomallei* in soil and surface water. Amer. J. trop. Med. *18*, 698-702 (1969 b)
49. Strauss, J. M.; S. Jason; H. Lee, and E. Gan: Melioidosis with spontaneous remission of osteomyelitis in a macaque *(Macaca nemestrina).* J. Amer. vet. med. Ass. *155*, 1169-1175 (1969 c)
50. Strobert, E. A. and R. B. Swenson: Treatment regimen for air sacculitis in the chimpanzee *(Pan troglodytes).* Lab. Anim. Sci. *29*, 387-388 (1979)
51. Struwe, F. E.: Beitrag zur Kenntnis der Infektionen mit *Pseudomonas aeruginosa (B. pyocyaneum)* im Säuglingsalter. Helv. paediat. Acta *19*, 127-136 (1964)
52. Sutter, R. L.; V. Hurst, and J. Fennell: A standardized system for phage-typing *Pseudomonas aeruginosa.* Hlth Lab. Sci. *2*, 7-16 (1965)
53. Tammemagi, L. and L. A. Y. Johnston: Melioidosis in an orangutan in North Queensland. Aust. vet. J. *39*, 241-242 (1963)
54. Verder, E. and J. Evans: A proposed antigenic schema for the identification of strains of *Pseudomonas aeruginosa.* J. infect. Dis. *109*, 183-193 (1961)
55. Wahba, A. H.: Überlebensdauer von Pseudomonas aeruginosa im Staub. Arch. Hyg. Bakt. *149*, 187-190 (1965)
56. Whitmore, A.: An account of a glanders-like disease occurring in Rangoon. J. Hyg. *13*, 1-34 (1913)
57. Yabuuchi, E.; N. Miyajima; H. Hotta, and Y. Furu: Serological typing of 31 achranogenic and 40 melanogenic *Pseudomonas aeruginosa* strains. Appl. Microbiol. *22*, 530-533 (1971)
58. Yow, E. H.: Development of Proteus and Pseudomonas infections during antibiotic therapy. J. Amer. vet. med. Ass. *49*, 1184-1188 (1952)

3.4 Brucella Spp.

Brucella spp. are zoonotic agents whose reservoir hosts are usually ruminants, swine or canids depending on the *Brucella* species involved. Man can be infected by *B. melitensis, B. abortus, B. suis* and rarely by *B. canis;* all of which cause serious human diseases.

In very rare cases evidence of *Brucella* sp. infections have been found in nonhuman primates. Pinkerton (1967) reported a case of brucellosis - tentatively identified as *B. melitensis* - in a wild baboon in Kenya. There remain questions as to the specificity of the finding, because the growth characteristics of *Bordetella bronchiseptica* resemble those of *Brucella* sp. and cross reactions in agglutination tests are extensive (Stoll, 1968). In other studies very few seropositive baboons were found in Kenya (Kalter et al., 1967; 1968). 50 sera of wild caught *Callithrix penicillata* proved to be anti-brucella-antibody negative (Ricciardi et al., 1976).

Macaques can experimentally be infected with *B. melitensis, B. abortus, B. suis,* and *B. canis* (see Percy et al., 1972) leading to focal granulomatous lesions in several organs.

The scarcity and questionable specificity of spontaneous infections of nonhuman primates justify the omission of further description of the agents and diseases caused in both man and animals.

References

1. Kalter, S.S.; R.E. Kuntz; Y. Al-Doory, and A.A. Katzberg: Collection of biomedical study materials from baboons in East Africa; Preliminary report. Lab. Anim. Care *16*, 161-177 (1967)
2. Kalter, S.S.; R.E. Kuntz; B.J. Myers; A.K. Eugster; A.R. Rodriguez; M. Benke, and G.V. Kalter: The collection of biomedical specimen from baboons *(Papio sp.)* Kenya. Primates *9*, 121-139 (1968)
3. Percy, D.H.; I.N. Egwu, and A.M. Jonas: Experimental *Brucella canis* infection in the monkey *(Macaca arctoides).* Canad. J. comp. Med. *36*, 221-225 (1972)
4. Pinkerton, M.E.: Bacteremia in wild baboons. Bact. Proc. *67*, 67 (1967)
5. Ricciardi, I.D.; M.P. Nunes; C.M. Andrade, and A.G. da Silva: Anti-brucella agglutinins in bats and "Callithrix" monkeys. J. Wildlife Dis. *12*, 52-54 (1976)
6. Stoll, L.: Bordetella bronchiseptica-Infektion bei einem Rhesusaffen. X. Internat. Symp. Erkr. Zootiere Salzburg, 55-57 (1968)

3.5 Bordetella Sp.

Bordetella Bronchiseptica

B. bronchiseptica is the only *Bordetella* species isolated from nonhuman primates. It infects various animal species and occasionally man after contact with infected animals. It causes bronchopneumonias in rodent-colonies, kennel-cough in dogs and in rare instances serious respiratory disease in man (Winsser, 1960; Lautrop and Lacey, 1960 - cit. from Stoll, 1968; Krepler and Flamm - cit. from Stoll, 1968; Thompson et al., 1976). In nonhuman primates *B. bronchiseptica* has been isolated from *M. mulatta* (Stoll, 1968), *C. aethiops* (Montano et al., 1966; Graves, 1968; 1970), *Callicebus* sp. (Seibold et al., 1970), *Callithrix jacchus* (Hill et al., 1978; Baskerville et al., 1983) and *Galago senegalensis* (Kohn and Haines, 1977).

Agents

B. bronchiseptica are aerophilic, thin, gram-negative rods with tapered ends, occurring singly or in pairs. They are motile by means of peritrichous flagella in contrast to the closely related *Brucella*- and *Alcaligenes* spp.

Growth is obtained on Bordet-Gengou-agar, brucella agar or blood agar, with most strains also growing on SS-agar (Bemis et al., 1977). The colonies appear in 3 forms:

1. as small (<1 mm diameter), pearl-like, opaque colonies with smooth surfaces and round borders which, when grown on blood-agar, cause haemolysis;
2. medium (1 to 2 mm diameter), convex, smooth, opaque to translucent colonies without haemolysis on blood agar;
3. large, umbronate, rough, translucent colonies with undulate margins, also non-haemolytic.

The biochemical characteristics of *B. bronchiseptica* are (Graves and Shah, 1972; Bemis et al., 1977):

Table 12. Biochemical Characteristics of *Bordetella bronchiseptica*

Glucose	Ø	Oxidase	+
Sucrose	Ø	Catalase	+
Lactose	Ø	Urea	+
Maltose	Ø	Citrate	+ (−)
Salicin	Ø	Nitrate	+
Gelatin	Ø	Tetrazolium reduction	+
Methyl red	Ø	Voges Proskauer	Ø
Indule	Ø		

For seroepidemiological studies haemagglutination tests are most useful (Graves and Shah, 1972).

Epidemiology

B. bronchoseptica has been isolated from a great variety of domesticated and wild living animals such as dogs, pigs, cats, guinea pigs, rabbits, rats, ferrets, hedgehogs and voles (Goodnow, 1980), but it is not known which of these species can serve as a reservoir of *B. bronchoseptica*.

No agglutinins were found in sera from free-living monkeys, but 12% of the free-ranging rhesus monkeys of the Cayo Santiago colony and 30% of a laboratory colony at Puerto Rico possessed antibodies to *B. bronchiseptica* (Graves and Shah, 1972) suggesting the infection being present in the human environment. Once a monkey becomes infected, the organisms remain viable in the upper respiratory tract of healthy animals for quite some time (Winsser, 1960). Transmission occurs probably via droplet infection.

Clinical Symptoms

Nonhuman primates: In marmosets no other symptoms except bilateral mucopurulent nasal discharge, dyspnea and elevated body temperatures have been noted (Baskerville et al., 1983). A rhesus monkey which died of the infection had exhibited a distinctive cough (Stoll, 1968). A galago suffering from otitis media due to *B. bronchoseptica* showed torticollis and epileptiform seizures, others were found dead without any preceeding symptoms (Kohn and Haines, 1977).

Man: On rare occasions *B. bronchiseptica* causes a pertussis-like disease in children (Brown, 1926 - cit. from Köhler and Mochmann, 1975; Finegold and Martin, 1982).

Pathology

Nonhuman primates: The most common and often fatal type of lesions in nonhuman primates consists of a focal to widespread purulent bronchopneumonia (Winsser, 1960; Montano et al., 1966; Stoll, 1968; Graves; 1968, 1970; Seibold et al., 1970; Kohn and Haines, 1977; Baskerville et al., 1983). The otitis media in a galago was also of the purulent type. It was associated with meningitis, venous thrombosis and focal cerebral infarction (Kohn and Haines, 1977).

Man: Acute fatal *B. bronchiseptica* infections are associated with lung congestion and edema and multifocal bronchopneumonia (Winsser, 1960).

Treatment

All *B. bronchiseptica* strains tested were sensitive to polymyxin B, chloramphenicol, and tetracyclin. Most strains also responded to gentamycin and kanamycin. Nalidixic acid, cephalothin, ampicillin and sulfonamide-trimethoprim were less effective. Streptomycin and penicillin did not inhibit the growth (Bemis et al., 1977).

Marmosets infected with oxytetracyclin-sensitive *B. bronchiseptica* responded well enough to i.m. inoculation of oxytetracyclin, but still carried the microorganisms 8 weeks later in nasal swabs (Baskerville et al., 1983).

References

1. Baskerville, M.; M. Wood, and A. Baskerville: An outbreak of *Bordetella bronchiseptica* pneumonia in a colony of common marmosets *(Callithrix jacchus)*. Lab. Anim. *17*, 350–355 (1983)
2. Bemis, D. A.; H. A. Greisen, and M. J. Appel: Bacteriological variation among *Bordetella bronchiseptica* isolated from dogs and other species. J. clin. Microbiol. *5*, 471–480 (1977)
3. Finegold, S. M. and W. J. Martin: Bailey and Scott's Diagnostic Microbiology. C. V. Mosby, St. Louis (1982)
4. Goodnow, R. A.: Biology of *Bordetella bronchiseptica*. Microbiol. Rev. *44*, 722–738 (1980)
5. Graves, I. L.: *Bordetella bronchiseptica* isolated from a fatal case of bronchopneumonia in an African green monkey. Lab. Anim. Care *18*, 405–406 (1968)
6. Graves, I. L.: Agglutinating antibodies for *Bordetella bronchiseptica* in sera before, during, and after an epizootic of pneumonia in caged monkeys. Lab. Anim. Care *20*, 246–250 (1970)
7. Graves, I. L. and K. V. Shah: *Bordetella bronchiseptica* in monkeys: Agglutinin survey and isolation from asymptomatic individuals. J. med. Primatol. *1*, 344–353 (1972)

8. Hill, A.C.; J.A. Turton, and J. Bleby: Bacterial and mycoplasma flora of a laboratory colony of the common marmoset *(Callithrix jacchus)*. Vet. Rec. *103*, 284-287 (1978)
9. Kohn, D.F. and D.E. Haines: *Bordetella bronchiseptica* infection in the lesser bushbaby *(Galago senegalensis)*. Lab. Anim. Sci. *27*, 279-280 (1977)
10. Köhler, W. und H. Mochmann: Grundriß der Medizinischen Mikrobiologie. Gustav Fischer, Jena (1975)
11. Montano, A.; J. Fernandes Gomes, et R. Oliveira e Silva: Bronchopneumonia contagiosa por Bordetella bronchiseptica em simios importados. Histologia comparada com a especie humana. Med. Contemp. *84*, 303-315 (1966)
12. Seibold, H.R.; E.A. Perrin, jr. and A.C. Garner: Pneumonia associated with *Bordetella bronchiseptica* in *Callicebus* species primates. Lab. Anim. Care *20*, 456-461 (1970)
13. Stoll, L.: Bordetella bronchiseptica - Infektion bei einem Rhesusaffen. X. Internat. Symp. Erkr. Zootiere, Salzburg, 55-57 (1968)
14. Thompson, H.; I.A.P. McCandish, and N.G. Wright: Experimental respiratory disease in dogs due to *Bordetella bronchiseptica*. Res. Vet. Sci. *20*, 16-23 (1976)
15. Winsser, J.: A study of *Bordetella bronchiseptica*. Proc. Anim. Care Panel *10*, 87-104 (1960)

3.6 Enterobacteriaceae

3.6.1 Escherichia Coli

E. coli is a normal commensal organism of the large intestine of vertebrates. More than hundred serotypes of *E. coli* have been recorded. Some serotypes have only a restricted host relationship, whilst others occur in a wide variety of hosts, occasionally being pathogenic to their host (Rolle and Mayr, 1978).

Some of the enteropathogenic *E. coli*-strains of man have been isolated also from nonhuman primates, indicating the cross-species transmissibility of certain *E. coli* serotypes from man to animal, and possibly vice versa.

Agents
E. coli, like all Enterobacteriaceae, is a gram-negative, aerobic to facultatively anaerobic, plump rod with rounded ends. Most *E. coli* strains are motile, with some possessing mucopolysaccharide-containing capsules.

E. coli has no special growth-requirements. For practical purposes selective media such as EMB-, Kligler-, McConkey-, SS-agar etc. are most useful. The colonies growing on the surface of plated or slanted agar are 1 to 2 mm in diameter, greyish-white or creamy, smooth or rough, convex to flat, sometimes mucoid. On selective media the colonies might be tinged by the indicators contained in the agars.

Some *E. coli* strains are haemolytic on blood agar, an either genome - determined characteristic and thus stabil (e.g. strains 0138 or 0141) or a plasmid-dependent property (Hyl-plasmid) and thus variable.

Serotyping of *E. coli* is associated with the presence of 0-, H- and K-antigens. 0-antigens are heat-stable, somatic antigens; H-antigens are of flagellar origin and are heat labile (destroyed by 100 °C); whilst K-antigens are somatic antigens occurring as envelopes or capsules which inhibit the agglutination of living bacteria in 0-antisera. The K-antigens are formed by a class of at least 3 varieties, designated L, A and B (Ewing and Martin, 1974). So far more than 100 serotypes have been described.

The normal biochemical reactions of *E. coli* are:

Table 13. Biochemical Characteristics of *Escherichia coli*

Glucose (acid and gas)	+	Raffinose	v
Lactose	+	Sorbitol	+
Sucrose	v	Arabinose	+
Salicin	v	Rhamnose	+
Indole	+	Xylose	v
H_2S	∅	Trehalose	+
Lysine decarboxylase	v	Maltose	+
Ornithine decarboxylase	v	Mannitol	+
Arginine	v	Catalase	+
Esculin	v	Methyl red	+
Dulcitol	v	Voges Proskauer	∅
Inositol	∅	Citrate	∅
Adonitol	∅	Urea	∅
		Gelatin	∅

Deviations from this pattern are not unusual. Enteroinvasive E. coli strains, for example, are commonly lactose negative or are slow fermenters of lactose. Certain strains of *E. coli* acquire the ability to produce hydrogen sulfide by recombination, in which an episome or plasmid carrying genetic material for that character is involved (Lautrop et al., 1971; Stoleru et al., 1972).

Some *E. coli* strains produce plasmid mediated cholera-toxin-like heat labile (LT) or heat stabile (ST) enterotoxins after entering the small intestine (Gross et al., 1983). At least 2 subtypes of ST are known: one is ST h (ST 1b), produced by human strains; the other is ST p (ST 1a), produced by bovine, porcine, and human strains (Hemelhof et al., 1984). For the detection of ST the infant mouse test (Dean et al., 1972) or ligated rabbit loop models (DuPont and Pickering, 1980) can be used. LT can be demonstrated in YI adrenal cell (Donta et al., 1974) or Chinese hamster ovary (CHO) cell tests (Guerrant et al., 1974).
Other *E. coli* strains are enteroinvasive, a property not normally present in commensal organisms. Most of the known enteroinvasive *E. coli*-strains have O-antigens identical or closely related to *Shigella dysenteriae* and *Sh. boydii* (Edwards and Ewing, 1972) with relevant cross reactions in serotests. Enteroinvasive *E. coli* possess 120 to 140 megadalton virulence plasmids antigenically related also to the similar plasmids of *Shigella* spp. (Silva et al., 1982; Pal et al., 1983; Sansonetti et al., 1983).

Enteroinvasiveness is checked for by the Sereny-test (Sereny, 1955, 1957), whereby positive strains produce keratoconjunctivitis after instillation into the guinea pig conjunctival sac (Toledo et al., 1980).

Epidemiology

Nonpathogenic and pathogenic *E. coli* are probably transmissible between different host species, although irrefutable evidence of transmission of enteropathogenic *E. coli* from animals to man is lacking (Myers, 1980). Strains that are pathogenic in one host species are not necessarily equally pathogenic in other host species. *E. coli* strains enteropathogenic for pigs, calves and lambs, when symptoms were at all observed, caused only mild diarrhea in some human volunteers (Myers, 1980).

In man, certain serotypes are generally accepted as etiological agents of diarrheal disease in infants, being responsible for approximately 15 % of all *E. coli*-related gastrointestinal diseases. These serogroups are as follows (Ewing and Martin, 1974):

026: K60 (B6)
055 a: K59(B5)
086 a: K61 (B7)
0111 a, 111 b: K58 (B4)
0119: K69 (B14)
0125 a, 125 b: K70 (B15)
0125 a, 125 c: K70 (B15)
0126: K71 (B16)
0127 a: K63 (B8)

The serogroups 026 ac
0112 ac
0124
0136
0143
0144
0152
0164

have been commonly found as enteroinvasive agents.

In nonhuman primates (*Macaca mulatta, M.fascicularis, Cercopithecus aethiops, Hylobates concolor, Pan troglodytes, Pongo pygmaeus, Presbytis cristatus, Alouatta villosa, Ateles fusciceps, A.geoffroyi, Aotus trivirgatus, Callithrix jacchus, Saguinus geoffroyi, Tupaia glis*) the following serotypes are found most frequently:

026: B6
055: B55
0111: B4
0111: B5
0119: B14
0124: B17

Other serotypes found occasionally are 06: H1; 015: H33; 044: K74; 064: H8; 078: K80; 086: B7; 0101: K30; 0111: B17; 0114: B14; 0117: H26; 0119: B14; 0125: K11; 0127: B8; 0128: B12; 014: K89; 0153: H40; 0162: H-. (Kourany and Porter, 1969; McClure et al., 1972; Schiff et al., 1972; Mohan et al., 1973, 1975; Walker et al., 1974; Brede et al., 1977; Panaitescu and Potorac, 1982; Selbitz et al., 1983).

Pathogenesis

E. coli may develop pathogenic properties in different ways:

1. by accidental contamination of wounds;
2. by acquiring the ability to invade intestinal or genito-urinary tissues, which is associated with the presence of certain fimbriae correlated to the exhibition of a

mannose-resistant haemagglutination (Nimmich et al., 1984; Saxena and Yadava, 1985)

3. by toxin production.

Ad 1. Accidental contamination of wounds is not associated to any specific properties of the bacteria. They probably can therefore affect different host species.

Ad 2. *E. coli* as a member of the Enterobacteriaceae is normally abundant in the colonic contents of most mammals. It participates in the establishment of the proper redox potential necessary to maintain a balanced aerobic: anaerobic flora in the gut. The intact bacterial flora in the large intestine have a great biochemical versatility, performing many fundamental reactions which their hosts are incapable of. No mammalian tissue is able to break down cellulose, hydrolyse urea, or obtain its nitrogenous needs entirely from ammonia-reactions critical to the life cycle on earth (Wrong et al., 1981).

Certain *E. coli*-strains share with the *Shigella* strains the ability to penetrate the epithelial cells of the large intestine and to multiply intracellularly, thereby destroying the affected cells (Gross et al., 1983; Pal et al., 1983) (Fig. 7).

After gaining access to the vascular system the invading bacteria are carried to other internal organs not normally accessible to *E. coli.* Destruction of the bacteria liberates shock-inducing endotoxins (Smith, 1975), which cause intravascular clotting with deposition of the resulting thrombi in the renal glomerular capillaries, consumption coagulopathy and haemodynamic alterations (Dinbar et al., 1971; Hammon et al., 1978).

Enteroinvasive *E. coli* possess a special plasmid of 120 to 140 megadalton size, which is required particularly in the initial steps of the invasion (Silva et al., 1982).

The most frequent extraintestinal *E. coli* infections not preceeded by septicemia are those of the genito-urinary tract. *E. coli* is one of the most common bacterial causes of pyelonephritis in man and animals. Pyelonephritogenic *E. coli* possess type I-fimbriae binding to mannose receptors of the uromucoids, thus impairing its normal clearance by the flow of urine (Roberts et al., 1984). The type I fimbriae do not adhere to the epithelial surface but to the glycoproteins contained in the mucus (Orskov et al., 1980 – cit. from Ketyi et al., 1983). In urinary infections (and in infantile meningitis) a predominance of K-antigen producing *E. coli* strains has been observed by many investigators (Ketyi et al., 1983). In human urinary *E. coli* infections certain additional virulence factors are known. The human uroepithelial cell receptors recognized by fimbriae of pyelonephritogenic *E. coli* are related to the human P-blood group system (Källenius et al., 1983). An additional factor is the alpha haemolysin produced by the bacteria (Ketyi et al., 1983). The influence of such fimbriae and cofactors in spontaneous urinary tract infections of animals are unknown.

Invasive *E. coli* often show a typical biochemical reaction, e.g. slow lactose fermentation and they usually do not decarboxylate lysine (Silva et al., 1980).

Ad 3. Enterotoxin production by *E. coli* induces intestinal hypersecretion in two different ways depending on the kind of enterotoxin (heat labile = LT, or heat stable = ST). LT consists of 2 subunits; the B polymer of 5 polypeptides which binds to GM 1-gangliosides of the enterocytes and the A subunit which catalyses ADP-ribosilation of a regulatory subunit of adenylate cyclase (Moss et al., 1981) and in that way enhances the cellular cyclic AMP. The cAMP in turn, inhibits the NaCl co-transportation through the brush border of the enterocytes. It also stimulates the CL-secretion in the intestinal crypts , and finally enhances the permeability of enterocyte membranes by phosphorylation of membrane fraction by activation of protein kinases (Shlatz and Kimberg, 1980). The result is an enhanced net-secretion from the small intestinal epithelium (Carpenter, 1980; Field, 1980; Binder, 1981; Heintze, 1981).

ST, on the other hand, activates guanylate cyclase and thus the cGMP (Carpenter, 1980; Field, 1980; Binder, 1981) without binding to the GM 1-ganglioside. ST stimulates only the intestinal guanylate cyclase in contrast to LT which activates the adenylate cyclase of both intestinal and nonintestinal cells (Frantz et al., 1984). ST acts like LT, but faster; it also disappears quicker.

Both types of enterotoxins require the attachment of the bacteria to the small intestinal wall. Normally the inhibitants of the large intestine will be flushed from the small intestines by the peristaltic waves unless they possess special attachment plasmids or fimbriae of 55 to 70 megadalton size (colonization factor antigens (CFAs)), which have been demonstrated in enterotoxic *E. coli* of human origin (Nataro et al., 1985). They serve as ligands in an intestinal binding process. Similar adhesion factors have been found in porcine strains as K88 surface antigen or as K99 surface antigen in strains of calf or lamb origin (DuPont and Pickering, 1980). CFA/I is transferable to many members of the family Enterobacteriaceae (Yamamoto et al., 1984).

In man several *E. coli* strains caused enteritis in children by colonization of the duodenum and upper jejunum without either enteroinvasiveness or the classical enterotoxinproducing ability. The pathogenesis of this type of diarrhea is not well understood. It appears to be caused by *E. coli* capable of attachment to the denuded surface of intestinal epithelial cells, with the production of small amounts of *Shigella dysenteriae*-like enterotoxin (Myers, 1980). These are called "enteropathogenic" *E. coli.* The serotypes most commonly associated with this type of disease are: 026; 055; 0111; 0119; 0127 and 0128.

In nonhuman primates the *E. coli* problem is far less well classified. Although the term "enteropathogenic" *E. coli* is commonly used in connection with diarrhoic or healthy monkeys (Fiennes et al., 1972; McClure et al., 1972; Schiff et al., 1972; Mohan et al., 1975; McClure, 1976) this term usually applies only to the isolation of serotypes known to be enterotoxic, enteroinvasive, or enteropathogenic in man. The ability of enterotoxin production or enteroinvasiveness has not been tested in *E. coli* strains of simian origin.

Clinical Symptoms

Nonhuman primates: The symptoms observed particularly in young animals vary from lethargy and mild to moderately severe watery diarrhea with or without mucus and/or blood added to the stools, which lasts for 2 to 10 days, to a rapidly fatal disease (McClure et al., 1972; McClure, 1976). Slow lactose fermenting (not serotyped) *E. coli* caused a watery-mucoid diarrhea of usually 2 to 4 days duration in *E. patas* (Wolf et al., 1971).

In experimental intravenous *E. coli* infections of baboons, in an attempt to simulate *E. coli* septicemia, the following symptoms were observed: marked tachycardia, profound fall of mean arterial pressure, decreased renal blood flow, increased plasma norepinephrine levels, acidosis, fall of platelet counts and haematocrit, and elevation of both partial thromboplastin time and thrombin time (Cavanagh et al., 1970; Hammon et al., 1978; Coalson et al., 1979). Similar effects were caused by inoculation of *E. coli*-endotoxins into *M. mulatta* (Dinbar et al., 1971).

Extraintestinal infections caused fatal pyometritis in a de Brazza's monkey *(Cercopithecus neglectus)* (Cooper, 1972) and airsacculitis, pneumonia, and abortion in an orangutan (Cambre et al., 1980).

Man: Enteroinvasive *E. coli* infections resemble those of bacillary dysentery in children and adults, with vomiting and diarrhoic stools which frequently contain blood and mucus (Toledo et al., 1980).

Enterotoxic strains stimulate the loss of water and electrolytes, causing watery diarrhea (Gross et al., 1983). They are a major cause of traveller 's diarrheas particularly in the tropics and subtropics.

Pathology

Nonhuman primates: In two young chimpanzees the most conspicious lesions were: distention of the gastrointestinal tract with bloody fluids, petechial and ecchymotic haemorrhages or haemorhagic necrosis of the intestinal wall, lung edema or haemorrhages and dehydration (McClure et al., 1972; McClure, 1976). In a baby gorilla acute haemorrhagic enterocolitis was caused by an overwhelming *E. coli* infection (Simmons, 1973). Similar lesions were followed by septicemia, haemorrhagic diathesis and haemorrhagic-purulent meningo - encephalitis in a newborn baboon (Brack et al., 1975).

Microthrombosis in experimental intravenous infections of baboons lead to fibrin deposits in renal glomerular capillaries, and occasionally, to tubular necrosis (Hammon et al., 1978; Coalson et al., 1979).

Uterine infection caused a purulent endo- and myometritis in a de Brazza 's monkey, but there was no indication of an extension of the infection to the urinary tract (Cooper, 1972).

Man: A distinctive feature of neonatal *E. coli* diarrhea is the frequent development of shock-like toxic states with haemorrhagic diathesis. Septicemia may occur with subsequent development of pneumonia (Cooper et al., 1956; Lindberg and Young, 1956; Drucker et al., 1970). The local lesions of the intestinal tract consist of edema, ulceration, haemorrhages, acute inflammation and often intravascular clotting.

Treatment

In antibiotic treatment drug resistance is easily transferred between the Enterobacteriaceae by plasmid mediated R-factors (Kawatomari, 1972; Fox et al., 1973; Platt et al., 1984), therefore antibiograms are required in each case.

Hypersecretion induced by ST can be inhibited by agents blocking the Ca uptake and the prostaglandin synthesis (Greenberg et al., 1980, 1982; Thomas and Knoop, 1982). LT-action can possibly be blocked by chlorpromazine, which is an adenylate-cyclase antagonist. The diarrheas of piglets suffering from Coli-enterotoxemia stopped immediately after treatment with 2 mg/kg chlorpromazine, and also in cholera of man 1.6 to 4 mg/kg chlorpromazine were equally efficient (Holmgren and Greenough, 1980).

3.6.2 Salmonella Spp.

Salmonella infections are one of the classical food-borne zoonoses. They also account for the majority of isolates from human gastroenteritis in some countries (Joseph et al., 1978). *Salmonella* sp. infections are common in exotic animals particularly after transportation and consequently are occasionally encountered in nonhuman primates (Tauraso, 1973; Ford et al., 1973; Tanaka and Katsube, 1981). Most isolations from nonhuman primates are made in summer and early fall (Schneider et al., 1960; Schröder, 1964), which has been ascribed to environmental conditions favouring the maintenance and spread of the bacteria in soil etc. Because the very same *Salmonella* spp. infect both man and nonhuman primates the possibility of cross species infections has been suggested by many authors, but only one such case (Anderson et al., 1971) could be verified.

Agents

Salmonella spp., like other Enterobacteria, are gram-negative, nonspore-forming rods that are aerobic or facultatively anaerobic. Most strains are motile (exception: *S. gallinarum-pullorum*) using peritrichous flagella. *Salmonellae* are killed at 55 °C in 1 h or at 60 °C in 15 to 20 min. Most desinfectants also destroy the bacteria.

Salmonella spp. grow in the presence of malachite or brillant green dyes which kill or inhibit competitive bacteria in primary cultures from fecal materials etc.

Salmonella spp. do not have special nutritional requirements, for practical purposes selective media such as McConkey-, EMB-, SS-, desoxycholate-, brillant green-, phenol red agar etc. in combination with enrichment media (selenite-, tetrathionate-, GN-broth etc.) are mostly used.

Salmonella colonies on the nutrient media's surface are usually translucent, convex, but deviations (R-forms etc.) are not uncommon.

The general biochemical characteristics of Salmonellae are given in Table 14:

Salmonella spp. possess thermostable O-, and thermolabile H-antigens, the latter in phase 1 and 2. *S. typhi* and *S. paratyphi* C also have K-antigens (Vi-antigens), which interfere with the O-agglutination. Serogrouping of the more than 1800 species depends on the spectrum of over 60 O-antigens for group determination and additional H-antigens demonstration (more than 90 H-antigens are presently

Table 14. Biochemical Characteristics of *Salmonella spp.*

Glucose	+ (gas +)	Lysine decarboxylase	+
Saccharose	∅	Ornithine decarboxylase	+
Salicin	∅	Arginine dihydrolase	+
Adonitol	∅	Hydrogen sulfide	∅
Sorbitol	+	Urea	∅
Maltose	+	Indole	∅
Arabinose	+	Methyl red	∅
Xylose	+	Voges Proskauer	∅
Rhamnose	+	Citrate	+
		KCN	−
		Phenylalanine deaminase	∅
		Mannitol	+
		Nitrate reduction	+

known) for species differentiation. It is clearly beyond the scope of the present article to go into more details on the classification of *Salmonella* spp.

Epidemiology

In captive nonhuman primates members of virtually all serogroups have been isolated except *S. paratyphi* A (Serogroup A) and *S. typhi* (Serogroup D) - the two most important human *Salmonella* species.

Nonhuman primates living in the wild were mostly free of Salmonella infections (Kourany and Porter, 1969; Rowe, 1969). Only Brede (1972) reported 13,75% of 851 *P. ursinus* to be *Salmonella*-carriers soon after capture, with *S. sundsvall* being the most frequently encountered species.

This contrasts sharply with the rate of infection found in newly imported, conditioned or quarantined monkeys and apes in some studies: 12,2% of newly imported cynomolgus monkeys (Tanaka and Katsube, 1981); 22% of newly imported macaques and chimpanzees (Schneider et al., 1960; Ford et al., 1972); and 27 % of conditioned macaques, *Cercopithecus* spp. and *Presbytis* spp. (Good et al., 1969) were found to be *Salmonella*-infected. Even higher infection rates have been reported, according to Tauraso (1973) the incidence of salmonellosis reached 78% in some shipments.

These findings suggest the infection of monkeys by *Salmonella* spp. in the human environment, either when rummaging through settlements or shortly after capture resp. later in captivity. The source of infections is usually food contaminated by feces, bird droppings, excretions from infected cage mates (especially at or after transportation according to Schröder, 1964); in Zoological gardens also from other animal species, particularly reptiles, which have even higher infection rates. Cockroaches and flies also have been shown to carry *Salmonella*-spp., but only in a small percentage (Müller, 1965). It has to be remembered that Salmonellae can exist in contaminated soil, vegetation and food, which facilitates indirect transmission (Fiennes, 1967).

The infectivity of Salmonellae to a certain host is enhanced by the presence of low gastric acidity or debilitating diseases (DuPont and Pickering, 1980), and in nonhuman primates mainly by transportation stresses.

The spread of the infection within a monkey or ape colony is favoured by the behavior of nonhuman primates which is fairly well below normal hygienic standards. It is also favoured by the development of clinically healthy carrier stages (Mohan et al., 1975; Tanaka and Katsube, 1981) which excrete the bacteria only inconsistently and only in small numbers.

Transmission of Salmonellae *(S. oranienburg)* from nonhuman primates to man (Fox, 1975), although often suspected, so far has only once been proven, resulting in diarrheic disease of a youngster (Anderson et al., 1971).

Pathogenesis

Salmonella spp. cause infection of the entire gastrointestinal tract. They invade and penetrate the intestinal epithelial barrier (Kent et al., 1966), then they usually multiply in the *lamina propria* and the submucosa of the infected bowel, causing heavy inflammation in the intestinal wall and haematogenous spread to other organs (Fig. 7). 12 hs p.i. in experimental infections, numerous *Salmonella*-organisms could be demonstrated in the brush border and within the villous epithelia of the ileum. The bacteria first cause a degeneration of the microvilli with disappearance of the terminal web, followed by elongation, fusion and swelling of the microvilli. The enterocytic apical cytoplasm then degenerates with formation of a cavity (vacuole) around the advancing bacteria, the cytoplasm may bulge and project into the gut lumen. *Salmonella* penetrating through the intercellular junctional complex are enclosed in vacuoles formed by the intercellular junctional complex of the adjoining cells (Takeuchi, 1975). As the penetration of the epithelium progresses neutrophilic granulocytes infiltrate the propria and the epithelium, in an inflammatory response seemingly out of proportion considering the number of bacteria causing it (Kent et al., 1966). Stimulation of the mucosal adenylate cyclase activity by the nonenterotoxin elaborating *Salmonella* spp. inhibits the active sodium absorption and stimulates the chloride secretion, resulting in increased fluid excretion (Giannella et al., 1975). Filtration loss through a "leaky" epithelium is unlikely to contribute to the diarrhea (Kinsey et al., 1976). In experimentally infected rhesus monkeys the fluid loss was greater in the ileum than in the jejunum (Rout et al., 1974).

In *Salmonella*-sepsis depression of the renal water clearance, elevation of the serum sodium and chloride concentration and kaluresis occur (Nichols et al., 1981).

Clinical Symptoms

Nonhuman primates: The clinical symptoms (if at all observed) are not very striking - diarrhea with or without fever, which by itself is very common in nonhuman primates especially after stress (Kent et al., 1966). In orangutans also vomiting and icterus have been observed (Meier and Sanborn, 1982).
Septic salmonellosis *(S. heidelberg)* caused abortion in a *Hylobates lar* (Thurman et al., 1983) or osteomyelitis (*S. typhimurium* in a rhesus monkey (Klumpp et al., 1986).

Man: Because *S. typhi* has not been isolated from nonhuman primates so far the description of typhoid fever in man will be omitted.

The paratyphoid Salmonellae cause acute gastroenteritis in man. Clinical symptoms usually begin 8 to 48 hrs. after the ingestion of contaminated food.

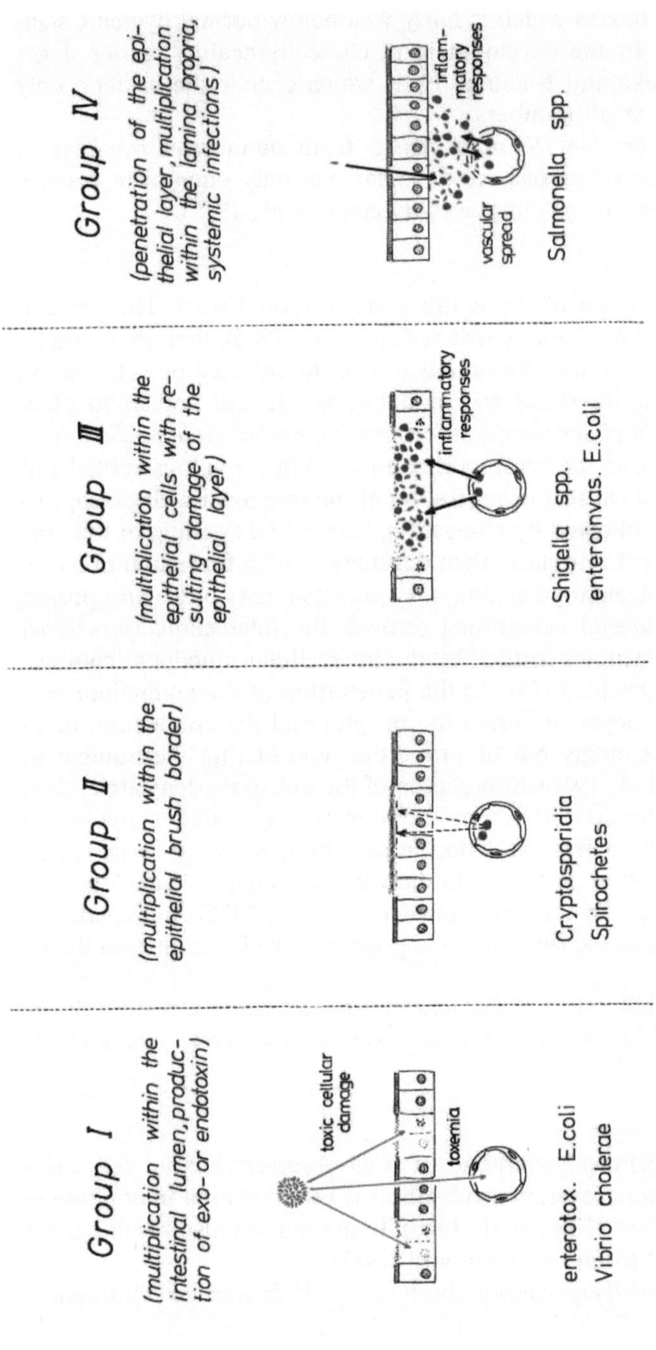

Fig. 7. Pathways of enteric bacterial infections

They consist of fever, headache, abdominal pain, diarrhea, and vomiting, sometimes also of respiratory symptoms (coughing). The diarrheic stools usually are watery to mucoid, but rarely contain any blood. *Salmonella*-gastroenteritis is usually mild, but fatalities can occur in infants, or in debilitated or elderly adults (Dupont and Pickering, 1980). Following acute *Salmonella* gastroenteritis *Salmonella*-excretion may continue for several months (Williams, 1980).

The second type of human salmonellosis (other than typhoid fever) is enteric fever or septic syndrome. It may follow gastroenteritic episodes or develop independently, it is seen particularly in human *S. cholerae suis*-infections.

The third syndrome is that of localized infection often following initial bacteremia. It leads to local inflammation or abscesses in any organ or tissue, resulting in bronchopneumonia, endocarditis, pyelonephritis, osteomyelitis, arthritis or meningitis (Williams, 1980).

Pathology

Nonhuman primates: The morphological alterations in the duodenum and jejunum are usually minimal with hyperplasia of crypt epithelial cells and reversal of the crypt: villous ratio, dilation of the crypts, reduced goblet cell mucus, slight inflammatory (polymorphonuclear) infiltration of the propria and finally villous atrophy.

The lesions in the ileum and especially the colon are much more pronounced as necrotizing enterocolitis (Hunt et al., 1978; Okoh and Onazi, 1980) with fulminating, acute, diffuse inflammation of the propria and submucosa, crypt abscesses and bacterial invasion of the epithelial cells and the *lamina propria* (Kent et al., 1966; Kinsey et al., 1976). With prolonged illness the leukocytes are replaced by large histiocytes.

Focal lesions also appear in the stomach as focal collections of polymorphonuclear leukocytes. Extraintestinal salmonellosis *(S. typhimurium)* was observed following a diarrheic episode in a black howler monkey *(Alouatta villosa)* as subcutaneous abscesses in both maxillas (Kourany and Rossan, 1971) or as severe osteomyelitis of the femur in an adult rhesus monkey (Klumpp et al., 1986).

It has to be remembered, however, that many *Salmonella*-infections in nonhuman primates are symptomless and without characteristic lesions.

Treatment

The treatment of the infection as well as the disease requires antibiotics, but care has to be taken because of existing or developing drug resistance. Antibiotic resistance is plasmid mediated by extrachromosomal DNA-containing genetic material (R-factor) consisting of 2 subunits, the resistance transfer factor (RTF) which is responsible for the production of pili and the actual transfer of the genetic material and the resistant determinant (RD) coding for drug resistance (Anderson and Lewis, 1965; Fox, 1975; Fleming, 1977). R-factors normally replicate in synchrony with the bacterial chromosome at cell division, but bacteria carrying the R-factor can transmit the antibiotic resistance to other susceptible *Enterobacteriaceae, Pseudomonadaceae* or *Vibrionaceae* (Harada et al., 1960; Baron and Falkow, 1961; Roe et al., 1971). Transfer of the R-factor occurs when a host and a suitable recipient cell touch each other, the duplicated R-factor, then passes through a pilus into the recipient bacterial cell. R-factors can be directed towards one or more antibiotics. They have been found against streptomycin, tetracycline, chloramphenicol, kanamycin, sulphonamide, trimethoprim etc.

If no resistance exists the following dosages are recommended in nonhuman primates:

chloramphenicol palmitate (orally):	25–75 mg/kg/day
streptomycin (i. m.)	: 500–750 mg/day
kanamycin (i. m.)	: 10 mg/kg/day
chlortetracycline (orally)	: 50 mg/kg
oxytetracycline (i. m.)	: 100–500 mg/day
furazolidone (Furoxon)	: 5–10 mg/day
trimethoprim	: 4.4 mg/kg/day
sulfamethoxazole	: 17.6 mg/kg/day

With prolonged vomitus or diarrhea supportive treatment (fluid, electrolytes) has to be also administered.

3.6.3 Shigella Spp.

Infections by *Shigella* spp., like those by *Salmonella* spp. are of great concern to public health authorities and to everyone responsible for the maintenance of primate colonies because of:

1. their transmissibility from man to nonhuman primates and vice versa;
2. their ability to induce severe, even fatal diseases in both man and nonhuman primates. In that respect they are more dangerous than the Salmonellae;
3. the practical difficulties in the eradication of the infection in nonhuman primates.

Shigellae, in contrast to the widely distributed Salmonellae tend to be limited to primates, infections of other genera have rarely been observed as *Sh. flexneri*- and *Sh. dysenteriae*-isolations from dogs in China, *Sh. sonnei* isolations from feces of a dog in Mexico and of *Shigella* spp. from tigers (Boro et al., 1980; Zaki et al., 1980). They have also been found in Colombian bats (Arata et al., 1968).

Shigella spp. differ from salmonellae also in their very limited ability to survive outside of their host 's body, although they have been isolated from egg albumin, checked and frozen eggs and from the fluid contents of a plastic ice cube (Eichner et al., 1968).

Shigellosis in nonhuman primates has first been reported by Ravaut and Dopter (1909), since that time a vast number of *Shigella*-infections or clinical shigelloses in simian primates have been described.

Agents

Shigella spp. resemble other Enterobacteriaceae in many respects. They are gram-negative rods of 2–3 × 0,5–0,7 μm size. They are non-motile, non-encapsulated, can not use citrate or malonate as sole C source, and their growth is inhibited by KCN and by bismuth sulfite. Shigellae have no special growth requirements, but the use of selective media is usually recommended in order to suppress other intestinal bacteria. For cultivation of Shigellae differential media without special inhibitors to other bacteria such as EMB- or McConkey agar can be employed. For suppression of other coliform Enterobacteriae differential media containing special inhibitors such as deoxycholate citrate agar, deoxycholate citrate lactose sucrose agar, xylose lysine deoxycholate agar and *Salmonella-Shigella*-(S-S) agar are sometimes preferred. These, however, might also inhibit growth of a significant proportion of *Shigella*-strains (Fleming, 1978). *Salmonella-Shigella*-agar, despite its name, only supported growth of half or less strains of Shigellae (Lindsey et al., 1971). In triple sugar iron (TSI) or KIA the Shigellae show an acid butt, no hydrogen sulfide, and an alkaline slant (Finegold and Martin, 1982).

Practical difficulties isolating Shigellae are caused by:

a) the smaller colony size compared to other coliforms or Salmonellae
b) the rapid decline in survival parallel to the decline of the environmental temperature. As Shigellae do not form any spores and the vegetative Shigellae are killed very rapidly by other Enterobacteriae, esp. Klebsiella spp. (Fincham and Seier, 1981) the best chances to obtain Shigellae are by using fresh faeces. At necropsy the chances of recovering *Shigella* spp. were greatly increased by brushing the mucosa down to the vascular bed to remove all luminal contents before swabbing (Hentges, 1967; Lindsey et al., 1971).
c) In carriers the intracellular maintenance of the Shigellae might be an obstacle to the *in vitro* isolation.

The biochemical reactions of Shigellae are usually:

Table 15. Biochemical Characteristics of *Shigella* spp.

Glucose	+ (no gas)	Christensen's citrate	Ø
Lactose	Ø to v	Simmon's citrate	Ø
Sucrose	0	Sodium acetate	Ø
Salicin	Ø	Mucate	Ø
Indole	Ø or +	Hydrogen sulfide	Ø
Adonitol	Ø	Methyl red	+
Mannitol	v	Voges Proskauer	Ø
Esculin	Ø	Phenylalanine deaminase	Ø
Lysine decarboxylase	Ø		
Arginine dihydrolase	d		
Ornithine decarboxylase	Ø or +		
Catalase	+ or Ø		
Urease	Ø		

The Shigellae can be grouped according to their biochemical reactions or they can be serogrouped using the presence of the O-antigen. Shigellae lack the H-antigens, but some strains do possess K-antigens which inhibit the O-agglutination of live bacteria. Suspicious, but not agglutinating colonies should be inoculated onto polytrophic media (e.g. Kligler 's medium), and the agglutination repeated with bacteria rinsed off and boiled for 1 hr in order to destroy the thermolabile K-antigens (Ogawa and Takasaka, 1966; Kaufmann, 1972; Takasaka et al., 1973).

So far, the following groups are distinguished: (see Tab. 16)

Another way of separating Shigellae is by colicine typing, which can be done in certain laboratories especially with respect to *Sh. sonnei.* 15 colicine types of *Sh. sonnei* are currently recognized.

Table 16. *Shigella* spp. - Serogroups

Group A:	Sh. dysenteriae 1 (Shiga-Kurse)	
	Sh. dysenteriae 2 (schmitzii)	
	Sh. dysenteriae 3-7 (10) (Large-Sachs-group)	
Group B:		antigenic formula
	Sh. flexneri 1a	I:4
	Sh. flexneri 1b	I:(4), 6
	Sh. flexneri 2a	II:3, 4
	Sh. flexneri 2b	II:7, 8
	Sh. flexneri 3a	III: (3, 4) 6
	Sh. flexneri 3b	III: (3, 4) 6, 7, 8
	Sh. flexneri 4a	IV: (3, 4)
	Sh. flexneri 4b	IV: (3, 4), 6
	Sh. flexneri 5a	V:(4)
	Sh. flexneri 5b	V:7
	Sh. flexneri var x	-:7, 8
	Sh. flexneri var y	-:3, 4

(Sh. flexneri serovar 6 has been transferred to Sh. boydii because of its fermentative properties, antigenic spectrum and genetic characterisation according to Petrovskaya and Khomenko, 1979; Goullet et al., 1983)

Group C:	Sh. boydii serovars 1-18
Group D:	Sh. sonnei (Kruse-Sonne)

Epidemiology

Shigellae have been found in 19,38% of *P. ursinus* soon after capture (Brede, 1972), but so far not in wild living nonhuman primates (Carpenter and Cooke, 1965; DeVos et al., 1973; Takasaka et al., 1973 a). They are commonly found in newly arrived animals, during quarantine or conditioning. The infection rate in newly arrived animals has been shown to range from 5,5 % (Hartley, 1975) to 20% or more (Takasaka et al., 1964; Good et al., 1969; Morris et al., 1971; Mohan et al., 1975). Subsequently quite often the infection rate increases, occasionally reaching more than 80% (Schneider et al., 1960; Caraway et al., 1968; Morris et al., 1971; Weil et al., 1971).

It has therefore been suggested that monkeys and apes acquire the infection on their initial contact with man after capture (Cook, 1969).

Shigella-infections, like those by *Salmonella*-spp. frequently induce latent (silent) carrier states. Transportation stresses and other types of stress convert silent infections easily to overt disease (Cook, 1969).

Transmission between simian primates depends primarily on the direct fecal-oral route. Fecal excretion of *Shigella*-bacteria lasts for 1 to 4 weeks if not treated (DuPont and Pickering, 1980), with long term excretion of more than one year occurring in carriers. Fecal-oral transmission in simian primates is favoured by the dirty habits of nonhuman primates, but also by shortcomings in the housing or feeding of the animals enabling food to be picked up from the ground. Spread by vapors or droplets created by high-pressure hosing of contaminated floors also have been considered to be a posssible way of transmission (Fincham and Seier,

1981). Shigella strains remain viable in water at room temperature for periods up to 6 months (Felsen, 1945) unless they are killed by other bacteria.

It is worth mentioning that the primary human pathogen, *Sh. dysenteriae* type 1 (Shiga-Kruse) so far has not been found in a nonhuman primate (Pinkerton et al., 1967; Göldenboth, 1967; Hartley, 1975), but transmission of other serotypes from nonhuman primates to man usually results in serious disease (Bach et al., 1931; Rewell, 1949; Carpenter and Sandiford, 1952; Haas and Peterson, 1959; Robinson et al., 1965, 1966; Anderson and Peterson, 1971; Anon, 1971; Lamm et al., 1971 a, b; Dertinger et al., 1972; Toft and Dunagan, 1973; Fox, 1975; Zuidema and Smits, 1979). Some have resulted in a fatal outcome in children or elderly persons (Bach et al., 1931; Lamm et al., 1971 a, b).

Pathogenesis

Most Shigellae, other than Salmonellae, are primarily intracellular parasites of the enterocytes. Entero- and neurotoxin-production (Shiga-toxin) is restricted to *Sh. dysenteriae* 1 (Branham et al., 1949; Gemski et al., 1972; Levine et al., 1973; Fernandez et al., 1984). The primary virulence determinant of Shigellae is the entero-invasiveness, which is both chromosomally and plasmid mediated (DuPont and Pickering, 1980; Sansonetti et al., 1982). In the early phase of shigellosis the initial multiplication of Shigellae takes place in the lumen of the small intestines (Keusch, 1978) which might cause watery diarrhea by a net increase in jejunal secretion due to an as yet unknown pathogenesis (Giannella et al., 1974; Rout et al., 1975). 12 hrs after ingestion the bacteria transiently multiplied in the small intestines to concentrations of 10^7 to 10^9 bacteria/ml of luminal content (DuPont and Pickering, 1980).

Sh. flexneri and *Sh. sonnei* both produce a cytotoxin (Keusch and Jacewicz, 1977; O 'Brien et al., 1977, 1982) which is considered to cause an outpouring of fluid and electrolytes from the enterocytes. The exotoxins released by virulent *Shigella*-strains are also highly lethal to phagocytes, explaining the resistance of virulent *Shigella*-strains to phagocytosis.

Overt clinical dysentery results from proliferation of the pathogens within ileocolonic enterocytes (Hornic, 1977). The pathogenic potential of Shigellae is directly correlated to their ability to invade and multiply in the colonic mucosa (Sansonetti et al., 1982). Virulent (phase I) strains of Shigellae possess a 120 to 140 mdal virulence plasmid (Sansonetti et al., 1983) encoded for adherence to the epithelium, whose loss in transition to phase II is associated with a loss in virulence (Kopecko et al., 1980; Sansonetti et al., 1980).

Another virulence-associated factor might be the production of glycosidases which uncover receptor sites from the intestinal surface glycoproteins (Prizont, 1982) and the chemical composition of the surface-lipopolysaccharide-O-repeat polymer (Gemski and Formal, 1975; Formal and Gemski, 1977; Edebo et al., 1983). Virulent phase I *Sh. sonnei* possess a complete lipopolysaccharide with 2-amino-2-deoxyl-1-altruronic acid as a constituent of the O-repeat polymer (Kontrohr, 1977). This O-repeat polymer is lost in phase II bacteria (Romanowska et al., 1974). A similar situation seems to exist in virulent versus avirulent *Sh. flexneri*-strains (Formal and Gemski, 1977).

The Shigellae gain access to the epithelial cells by an endocytic process (Hale et al., 1979), whereby the cellular plasma membrane becomes ruffled and enfolded. Intracellular Shigellae are often found within phagosome-like membrane structures (Takeuchi et al., 1965, 1968). Bacterial invasion unevenly affects the epithelial cells, resulting in extrusion of epithelial cells and disappearance of the goblet cells which are killed by the intracellularly multiplying Shigellae. Microulcers are formed at the luminal surface by horizontal spread of the Shigellae. By confluence of the lesions, larger ulcerated areas might be generated, invasion of deeper structures and septicemias although not as common as in Salmonellae, nevertheless occur (Swenson and McClure, 1974).

Clinical Symptoms

In clinical shigellosis one has to distinguish between diarrhea (watery feces) and dysentery (mucoid, bloody feces).

Nonhuman primates: The symptoms in nonhuman primates are frequently more severe than those observed in man (Cook, 1969). Clinical shigellosis is much more common in Old World monkeys and apes than it is in New World monkeys (Hunt et al., 1978). Silent carriers are also common in nonhuman primates being an epidemiological risk but they are also prone to conversion to clinical disease after stressful events.

Sick monkeys are seen typically sitting in a hunched and cramped posture. In the initial stages they might vomit and pass watery feces, followed by bloody-mucoid stools containing leukocytes and erythrocytes in considerable numbers in a matrix of intestinal epithelia (David and Schirl, 1939; Mulder, 1971; Cooper and Needham, 1976; Fincham and Seier, 1981). They are anorectic, their faces drawn. Other general symptoms are fever, depression, tachypnea, marasmus, dehydration, and low haematocrit (Barnes, 1970; Simmons, 1973; Lemen et al., 1974; Takeuchi, 1982). The death rate attributable to shigellosis averaged 12,5% in *M. mulatta* and *M. fascicularis* (Hartley, 1975).

In experimental infections of *M. fascicularis* using 10^{10} viable cells bloody mucoid dysentery was noticed already 1 to 4 days p.i. (Honjo et al., 1964; Takasaka et al., 1969).

Extraintestinal shigellosis in nonhuman primates has been reported as gingivitis in rhesus monkeys, most prominent in the molar and premolar regions (McClure et al., 1976; Armitage et al., 1982, 1983); as air sacculitis (McClure et al., 1976), as abortion (Swenson and McClure, 1974; McClure et al., 1976), or as epicarditis and meningitis (Barnes, 1970).

Man: Bacillary dysentery of man is a biphasic illness. Volunteer experiments showed a transient multiplication of the bacteria in the small intestines 12 hrs p.i., associated with abdominal cramps, fever and watery diarrhea. After 1 to 3 days the temperature dropped, the pain and abdominal tenderness shifted to the lower abdominal quadrants and finally dysentery characterized by the painful, frequent passage of stools containing blood and mucus developed (DuPont and Pickering, 1980). Children may come down with the early form of the disease without clearly recognizable enteric symptoms.

Extraintestinal infections have been reported as upper respiratory symptoms, chest pain, pneumonia, seizures, skin rash, furuncles, conjunctivitis, vulvovaginitis, acute and chronic nephritis (Barrett-Connor and Connor, 1970; Davis, 1975; Stoll, 1986).

Pathology

Nonhuman primates: The watery diarrhea in early stages of the infection due to jejunal net secretion apparently is not accompanied by any gross or microscopical lesions. Only the ileocecal valve was edematous and at times haemorrhagic at 2 hrs p.i. (Takeuchi et al., 1968; Takeuchi, 1982). Some references, however, point to a concentration of lesions in the terminal ileum (see Fincham and Seier, 1981), or to lesions in the gastric mucosa, described as haemorrhages and ulcers in the fundic and antral mucosa (Kent et al., 1967; Takeuchi et al., 1975).

Usually the lesions in clinical shigellosis are confined to the ascending colon and cecum, as catarrhalic to ulcerative inflammation diminishing towards the rectum (Ogawa and Takasaka, 1966; Lindsey et al., 1971; Cooper and Needham, 1976; Takeuchi, 1982).

Three types of colitis have been separated by Ogawa et al., 1964:

a) catarrhalic colitis
b) diphtheric colitis
c) ulcerative colitis

with the diphtheric form being the most common one.

Catarrhalic colitis is characterized by flattening, pseudostratification, desquamation and necrosis of enterocytes, and infiltration predominantly with polymorphonuclear leukocytes of the edematous proprial and submucosal layers. The goblet cells disappeared. In many cases vascular fibrin thrombi were present (Takeuchi et al., 1968, 1975). Fat droplets were noted in surface and crypt cells. The surface coating of the epithelium became obscured corresponding to severe degeneration of the microvilli, terminal web and the extraneous coating layer.

In *diphtheric* colitis the mucosa is grossly swollen and haemorrhagic, the mucous membranes are covered with a pseudomembrane composed of nuclear debris, desquamated mucosa and fibrin. The edema is less pronounced than in catarrhalic inflammation. The inflammatory substrate gradually changes to histocytes, lymphocytes and fibroblasts. Epithelial regeneration can be observed in some areas. Microulcers were characterized by focal destruction of the basement membrane, covered by fibrinous exudate or by leukocytes and progressively extending into the lamina propria (Takeuchi et al., 1968).

Ulcerative colitis: Based on catarrhalic or diphtheric colitis ulcers develop usually at the site of lymphatic nodules. The ulcers extend down to the submucosal layer being edematous around the ulcers. Blood vessels are frequently filled with thrombi. In polypous ulcerative lesions perivascular histiocytic and lymphocytic infiltration was prominent (Ogawa et al., 1964).

The mesenteric lymphnodes usually show signs of leukocytic lymphadenitis (Mulder, 1971).

By means of fluorescent antibody staining, Shigella organisms can be identified mainly in the surface epithelium and the more luminal crypt cells; less frequently they can be found in the deeper crypt cells (Takeuchi et al., 1968).

As a secondary sequence of bacillary dysentery rectum prolapse sometimes has been ascribed to shigellosis (Cass, 1952; Ruch, 1959).

Shigella flexneri infected periodonal plaques were characterized by ulceration of the pocket wall and heavy lymphocytic and plasma cellular infiltration of the gingival connective tissues (Armitage et al., 1982). Aborted fetuses due to inapparent *Shigella*-infections were associated with petechial haemorrhages all over the body (Swenson and McClure, 1974), in other extraintestinal infections serofibrinous pericarditis, peritonitis and septic-embolic suppurative meningitis were observed (Barnes, 1970).

Man: The typical lesions of bacillary dysentery in man are catarrhalic colitis, reduction of the crypts and infiltration of the propria and submucosa by histiocytes and leukocytes (Ogawa et al., 1964). In fatal cases shortly after onset of clinical illness only edema and hyperemia of the mucosa and hypertrophy of lymphoid follicles have been noted. After a few days of illness, shallow mucosal ulcerations and exudation of fibrinous, purulent pseudomembranes occur. In prolonged disease segments of the membrane are sloughed, leaving a raw, beefy, shaggy mucosa and islands of hyperplastic lymphoid tissue (Nelson, 1981).

Treatment

The rapid emergence of single or multiple drug resistant Shigella strains has become one of the most serious problems in clinical medicine and in the husbandry of simian primates (Lindsey et al., 1971; Cooper and Needham, 1976). The drug resistance of Shigellae is R-plasmid mediated as it is in other *Enterobacteriaceae* and transmissible between different bacterial species and genera (Farrar et al., 1972; Kawatomari, 1972). As an extrachromosomal mechanism it is an example of infective heredity (Fox, 1975). R-factors have been observed particularly to tetracycline, streptomycin, sulphonamides, and ampicillin (Hardy et al., 1968; Farrar et al., 1972; Fleming, 1977; Frost and Rowe, 1983; Chun et al., 1984); and less frequently to trimethoprim or erythromycin (Greenberg et al., 1982; Chun et al., 1984). Quite often the drug resistance reflects the availability of the drugs over the counter or their uncritical use.

Streptomycin is contraindicated in shigellosis also for another reason: many *Sh. flexneri* 2 a strains actually depend on streptomycin for growth (Mel et al., 1970).

Unless drug resistance renders them useless, the following dosages of antibiotics have been recommended in nonhuman primates (Mulder, 1971):

	monkeys	chimpanzees
chloramphenicol	250 mg orally 3-5 days	0,5-2 gr orally 3-5 days
chlortetracycline	50 mg/kg orally	50 mg/kg orally
oxytetracycline	100-500 mg i.m. daily	0,5-2 gr i.m. daily
sulfadiazine	200 mg/kg added to 15 ml kaopectate plus 200 mg neomycin	

triple sulfonamides 150 mg/kg not to exceed 6 gr daily

For elimination of carrier states trimethoprim-sulfamethoxazole is recommended, given 2 × daily in dosages of 8 mg trimethoprim and 40 mg sulfamethoxazole perorally in 3 ml of sterile water (Pucak et al., 1977).

Rifampicin showed remarkable therapeutic effects on bacillary dysentery in *M. fascicularis* at dosages of 200 mg/animal at 3 consecutive days, but was uneffective in diarrhoic cases (Takasaka et al., 1973b).

Another method of treatment would be the competitive displacement of the pathogens by other, nonpathogenic microorganisms such as *Saccharomyces boulardi*. This should be considered as an alternative to antibiotic treatment (Fincham and Seier, 1981).

3.6.4 Klebsiella Spp.

Klebsiella spp. are often found as saprophytic organisms of the digestive and respiratory tracts in both human beings and nonhuman primates. Nevertheless, under certain conditions they may become true pathogens. The *Klebsiella* species most often encountered in nonhuman primates is *K. pneumoniae*, although *in praxi* not every *Klebsiella* strain can be speciesdifferentiated.

Klebsiella Pneumoniae
K. pneumoniae types 1 and 2 are among the most common causes of fatal pneumonia in captive baboons (Pinkerton, 1972) but they can also be isolated from the normal oral, fecal or upper respiratory flora in baboons (Pinkerton et al., 1967; Brede and Murphy, 1968). Similarly *K. pneumoniae* has been found either as a pathogen or a commensal organism in marmosets (Gebauer and Raethel, 1958; Deinhardt et al., 1967; own results), rhesus monkeys (Hunt et al., 1968; Houser et al., 1970); a gibbon (Kageruka et al., 1971); mangabey (Kageruka et al., 1971); night monkeys (Giels et al., 1974); and chimpanzees (Schmidt and Butler, 1971; Butler, 1972; Eichberg, 1985).

Agent
Klebsiellae are short, gram-negative, encapsulated, nonmotile rods of 5 μm length. They grow in aerobic conditions without special nutrient requirements. After 24 hrs of incubation at 37 °C, glistening, mucoid colonies stained pink to violet appear on McConkey- and EMB agar. Single colonies are half-spherical, but due to a distinct confluent tendency soon form large, confluent lines along the incubation streaks if present in larger numbers.

The biochemical characteristics of the *Klebsiella*-genus are given in Table 17 according to Ewing and Martin (1974) and Finegold and Martin (1982).

Epidemiology
Klebsiella pneumoniae, being present on the mucous membrane surfaces of the healthy respiratory and digestive tracts may form overwhelming infections after viral or parasitic infections, stress induced breakdown of the immunity or, probably most commonly, after elimination of the normal respiratory or intestinal flora by indiscriminate use of antibiotics. Klebsiella usually are much more resistant to antibiotics than the other Enterobacteria are. Once they are able to invade the tissues they cause cellular destruction, purulent inflammation or abscess formation.

Transmission between animals and possibly from animal to man is probably by direct contact. It does not necessarily lead to clinical disease, but antibiotic resistance may be so transmitted.

Table 17. Biochemical Differentiation of *Klebsiella* spp.

Test or substrate	*K. pneumoniae*	*K. ozaenae*	*K. rhinoscleromatosis*	*K. oxytoca*
Glucose	+ (gas)	V	–	+ (gas)
Lactose	+	d (V)	d (V)	+
Saccharose	+			
Ornithin decarboxylase	–	–	–	–
Lysine decarboxylase	+	V	–	+
Arginine dihydrolase	–	–	–	–
Indol	V	V	+	+
H_2S	+			
Adonitol	V	+	+	V
Dulcitol	V	–	–	V
Mucate	+	V	–	
Malonate	+	–	V	
Phenylalanine	V	–	–	–
Citrate (Simmons)	+	–	–	+
Urease	+	d	–	+
Methyl red	V	+	+	V
Voges Proskauer	+	–	–	+
Inositol	+	V	V	+
Mannitol	+	+	+	+
Salicin	+	+	+	+

Clinical Symptoms

Nonhuman primates: *K. pneumoniae* has been shown to induce bronchopneumonia, air sacculitis, meningitis, arthritis, peritonitis and cystitis in monkeys and apes (Houser et al., 1970; Snyder et al., 1970 - cit. from Schmidt and Butler, 1971; Schmidt and Butler, 1971).

Klebsiella-bronchopneumonia is characterized by loss of appetite, refusal to drink, depression and respiratory difficulties (Kageruka et al., 1971; Schmidt and Butler, 1971; Butler, 1972). Urinary tract infection similarly results in depression, loss of appetite and a viscous urine (Houser et al., 1970). Air sacculitis, if at all recognized during life, induces a purulent nasal discharge and apparent swelling in the cervical and intermandibular areas (Giles et al., 1974).

Man: The most common serotype in man is *K. pneumoniae* I, which causes bronchopneumonia, pulmonary abscesses and a thick, gelatinous sputum (Falls, 1964). *K. pneumoniae* causes also meningitis in man (Köhler and Mochmann, 1975).

Pathology

Nonhuman primates: The principal lesions of *K. pneumoniae* infections in nonhuman primates are purulent to fibrinopurulent inflammations. In the lung tissues these appear as purulent bronchopneumonia, sometimes accompanied by alveolar haemorrhages or alveolar microabscesses and frequently associated with extrapulmonary secondary manifestations. Concomitant *K. pneumoniae* lesions are often found as focal purulent hepatitis, myocarditis or as a serous to fibrinous-purulent peritonitis, pleuritis or pericarditis (Kageruka et al., 1971; Schmidt and Butler, 1972). *Klebsiella*-airsacculitis was expressed by a variety of lesions ranging from diffuse hyperemia of the air sac mucosa, petechial haemorrhages, thickening and edema of the mucosa and deeper surrounding tissues to suppurative inflammation of all parts of the mucosa with distension of the air sac, accompanied by thromboemboly of subepithelial vessels and a suppurative cervical lymphadenitis (Giles et al., 1974). Urinary tract infections in young rhesus monkeys resulted in abscesses of the urinary bladder and diffuse fibrinopurulent peritonitis. Systemic infection in this species proceded to polyarthritis, meningitis and diffuse purulent peritonitis (Houser et al., 1970).

Man: The most common pathological picture of *K. pneumoniae* infections of man is characterized by a severe broncho-pneumonia which spreads rapidly, sometimes mimicing lobar pneumonia, and frequently associated with pleural involvement and empyema. The cut surface of the lung is extremely friable, the exudate is singularly tenacious, gelatinous or slimy (Falls, 1964; Sanford and Luby, 1981).

Treatment

There is a natural or readily induced resistance of *K. pneumoniae* to most antibiotics, especially to penicillin (Houser et al., 1970; Pinkerton, 1972). Hunt et al., 1968 used a weekly succession of 75 mg/kg streptomycin sulfate, 200 mg/kg cephalotin and 50 mg/kg kanamycin all given twice daily. Other authors have used sulfonamides, streptomycin and broad-spectrum antibiotics, but care has to be taken to not destroy the normal, balancing intestinal flora. In *K. pneumoniae*-pneumonia a rapid diagnosis is a necessity and a supportive treatment especially recommended (Pinkerton, 1972).

3.6.5 Klebsiella Oxytoca

K. oxytoca, whose validity as a separate species, distinct from *K. pneumoniae* is debated (Blobel and Schliesser, 1981) has been described as a causative agent of catheter associated septicemia in man. It also caused septicemia in 8 out of 20 rhesus monkeys in one laboratory, which originated from chronic indwelling catheters (DaRif and Rush, 1983).

K. oxytoca does not grow at 44.5 °C or above. It produces a dark-brown pigment on gluconate and ferric citrate containing media (Orskov, 1974). Other characteristics are given in Table 17. *K. oxytoca* is indole-, gelatin- and pectate-positive in contrast to the classical *K. pneumoniae*-strains (Lautrop, 1956; von Riesen, 1976).

K. oxytoca has been isolated from contaminated food. In this laboratory it has been repeatedly isolated from normal and diarrhoic marmoset faeces.

For treatment of catheter induced septicemia the removal of the infected catheters is required, in additon to proper antibiotic treatment.

3.6.6 Yersinia Spp.

The genus *Yersinia*, formerly included in the genus *Pasteurella*, was proposed in 1944 (van Loghem, 1944). It is currently included in the family Enterobacteriaceae (Mollaret and Thal, 1974) because of its having a common antigen with the other Enterobacteriaceae. It is motile with peritrichous flagella; fermentates glucose, reduces nitrate and has a negative oxidase-reaction (Le Minor, 1979). Two species, *Y. pseudotuberculosis* and *Y. enterocolitica* are zoonotic agents occurring naturally in nonhuman primates.

3.6.6.1 Yersinia Pseudotuberculosis (Pasteurella Pseudotuberculosis)

Y. pseudotuberculosis, first described by Malassez and Vignal (1883, 1884), apparently originated in northern and central Europe possibly due to a transmutation of the medieval plague bacilli (Stovell, 1980). It is still most prevalent in northern and central Europe, but is rarely seen in Italy, Greece, Spain, Portugal, the Near- and Middle East. The African continent seems to be free with the exception of Algeria and Morocco. The infection has only been seen once in Southeast Asia and India, but has been noted several times in Japan (Tsubokura et al., 1982), Australia, and New Zealand (Mair, 1975).

A difficulty in the evaluation of case reports on yersinial pseudotuberculosis is the inclusion of *Y. enterocolitica* into the species *Y. pseudotuberculosis* until 1964 when it was proposed to be a different species on its own (Frederiksen, 1964). *Y. pseudotuberculosis* in its present definition infects a wide variety of birds and mammals including man and nonhuman primates.

Y. pseudotuberculosis induced diseases in the nonhuman primates are not restricted to any particular host species as shown in Table 18, which summarizes reports published after 1970. Earlier reports have not been included because of the taxonomical problems mentioned above.

In animals pseudotuberculosis usually is a serious infection of sometimes endemic dimensions, whereas in man it usually has a more benign, sporadic nature.

Agent

Y. pseudotuberculosis is generally a bipolar, gram-negative, coccobacillary to pleomorphic, round ended, noncapsulated and nonspore-forming bacterium. The bipolar appearance can be better revealed using Wayson 's stain. It is usually straight of axis, but in some strains it forms short chains or longer filaments; the single bacterium might be curved. Most cells are coccobacillary or ovoid and usually of less than 1,0 µm in length, but nevertheless in longer filaments they may reach up to 6,0 µm. *Y. pseudotuberculosis* is motile at 22 to 27 °C by means of 1 to 6 peritrichous flagella or 1 to 2 parapolarly located flagella. In india ink preparations an envelope may be visible.

Y. pseudotuberculosis is aerobic or facultatively anaerobic. Optimum growth, virulence preservation and expression of smooth somatic antigenic components

Table 18. *Yersinia pseudotuberculosis* in Nonhuman Primates (1970-1984)

Primate species	Y. pseudotuberculosis-Serotype	Reference
Old World monkeys:		
Macaca fascicularis	I B, III	Bronson et al. (1972); Rosenberg et al. (1980); Strickland et al. (1982); MacArthur and Wood (1983)
Macaca radiata	III	Bronson et al. (1972)
Macaca nigra		Gillespie and Arnold (1982)
Erythrocebus patas	I, I B	Keymer (1972); Hirai et al. (1974); Kageruka et al. (1976)
Cercocebus fulliginosus	I B	Bronson et al. (1972)
Cercopithecus aethiops	II	Keymer (1972); Kageruka et al. (1976)
Cercopithecus cephus	I, II	Kageruka et al. (1976); Fukuda and Asakura (1960); Jentzsch et al. (1969)
Cercopithecus ascanius	I	Kageruka et al. (1976)
Cercopithecus erythrotis	II	Kageruka et al. (1976)
Cercopithecus mona		Frolka (1980)
Cercopithecus diana		Frolka (1980)
Cercopithecus sabaeus	I	Kageruka et al. (1976)
Cercopithecus spp.		Schröder (1980)
Colobus polykomos		King (1977); Arnold (1981)
New World monkeys:		
Lagothrix sp.		Keymer (1972)
Ateles sp.		Pierce et al. (1973)
Cebus sp.		Keymer (1972)
Saimiri sciureus	I, III	Kageruka et al. (1976); Buhles et al. (1981)
Callimico goeldii		Arnold (1981); Gillespie and Arnold (1982)
Callithrix jacchus	II	Kageruka et al. (1976)
Callithrix argentata argentata		Arnold (1981)
Callithrix argentata melanura		Gillespie and Arnold (1982)
Saguinus labiatus	II B	Taffs and Dunn (1983)
Prosimians:		
Galago crassicaudatus	I	Kageruka et al. (1976); Chang et al. (1980)

are obtained at 27° to 30 °C, although much lower or higher temperatures up to 43 °C are tolerated. It grows easily on meat-extract media. In broth media the S-phase produces a uniform turbidity and after several days a sediment, whereas the R-phase sediments out without previous turbidity being apparent. *Y. pseudotuberculosis* also grows on all usual media such as EMB-, McConkey-, desoxy-cholate-citrate-, Endo-agar etc., achieving maximal and characteristic growth in less than 48 hrs. However, isolation of Yersinae from fecal specimens can be difficult because of overgrowth by normal intestinal flora (McClure and King, 1984). On solid media containing serum or blood smooth to slimy, light, transparent colonies with a diameter of 2 to 3 mm develop within 48 hs at 30 °C (at 37 °C the colonies are thin, irregular with rough edges). Later on, the colonies develop from low convex to umbonate, they become granular, greyish-yellow and opaque. The edges become effusive with radial striations that stretch back to the umbronate center. Haemolysis is not produced, the characteristics on TSI-agar are: alkaline slant and acid butt. The biochemical reaction of both *Y. pseudotuberculosis* and *Y. enterocolitica* are compared in Table 19.

Serotyping by tube agglutination (Mair et al., 1960) distinguishes 6 main serotypes with four subtypes based on 15 somatic (O) antigens (Table 20). Two new serotypes (VII and VIII) have been added by Tsubokura et al. (1984).

The most common serotype in man and animals is type I (64 %) followed by type II (17%), type III (12%), type IV (7%), type V (2%), and type VI (1%) with type I probably being virulent for all species (Stovell, 1980).

Table 19. Biochemical Differentiation of *Y. pseudotuberculosis* and *Y. enterocolitica*

Substrate	Y. pseudotuberculosis	Y. enterocolitica
Glucose	+ (no gas)	+ (no gas)
Lactose	−	− (+)
Sucrose	−	+
Maltose	+	d
Rhamnose	+	−
Xylose	+	d
Cellobiose	−	+
Raffinose	−	−
Arabinose	+	+
Melibiose	+	−
Trehalose	+	d
Mannitol	+	+
Sorbitol	−	+
Adonitol	±	−
Indole	−	+ or −
Salicin	+	− (+)
Citrate (Simmons)	−	−
H_2S	−	−
Urease	+	+
Phenylalanine deaminase	−	−
Oxidase	+	+
Catalase	+	+
β-Galactosidase	+	+
Nitrate reduction	+	+ (−)
Methyl red	+	+
Voges Proskauer	−	− (+)
Lysine decarboxylase	−	−
Arginine dihydrolase	−	+
Ornithin decarboxylase	−	+
Esculin hydrolysis	+	− (+)
Gelatin hydrolysis	−	−

Table 20. Serotypes of *Yersinia pseudotuberculosis*

O-group	O-Subgroup	R-antigen	Thermostable O-antigens	Thermolabile H-antigens
I	A	(I)	2,3	a, c
	B	(I)	2,4	a, c
II	A	(I)	5,6	a, d
	B	(I)	5,7	a, d
III	-	(I)	8	a
IV	A	(I)	9,11	b or a, b
	B	(I)	9,12	a, b, d
V	A	(I)	10,14	a, e (b)
	B		10,15	a
VI	-	(I)	13?	a
VII			19	ND
VIII			20	ND

Y. pseudotuberculosis is highly resistant to adverse environmental conditions, surviving in soil for 11 to 18 months (Mollaret, 1963); in water for several months and in meat at 4 °C for up to 145 days (Serov et al., 1968 - cit. from Mair, 1975). In fecal deposits it survives in cold weather for several days (Paterson and Cook, 1963). Transmission is mainly by contamination of food or by smear infections (Fig. 8).

Percutaneous infection by contaminated arthropods has been suspected, as in one disease outbreak, the agent was isolated from ixodid ticks removed from infected cattle (Ul 'Yanova, 1961) and as there is an association between the occurrence of an unusual epididymo-orchitis of rams and the tick season in England (Jamieson and Soltys, 1947 - cit. from Stovell, 1980). Inhalatory infection may play some role in avians as suggested by the large number of respiratory tract lesions found in birds (Steele, 1975 - cit. from Stovell, 1980).

Pathogenesis

Y. pseudotuberculosis probably enters the intestinal drainage system in the ileal region (Daniels, 1961), allowing the development of focal lymphatic abscesses. The first step might be a focal, rapidly healing intestinal lesion parallel to the established pathways of *Y. enterocolitica* (Stovell, 1980). The next step is involvement of the regional lymph nodes. Haematogenous spread and septicemia may occur, but these are not part of the normal sequence of the infection.

Clinical Symptoms

Nonhuman primates: Pseudotuberculosis in nonhuman primates may appear as a sporadic disease (Bronson et al., 1972; Keymer, 1972; Pierce et al., 1973; King, 1977; Chang et al., 1980; Schröder, 1980; Arnold, 1980, 1981; Gillespie, 1982), but it often follows an enzootic course, sometimes with mortality rates of more than 10% of the animals at risk (Kageruka et al., 1976; Rosenberg et al., 1980; Buhles et

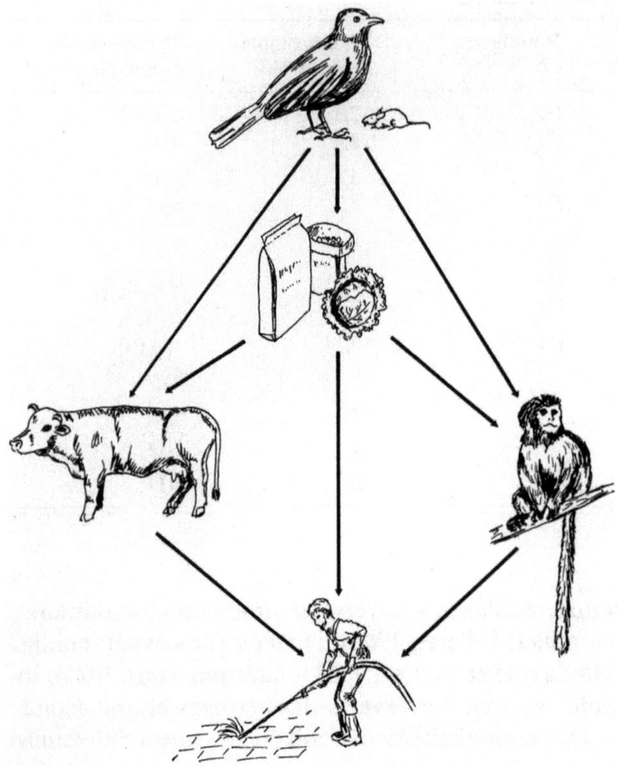

Fig. 8. Yersiniosis-Cycle

al., 1981; Strickland et al., 1982; MacArthur and Eood, 1983; Taffs and Dunn, 1983).

The evolution of pseudotuberculosis in simian primates usually follows an insidious course without pathognomonic symptoms (Kageruka et al., 1976). The animals are either found dead or suffer from only nonspecific symptoms such as weakness, depression, watery or haemorrhagic diarrhea, dehydration, fever or hypothermia, hyponatremia and hypochloremia (Bronson, 1972; Hirai et al., 1974; Rosenberg et al., 1980; Buhles et al., 1981; Strickland et al., 1982; Taffs and Dunn, 1983; McClure and King, 1984).

Increased abortion- and stillborn rates have been observed during outbreaks in Java-macaques (Rosenberg et al., 1980) and squirrel monkeys (Buhles et al., 1981).

Man: Probably most people develop transitory subclinical or latent infections after being exposed to the causative agent. Children, aged or debilitated persons are most susceptible to clinical disease (Stovell, 1980). A higher prevalence among male adolescents is sometimes remarkable (Mair, 1975). The human disease, similar to animal infections, is more prevalent in northern and central Europe than in the U.S. or in southern Europe (Stovell, 1980).

Incubation periods of 10 days or less have been reported (Mair, 1975; Stovell, 1980), in other cases circumstantial evidence has pointed to prolonged incubations.

Clinical pseudotuberculosis in man appears as:

1. Pseudoappendicitis, which can be caused by
   a) acute mesenteric lymphadenitis (retikulär - abszedierende Lymphadenitis Masshoft und Dölle) or
   b) acute terminal ileitis
2. scarlatiniform fever in the USSR
3. erythema nodosum
4. polyarthritis
5. septicemia

Ad 1. Pseudoappendicitis is marked by abdominal pains primarily of the right lower quadrant, nausea, vomiting, and a rapidly developing fever up to 40° C. Mesenteric lymphadenitis is most common in children and young adults and is usually benign.

Ad 2. Scarlatiniform fever: *Y. pseudotuberculosis* has caused an acute generalized infection in the USSR, characterized by fever, scarlatiniform eruptions and acute polyarthritis with arthralgia. Occasionally icterus and abdominal pain are also seen.

Ad 3. Erythema nodosum is a secondary immunologically mediated complication occurring in perhaps 15 to 20% of cases of acute infection. After a prodromal phase of approximately 2 weeks, one to three crops of eruption develop, accompanied by fever and fatigue, which reach their maximum values in a few days. The disease disappears in 2 to 3 weeks, relapses do not occur.

Ad 5. Septicemia occurs particularly in older people with liver damage or in patients with blood disease. The clinical picture is dominated by a severe septicemic typhoidal course with mortality of up to 50% even with antibiotic treatment.

Pathology

Nonhuman primates: The gastrointestinal tract is involved in virtually all cases as an acute ulcerative enterocolitis with villous tip necrosis, erosions or ulcerations in the small and large intestines. The mucosa and submucosa are edematous, whilst the mucosal surface is covered by fibrinous pseudomembranes. The discrete ulcerations often lie over the lymphoid follicles, they rarely extend down to the submucosa.

The mesenteric lymphnodes are usually swollen and hyperemic. Lymphadenitis simplex or purulenta may be present or absent. Focal necroses or necrotic abscesses develop particularly in the marginal sinuses.

Multifocal miliary necrotic foci (1,5 to 2 mm in diameter) also develop in the liver and spleen, which are sometimes, but not always, swollen. In *Saguinus labiatus* and *Galago crassicaudatus* the hepatic and splenic necroses were the most common lesions (Taffs and Dunn, 1983). In the liver the intralobular necroses were filled by the emigration of variable numbers of polymorphonuclear leukocytes, swollen endothelial cells and fibrin exudation.

Uterine lesions, if present, were characterized by acute, erosive endometritis with focal necroses, haemorrhages and myometritis (Buhles et al., 1981).

In the galagos, a nonsuppurative pneumonitis was scattered throughout the lungs of all 3 animals.

Man: The lesions in human pseudotuberculosis vary slightly depending on the type of disease.

In pseudoappendicitis they are more or less confined to the mesenteric lymph nodes as swelling and redness. The cut lymphonodular surfaces show follicular hyperplasia and miliary abscessation - the characteristic necrotizing abscesses of acute mesenteric lymphadenitis Masshoff und Dölle. The appendix itself is usually not affected. However, the terminal ileal and caecal mucosa may be thickened due to hyperemia and edema (Otto et al., 1976).

In the enteric form the lymphoid follicles of the terminal ileum are hyperplastic and edematous. Irregular ulcers and fibrinous-serous peritonitis accompany the lesions (Otto et al., 1976).

The lesions in the septicemic form are profound, affecting many organs as nodular caseous or abscessed necrotic foci of 1 to 10 mm diameter particularly in the enlarged liver, spleen, mesenteric lymph nodes and pancreas.

Treatment

The usefulness of antibiotics depends on the development of antibiotic resistance. Ippen and Stoll (1959), for that reason, advised the avoidance of streptomycin and penicillin. Bronson et al. (1972) found all isolates to be sensitive to tetracycline, neomycin and cephalotin, whilst some strains were resistant to chloramphenicol, streptomycin, furazolidone, penicillin and polymyxin. McClure and King (1984) found all strains resistant to lincomycin, methicillin and erythromycin, with some strains also resistant to polymyxin or tetracycline. All strains were susceptible to ampicillin, penicillin, cephalotin, chlormycetin, kanamycin, gentamycin and trimethoprim-sulfamethoxazole. Garlt (1965) successfully used 50 mg/day oxytetracycline per os in Cercopithecids.

3.6.6.2 Yersinia Enterocolitica (Pasteurella X)

Y. enterocolitica was formerly contained within the species *Pasteurella pseudotuberculosis* until 1939, when Schleifstein and Coleman separated the species based on their biochemical differences.

Y. enterocolitica, first named *Bacterium enterocoliticum*, has also been known as *Pasteurella* X, *Pasteurella* Y or *P. pseudotuberculosis*-like (Staley, 1980). Frederiksen (1964) finally proposed the adoption of the current terminology *Yersinia enterocolitica*.

Today *Y. enterocolitica* is known to cause a number of syndromes in man and animals including nonhuman primates. During the past 10 years the incidence of human and animal *Y. enterocolitica* infections has shown a marked increase in Europe and in North America (Chiesa et al., 1984). The cases reported in nonhuman primates since 1970 are summarized in the appendix.

Agent

The general characteristics of *Y. enterocolitica* are similar to those of *Y. pseudotuberculosis* and have been described in the previous chapter. The biochemical differences between the two species have also been described in the Table 19 of that chapter.

Y. enterocolitica strains, unlike *Y. pseudotuberculosis*, are biochemically heterogenous and are also influenced by the incubation temperature (Brenner, 1981). These biochemical differences between the different *Y. enterocolitica* strains led to the establishment of several biotyping schemes.

Each biotype is composed of one or more serotypes (Staley, 1980), the serotyping system utilizing O- or H-antigens or combinations of both. With respect to serotyping it is noteworthy that there are considerable cross reactions between *Y. enterocolitica* and other bacteria (Corbel, 1979; Thomas et al., 1983), especially between anti-*Brucella abortus* antisera and *Y. enterocolitica* serotype 9 (Sonnenwirth, 1974). *Y. enterocolitica*, like other Enterobacteriaceae, posses lipopolysaccharide determinants corresponding to O-antigens which are heat-stable and alcohol-resistant. The demonstration of O-antigens depends on the use of S-colonies. O-serotyping is often hampered by the presence of R-agglutinins in most antisera, by envelope - (K-)antigens (K1 to K6) and by the marked tendency of *Y. enterocolitica* to change to R-forms. In order to select S-forms for serotyping the use of SS-, McConkey- or desoxycholate-citrate-lactose agar for cultivation is recommended (Wauters, 1981).

So far at least 57 O-factors and a great number of H-antigens have been found in various combinations (Wauters, 1981). For practical purposes 34 serotypes of *Y. enterocolitica* are recognized (Staley, 1980). Serotyping of *Y. enterocolitica* is not a tool for classification but rather of epidemiological value. The most common serotypes in man in North America are 03, 08, 09 and 05,27 (Bottone, 1977; Schiemann, 1984); that in India 04 (Varghese et al., 1984). Recently a new serotype (013a, 13b) was isolated from man and nonhuman primates in the USA (Toma et al., 1984). *Y. enterocolitica* possess a 40- to 82-megadalton virulence-plasmid associated with specific outer membrane polypeptides (Chang and Doyle, 1984; Skurnik, 1984). Virulence of *Y. enterocolitica*, like that of *Y. pestis*, is associated with the capacity to take up Congo-red, and therefore a special Congo-red agar for the differentiation between virulent and avirulent *Y. enterocolitica*-isolates has been developed (Prpic et al., 1983).

Y. enterocolitica 03-strains have been shown to produce heatstable enterotoxins in rabbit and mice gut loop tests (Sakazaki et al., 1974; Pai and Mors, 1978). Their chemical structures resemble those of *E. coli*-ST (Takao et al., 1984) and their genetic control mechanisms are still debated (Velin and Emödy, 1982). Heat-labile enterotoxins could not so far be detected, but all strains tested showed invasive capacities (Feeley et al., 1979).

Epidemiology

Y. enterocolitica seems to be a ubiquitous organism, being demonstrated largely in terrestrial and aquatic ecosystems (Bercovier et al., 1978; Saari and Jansen, 1979) and infecting a wide variety of mammals (Hurvell, 1981), especially chinchillas (Beht, 1962; Daniels, 1963; Knapp and Thal, 1963; Brack, 1966), pigs (Dickinson and Mocquot, 1961) and hares (Mollaret and Lucas, 1965).

Y. enterocolitica has a wider geographical range than *Y. pseudotuberculosis*, being found also in the Americas (Mollaret et al., 1970; Pizsolitto et al., 1979), the Near and Middle East and several African countries (Mair, 1975; Mollaret et al., 1979; Robins-Browne et al., 1981). The highest concentration has, however, also been found in northern Europe.

The seasonal influences by cold weather are similar to, but less pronounced than in *Y. pseudotuberculosis* infections.

Human yersiniosis enterocolitica is due to strains circulating from man to man or to contamination of food (Mair, 1975; Bercovier, 1980; Lee et al., 1981; Shayegani, 1981) or water (Highsmith et al., 1977; Kapperud, 1977). Transmission from animal to animal or from animal to man has not been established (Mair, 1975), but animals may be the reservoir and foods may act as the vehicle of infection (Staley, 1980). With respect to nonhuman primates it should be noted that quite a number of nonhuman primate infections were caused by serotype 3,9 or 5,27 (see appendix), the most frequent human pathogens.

Pathogenesis

Y. enterocolitica-strains isolated from enteritis have been found to be enteroinvasive and also to have positive Sereny tests, indicating the importance of enteroinvasiveness in the pathogenesis of Yersiniosis enterocolitica (Olsson et al., 1980). Enteroinvasiveness is plasmid controlled (Heesemann et al., 1983; Robins-Browne et al., 1985).

Y. enterocolitica also produces a heat-stable enterotoxin, whose properties may closely resemble those of enterotoxic *E. coli.* The enterotoxin gives a positive rabbit skin permeability test like *E. coli* LT (Olsson et al., 1980), while culture supernatants give positive rabbit loop reactions (Sakazaki et al., 1974). Enterotoxin-production of *Y. enterocolitica* is controlled by a chromosomal gene (Robins-Browne et al., 1985).

Intragastric inoculation into mice is followed in less than 24 hrs by neutrophilic infiltration in the Peyer 's patches of the ileum. By 5 days the infection spreads to the mesenteric lymph nodes causing abscess formation, thereafter the infection becomes generalized (Wormser and Keusch, 1981).

Clinical Symptoms

Nonhuman primates: The clinical symptoms of *Y. enterocolitica* infections in nonhuman primates, although principally similar to those of *Y. pseudotuberculosis*, are usually more dramatic. The animals were found dead or succumb to a syndrome of hypothermia, collapse, abdominal pain (Breshnahan et al., 1984) and coma with or without diarrhea within hours (Mair, 1970; McClure et al., 1971; Pattyn et al., 1978; Skavlen et al., 1983). The diarrhea may be bloody or watery (McClure et al., 1971).

Man: The majority of patients are young. The most common form in children less than 5 years of age is a self-limiting gastroenteritis (DuPont and Pickering, 1980; Varghese et al., 1984). The stools may contain blood and mucus or may appear as profuse watery diarrhea. Other symptoms include vomiting, abdominal pain and fever. The duration of the illness is usually two weeks, but chronic diarrhea may occur (Wormser and Keusch, 1981).

Pseudoappendicitis, which occurs in children over 5 years of age results in fever (39 °C), elevated pulse rate, rebounding tenderness in the right iliac fossa, nausea, vomiting and diarrhea together with elevated WBC (Gurry, 1974 - cit. from Staley, 1980).

Recovery from gastrointestinal *Y. enterocolitica* infections is occasionally followed by a nonsuppurative arthritis, which has an acute onset and usually affects several joints. It is mostly self-limiting, but may persist for several months. Some patients may also experience a benign form of carditis, Reiter 's syndrome or prolonged arthralgias (Aho et al., 1981).

Other extraintestinal signs are erythema nodosum, septicemia and pyuria (Winblad, 1981).

Erythema nodosum, which is associated with a variety of infections is a common complication of *Y. enterocolitica* infection. The skin lesions resemble other eruptions, but they are sometimes not restricted to the legs or arms, and occur also on hands, feet or neck.

Pathology

Nonhuman primates: The lesions in nonhuman primates which died from *Y. enterocolitica*-infections were dominated by the triad of necrotic foci in liver and/or spleen and ulcerative enteritis, which was typically most pronounced in the terminal ileum, caecum and anterior colon (Mair, 1970; McClure et al., 1971; Baggs et al., 1976; van Damme et al., 1978; Pattyn et al., 1979; Skavlen et al., 1983; Breshnahan et al., 1984). The intestinal necrosis sometimes involves the entire thickness of the gut wall.

Man: The pathologic lesions of the gastrointestinal tract infections include ulceration of the gut (occasionally including the stomach), microabscesses within the Peyer 's patches and abscesses in the liver and spleen, which are occasionally also found in the lungs. The yellowish-white abscesses are dominated by neutrophils except in the lung lesions where plasma cells may predominate (Rabson et al., 1972; Sternby, 1973 - both cit. from Staley, 1981). The extent and distributions of these lesions may slightly depend on the type of disease.

Biopsies taken from sigmoids in the cases of gastroenteritis show signs of chronic, nonspecific inflammation (mononuclear cells and lymphocytes) and crypt abscesses (Wormser and Keusch, 1981).

In pseudoappendicitis mesenteric lymphadenitis and terminal ileitis predominate. Sometimes large conglomerate masses are formed by the confluent, enlarged mesenteric lymph nodes, usually appearing in the right lower quadrant. Histopathologically the lymph nodes show proliferation of large pyroninophilic cells and focal necrosis.

In terminal ileitis the most distal part of the ileum is edematous and inflamed with a hardened mesentery. Superficial ulceration may be present along the mucosal surface down to the muscularis. The diseased part of the ileum is usually sharply demarcated from the adjacent healthy intestine.

Bacteremia with *Y. enterocolitica*, usually a sequela of other predisposing diseases is highly fatal and often associated with hepatic or splenic abscesses or infections of meninges, synovium, skin, lung, bone or kidney.

The arthritis associated with *Y. enterocolitica* is of a nonpurulent type and has been especially reported in Scandinavian countries (Aho et al., 1981).

Treatment

Y. enterocolitica, in contrast to *Y. pseudotuberculosis* produces ß-lactamase and therefore is not or only little affected by antibiotics of the ß-lactam group (Penicillin, cephalothin, ampicillin, carbenicillin). The ß-lactamase is chromosomally mediated, other plasmid-mediated R-factors can be acquired from other enterobacteriae (Cornelis, 1981).

Y. enterocolitica generally responds satisfactorily to tetracycline, chloramphenicol, neomycin, gentamycin and streptomycin.

References

1. Aho, K.; P. Ahvonen; O. Laitinen, and M. Leirisalo: Arthritis associated with *Yersinia enterocolitica* infection; pp. 113–124 in "Yersinia enterocolitica" (Bottone, E.J., ed.) CRC Press, Boca Raton (1981)
2. Anderson, E.S. and J.J. Lewis: Drug resistance and its transfer in *Salmonella typhimurium.* Nature *206,* 579–583 (1965)
3. Anderson, H.W. and D.R. Peterson: Shigellosis and salmonellosis from a spider monkey. CDC Morbid. Mortal. Weekly Rep. *20* (33), 191–292 (1971)
4. Anon: Pet spider monkey transmits shigellosis and salmonellosis to humans. Lab. Primate Newsl. *10* (4), 12 (1971)
5. Appleby, E.C.; O. Graham-Jones, and S.A. Keeble: Primate diseases infectious to man. Vet. Rec. *75,* 81–86 (1963)
6. Arata, A.A.; J.B. Vughn; K.W. Newell; R.A.J. Barth, and M. Gracian: *Salmonella* and *Shigella* infections in bats in selected areas in Columbia. Amer. J. trop. med. Hyg. *17,* 92–95 (1968)
7. Armitage, G.S.; T.A. Banks; E. Newbrun; J.S. Greenspan; C.I. Hoover, and H.H. Anderson: Immunologic observations in macaques with Shigella-associated periodontal disease. J. Periodont. Res. *18,* 139–148 (1983)
8. Armitage, G.C.; E. Newbrun; C.I. Hoover, and J.H. Anderson: Periodontal disease associated with Shigella flexneri in rhesus monkeys. Clinical, microbiologic and histopathologic findings. J. Periodont. Res. *17,* 131–144 (1982)
9. Arnold, L.: Pathology report. Dodo J. Wildl. Preserv. Trust *17,* 96–104 (1980)
10. Arnold, L.: Pathology report. Dodo J. Wildl. Preserv. Trust *18,* 86–94 (1981)
11. Bach, F.W.; J.H. Fülscher und Harnack: Über eine durch Affen verursachte Ruhrepidemie. Veröffentlich. Geb. Medizinalverw. *34,* 62–73 (1931)
12. Baggs, R.B.; R.D. Hunt; F.G. Garcia; E.M. Hajema; B.J. Blake, and C.E.O. Fraser: Pseudotuberculosis *(Yersinia enterocolitica)* in the owl monkey *(Aotus trivirgatus).* Lab. Anim. Sci. *26,* 1079–1083 (1976)
13. Barnes, D.M.: Septicemic shigellosis in two woolly monkeys. Primate Zoonos. Surveil. Rep. *2,* 7–8 (1970)
14. Baron, L.S. and S. Falkow: Genetic transfer of episomes from *Salmonella typhosa* to *Vibrio cholerae.* Rec. Genet. Soc. Am. *30,* 59 (1961)
15. Barrett-Connor, E. and J.D. Connor: Extraintestinal manifestation of shigellosis. Amer. J. Gastroenterol. *53,* 234–245 (1970)
16. Becht, H.: Untersuchungen über die Pseudotuberkulose beim Chinchilla. Dtsch. tierärztl. Wschr. *69,* 626–627 (1962)
17. Ben-Gurion, R. and A. Schaferman: Essential virulence determinants of different Yersinia species are carried on a common plasmid. Plasmid *5,* 183–187 (1981)
18. Bercovier, H.; J. Brault; N. Barre; H. Treignier; J.M. Alonso, and H.H. Mollaret: Biochemical, serological and phage typing characteristics of 459 *Yersinia* strains isolated from a terrestrial ecosystem. Curr. Microbiol. *1,* 353–357 (1978)
19. Binder, H.J.: Diarrhea, intestinal secretion and cyclic AMP, pp. 177–181 in "Diarrhea in Disorders of Intestinal Transport" (Ruppin, H.; W. Domschke, and K.H. Soergel, eds.) Thieme, Stuttgart (1981)
20. Blobel, H. und T. Schliesser: Handbuch der bakteriellen Infektionen bei Tieren, Bd. III. Gustav Fischer, Stuttgart (1981)

21. Boro, B.R.; G.Sarma, and A.K. Sarmah: Bacteriological investigation of enteric infections in zoo animals. Indian J.Anim. Hlth *19*, 39-41 (1980)
22. Bottone, E.J.: *Yersinia enterocolitica*: a panoramic view of a charismatic microorganism. Crit. Rev. Microbiol. *5*, 211-240 (1977)
23. Brack, M.: Beitrag zu den Erkrankungen des Chinchilla an Pasteurella X. Das Pelzgewerbe *17*, 41-43 (1966)
24. Brack, M.; L.H. Boncyk; G.T. Moore, and S.S. Kalter: Bacterial meningo-encephalitis in newborn baboons *(Papio cynocephalus)*. Contemporary Primatology; 8th int. Congr. Primat. Nagoya, pp.493-501, Karger, Basel (1975)
25. Branham, S.E.; K.Habel, and R.D. Lillie: Studies with *Shigella dysenteriae*(Shiga). III. Infection and intoxication in *Macacus mulatta* monkeys. J.Infect. Dis. *85*, 295-303 (1949)
26. Brede, H.D.: Pathogenic enterobacteriaceae of free-living baboons *(Papio ursinus)*. Medical Primatology 1972. Proc. 3rd. Conf. exp. Med. Surg. Primates, Lyon 1972, part III, pp.35-38. Karger, Basel (1972)
27. Brede, H.D.; L.Heber, and A.Steinbach: Comparative bacteriological survey of imported experimental monkeys. Arb. Paul-Ehrlich-Inst. *71*, 113-117 (1977)
28. Brede, H.D. and G.P. Murphy: Bacteriological aspects of renal allotransplantation or homotransplantation. S.Afr. J.Med. (Suppl.) *17*, 83-87 (1968)
29. Brenner, D.J.: Classification of *Yersinia enterocolitica*, pp. 1 - 8 in "Yersinia enterocolitica" (Bottone, E.J. ed.), CRC Press, Boca Raton (1981)
30. Breshnahan, J.F.; U.G. Whitworth; Y.Hayes; E.Summers, and J. Pollock: *Yersinia enterocolitica* infection in breeding colonies of ruffed lemurs. J.Amer. vet. med. Ass. *185*, 1354-1356 (1984)
31. Bronson, R.T.; B.D. May, and B.H. Ruebner: An outbreak of infection by Yersinia pseudotuberculosis in nonhuman primates. Amer. J.Path. *69*, 289-303 (1972)
32. Buhles, W.C., jr.; J.E. Vanderlip; S.W. Russell, and N.L. Alexander: *Yersinia pseudotuberculosis* infection: study of an epizootic in squirrel monkeys. J.Clin. Microbiol. *13*, 519-525 (1981)
33. Butler, T.M.: Reproductive history of the Holloman chimpanzee colony. J.med. Prim. *1*, 51-57 (1972)
34. Cambre, R.C.; H.L. Wilson; T.R. Spraker, and B.E. Favara: Fatal airsacculitis and pneumonia, with abortion, in an orang utan. J.Amer. vet. med. Ass. *177*, 822-824 (1980)
35. Caraway, B.L.; R.R. Estes; P.M. Neda, and O.W. Thornton: Results of screening for shigella and other bacteria in rhesus monkeys during quarantine. Lab. Anim. Care *18*, 568-569 (1968)
36. Carpenter, C.C.J.: Clinical and pathophysiologic features of diarrhea caused by *Vibrio cholerae* and *Escherichia coli*; pp.67-83 in "Secretary Diarrhea " (Field, M.; J.S. Fordtran, and S.G. Schultz, eds.) Amer. Physiol. Soc., Bethesda (1980)
37. Carpenter, K.P. and E.R.N. Cooke: An attempt to find Shigellae in wild primates. J.Comp. Path. *75*, 201-204 (1965)
38. Carpenter, K.P. and B.R. Sandiford: Epidemiology of a human case of bacillary dysentery due to infection by Shigella flexneri 103Z. Brit. Med. J. *I*, 142-143 (1952)
39. Cass, J.S.: Enteric infection in monkeys. Proc. Anim. Care Panel *3*, 14-22 (1952)
40. Cavanagh, D.; P.S. Rao; D.M.C. Sutton; B.der Bhagat, and F.Bachmann: Pathophysiology of endotoxin shock in the primate. Amer. J.Obstet. Gyncol. *108*, 705-722 (1970)
41. Chang, J.; J.L. Wagner, and R.W. Kornegay: Fatal *Yersinia pseudotuberculosis* infection in captive bushbabies. J.Amer. vet. med. Ass. *177*, 820-821 (1980)
42. Chang, M.T. and M.P. Doyle: Identification of specific outer membrane polypeptides associated with virulent *Yersinia enterocolitica*. Infect. Immun. *43*, 472-476 (1984)
43. Chiesa, C.; E.J. Bottone, and H.L. Hodes: Status of immunity of newborns against *Yersinia enterocolitica*: A preliminary report. Amer. J.med. Sci. *287*, 18-20 (1984)
44. Chun, D.; D.T. Cho; S.Y. Seol; M.H. Suh, and Y.C. Lee: R plasmids conferring multiple drug resistance from shigella isolated in Korea. J.Hyg. *92*, 153-160 (1984)
45. Coalson, J.J.; L.T. Archer; B.A. Benjamin; B.K. Beller-Todd, and L.B. Hinshaw: A morphologic study of live *Escherichia coli* organism shock in baboons. Exp. molec. Path. *31*, 10-22 (1979)
46. Cook, R.: Simian shigellosis; pp. 83-91 in "Hazards of Handling Simians". Laboratory Ani-

mal Handbooks No. 4 (Perkins, F.T. and P.N. O 'Donoghue, eds.), Lab. Anim. Ltd., London (1969)
47. Cooper, J.E.: A case of pyometritis in an East African de Brazza 's monkey *(Cercopithecus neglectus).* Vet. Rec., 457-458 (1972)
48. Cooper, J.E. and J.R. Needham: An outbreak of shigellosis in laboratory marmosets and tamarins (Family: Callitricidae). J. Hyg. (Camb.) *76*, 415-424 (1976)
49. Cooper, M.L.; E. Walters, and H.M. Keller: *Escherichia coli* associated with infantile diarrhea. Ann. N.Y. Acad. Sci. *66*, 18 - 89 (1956)
50. Corbel, M.J.: The relationship between the protective and cross-reacting antigens of *Brucella* spp., *Yersinia enterocolitica* 0:9 and *Salmonella* serotypes of Kauffmann-White group N., pp. 50-63 in "Yersinia enterocolitica. Biology, Epidemiology and Pathology." (Carter, P.B., L. Lafleur and S. Thoma, eds.) Contrib. Microbiol. Immunol. 5. Karger, Basel (1979)
51. Cornelis, G.: Antibiotic resistance in *Yersinia enterocolitica*; pp. 55-71 in "Yersinia enterocolitica" (Bottone, E.J. ed.), CRC Press, Boca Raton (1981)
52. Daniels, J.J.H.M.: Enteral infection with Pasturella pseudotuberculosis. Brit. Med. J. ii, 997 (1961)
53. Daniel, J.J.H.M.: Untersuchungen an als *Pasteurella pseudotuberculosis* diagnostizierten Stämmen von Chinchillas. Zbl. Vet. Med. *10*, 413-417 (1963)
54. DaRif, C.A. and H.G. Rush: Management of septicemia in rhesus monkeys with chronic indwelling venous catheters. Lab. Anim. Sci. *33*, 90-94 (1983)
55. David, H. und A. Schirl: Über Ruhrerkrankungen bei Menschenaffen. Zbl. Bakt. I. Orig. *144*, 43-45 (1939)
56. Davis, T.C.: Chronic vulvovaginitis in children due to *Shigella flexneri.* Pediatrics *56*, 41-44 (1975)
57. Dean, A.G.; Y.-C. Ching; R.G. Williams, and L.B. Harden: Test for *Escherichia coli* enterotoxin using infant mice: application in a study of diarrhea in children in Honolulu. J. infect. Dis. *125*, 407-411 (1972)
58. Deinhardt, J.B.; J. Devine; M. Passovoy; R. Pohlman, and F. Deinhardt: Marmosets in the laboratory. Pathology and outline of statistical evaluation of data. Lab. Anim. Care *17*, 11-29 (1967)
59. Dertinger, L.; H.W. Anderson; R.B. Watkins; J.P. Pflueger; E.D. Ploeger; J. Allard, and B.J. Francis: Epidemiologic notes and reports. Monkey-associated gastroenteritis in Washington. CDC Morbid. Mortal. Wkly. Rep. *21*, 299-304 (1972)
60. DeVos, V.; C.A.W.J. van Niekerk, and E.E. McConnell: A survey of selected bacteriological infections of the chacma baboon *Papio ursinus* from the Kruger National Park. Koedoe *16*, 1-10 (1973)
61. Dickinson, A.B. and G. Mocquot: Studies on the bacterial flora of the alimentary tract of pigs. I. Enterobacteriaceae and other gram-negative bacteria. J. appl. Bact. *24*, 252-284 (1961)
62. Dinbar, A.; S.I. Rapaport; M.J. Patch; W. Grant, and E.W. Fonkalsrud: Hematologic effects of endotoxin on the macaque monkey. Surgery *70*, 596-603 (1971)
63. Donta, S.T.; H.W. Moon, and S.C. Whipp: Detection of heat-labile *Escherichia coli* enterotoxin with the normal use of adrenal cells in tissue culture. Science *183*, 334-336 (1974)
64. Drucker, M.M.; A. Polliak; R. Yeivin, and T.G. Sacks: Immunofluorescent demonstration of enteropathogenic *Escherichia coli* in tissues of infants dying with enteritis. Pediatrics *46*, 855-864 (1970)
65. Dupont, H.L. and L.K. Pickering: Infection of the Gastrointestinal Tract. Microbiology, Pathophysiology and Clinical Features. Plenum Med. Book Co. New York (1980)
66. Edebo, L.; K.E. Magnusson, and O. Stendahl: Physico-chemical surface properties of *Shigella sonnei.* Acta path. microbiol. immunol. scand. B. *91*, 101-106 (1983)
67. Edwards, P.R. and W.H. Ewing: Identification of Enterobacteriaceae. 3rd ed. Burgess. Publ. Co., Minneapolis (1972)
68. Eichberg, J.: Changes in infant chimpanzee husbandry in response to a *Klebsiella pneumoniae* outbreak; pp. 43-45 in "Clinical Management of Infant Great Apes" (Graham, C.E. and J.A. Bowens, eds.) Liss Inc., New York (1985)
69. Eichner, E.R.; E.J. Gangarosa, and J.B. Goldsby: The current status of Shigellosis in the United States. Amer. J. Publ. Hlth *58*, 753-763 (1968)
70. Ewing, W.H. and W.J. Martin: Enterobacteriaceae; pp. 189-221 in "Manual of Clinical

Microbiology" (Lennette, E. H., E. H. Spaulding, and J. P. Truant, eds.) 2nd ed. Amer. Soc. Microbiol., Washington (1974)

71. Farrar, W. E., jr.; M. Eidson; P. Guerry; S. Falkow; L. M. Drusin, and R. B. Roberts: Interbacterial transfer of R factor in the human intestine: in- vivo acquisition of R-factor-mediated kanamycin resistance by a multiresistant strain of *Shigella sonnei*. J. Infect. Dis. *126*, 27-33 (1972)
72. Feely, J. C.; J. G. Wells; T. F. Tsai, and N. D. Puhr: Detection of enterotoxigenic and invasive strains of *Yersinia enterocolitica*; pp. 329-334 in "Yersinia enterocolitica. Biology, Epidemiology and Pathology." (Carter, P. B., L. Lafleur and S. Toma, eds.). Contrib. Microbiol. Immunol. *5*, Karger, Basel (1979)
73. Felsen, J.: Bacillary Dysentery, Colitis and Enteritis. W. B. Saunders, Philadelphia (1945)
74. Fernandez, A.; C. A. Sninsky; A. D. O 'Brien; M. H. Clench, and J. R. Mathias: Purified *Shigella* enterotoxin does not alter intestinal motility. Infect. Immun. *43*, 477-481 (1984)
75. Field, M.: Regulation of small intestinal ion transport by cyclic nucleotides and calcium; pp. 21-30 in "Secretary Diarrhea" (Field, M.; J. S. Fordtran, and S. G. Schultz, eds.) Amer. Physiol. Soc., Bethesda (1980)
76. Fiennes, R.: Zoonoses of Primates. The Epidemiology and Ecology of Simian Diseases in Relation to Man. Cornell Univ. Press, Ithaca (1967)
77. Fiennes, R. N. T-W.; M. Pinkerton, and E. K. Dshikidze: Enteropathogenic organisms; pp. 263-276 in "Pathology of Simian Primates", Vol. 2 (Fiennes, R. N. T-W., ed.), Karger, Basel (1972)
78. Fincham, J. E. and J. V. Seier: Endemic enteric disease in vervet monkeys. J. S. Afr. Vet. Ass. *52*, 177-179 (1981)
79. Finegold, S. M. and W. J. Martin: Diagnostic Microbiology. 6 th ed., C. V. Mosby Co., St. Louis (1982)
80. Fleming, M. P.: Infectious drug resistance and disease control in imported simian primates. Primate Supply *2* (2), 21-22 (1977)
81. Fleming, M. P.: Bacillary dysentery (Shigellosis) in simian primates. Primate Supply *3* (1), 6-7 (1978)
82. Ford, A. C.; T. M. Speltie, and W. D. H. Hendriks: Studies on the prevalence of *Salmonella* serotypes in non-human primates. Lab. Anim. Sci. *23*, 649-652 (1973)
83. Ford, A. C.; T. M. Speltie, and D. van der Waaij: Incidence and therapy of Salmonella in newly arrived non-human primates, pp. 157-159, Ann. Rep. Radiobiol. Inst. TNO, Inst. Exp. Geront. TNO, Primate Center TNO, Rijswijk (1972)
84. Formal, S. B. and P. Gemski, jr.: Bacterial enterocolitis; pp. 31-35 in "The Gastrointestinal Tract. International Academy of Pathology Monograph 18". (Yardley, J. H., B. C. Morson, and M. R. Abell, eds.). Williams and Wilkins Co., Baltimore (1977)
85. Fox, J. G.: Transmissible drug resistance in *Shigella* and *Salmonella* isolated from pet monkeys and their owners. J. med. Primatol. *4*, 165-171 (1975)
86. Fox, J. G.; S. B. Snyder, and O. A. Soave: Transmissible drug resistance in enterobacteriaceae isolated from healthy nonhuman primates. Amer. J. vet. Res. *34*, 965-970 (1973)
87. Frantz, J. C.; L. Jaso-Friedman, and D. C. Robertson: Binding of *Escherichia coli* heat-stable enterotoxin to rat intestinal cells and brush border membranes. Infect. Immun. *43*, 622-630 (1984)
88. Frederiksen, W.: A study of *Yersinia pseudotuberculosis*-like bacteria (*Bacterium enterocoliticum* and *Pasteurella* X). Proc. 14th Scand Congr. Pathol. Microbiol. Oslo, Abstr. No. *47*, 103-105 (1964)
89. Frollka, J.: Zum Vorkommen von Pseudotuberkulose bei Meerkatzen im Zoo Lesna: XXII. Internat. Symp. Erkr. Zootiere, Arnhem, pp. 117-123 (1980)
90. Frost, J. A. and B. Rowe: Plasmid-determined antibiotic resistance in *Shigella flexneri* isolated in England and Wales between 1974 and 1978. J. Hyg. *90*, 27-32 (1983)
91. Fukuda, N. and S. Asakura: Diseases of monkeys in the Ueno zoo. 2. Internat. Symp. Zoo Anim. Path. Warszaw, pp. 191-194 (1960)
92. Garlt, C.: Seuchenhaftes Auftreten von Pseudotuberkulose bei Affen. VII. Internat. Symp. Erkr. Zootiere, Zürich - Basel, pp. 139-145 (1965)
93. Gebauer, B. and H. S. Raethel: Klebsiellose bei südamerikanischen Krallenäffchen (Callitrichidae). Dtsch. tierärztl. Wschr. *65*, 383-386 (1958)

94. Gemski, P. and S. B. Formal: Shigellosis: an invasive infection of the gastrointestinal tract; pp. 165-169 in "Microbiology 1973" (Schlesinger, D., ed.), Amer. Soc. Microbiol., Washington D.C. (1975)
95. Gemski, P.; J.R. Lazere; T. Casey, and J.A. Wohlhieter: Presence of a virulence-associated plasmid in *Yersinia pseudotuberculosis.* J. Infect. Immun. *28*, 1044-1047 (1980)
96. Gemski, P., jr.; A. Takeuchi; O. Washington, and S. B. Formal: Shigellosis due to *Shigella dysenteriae* 1: Relative importance of mucosal invasion versus toxin production in pathogenesis. J. Infect. Dis. *126*, 523-530 (1972)
97. Giannella, R.A.; R.E. Gots; A.N. Charney; W.B. Greenough, III, and S.B. Formal: Pathogenesis of salmonella-mediated intestinal fluid secretion. Activation of adenylate cyclase and inhibition by indomethacin. Gastroenterology *69*, 1238-1245 (1975)
98. Giannella, R.A.; W.R. Rout; S.B. Formal, and G.J. Dammin: Pathophysiology of shigella diarrhea and dysentery in rhesus monkey. Gastroenterology *66*, 698 (1974)
99. Giles, R.C.; P.K. Hildebrandt, and C. Tate: Klebsiella air sacculitis in the owl monkey *(Aotus trivirgatus).* Lab. Anim. Sci. *24*, 610-616 (1974)
100. Gillespie, P. and L. Arnold: Pathology report. Dodo J. Jersey Wildl. Preserv. Trust *19*, 81-94 (1982)
101. Göldenboth, R.: Ein Ausbruch von bakterieller Ruhr bei Menschenaffen. Kleintierpraxis *12*, 79-82 (1967)
102. Good, R.C.; B.D. May, and T. Kawatomari: Enteric pathogens in monkeys. J. Bacteriol. *97*, 1048-1055 (1969)
103. Goullet, P.; B. Picard, et M. Toucas: Différenciation du serotype 6 des serotypes 1-5 de *Shigella flexneri* par électrophorèse et focalisation isoélectrique de leurs esterases. Ann. Microbiol. (Inst. Pasteur) 134 B, 269-276 (1983)
104. Greenberg, R.N.; R.L. Guerrant; B. Chang; D.C. Robertson, and F. Murad: Inhibition of *Escherichia coli* heat-stable enterotoxin effects on intestinal guanylate cyclase and fluid secretion by quinacrine. Biochem. Pharmacol. *31*, 2005-2009 (1982)
105. Greenberg, R.N.; F. Murad; B. Chang; D.C. Robertson, and R.L. Guerrant: Inhibition of *Escherichia coli* heat-stable enterotoxin by indomethacin and chlorpromazine. Infect. Immun. *29*, 908-913 (1980)
106. Greenberg, R.N.; R.D. Pearson; D.J. Innes; K.T. Sauer; L.D. Halterman, and R.L. Guerrant: Susceptibility of *Shigella* species to erythromycin. Antimicrob. Agents Chemother. *21*, 347-348 (1982)
107. Gross, R.J.; L.V. Thomas; T. Cheasty; N.P. Day; B. Rowe; M.R.F. Toledo, and L.R. Trabulsi: Enterotoxigenic and enteroinvasive *Escherichia coli* strains belonging to a new 0 group, 0167. J. clin. Microbiol. *17*, 521-523 (1983)
108. Guerrant, R.L.; L.L. Brunton; T.C. Schnaitman; L.I. Rebhuhn, and A.G. Gilman: Cyclic adenosine monophosphate and alteration of Chinese hamster ovary cell morphology: a rapid sensitive in vitro assay for the enterotoxins of *Escherichia coli.* Infect. Immun. *10*, 320-327 (1974)
109. Haas, R. und K.F. Petersen: Shigella- und Salmonella-Befunde bei Affen. Ein Beitrag zum Problem der Einschleppung pathogener Darmbakterien. Med. Klin. *54*, 1947-1949 (1959)
110. Hale, T.L.; R.E. Morris, and P.F. Bonventre: Shigella infection of Henle intestinal epithelial cells: Role of the host cell. Infect. Immun. *24*, 887-894 (1979)
111. Hammon, J.W.; J.L. Kitzmiller; R. Virgilio; L.G. Getz, and W.E. Lucas: The role of fibrin formation in the pathogenesis of bacteremic shock in the primate. Surg. Gynec. Obstet. *146*, 535-540 (1978)
112. Harada, K.M.; M. Suzuki; M. Kameda, and S. Mitsuhashi: On the drug resistance among enteric bacteria. II. Transmission of the drug resistance among *Enterobacteriaceae.* Jap. J. exp. Med. *30*, 289-299 (1960)
113. Hardy, P.H.; J.R. Lindsey, and E.C. Melby: Transferable multiple drug resistance in enterobacteria from non-human primates. I. Development and characteristics of resistance in *Shigella flexneri.* Johns Hopkins Med. J. *123*, 29-37 (1968)
114. Hartley, E.G.: The incidence and antibiotic sensitivity of *Shigella* bacteria isolated from newly imported macaque monkeys. Brit. vet. J. *131*, 205-212 (1975)
115. Heesemann, J.; C. Keller; R. Morawa; N. Schmidt; H.J. Siemens, and R. Laufs: Plasmids of

human strains of *Yersinia enterocolitica*: Molecular relatedness and possible importance for pathogenesis. J. infect. Dis. *147*, 107-115 (1983)
116. Heintze, K.: Physiology of intestinal electrolyte and water transport; pp. 2-14 in "Diarrhea in Disorders of Intestinal Transport" (Ruppin, H., W. Domschke, and K. H. Soergel, eds.) Thieme, Stuttgart (1981)
117. Hemelhof, W.; P. Retore; P. de Mol; J. P. Butzler; T. Takeda; T. Miwatani, and Y. Takeda: Production of a monoclonal antibody against heat-stable enterotoxin produced by human strain of enterotoxigenic Escherichia coli. Lancet I, 1011-1012 (1984)
118. Hentges, D. J.: Inhibition of *Shigella flexneri* by the normal intestinal flora. I. Mechanisms of inhibition by *Klebsiella*. J. Bacteriol. *93*, 1369-1373 (1967)
119. Highsmith, A. K.; J. C. Feeley; P. Skally; J. G. Wells, and B. T. Wood: Isolation of *Yersinia enterocolitica* from well water and growth in distilled water. Appl. environ. Microbiol. *34*, 745-750 (1977)
120. Hirai, K.; Y. Suzuki; N. Kato; K. Yagami; A. Miyoshi; Y. Mabuchi; H. Nigi; H. Inagaki; K. Otsuki, and M. Tsubokura: *Yersinia pseudotuberculosis* infection occurred spontaneously in a group of patas monkeys *(Erythrocebus patas)*. Jap. J. vet. Sci. *36*, 351-355 (1974)
121. Holmgren, H. and W. B. Greenough: Reversal of enterotoxin diarrhea by chlorpromazine and related drugs; pp. 211-218 in "Secretary Diarrhea" (Field, M.; J. S. Fordtran and S. G. Schultz, eds.), Amer. Physiol. Soc., Bethesda (1980)
122. Honjo, S.; M. Takasaka; T. Fujiwara; M. Nakagawa; K. Andoo; H. Ogawa; R. Takahashi, and K. Imaizumi: Shigellosis in cynomolgus monkeys *(Macaca irus)*. II. Experimental infection with *Shigella flexneri* 2a with special references to clinical and bacteriological findings. Jap. J. Med. Sci. Biol. *17*, 307-319 (1964)
123. Hornic, R. B.: Shigellosis; pp. 549-555 in "Infectious Diseases" (Hoeprich, D., ed.), Harper and Row, Hagerstown (1977)
124. Houser, W. D.; D. H. Norback, and W. L. Ragland: Atypical *Klebsiella* infections in infant monkeys. Mainly Monkeys *1*, 20 (1970)
125. Hunt, D. E.; R. F. Pitillo; G. A. Deneau; F. M. Schabel, jr., and L. B. Mellett: Control of an acute Klebsiella pneumoniae infection in a rhesus monkey colony. Lab. Anim. Care *18*, 183-185 (1968)
126. Hunt, R. D.; M. P. Anderson, and L. V. Chalifoux: Spontaneous infectious disease of marmosets. Prim. Med. *10*, 239-253 (1978)
127. Hurvell, B.: Zoonotic *Yersinia enterocolitica* infection: Host range, clinical manifestations and transmission between animals and man; pp. 145-159 in "Yersinia enterocolitica" (Bottone, E. J., ed.) CRC Press, Boca Raton (1981)
128. Ippen, R. und L. Stoll: *Pasteurella pseudotuberculosis* bei einem Virginiahirsch. Berl. Münch. Tierärztl. Wschr. *72*, 439-440 (1959)
129. Jentzsch, K. D.; H. D. Schroeder und E. Karasek: Beitrag zum Vorkommen von Zooanthroponosen bei Zootieren. XI. Internat. Symp. Erkr. Zootiere, Zagreb, pp. 7-17 (1969)
130. Joseph, P. G.; M. Anwar, and M. Jegathesan: Animal salmonellosis in peninsular Malaysia. II. Animal and zoological distribution of *Salmonella* serotypes over the 10-year period 1966-1975. Amer. J. trop. Med. Hyg. *27*, 562-566 (1978)
131. Källenius, G.; S. B. Svenson; H. Hultberg; R. Möllby; J. Winberg, and J. A. Roberts: P-fimbriae of pyelonephritogenic Escherichia coli: Significance for reflux and renal scarring - a hypothesis. Infection *11*, 73-76 (1983)
132. Kageruka, P.; H. H. Mollaret et J. Vercruysse: Pseudotuberculose d 'un singe (Cercopithecus cephus, Lin) à Yersinia enterocolitica. Acta zool. path. antverp. *56*, 31-36 (1972)
133. Kageruka, P.; J. Mortelmans, and J. Vercruysse: Klebsiella pneumoniae infection in monkeys. Acta zool. path. antverp. *52*, 83-88 (1971)
134. Kageruka, P.; J. Mortelmans; J. Vercruysse, and C. Beernaert-Declercq: Pseudotuberculosis in the Antwerp Zoo. Acta zool. path. antverp. *66*, 111-120 (1976)
135. Kapperud, G.: *Yersinia enterocolitica* and *Yersinia* like microbes isolated from mammals and water in Norway and Denmark. Acta path. microbiol. scand., Sect. B *85*, 129-135 (1977)
136. Kaufmann, A. F.: The effects of spontaneous disease during quarantine, conditioning, and production of nonhuman primates for scientific use following importation. Pan Amer. Hlth Org. sci. Publ. *235*, 39-48 (1972)
137. Kawatomari, T.: Enteropathogenic bacteria of nonhuman primates. *In vivo* and *in vitro*

transfer of R-factors from *Escherichia coli* to *Shigella flexneri.* J. med. Primatol. *1*, 354 - 362 (1972)
138. Kent, T.H.; S.B. Formal, and E.H. Labrec: *Salmonella* gastroenteritis in rhesus monkeys. Arch. Path. *82*, 272-279 (1966)
139. Kent, T.H.; S.B. Formal; E.H. LaBrec; H. Sprinz, and R.M. Maenza: Gastric shigellosis in rhesus monkeys. Amer. J. Path. *51*, 259-267 (1967)
140. Ketyi, I.; G. Naumann, and W. Nimmich: Urinary tract infectivity of R-strains of *Escherichia coli* carrying various virulence factors. Acta microbiol. Acad. Sci. hung. *30*, 155-161 (1983)
141. Keusch, G.T.: Ecological control of the bacterial diarrheas: a scientific strategy. Amer. J. Clin. Nutr. *31*, 2208-2218 (1978)
142. Keusch, G.T. and M. Jacewitz: The pathogenesis of shigella: VI. Toxin and antitoxin in *Shigella flexneri* and *Shigella sonnei* infections in humans. J. Infect. Dis. *138*, 552-556 (1977)
143. Keymer, I.F.: Report of the pathologist, 1969 and 1970. J. Zool. *166*, 515-540 (1972)
144. King, G.: Parasite and post mortem reports. Dodo J. Jersey Wildl. Preserv. Trust *14*, 1-8 (1977)
145. Kinsey, M.D.; G.J. Dammin; S.B. Formal, and R.A. Giannella: The role of altered intestinal permeability in the pathogenesis of salmonella diarrhea in the rhesus monkey. Gastroenterology *71*, 429-434 (1976)
146. Klumpp, S.A.; D.S. Weaver; C.P. Jerome, and M.P. Jokinen: Salmonella osteomyelitis in a rhesus monkey. Vet. Pathol. *23*, 190-197 (1986)
147. Knapp, W. und E. Thal: Untersuchungen über die kulturell-biochemischen, serologischen, tierexperimentellen und immunologischen Eigenschaften einer vorläufig "Pasteurella X" benannten Bakterienart. Zbl. Bakt., I. Abt. Orig. *190*, 472-484 (1963)
148. Köhler, W. and H. Mochmann: Grundriß der Medizinischen Mikrobiologie. VEB Gustav Fischer Verlag Jena (1975)
149. Kontrohr, T.: The identification of 2-amino-2-deoxy-L-altruronic acid as a constituent of Shigella sonnei phase I lipopolysaccharide. Carbohydr. Res. *58* 498-500 (1977)
150. Kopecko, D.J.; O. Washington, and S.B. Formal: Genetic and physical evidence for plasmid control of *Shigella sonnei* form I cell surface antigen. Infect. Immun. *29*, 207-214 (1980)
151. Kourany, M. and R.N. Rossan: A subcutaneous abscess associated with *Salmonella typhimurium* in black howler monkey *(Alouatta villosa).* Lab. Anim. Sci. *21*, 412-414 (1971)
152. Kourany, M. and J.A. Porter, jr.: A survey for enteropathogenic bacteria in Panamanian primates. Lab. Anim. Care *19*, 336-341 (1969)
153. Lamm, S.H.; M. Nadel; R.M. Jakowski, and J.C. Hart: A fatal case of human shigellosis associated with a pet spider monkey. Primate Zoonoses Surveil. Rep. *4*, 12-13 (1971 a)
154. Lamm, S.H.; M. Nadel; R.M. Jakowski, and J.D. Hart: A fatal case of human shigellosis associated with a pet spider monkey. Primate Zoonoses Surveil. Rep. *5*, 15 (1971 b)
155. Lautrop, H.: Gelatin-liquefying *Klebsiella* strains (*Bacterium oxytocum* Flugge). Acta pat. microbiol. scand. *39*, 375-384 (1956)
156. Lautrop, H.; I. Oerskov, and K. Gaarslev: Hydrogensulfide producing variants of *Escherichia coli.* Acta path. microbiol. scand., Sect. B *79*, 641-650 (1971)
157. Lee, W.H.; C. Vanderzant, and N. Stern: The occurrence of *Yersinia enterocolitica* in foods; pp. 161-171 in "Yersinia enterocolitica" (Bottone, E.J., ed.), CRC Press, Boca Raton (1981)
158. Lemen, R.; S. Lemen; R. Morrish, and W. Tooley: Marasmus and shigellosis in two infant gorillas. J. med. Primatol. *3*, 365-369 (1974)
159. LeMinor, L.: Common enterobacterial antigen and ONPG test in the taxonomy of the genus *Yersinia*, pp. 1-7 in "Yersinia enterocolitica. Biology, Epidemiology, and Pathology." (Carter, P.B.; L. Lafleur, and S. Toma, eds.). Contrib. Microbiol. Immunol. *5* (1979)
160. Levine, M.M.; H.L. Du Pont; S.B. Formal; R.B. Hornick; A. Takeuchi; E.J. Gangarosa; M.J. Snyder, and J.P. Libonati: Pathogenesis of *Shigella dysenteriae* 1 (Shiga) dysentery. J. Infect. Dis. *127*, 261-270 (1973)
161. Lindberg, R.B. and V.M. Young: Observation on enteropathogenic *Escherichia coli.* Ann. N.Y. Acad. Sci. *66*, 100-107 (1956)
162. Lindsey, J.R.; P.H. Hardy, jr.; H.J. Baker, and E.C. Melby, jr.: Observations on shigellosis and development of multiply resistant shigellosis in *Macaca mulatta.* Lab. Anim. Sci. *21*, 832-844 (1971)
163. Mair, N.S.: Diseases in free-living wild animals. Academic Press, London (1969)

164. Mair, N.S.: Yersiniosis (Infections due to *Yersinia pseudotuberculosis* and *Yersinia enterocolitica*); pp. 174-185 in "Diseases Transmitted from Animals to Man", 6th ed. (Hubbert, W.T., W.F. McCulloch and P.R. Schnurrenberger, eds.). C.C. Thomas, Springfield (1975)
165. Mair, N.S.; H.J. Mair; E.M. Stirk, and J.E. Corson: Three cases of acute mesenteric lymphadenitis due to *Pasteurella pseudotuberculosis*. J. Clin. Path. *13*, 432-439 (1960)
166. Mair, N.S.; G.D. White; F.K. Schubert, and J.F. Harbourne: Yersinia enterocolitica infection in the bush-baby (Galago). Vet. Rec. *86*, 69-71 (1970)
167. MacArthur, J.A. and M. Wood: Yersiniosis in a breeding unit of *Macaca fascicularis* (cynomolgus monkeys). Lab. Anim. *17*, 151-155 (1983)
168. Malassez, L. and W. Vignal: Tuberculose zoogleique (forme ou espèce de tuberculose sans bacilles). Arch. Physiol. Norm. Pathol. *2* (ser. 3), 369-412 (1883)
169. Malassez, L. and W. Vignal: Sur le microorganisme de la tuberculose zoogleique. Arch. Physiol. Norm. Pathol. *4* (ser. 3), 80-105 (1884)
170. McClure, H.M.: Enteropathogenic *Escherichia coli* infection. Comp. Path. Bull. *8* (4), 1-3 (1976)
171. McClure, H.M.; P. Alford, and B. Swenson: Nonenteric shigella infections in nonhuman primates. J. Amer. vet. med. Ass. *189*, 938-939 (1976)
172. McClure, H.M. and F.A. King: Yersiniosis: A review and report of an epizootic in nonhuman primates; pp. 217-241 in "One Medicine" (Ryder, O.A. and M.L. Bird, eds.). Springer, Berlin (1984)
173. McClure, H.M.; L.M. Strozier, and M.E. Keeling: Enteropathogenic *Escherichia coli* infection in anthropoid apes. J. Amer. vet. med. Ass. *161*, 687-689 (1972)
174. McClure, H.M.; R.E. Weaver, and A.F. Kaufmann: Pseudotuberculosis in nonhuman primates: Infection with organisms of the *Yersinia enterocolitica* group. Lab. Anim. Sci. *21*, 376-382 (1971)
175. Meier, J.E. and W. Sanborn: A preliminary report on the management and treatment of salmonellosis with trimethoprim - sulfamethoxazole in an exotic animal nursery. J. Zoo Ass. Med. *13*, 26-29 (1982)
176. Mel, D.M.; B. Cvjetanovic, and O. Felsenfeld: Studies on vaccination against bacillary dysentery. 5. Studies in *Erythrocebus patas*. Bull. Wld Hlth Org. *43*, 431-437 (1970)
177. Mohan, K.; S.C. Pal; J.N. Ghosh; A.K. Chakravarty; S. Pattanayak, and S.C. Arya: Enteric infections in captive monkeys. Indian J. med. Res. *61*, 383-388 (1973)
178. Mohan, K.; J.C. Suri, and M. Madan: Enteric infections in captive monkeys. II. Symptomless shedder of enteropathogenic bacteria. Indian J. med. Res. *63*, 22-26 (1975)
179. Mollaret, H.H.: Conservation expérimentale de la peste dans le sol. Bull. Soc. Pathol. Exot. *56*, 1168-1182 (1963)
180. Mollaret, H.H.; H. Bercovier, and J.M. Alonso: Summary of the data received at the WHO Reference Center for *Yersinia enterocolitica*, pp. 174-184 in "Yersinia enterocolitica. Biology, Epidemiology and Pathology." (Carter, P.B., L. Lafleur, and S. Toma, eds.). Contrib. Microbiol. Immunol. *5*, Karger, Basel (1979)
181. Mollaret, H.H.; W. Giorgi; A. Matera; A.F. Pestana de Castro et J.C. Guillon: Isolement de *Yersinia enterocolitica* chez le singe callitriche au Brésil. Rec. Med. Vet. *146*, 919-924 (1970)
183. Mollaret, H.H. et A. Lucas: Sur les particularités biochimiques des souches de *Yersinia enterocolitica* isolées chez les lièvres. Ann. Inst. Pasteur *108*, 121-125 (1965)
184. Mollaret, H.H. and E. Thal: Yersinia; pp. 330-332 in "Bergey 's Manual of Determinative Bacteriology" 8 th ed. (Buchanan, R.E. and N.E. Gibbons, eds.). William and Wilkins Co., Baltimore (1974)
185. Morris, G.K.; J.A. Koehler, and E.J. Snow, jr.: Shigella isolates from recently imported rhesus monkeys. Primate Zoonoses Surveil. Rep. *3*, 14-15 (1971)
186. Moss, J.; J.C. Osborne, jr.; P.H. Fishman; S. Nakaya, and D.C. Robertson: *Escherichia coli* heat-labile enterotoxin, ganglioside specificity and ADP-ribosyltransferse activity. J. biol. Chem. *256*, 12861-12865 (1981)
187. Müller, G.: Wie weit sind niedere Tiere an der Verbreitung von Salmonellen beteiligt? Zbl. Bakt., I. Abt. Orig. *197*, 295-298 (1965)
188. Mulder, J.B.: Shigellosis in nonhuman primates: A review. Lab. Anim. Sci. *21*, 734-738 (1971)
189. Myers, L.L.: Zoonotic Potential of Animal Strains of *Escherichia coli*, pp. 38-45 in "CRC

Handbook Series in Zoonoses" (Steele, J.H., Series ed.); Section A: Bacterial, Rickettsial and Mycotic Diseases. Vol. II (Stoenner, H.; W. Kaplan and M. Torten, Sect. eds.), CRC Press, Boca Raton (1980)
190. Nataro, J.P.; I.C.A. Scaletsky; J.B. Kaper; M.M. Levine, and L.R. Trabulsi: Plasmid-mediated factors conferring diffuse and localized adherence of enteropathogenic *Escherichia coli*. Infect. Immun. *48*, 378-383 (1985)
191. Nelson, J.D.: Shigellosis; pp. 254-259 in "Infectious Diseases" (Sanford, J.P. and J.P. Luby, eds.), Grune and Stratton, New York (1981)
192. Nichols, B.L.; G.L. Bilbrey; C.F. Hazlewood; S.L. Kimzey; C.T. Liu, and W.R. Beisel: Electrolyte metabolism in rhesus monkeys with experimental salmonella sepsis. Amer. J. clin. Nutr. *34*, 1362-1371 (1981)
193. Nimmich, W.; G. Zingler, and I. Orskov: Fimbrial antigens of Escherichia coli O1: K1: H7 and O1: K1: H- strains isolated from patients with urinary tract infections. Zbl. Bakt. Hyg. A *258*, 104-111 (1984)
194. O'Brien, A.D.; G.D. LaVeck; M.R. Thompson, and S.B. Formal: Production of *Shigella dysenteriae* type 1-like cytotoxin by *Escherichia coli*. J. Infect. Dis. *146*, 763-769 (1982)
195. O 'Brien, A.D.; M.R. Thompson; P. Gemski; B.P. Doctor, and S.B. Formal: Biological properties of *Shigella flexneri* 2a toxin and its serological relationship in *Shigella dysenteriae* 1 toxin. Infect. Immun. *15*, 796-798 (1977)
196. Obwolo, M.J.: A review of yersiniosis (*Yersinia pseudotuberculosis* infection). Vet. Bull. *46*, 167-171 (1976)
197. Ogawa, H. and M. Takasaka: Bacterial infections among monkeys used as experimental animals at N.I.H.. Jap. J. Med. Sci. Biol. *19*, 219 (1966)
198. Ogawa, H.; R. Takahashi; S. Honjo; M. Takasaka; T. Fujiwara; K. Ando; M. Nakagawa; T. Muto, and K. Imaizumi: Shigellosis in cynomolgus monkeys *(Macaca irus)*. III. Histopathological studies on natural and experimental *Shigellosis*. Jap. J. Med. Sci. Biol. *17*, 321-332 (1964)
199. Okoh, A.E.J. and M. Onazi: Notes on Salmonellae isolated from wildlife in Kano Zoological Gardens. J. Wildlife Dis. *16*, 7-10 (1980)
200. Olsson, E.; K. Krovacek; B. Hurvell, and T. Wadström: Enterotoxigenic porcine and human isolates of *Yersinia enterocolitica* in Sweden. Current Microbiol. *3*, 267-271 (1980)
201. Orskov, I.: Klebsiella; pp. 321-324 in "Bergey 's Manual of Determinative Bacteriology". 8th ed. (Buchanan, R.E. and N.E. Gibbons, eds.), Williams and Wilkins, Baltimore (1974)
202. Otsuki, K.; M. Tsubokura; K. Itagaki; K. Hirai, and H. Nigi: Isolation of *Yersinia enterocolitica* from monkeys and deers. Jap. J. vet. Sci. *35*, 447-448 (1973)
203. Otto, H.F.; M. Wanke, and J. Zeitlhofer: Darm und Peritoneum, "Spezielle pathologische Anatomie", Band 2, Teil 2 (Doerr, W., G. Seifert und E. Uehlinger, eds.) Springer, Berlin (1976)
204. Pai, C.H. and V. Mors: Production of enterotoxin by *Yersinia enterocolitica*. Infect. Immun. *19*, 908-911 (1978)
205. Pal, T.; S. Pacsa; L. Emody, and S. Voros: Antigenic relationship among virulent enteroinvasive Escherichia coli, Shigella flexneri and Shigella sonnei detected by ELISA. Lancet II, 102 (1983)
206. Panaitescu, M. and E. Potorac: The etiopathogeny of gastrointestinal infection in newly imported monkeys with special emphasis on associated parasitic and bacterial infections. Arch. roum. Path. exp. *40*, 273-285 (1981)
207. Paterson, J.S. and R. Cook: A method for the recovery of *Pasteurella pseudotuberculosis* from faeces. J. Path. Bact. *85*, 241-242 (1963)
208. Pattyn, S.R.; P. Piot; M. Asman; W. de Meurichy and G. Ghysels: Isolation of *Yersinia enterocolitica* 0:3 *Salmonella Antwerpen* n.sp. (1, 42 1, 42 2: c.e, n, z15, z17) from *Perodicticus potto* (Muller). Acta zool. path. antverp. *73*, 213-214 (1978)
209. Petrovskaya, V.G. and N.A. Khomenko: Proposals for improving the classification of members of the genus *Shigella*. Int. J. System. Bacteriol. *29*, 400-402 (1979)
210. Pierce, R.L.; M.W. Vorhies, and E.J. Bickwell: *Yersinia pseudotuberculosis* infection in a Siberian tiger and a spider monkey. J. Amer. vet. med. Ass. *163*, 547 (1973)
211. Pinkerton, M.: Miscellaneous organisms; pp. 283-313 in "Pathology of Simian Primates", Part II. (Fiennes, R.N.T-W., ed.), Karger, Basel (1972)

212. Pinkerton, M.E.; L.M. Boncyk, and J.A. Cline: Microbiological parameters of the baboon (*Papio* sp.): Bacteriology; pp.717-731 in "The Baboon in Medical Research", Vol. II (Vagtborg, H., ed.), Univ. Texas Press, Austin (1967)
213. Pizsolitto, A.C.; D.P. Falcao; M.T. Shimizu; S.H.M. Galvão, and W. Giraldini: The first isolation of human *Yersinia enterocolitica* in Brazil: Case report. Pp.169-173 in "Yersinia enterocolitica. Biology, Epidemiology and Pathology." (Carter, P.B., L. Lafleur, and S. Toma, eds.) Contr. Microbiol. Immunol. *5*, Karger, Basel (1979)
214. Platt, D.J.; J.S. Sommerville; C.A. Kraft, and M.C. Timbury: Antimicrobial resistance and the ecology of *Escherichia coli* plasmids. J. Hyg. Camb. *93*, 181-188 (1984)
215. Poelma, F.G.; G.H.A. Borst, and P. Zwart: *Yersinia enterocolitica* infections in non-human primates. Acta zool. path. antverp. *69*, 3-9 (1977)
216. Prizont, R.: Degradation of intestinal glycoproteins by pathogenic *Shigella flexneri*. Infect. Immun. *36*, 615-620 (1982)
217. Prpic, J.K.; R.M. Robins-Browne, and R. Brent Davey: Differentiation between virulent and avirulent *Yersinia enterocolitica* isolates by using Congo red agar. J. clin. Microbiol. *18*, 486-490 (1983)
218. Pucak, G.J.; R.P. Orcutt; R.J. Judge, and F. Rendon: Elimination of the *Shigella* carrier state in rhesus monkeys *(Macaca mulatta)* by trimethoprim - sulfamethoxazole. J. med. Primatol. *6*, 127-132 (1977)
219. Ravaut, P. et C. Dopter: Une épidémie de dysenterie bacillaire chez des macaques. Bull. Soc. Path. exot. *2*, 17-20 (1909)
220. Rewell, R.E.: Outbreak of *Shigella schmitzii* infection in man and apes. Lancet *256*, 220-221 (1949)
221. Roberts, J.A.; B. Kaack; G. Källenius; R. Möllby; J. Winberg, and S.B. Svenson: Reception for pyelonephritogenic Escherichia coli in primates. J. Urol. *131*, 163-168 (1984)
222. Robins-Browne, R.M.; A.R. Rabson, and H.J. Koornhof: *Yersinia enterocolitica* in South Africa; pp. 193-203 in "Yersinia enterocolitica" (Bottone, E.J., ed.), CRC Press, Boca Raton (1981)
223. Robins-Browne, R.M.; S. Tzipori; G. Gonis; J. Hayes; M. Withers, and J.K. Prpic: The pathogenesis of *Yersinia enterocolitica* infection in gnotobiotic piglets. J. med. Microbiol. *19*, 297-308 (1985)
224. Robinson, D.T.; E.C. Armstrong, and S.P. Carpenter: Outbreak of dysentery due to contact with a pet monkey. Brit. med. J. *I*, 903-905 (1965)
225. Robinson, D.T.; E.C. Armstrong, and P. Carpenter: Visiting pet monkey gives children dysentery. Lab. Primate Newsl. *4*, (4), 12 (1966)
226. Roe, E.; R.J. Jones, and E.J.L. Lowbury: Transfer of antibiotic resistance between *Pseudomonas aeruginosa, Escherichia coli* and other gramnegative bacilli in burns. Lancet I, 149-152 (1971)
227. Rolle, M. and A. Mayr: Mikrobiologie, Infektions- und Seuchenlehre für Tierärzte, Biologen und Agrarwissenschaftler. Enke, Stuttgart (1978)
228. Romanowska, E.; C. Lugowski, and M. Mulcyzyk: Studies on Shigella sonnei phase I and phase II lipopolysaccharides. J. Hyg. *18*, 391-396 (1974)
229. Rosenberg, D.P.; N.W. Lerche, and R.V. Henrickson: *Yersinia pseudotuberculosis* infection in a group of *Macaca fascicularis*. J. Amer. vet. med. Ass. *177*, 818-819 (1980)
230. Rout, W.R.; S.B. Formal; R.A. Giannella, and G.J. Dammin: Pathophysiology of Shigella diarrhea in the rhesus monkey: intestinal transport, morphological and bactèriological studies. Gastroenterology *68*, 270-278 (1975)
231. Rowe, B.: Salmonellosis in simians and other non-human primates. Lab. Anim. Handb. *4*, 63-75 (1969)
232. Ruch, T.C.: Diseases of Laboratory Primates. W.B. Saunders, Philadelphia (1959)
233. Saari, T.M. and G.P. Jansen: Waterborne *Yersinia enterocolitica* in the midwest United States, pp. 185-195 in "Yersinia enterocolitica. Biology, Epidemiology and Pathology" (Carter, P.B., L. Lafleur, and S. Toma, eds.) Contr. Microbiol. Immunol. *5*, Karger, Basel (1979)
234. Sakazaki, R.; K. Tamura; A. Nakamura, and T. Kurata: Enteropathogenic and enterotoxigenic activities on ligated gut loups in rabbits of *Salmonella* and some other enterobacteria isolated from human patients with diarrhea. Jap. J. med. Sci. Biol. *27*, 45-48 (1974)
235. Sanford, J.P. and J.P. Luby: Infectious Diseases. Grune and Stratton, New York (1981)

236. Sansonetti, P.J.; M. David, and M. Toucas: Correlation entre la perte d 'ADN plasmidique et le passage de la phase 1 virulence à la phase II avirulente chez *Shigella sonnei*. C.R. Acad. Sci. D, *290*, 879-882 (1980)
237. Sansonetti, P.J.; H.d 'Hauteville; C. Ecobichon, and C. Purcel: Molecular comparison of virulence plasmids in *Shigella* and enteroinvasive *Escherichia coli*. Ann. Microbiol. *134* A, 295-318 (1983)
238. Sansonetti, P.J.; D.J. Kopecko, and S.B. Formal: Involvement of a plasmid in the invasive ability of *Shigella flexneri*. Infect. Immun. *35*, 852-860 (1982)
239. Saxena, V.K. and J.N.S. Yadava: Haemagglutinins and enterotoxins in *Escherichia coli* strains from human urinary tract infections. Indian J. Med. Res. *81*, 35-40 (1985)
240. Schiemann, D.A.: Antigenic identity of *Yersinia enterocolitica* serotypes 0: Tacoma and 021. J. clin. Microbiol. *20*, 831-832 (1984)
241. Schiff, L.J.; P.W. Barbera; C.D. Port; H.M. Yamashiroya; A.M. Sheffner, and S.M. Poiley: Enteropathogenic *Escherichia coli* infections: increasing awareness of a problem in laboratory animals. Lab. Anim. Sci. *22*, 705-708 (1972)
242. Schleifstein, J. and M.B. Coleman: An unidentified microorganism resembling *B. lignieri* and *Past. pseudotuberculosis* and pathogenic for man. N.Y. St. J. Med. *39*, 1749-1753 (1939)
243. Schmidt, R.E. and T.M. Butler: *Klebsiella-Enterobacter* infections in chimpanzees. Lab. Anim. Sci. *21*, 946-949 (1971)
244. Schneider, N.J.; E.C. Prather; A.L. Lewis; J.E. Scatterday, and A.V. Hard: Enteric bacteriological studies in a large colony of primates. Ann. N.Y. Acad. Sci. *85*, 935-941 (1960)
245. Schröder, H.D.: Zum Vorkommen und zur Bedeutung von Yersinia-Infektionen bei Affen. XXII. Internat. Symp. Erkr. Zootiere, Arnhem, pp. 105-110 (1980)
246. Schröder, H.D.: Über das Vorkommen von Salmonellen bei Zootieren. VI. Internat. Symp. Erkr. Zootiere, Wien, 8.-10.5.64, pp. 130-133 (1964)
247. Schneider, N.J.; E.C. Prather; A.L. Lewis; J.E. Scatterday, and A.V. Hardy: Enteric bacteriological studies in a large colony of primates. Ann. N.Y. Acad. Sci. *85*, 935-941 (1960)
248. Selbitz, H.J.; K. Elze, and G. Seifert: Nachweis und klinische Bedeutung human- und tierpathogener Serotypen von Escherichia coli bei Affen. 25. Internat. Symp. Erkr. Zootiere, Wien, pp. 275-279 (1983)
249. Sereny, B.: Experimental *Shigella* conjunctivitis. Acta microbiol. Acad. Sci. hung. *2*, 293-296 (1955)
250. Sereny, B.: Experimental keratoconjunctivitis shigellosa. Acta microbiol. Acad. Sci. hung. *4*, 367-376 (1957)
251. Shayegani, M.: Epidemiological aspects of *Yersinia enterocolitica* in New York state with emphasis on a recent food-borne outbreak; p. 173-182 in "Yersinia enterocolitica" (Bottone, E.J., ed.) CRC Press, Boca Raton (1981)
252. Shlatz, L.J. and D.V. Kimberg: Membrane polarity in regulation of ion transport in the small intestine; pp. 45-60 in "Secretary Diarrhea" (Field, M., J.S. Fordtran and S.G. Schultz, eds.) Amer. Physiol. Soc., Bethesda (1980)
253. Silva, R.M.; M.R.F. Toledo, and L.R. Trabulsi: Biochemical and cultural characteristics of invasive *Escherichia coli*. J. clin. Microbiol. *11*, 441-444 (1980)
254. Silva, R.M.; M.R.F. Toledo, and L.R. Trabulsi: Correlation of invasiveness with plasmid in enteroinvasive strains of *Escherichia coli*. J. infect. Dis. *148*, 706 (1982)
255. Simmons, L.G.: Observations on three gastrointestinal problems encountered in baby gorillas. Pp. 273-277 in "American Association of Zoo Veterinarians: Annual proceedings." Topeka, Kansas, priv. publ. (1973)
256. Simmons, L.G.: Fatal shigellosis in a two year old gorilla. Amer. Ass. Zoo Vet. Ann. Proc. - privately publ., Topeka, Kansas (1973)
257. Skavlen, P.A.; H.F. Stills, jr. and C.C. Middleton: *Yersinia enterocolitica* septicemia in *Erythrocebus patas*. Lab. Anim. Sci. *33*, 481 (1983)
258. Skurnick, M.; I. Bölin; H. Heikkinen; S. Piha, and H. Wolf-Watz: Virulence plasmid-associated autoagglutination in *Yersinia* spp. J. Bact. *158*, 1033-1036 (1984)
259. Smith, H.: Gram-negative bacteria: The determinants of pathogenicity; pp. 8-15 in "Gram-Negative Bacterial Infections and Mode of Endotoxin Actions. Pathophysiological, Immunological and Clinical Aspects" (Urbaschek, B., R. Urbaschek, and E. Neter, eds.) Springer, Wien - New York (1975)

260. Sonnenwirth, A.C.: Yersinia; pp.222-229 in "Manual of Clinical Microbiology". 2nd ed. (Lennette, E.H., E.H. Spaulding, J.P. Truant) Amer. Soc. Microbiol. Washington D.C. (1974)
261. Staley, L.G.: Yersiniosis *(Yersinia enterocolitica)*; pp.257-271 in "CRC Handbook Series in Zoonoses" (Steele, J.H., Series ed.), Section A: Bacterial, Rickettsial and Mycotic Diseaes, Vol. II (Stoenner, H., W. Kaplan and M. Torten, Section ed.) CRC Press, Boca Raton (1980)
262. Stoleru, G.H.; G.R. Gerbaud; D.H. Bouanchaud, and L. LeMinor: Etude d'un plasmide transferable determinant la production d 'H2S et la resistance a la tetracycline chez *Escherichia coli.* Ann. Inst. Pasteur *123*, 743-754 (1972)
263. Stoll, D.M.: Cutaneous shigellosis. Arch. Dermatol. *122*, 22 (1986)
264. Stovell, P.L.: Pseudotubercular yersiniosis; pp.209-256 in "CRC Handbook Series in Zoonoses" (Steele, J.H., Series ed.), Section A: Bacterial, Rickettsial and Mycotic Diseaes. Vol. II (Stoenner, H., W. Kaplan and M. Torten, Section eds.). CRC Press, Boca Raton (1980)
265. Strickland, H.L.; B. Rader, and N.D.M. Lehner: An outbreak of infection with *Yersinia pseudotuberculosis* in a group of macaques. Lab. Anim. Sci. *32*, 426 (1982)
266. Swenson, R.B. and H. McClure: Septic abortion in a gorilla due to Shigella flexneri. Ann. Proc. Amer. Ass. Zoo Vet. 1974, 938-939 (1974)
267. Szita, J.; A. Svidro; M. Kubinyi; I. Nyomarkay, and I. Mihalyfi: Yersinia enterocolitica infection of animals and human contacts. Acta microbiol. Acad. Sci. hung. *27*, 103-109 (1980)
268. Taffs, L.F. and G. Dunn: An outbreak of *Yersinia pseudotuberculosis* infection in a small indoor breeding colony of red-bellied *(Saguinus labiatus)* tamarins. Lab. Anim. *17*, 311-320 (1983)
269. Takao, T.; N. Tominaga; Y. Shimonishi; S. Hara; T. Inoue, and A. Niyama: Primary structure of heat-stable enterotoxin produced by *Yersinia enterocolitica.* Biochem. Biophys. Res. Comm. *125*, 845-851 (1984)
270. Takasaka, M.; S. Honjo; T. Fujiwara; T. Hagiwari; H. Ogawa, and K. Imaizuni: Shigellosis in cynomolgus monkeys *(Macaca irus).* I. Epidemiological surveys on Shigella infection rate. Jap. J. Med. Sci. Biol. *17*, 259-265 (1964)
271. Takasaka, M.; S. Honjo, and K. Imaizumi: Shigella infection rate in imported cynomolgus monkeys and drug-sensitivity of the isolates. Exp. Anim. *22*, 377-387 (1973a)
272. Takasaka, M.; S. Honjo; K. Imaizumi; H. Ogawa; R. Nakaya; A. Nakamura, and K. Mise: Experimental infection with *Shigella sonnei* in cynomolgus monkeys *(Macaca irus).* Jap. J. Med. Sci. Biol. *22*, 389-393 (1969)
273. Takasaka, M.; S. Morota; T. Kasono; M. Abe, and S. Honjo: Incidence of diarrhea and antibiotic therapy in cynomolgus monkeys imported for medical use. Exp. Anim. *22*, 227-236 (1973 b)
274. Takeuchi, A.: Electron microscope observation on penetration of the gut epithelial barrier by *Salmonella typhimurium.* Microbiology 174-181 (1975)
275. Takeuchi, A.: Early colonic lesions in experimental Shigella infection in rhesus monkeys: Revisited. Vet. Path. *19*, (Suppl. 7), 1-8 (1982)
276. Takeuchi, A.; S.B. Formal, and H. Sprinz: Experimental acute colitis in the rhesus monkey following per oral infection with *Shigella flexneri*: an electron microscope study. Amer. J. Path. *52*, 503-520 (1968)
277. Takeuchi, A.; H.R. Jervis, and S.B. Formal: Animal model: Monkey shigellosis or dysentery. Amer. J. Path. *81*, 251-254 (1975)
278. Takeuchi, A.; H. Sprinz; E.H. LeBrec, and S.B. Formal: Experimental bacillary dysentery: an electron microscopic study of the response of the intestinal mucosa to bacterial invasion. Amer. J. Path. *47*, 1011-1014 (1965)
279. Tanaka, Y. and Y. Katsube: *Salmonella* carriers in the imported cynomolgus monkeys *(Macaca fascicularis).* Jap. J. vet. Sci. *43*, 787-789 (1981)
280. Tauraso, N.M.: Review of recent epizootics in nonhuman primate colonies and their relation to man. Lab. Anim. Sci. *23*, 201-210 (1973)
281. Thomas, L.V.; R.J. Gross; T. Cheasty; C.R. Shipp, and B. Rowe: Antigenic relationship among type strains of *Yersinia enterocolitica* and those of *Escherichia coli*, *Salmonella* spp. and *Shigella* spp. J. clin. Microbiol. *17*, 109-111 (1983)
282. Thomas, D.D. and F.C. Knoop: The effect of calcium and prostaglandin inhibitors on the intestinal fluid response to heat-stable enterotoxin of *Escherichia coli.* J. infect. Dis. *145*, 141-147 (1982)

283. Thurman, J.D.; R.J. Morton, and E.L. Stair: Septic abortion caused by *Salmonella heidelberg* in a white-handed gibbon. J. Amer. vet. med. Ass. *183*, 1325 (1983)
284. Toft, J.D., II, and J.R. Dunagan: Human shigellosis associated with exposure to a pet woolly monkey. Primate Zoonoses Surveil. Rep. *11*, 8-9 (1973)
285. Toledo, M.R.; M.H.L. Reis, and L.R. Trabulsi: New invasive *Escherichia coli* strains. J. clin. Microbiol. *11*, 422-423 (1980)
286. Toma, S.; G. Wauters; H.M. McClure; G.K. Morris, and A.S. Weissfeld: 0:13a, 13b, a new pathogenic serotype of *Yersinia enterocolitica*. J. Clin. Microbiol. *20*, 843-845 (1984)
287. Tsubokura, M.; K. Otsuki; Y. Kawaoka; H. Fukushima; K. Ikemura, and Y. Kanazawa: Addition of new serogroups and improvement of the antigenic designs of *Yersinia pseudotuberculosis*. Curr. Microbiol. *11*, 89-92 (1984)
288. Tsubokura, M.; K. Otsuki; Y. Kawaoka; T. Hongo; H. Otani; K. Kishida; M. Tanaka, and K. Sato: The first successful isolation and identification of *Yersinia pseudotuberculosis* serogroup II A. Japan. Microbiol. Immunol. *26*, 1187-1188 (1982)
289. Ul'Yanova, N.L.: The natural focal occurrence of pseudotuberculosis. J. Microbiol. Epidemiol. Immunobiol. *32*, 2268-2272 (1961)
290. Van Damme, L.R.; j. Vandepitte; P. Piot, and E. van Dijck: Isolation of *Yersinia enterocolitica* serotype 0:3 from captive monkeys and domestic pigs in Zaire. Ann. Soc. belge Med. trop. *58*, 341-345 (1978)
291. Van Loghem, J.J.: The classification of plague-bacillus. Antoine von Leuwenhoek. J. Microbiol. Serol. *10*, 15-16 (1944)
292. Varghese, A.; V.G. Ramachandran, and D.S. Agarwal: Isolation and characterization of *Yersinia enterocolitica* from patients of gastroenteritis. Indian J. Med. Res. *79*, 35-40 (1984)
293. Velin, D. and L. Emödy: The stability of enterotoxin production in *Yersinia enterocolitica* and the methanol solubility of heat-stable enterotoxin. Acta microbiol. Acad. Sci. hung. *29*, 227-233 (1982)
294. Von Riesen, V.L.: Pectinolytic, indole-positive strains of *Klebsiella pneumoniae*. Int. J. Sys. Bact. *26*, 143-145 (1976)
295. Walker, J.S.; F.C. Cadigan; A.R. Sirimanne, and J. Binabass: Silvered leaf-monkeys *(Presbytis cristatus)*. I. Colony management, caging, diet, and biologic values. Lab. Anim. Sci. *24*, 290-298 (1974)
296. Wauters, G.: Antigens of *Yersinia enterocolitica*; pp. 41-53 in "Yersinia enterocolitica" (Bottone, E.J. ed.), RCR Press, Boca Raton (1981)
297. Weil, J.D.; M.K. Ward, and R.O. Spertzel: Incidence of *Shigella* in conditioned rhesus monkeys *(Macaca mulatta)*. Lab. Anim. Sci. *21*, 434-437 (1971)
298. Wetzler, T.F.: Pseudotuberculosis; pp. 253-262 in "Infectious Diseases of Wild Mammals". 2nd ed. (Davis, J.W.; L.H. Karstad, and D.O. Trainer, eds.), Iowa State Univ. Press, Ames/Iowa (1981)
299. Williams, L.P., jr.: Salmonellosis; pp. 11-37 in "CRC Handbook Series in Zoonoses". (Steele, J.H., Series ed.), Section A: Bacterial, Rickettsial and Mycotic Diseases. Vol. II (Stoenner, H., W. Kaplan, and M. Torten, Section eds.), CRC Press, Boca Raton (1980)
300. Winblad, S.: Erythema nodosum associated with infection with *Yersinia enterocolitica*; pp. 125-133 in "Yersinia enterocolitica" (Bottone, E.J., ed.), CRC Press, Boca Raton (1981)
301. Wolf, R.H.; J.E. Haught; O. Felsenfeld, and A.D. Felsenfeld: Slow lactose fermenting *Escherichia coli* diarrhea in a group of patas monkeys *(Erythrocebus patas)*. Lab. Anim. Sci. *21*, 549-552 (1971)
302. Wormser, G.P. and G.T. Keusch: *Yersinia enterocolitica*: Clinical Observations, pp. 83-93 in "Yersinia enterocolitica" (Bottone, E.J., ed.), CRC Press, Boca Raton (1981)
303. Wrong, O.M.; C.J. Edmonds, and V.S. Chadwick: The Large Intestine: Its Role in Mammalian Nutrition and Homeostasis. MTP Press Ltd., Falcon House, Lancaster (1981)
304. Yamamoto, T.; T. Honda; T. Miwatani, and T. Yokota: A virulence plasmid in *Escherichia coli* enterotoxigenic for humans: Intergenetic transfer and expression. J. Infect. Dis. *150*, 688-698 (1984)
305. Zaki, S.; T.S. Nalini; M. Satyanarayana Rao, and B.S. Keshavamurthy: Isolation of *Shigella flexneri* from two tiger cubs. Curr. Sci. *49*, 288 (1980)
306. Zuidema, P.J. en G.M. Smits: Casuistische mededelingen. Chimpansee als huisdier: oorzak van een shigella-infectie. Ned. T. Geneesk. *123*, 664-666 (1979)

3.7 Gram Negative Bacteria of Uncertain Affiliation

3.7.1 Pasteurella Multocida (P. Septica)

P. multocida is normally a commensal of the respiratory tract in man and animals, but under certain conditions may become pathogenic in animals and man. It causes shipping fever in cattle, avian cholera in waterfowl, snuffles in rabbits, and haemorrhagic septicemias in cattle, buffalos, sheep and pigs (Rosen, 1975). *P. multocida* is a pathogen in man (Branson and Burkfeldt, 1967; Holloway et al., 1969; Tindall and Harrison, 1972) and has been reported as the cause of respiratory or generalized infections in squirrel monkeys, owl monkeys, capuchin monkeys, rhesus monkeys and a baboon (Cook, 1954; Smith, 1954; Greenstein et al., 1965; Benjamin and Lang, 1971; Gross, 1978).

Agent
Pasteurellae are gram-negative, non acid-fast coccobacilli to filamentous rods with a marked bipolar staining. They are nonmotile, mucoid strains are encapsulated.

P. multocida grows aerobically or facultatively anaerobically on most ordinary media (no growth on McConkey-, SS- and XLD-agar!), but better growth is obtained on blood - or serum agar with increased CO_2-tension at 37 °C. Haemolysin is not produced on blood agar, although the medium may show a brownish coloration. Growth is inhibited by bile which explains the failure of the above mentioned media in cultivating *P. multocida.*

P. multocida cultures have a characteristic musty or semenlike odor (Branson and Burkfeldt, 1967; Horne and Berlyne, 1958), they appear in 3 different forms:
a) highly virulent S-colonies of 1.0 to 1.5 mm diameter that are usually iridescent to obliquely transmitted light;
b) mucoid strains reaching 2 to 3 mm in diameter and of low virulence;
c) rough (R)-strains of similar size as the S-colonies but with a grainy surface. R-strains are avirulent in man.

The S-form contains a type-specific polysaccharide capsular antigen, the mucoid form possesses a thick hyaluronic acid-containing capsule and sometimes the type-specific capsular antigen, R-forms lack both the capsular and the mucoid antigens (Eisenberg and Cavanaugh, 1974). The biochemical characteristics are listed in Table 21.

Serotyping distinguishes 16 distinct serotypes (Finegold and Martin, 1982), but is complicated by common somatic antigens or by a blocking capsular antigen in mucoid strains (Eisenberg and Cavanaugh, 1974).

Epidemiology
P. multocida is a commensal in the upper and lower respiratory tract in man and animals (Smith, 1955; Hubbert and Rosen, 1970), it was also isolated from feces and urine of infected animals (Stahlheim and Heddleston, 1970). Observations suggest an aerogenous route as the normal way of natural infection and the dependence of diseases from the presence of carriers. Preceding an outbreak of

Table 21. Biochemical Characteristics of *P. multocida*

Catalase	+	Salicin	Ø
Oxidase	+	Glucose	+ (no gas)
Ornithin decarboxylase	+	Sucrose	+
Arginine dihydrolase	Ø	Fructose	V
Urease	V	Lactose	Ø
Nitrate reductase	+	Mannose	+
Indole	+	Mannitol	V
Methyl red	Ø	Maltose	V
Voges Proskauer	Ø	Sorbitol	V
Citrate	Ø	Xylose	V
Gelatin	Ø	Arabinose	V
Methylen blue reduction	+	Rhamnose	V
H_2S	+	Dulcitol	V
		Melibiose	Ø
		Trehalose	V

haemorrhagic septicemia in an animal population there is a rise in the normal carrier rate (Pinkerton, 1972).

Clinical Symptoms

Nonhuman primates: Pasteurellosis in nonhuman primates frequently followed transportation stress. The animals were either dead on arrival (Smith, 1954) or died within the first days (Cook, 1953; Benjamin and Lang, 1971). Clinical symptoms were rarely observed in acute respiratory disease since the disease is commonly acute or peracute with death within several hours to days and with a mortality rate of 70 to 100% (Greenstein et al., 1965; Pinkerton, 1972).

Intestinal forms are marked by the frequent presence of bloody mucus in the intestinal contents, air-sacculitis in a baboon became apparent by progressive air sac-distention due to accumulation of an initially clear, subsequently purulent fluid (Gross, 1978).

Man: Infection with *P. multocida* in man are increasing with age. According to Rosen (1975) human pasteurellosis can be distinguished into:

a) wound infection caused by animal bites and scratches appearing a few hours after infected animals bite as redness, swelling and excruciating pain. The infection may progress and penetrate the deeper tissues as tendovaginitis, or periostitis with subsequent necrosis. The body temperature usually remains within normal limits. A scratch on the cornea resulted in panophthalmitis and loss of the eye (Galloway and Robinson - cit. from Biberstein, 1979).
b) latent infection of the upper respiratory tract, which may fulminate following other favouring incidents and possibly result in conjunctivitis, sinusitis or meningitis;
c) lower respiratory tract infection such as pneumonitis, empyema, pneumonia or bronchiectasia. The incubation period in that form may be as short as a few hours after animal contact, but may be as long as several months as well;
d) cryptogenic infection of internal organs such as cystitis, pyelonephritis, appendicitis, peritonitis, metritis, vaginitis, otitis media, meningitis and fatal septicemia.

Pathology

Nonhuman primates: The predominant lesions of pasteurellosis are found as petechial haemorrhages in pleura, endocardium and kidney. Larger haemorrhagic areas may be seen in connective tissues, lymphatic glands, muscles and serous membranes (Greenstein, 1965). In pulmonary pasteurellosis the lungs are hyperemic and focally consolidated, the pleura and the pericardium are covered by a fibrinous exudate, the mucosal membranes of the trachea and bronchi are congested (Pinkerton, 1972). Histopathologically the lesions are characterized as purulent or fibrinopurulent pneumonia and focally severe acute interstitial pneumonitis with focal alveolar necrosis (Benjamin and Lang, 1971).

Intestinal forms were described as acute haemorrhagic enteritis (Pinkerton, 1972) or as purulent enterocolitis with focal mucosal necrosis (Benjamin and Lang, 1971).

Spread of the infection to internal organs resulted in focal hepatic, renal, cardiac and endometrial necrosis and in one case in severe, generalized purulent meningoencephalitis (Benjamin and Lang, 1971).

Man: Fatalities are very rare in wound infections and are usually attributed to other concurrent conditions (Biberstein, 1979).

In respiratory infections pneumonia, pneumonitis, pulmonary abscesses, chronic bronchitis, chronic and obstructive lung diseases, sometimes empyema, tonsillitis and sinusitis have been reported.

Spread of the infection resulted in otitis media, cerebellar abscess, peritonitis, enteritis, Bartholin abscess, cystitis, septicemia, panophthalmia, possibly even in septic arthritis (Biberstein, 1979).

Treatment

P. multocida is very susceptible to penicillin. Tetracycline and chloramphenicol are also useful therapeutic agents, whereas erythromycin, clindamycin and aminoglycosides are relatively ineffective (Finegold and Martin, 1982).

Abscesses, empyemas or air sacculitis sometimes have to be treated surgically as well as antibiotically.

3.7.2 Francisella Tularensis (Pasteurella Tularensis, Brucella Tularensis)

F. tularensis is the causative agent of a plague-like disease in lagomorphs and rodents and is comprised of two subspecies: *F. tularensis tularensis* (type A), and *F. tularensis palearctica* (type B). Type A is restricted to North America, occurring mainly in lagomorphs, e.g. cotton tail rabbits (*Sylvilagus* spp.) and jack rabbits *(Lepus californicus).* Type B is of holoarctic distribution, affecting mainly rodents, particularly water rats *(Arvicola terrestris).* Both subspecies are notoriously infectious to man (Bell, 1980).

As a few spontaneous *F. tularensis*-infections have been reported in nonhuman primates - *Saguinus nigricollis, Saimiri sciureus, Miopithecus talapoin* (Emmons et al., 1970; Nayar et al., 1979) the disease has to be mentioned.

Agent

The two subspecies of *F. tularensis* seem to be antigenically identical, but they differ in their virulence, their ability to ferment glycerol and in the possession of citrullineureidase. Type A, the American subspecies, is associated with a significantly higher mortality in man as compared to the Eurasian type B (Kadull et al., 1950 – cit. from Bell, 1980). Type A utilizes glycerol (Olsuf 'Jev and Emelyanova, 1962) and possesses the enzyme citrullineureidase (Marchette and Nicholes, 1961).

Both subtypes of *F. tularensis* are very pleomorphic, Gram-negative bacteria, which are sometimes even able to pass filters impermeable to most other bacteria. They may be coccobacillary but they often occur in groups or chains. *F. tularensis* is non-motile, although some controversy exists concerning their motility as well as the presence of a capsule.

F. tularensis is strictly aerobic. It does not grow on ordinary media presumably due to the requirement of a growth factor and the unusual sensitivity of *F. tularensis* to mild oxidants in the medium. The best results are obtained with cystine-glucose-blood agar (CGBA), since its growth is generally enhanced by cystine or cysteine with production of H_2S in cystine containing media. On CGBA, growth is usually abundant within 24 hs at 37 °C (no growth at 10 °C!). The mucoid, gray-colored colonies may reach a diameter of 4 mm, the blood-containing agar becomes greenish.

F. tularensis remains viable for weeks or months in mud, soil, water and fomites, especially at lower temperatures, but it is easily killed by chlorine (1.5 mg/l) or by exposure to 56 °C for 10 min.

Epidemiology

The epidemiology of tularemia is very diverse including foodborne, water-borne, arthropod-vectored and contact infections depending on the geographical region, the bacterial subspecies involved, occupational exposure, carnivorism etc.

Type A occurs solely in North- and Central America as far south as Mexico. Its reservoir hosts are lagomorphs (*Sylvilagus* spp., *Lepus californicus*) and beavers *(Castor canadensis)*. It is transmitted by ticks (*Haemaphysalis, Amblyomma* and *Dermacentor* spp.) and deer flies *(Chrysops discalis)* among rabbits, or during the lambing or shearing season even to sheep. In the tick transovarian infection is possible (Pinkerton, 1972). Man is infected by contact with the infected rabbits or beavers mainly as an occupational hazard, or by tick bites. Type A has not been found to infect surface water (Bell, 1980).

Type B typically infects rodents, especially Cricetidae (*Arvicola terrestris* – water rat), microtines (voles), lagomorphs (*Lepus timidus* and *L. europaeus* – hares), muskrats, lemmings, and sometimes birds of several families (galliformes, anserinae, passerinae) in Europe, Asia and America. It frequently contaminates water, but it has not been isolated from sheep in contrast to type A. Transmission is possible via contaminated water, by carnivorism, during meat or fur processing as an occupational infection. In the USSR tabanids and *Culicoides* spp. have been identified as vectors (Bell, 1980).

Pathogenesis

Parenteral inoculation by arthropods, bone splinters, knives etc. are probably the most common causes of human infection but *F. tularensis* is also able to penetrate the unbroken skin or can take the enteric route.

At the port of entry the tissues become necrotic, forming a crater, which is sometimes covered by a scab. The ulcer is sharply demarcated by a wall of epithe-loid and plasma cells and lymphocytes. Repair of the ulcers is by formation of granulation tissue containing the above cells plus macrophages and mast cells. From the primary lesion the pathogen is initially lymphogenously transported, thereby inducing regional lymphadenopathy.

One major pathogenic stage of *F. tularensis* infection is the development of an allergic reaction resulting in early coagulation necrosis, with swelling and prolifera-tion of the RHS-cells of the sinuses. This reaction is more pronounced in lymph nodes more remote from the port of entry.

The infection, if it becomes generalized, spreads to other organs by lymphoge-nous or haematogenous route or by a combination of both. The most commonly affected organs are the spleen, liver and pleura, with the formation of focal necro-ses in the involved tissues.

Clinical Symptoms

Nonhuman primates: Three black and red tamarins (*Saguinus nigricollis*) in a Cana-dian Zoological Garden, which were naturally infected by *F. tularensis* suddenly col-lapsed and died or were found dead. A dwarf guenon from the same facilities, also infected by *F. tularensis*, had a severe tongue ulceration and as a result profuse saliva-tion, with ocular and nasal discharges (Nayar et al., 1979). A young squirrel monkey obtained from a San Francisco pet shop suffered from anorexia and weight loss. It became irritable and died 5 days later (Emmons et al., 1970).

Man: The clinical symptoms of tularemia in man are as variable as the mor-phology of the etiological agent. The ensuing disease may range from a prolonged disorder with a low-grade fever and adenopathy to a severe, toxic, overwhelming infection depending on the site of infection, infecting dose and the bacterial sub-species involved. Type A infections, for example, are usually more severe than those caused by type B (Bell, 1980).

The course of prolonged skin infections typically extends as ulceroglandular tularemia, the most common form, over a period of 2 or 3 months. The disease starts with sudden headaches, chills, vomiting, fever and prostration. At the site of entry an ulcer becomes punched out, whilst the regional lymph nodes enlarge and become tender and painful. Chronic adenitis may persist for months or years. Exanthema, which has been reported in several cases, usually appears during the first two weeks of illness as erythematous, urticarial or acneform eruptions which persist for 2 or 3 weeks.

Tularemic pneumonia, either after aerogenous infection or after spread from other sites is usually peracute or fulminant. It is marked by a dry cough, pleurisy and dyspnea.

In typhoidal tularemia the patients are extremely ill with fever, chills, head-aches, stupor, delirium, malaise, anorexia, abdominal pain, nausea, vomiting, diar-rhea, meningism and marked dehydration (Sanders and Torres-R, 1981).

Other forms of the disease include oculoglandular tularemia, which arises from transfer of the agent into the eye from the fingers, and oropharyngeal tularemia after oral infection.

Pathology
Pathological data in tularemia are rather scarce in both man and nonhuman primates.

Nonhuman primates: In all three tamarins the lesions consisted of slight increase in the abdominal fluid, fibrinous pleuritis and peritonitis, splenomegaly and tiny foci of caseation necrosis in the liver and spleen; lymphadenitis of the mesenteric lymphnodes and occasional mild intestinal haemorrhages (Nayar et al., 1979).

Man: Tularemic pneumonia is of multifocal necrotizing character, accompanied by a seropurulent pleuritis and necrotizing lymphadenitis (Morgenroth, 1983).

Treatment
Streptomycin, kanamycin and gentamicin are the antibiotics of choice, while tetracycline and chloramphenicol are less effective (Sanders and Torres-R, 1981).

Unclassified Bacteria:

3.7.3 Chromobacterium Violaceum

C. violaceum, generally considered as a saprophyte especially in tropical and subtropical climates (Johnson, 1971), has been reported on rare occasions to cause disease in man (Black and Shahan, 1938; Soule, 1939; Dauphinais and Robben, 1968; Ognibene and Thomas, 1970).

C. violaceum has also been reported in gibbons, a guenon monkey *(C. cephus)*, a stumptailed macaque *(Macaca arctoides)* and an Assam macaque *(M. assamensis)* (Audebaud et al., 1954 - cit. from McClure and Chang, 1976; Groves et al., 1969; Johnson et al., 1971; Joseph et al., 1971; McClure and Chang, 1976; Collins et al., 1985).

Agent
C. violaceum is a gram-negative, sometimes slightly curved rod. It grows aerobically or facultatively anaerobically on most standard nutrient media with the production of a violet, alcohol-soluble, water- and chloroform insoluble pigment (violacein) of antibiotic activity and of lethal action to certain protozoa (Sneath, 1960; Moss and Ryall, 1981; Finegold and Martin, 1982). In broth it forms a fragile pellicle, whilst on blood agar it is sometimes haemolytic.

C. violaceum is motile by means of one polar and 1 to 4 subpolar flagellae. It produces an alkaline slant and acid butt on triple sugar iron agar, other biochemical reactions are summarized in Table 22.

Table 22. Biochemical Characteristics of *Chromobacterium violaceum*

Oxidase	Ø	Xylose	Ø
Catalase	+	Ornithine decarboxylase	Ø
Glucose	+ (gas Ø)	Arginine dihydrolase	+
Lactose	Ø	Indole	Ø
Sucrose	+	H_2S	Ø
Arabinose	Ø	Citrate (Simmons)	+
Melibiose	Ø	Urea	Ø
Trehalose	+	Nitrate	+
Rhamnose	Ø	Gelatin	+
Mannitol	Ø	Methyl red	Ø
Sorbitol	Ø	Voges Proskauer	Ø

Epidemiology

The route of infection may be aerogenous, peroral, or by way of skin injuries (Sippel et al., 1954; Sneath, 1960; Johnson et al., 1971), and the organism may be transmitted from animals to man by any of these routes, in the latter case possibly by bite wounds.

Most human infections occur in southeast Asia and southeastern USA (Finegold and Martin, 1982); therefore care has to be taken particularly in handling animals imported from these areas. The infection in man frequently is fatal, the exact incubation time is unknown but might be prolonged.

Clinical Symptoms

Nonhuman primates: In the Assam macaque observed at Yerkes Primate Center no clinical signs prior to its sudden death were noted (McClure and Chang, 1976), the stumptailed macaque exhibited signs of anorexia, lethargy and depression, hydrops ascites and a mass in the dorsal abdomen (Collins et al., 1985).

Man: Infection in human beings usually involves abscess formation or bacteremia. Systemic (bacteremic) infection is almost uniformely fatal (Finegold and Martin, 1982).

Pathology

Nonhuman primates: The leading pathological lesions in the Assam macaques were hepatomegaly and extensive hepatic necrosis with only a minor inflammatory response, associated with other, but smaller focal necroses in the lungs, spleen, lymph nodes, adrenals and kidneys (McClure and Chang, 1976).

Man: The disease in man is characterized by hepatic necrosis and abscess formation (McClure and Chang, 1976).

Treatment

Most strains of *C. violaceum* produce beta-lactamase and are thus resistant to penicillin, cephalosporin, cephalotin, ampicillin and colistin (Farrar and O'Dell, 1976), but all isolates have been proven susceptible to chloramphenicol (Finegold and Martin, 1982).

3.7.4 Haemophilus Influenzae

Haemophilus spp. are strict parasites requiring accessory factors for growth. *Haemophilus influenzae* B is common especially in children up to 3 years of age, above that age, other serotypes increase in frequency.

Occasionally *Haemophilus influenzae* B also causes spontaneous infections and diseases in nonhuman primates. Macaques are very susceptible to experimental infections with man-derived strains.

Agent

Haemophilus influenzae are minute, Gram-negative, coccoid rods of 0.2 to 0.5 × 0.3 to 2.0 μm size. The formation of threadlike filaments and pleomorphism are common, bipolar staining is often observed. They are non-motile, aerobic, and need special conditions for successful growth. They do not grow on fresh human or sheep blood agar, because both agars contain heat labile inhibitors of *H. influenzae* (Young, 1974) which are destroyed in chocolate agar. They require the haeme-position of haemoglobin (X-factor) and nicotinamide-adenine-dinucleotide (V-factor) for growth. If cultivated on chocolate agar at 37 °C, small, dew-drop-like colonies with a bluish sheen in oblique light appear after 48 hs (Young, 1974; Burr and Kulshresta, 1982). If grown on other blood agars than the above mentioned (horse-blood, guinea pig etc.) *H. influenzae* colonies are haemolytic. R-forms may be flat, slightly wrinkled, granular and cohesive. For serologic demonstration of the capsular antigen, antibody-coated latex particle agglutination has been proven to be superior over countercurrent immunoelectrophoresis (Scheifele et al., 1979).

Pathology

Nonhuman primates: Gross lesions include serous to purulent sinusitis, acute rhinitis, congestion of the mucous membranes and tracheobronchitis. Lung involvement is characterized by purulent to lymphocytic bronchiolitis, peribronchiolitis and bronchopneumonia (Carlton and Hunt, 1978). The arthrosis that occurred in the cynomolgus monkey after it had survived the initial attack consisted of osteolysis, fibroblast proliferation, erosion and villous proliferation of the articular cartilage (Dawkins, 1982). In experimental infections of infant *M. mulatta, H. influenzae* caused pyarthrosis of the major joints. They were filled by an opaque, yellow-grey purulent material. Other lesions were fibrinopurulent pneumonia, panophthalmitis, pericarditis and purulent leptomeningitis (Scheifele et al., 1980).

Treatment

Most *H. influenzae*-strains are susceptible to penicillin, chloramphenicol, tetracyclines and sulfonamids. The most commonly used antibiotic in man is ampicillin (Kahn et al., 1967).

References

1. Alexander, H.E.: Bacterial and Mycotic Infections of Man. 4th ed. Lippincott, Philadelphia (1965)
2. Bell, J.F.: Tularemia; pp.161-193 in "CRC Handbook Series in Zoonoses" (Steele, J.H., Series ed.), Section A: Bacterial, Rickettsial and Mycotic Diseases. Vol. II (Stoenner, H., W.Kaplan and M.Torten, Section eds.) CRC Press, Boca Raton (1980)
3. Benjamin, S.A. and C.M. Lang: Acute pasteurellosis in owl monkeys *(Aotus trivirgatus).* Lab. Anim. Sci. *21,* 258-262 (1971)
4. Biberstein, E.L.: The pasteurellosis; pp.495-514 in "CRC Handbook Series in Zoonoses." (Steele, D.H., Series ed.); Section A: Bacterial, Rickettsial and Mycotic Diseases, Vol. I (Stoenner, H., W.Kaplan and M.Torten, Section eds.) CRC Press, Boca Raton (1979)
5. Black, M.E. and J.Shahan: *Bacillus violaceus* infection in a human being. J.Amer. med. Ass. *110,* 1270-1271 (1938)
6. Branson, D.and F.Burkfeldt: *Pasteurella multocida* in animal bites of humans. Amer. J.clin. Path. *48,* 552-555 (1967)
7. Burr, E.W. and S.B. Kulshrestha: Hemophilus influenza recorded from a sub-cutaneous abscess of a rhesus monkey in Uttar Pradesh, India. Indian Vet. J. *59,* 10-11 (1982)
8. Carlton, W.W. and R.D. Hunt: Bacterial Diseases. Pp.1366-1480 in "Benirschke, K., F.M. Garner, and T.C. Jones: Pathology of Laboratory Animals." Vol. II, Springer, New York (1978)
9. Collins, B.; G.Kollias; A.Battles, and A.Moreland: *Chromobacterium violaceum* infection in a stumptailed monkey *(Macaca arctoides).* Lab. Anim. Sci. *35,* 530-531 (1985)
10. Cook, R.: The naturally occurring diseases of animals and practical measures for the control of these diseases. J.med. lab. Technol. *11,* 30-35 (1953)
11. Dauphinais, R.M. and G.G. Robben: Fatal infections due to *Chromobacterium violaceum.* Amer. J.clin. Path. *50,* 592-597 (1968)
12. Dawkins, B.G.: *Haemophilus influenzae* in a cynomolgus monkey. J. Amer. vet. med. Ass. *181,* 1430-1431 (1982)
13. Eisenberg, G.H.G. and D.C. Cavanaugh: Pasteurella; pp.246-249 in "Manual of Clinical Microbiology" 2nd ed. (Lennette, E.H., E.H. Spaulding, and J.P. Truant, eds.) Amer. Soc. Microbiol., Washington, D.C. (1974)
14. Emmons, R.W.; J.D. Woodie; M.S. Taylor, and G.S. Nygaard: Tularemia in a pet squirrel monkey *(Saimiri sciureus).* Lab. Anim. Care *20,* 1149-1153 (1970)
15. Farrar, W.E., jr. and N.M. O'Dell: β-Lactamase activity in *Chromobacterium violaceum.* J.Infect. Dis. *134,* 290-293 (1976)
16. Finegold, S.M. and W.J. Martin: Diagnostic Microbiology. 6th ed., C.V. Mosby Co., St. Louis (1982)
17. Greenstein, E.T.; R.W. Doty, and K.Long: An outbreak of a fulminating infectious disease in the squirrel monkey *(Saimiri sciureus).* Lab. Anim. Care *15,* 74-80 (1965)
18. Gross, G.S.: Medical and surgical approach to laryngeal air sacculitis in a baboon caused by *Pasteurella multocida.* Lab. Anim. Sci. *28,* 737-741 (1978)
19. Groves, M.G.; J.M. Strauss; J.Abbas, and C.E. Davis: Natural infections of gibbons with a bacterium producing violet pigment *(Chromobacterium violaceum).* J.Infect. Dis. *120,* 605-610 (1969)
20. Hirschmann, J.V. and E.D. Everett: *Haemophilus influenzae* in adults: Report of nine cases and review of the literature. Medicine *58,* 80-94 (1979)
21. Holloway, W.J.; E.G. Scott, and Y.B. Adams: *Pasteurella multocida* infection in man. Report of 21 cases. Amer. J.clin. Path. *51,* 705-708 (1969)
22. Horne, W.I. and G.M. Berlyne: Empyema caused by *Pasteurella septica.* Brit. med. J. *2,* 896 (1958)
23. Howard, J.D. and C.L. Highgenboten: Residual effects of septic arthritis in infancy and childhood. J.Amer. vet. med. Ass. *236,* 732-735 (1976)
24. Hubbert, W.T. and M.N. Rosen: *Pasteurella multocida* infection in man unrelated to animal bite. Amer. J.publ. Hlth *60,* 1109-1117 (1970)
25. Johnson, W.M.; A.F.D. Salvo, and R.R. Steuer: Fatal *Chromobacterium violaceum* septicemia. Amer. J.clin. Path. *56,* 400-406 (1971)

26. Joseph, P.G.; R. Sivendra; M. Anwar, and S.F. Ong: *Chromobacterium violaceum* infection in animals. Kajian Vet., Malaysia-Singapore *3*, 55-66 (1971)
27. Kahn, W.; S. Ross, and E.A. Zaremba: Comparative inhibition of *Haemophilus influenzae* by eight antibiotics. Antimicrob. Agents Chemother. *1966*, 393-396 (1967)
28. Marchette, N.J. and P.S. Nicholes: Virulence and citralline ureidase activity of *Pasteurella tularensis*. J. Bacteriol. *82*, 26-32 (1961)
29. McClure, H.M. and J. Chang: *Chromobacterium violaceum* infection in a nonhuman primate *(Macaca assamensis)*. Lab. Anim. Sci. *26*, 807-810 (1976)
30. McCoy, G.W.: A plague-like disease of rodents. Publ. Hlth Bull. *43*, 53-71 (1911)
31. McCoy, G.W. and C.W. Chapin: Further observation on a plague-like disease of rodents with a preliminary note on the causative agent, *Bacterium tularense*. J. Infect. Dis. *19*, 61-72 (1912)
32. Morgenroth, K.: Lungenentzündungen; pp. 395-472 in "Spezielle pathologische Anatomie, Bd. 16/1: Pathologie der Lunge I" (Doerr, W., G. Seifert und E. Uehlinger, eds.) Springer, Berlin (1983)
33. Moss, M.O. and C. Ryall: The genus *Chromobacterium*; pp. 1355-1364 in "The Prokaryotes. A Handbook on Habitats, Isolation and Identification of Bacteria", Vol. II (Starr, M.P., H. Stolp, H.G. Trüper, A. Balows, and H.G. Schlegel, eds.) Springer, Berlin (1981)
34. Nayar, G.P.S.; G.J. Crawshaw, and J.L. Neufeld: Tularemia in a group of nonhuman primates. J. Amer. vet. med. Ass. *175*, 962-963 (1979)
35. Ognibene, A.J. and E. Thomas: Fatal infection due to *Chromobacterium violaceum* in Vietnam. Amer. J. clin. Path. *54*, 607-610 (1970)
36. Olsuf'Jev, N.G. and O.S. Emelyanova: Further studies of strains of tularaemic bacteria of the Old and New World. J. Hyg. Epidem. (Praha) *6*, 193-205 (1962)
37. Pinkerton, M.: Miscellaneous organisms; pp. 283-313 in "Pathology of Simian Primates." (Fiennes, R.N.T-W., ed.) Karger, Basel (1972)
38. Rosen, M.N.: Pasteurellosis *(P. multocida)*; pp. 129-138 in "Diseases Transmitted from Animals to Man" 6 th ed. (Hubbert, W.T., W.F. McCulloch, and P.R. Schnurrenberger, eds.) C.C. Thomas, Springfield (1975)
39. Sanders, C.V. and J.R. Torres-R: Tularemia; pp. 265-268 in "Infectious Diseases" (Sanford, J.P. and J.P. Luby, eds.), Grune and Stratton, New York (1981)
40. Scheifele, D.W.: *Haemophilus influenzae* bacteremia and meningitis in infant primates. J. Lab. clin. Med. *95*, 450-462 (1980)
41. Scheifele, D.W.; R.S Daum; V.P. Syriopoulou; D.R. Averill, and A.L. Smith: Haemophilus influenzae bacteremia and meningitis in infant primates. J. Lab. clin. Med. *95*, 450-462 (1980)
42. Scheifele, D.W.; R.S. Daum; V.P. Syriopoulou; G.R. Siber, and A.L. Smith: Comparison of two antigen detection techniques in a primate model of *Haemophilus influenzae* type 6 infection. Infect. Immun. *23*, 827-831 (1979)
43. Sippel, W.L.; G. Medina, and M.B. Atwood: Outbreaks of disease in animals associated with *Chromobacterium violaceum*. The disease in swine. J. Amer. vet. med. Ass. *124*, 467-471 (1954)
44. Smith, H.C.: Pasteurellosis in monkeys. J. Amer. vet. med. Ass. *124*, 147-148 (1954)
45. Smith, J.E.: Studies on Pasteurella septica: I. The occurrence in the nose and tonsils of dogs. J. comp. Path. *65*, 239-245 (1955)
46. Sneath, P.H.A.: Study of the bacterial genus *Chromobacterium*. Iowa State J. Sci. *34*, 243-500 (1960)
47. Soule, M.: A study of two strains of *B. violaceus* isolated from human beings. Amer. J. Path. *15*, 592-595 (1939)
48. Stahlheim, O.H.V. and K.L. Heddleston: Isolation of *Pasteurella multocida* from the urine of steers. Vet. Rec. *87*, 135-136 (1970)
49. Tindall, J.P. and C.M. Harrison: *Pasteurella multocida* infections following animal injuries, especially cat bites. Arch. Derm. *105*, 412-416 (1972)
50. Wiley, J.J. and G.A. Fraser: Septic arthritis in childhood. Canad. J. Surg. *22*, 326-330 (1979)
51. Young, V.M.: Haemophilus; pp. 302-307 in "Manual of Clinical Microbiology", 2nd ed. (Lennette, E.H., E.H. Spaulding, and J.P. Truant, eds.). Amer. Soc. Microbiol., Washington (1974)

3.8 Bacteroides Spp.

Bacteroides Melaninogenicus-Group (B. M. Subsp. Macacae)
Bacteroides spp. *(B. fragilis, B. melaninogenicus)* are important pathogens in man (Finegold and Martin, 1982). *Bacteroides* sp. have also been isolated from baboon feces (Brinkley and Mott, 1978) and a subspecies of *B. melaninogenicus* subsp. *macacae* was cultured from the gingival tissues of *Macaca arctoides* (Slots and Genco, 1980). Whether or not the *Bacteroides*-isolates from nonhuman primates are pathogenic in man remains to be determined.

Agent
B. m. subsp. *macacae* weakly ferments galactose, glucose, lactose, and mannose. It utilizes pyruvate, digests gelatin and produces indole and H_2S (Slots and Genco, 1980).

References

1. Brinkley, A. W. and G. E. Mott: Anaerobic fecal bacteria of the baboon. Appl. environ. Microbiol. *36*, 530-532 (1978)
2. Finegold, S. M. and W. J. Martin: Diagnostic Microbiology. 6th ed. C. V. Mosby, St. Louis (1982)
3. Slots, J. and R. J. Genco: *Bacteroides melaninogenicus* subsp. *macacae*, a new subspecies from monkey periodontopathic indigenous microflora. Int. J. System. Bact. *30*, 82-85 (1980)

3.9 Micrococcaceae

Staphylococcus Aureus
S. aureus, a well known pathogen in man and animals is also commonly found in simian primates particularly as part of the oral, nasal, vaginal or intestinal flora (Boncyk and Kalter, 1972; Blouse et al., 1976; Camporiondo and deMori, 1981; Dzhikidze et al., 1983). It may be transmitted to man by contamination of the human environment or via skin lesions, bite wounds etc. especially in animal handers, veterinarians etc.

Agent
S. aureus are gram-positive, spherical coccobacteria of about 0,5 to 0,8 μm diameter that occur in irregular clusters. They grow aerobically or facultatively anaerobic on most standard nutrient media, whilst media with a high (7,5%) salt content

such as mannitol salt agar are recommended as selective media. Animal strains usually produce β-haemolysis whilst human strains usually α-haemolysis on blood agar. They are coagulase-positive.

The biochemical characteristics are:

Table 23. Biochemical Characteristics of *S. aureus*

Catalase	+	Inositol	∅
Arginine dihydrolase	+	Inulin	∅
Glucose	+	Esculin	∅
Lactose	+	Indol	∅
Maltose	+	Nitrate reduction	+
Arabinose	∅		
Cellobiose	∅		
Raffinose	∅		

S. aureus produces a yellow-golden pigment of carotin nature; α-, β-, γ- and δ-toxin; leukocidin, dermonecrotic toxin and 6 antigenically distinct enterotoxins (Kienitz, 1962; Grün, 1964; Kienitz and Ritzerfeld, 1964; Bergdoll, 1972).

S. aureus is sensitive to phenols and their derivatives, halogens, surface active compounds and other desinfectants.

Epidemiology

Staphylococcal diseases can be caused either by infection or by toxic action. Nonhuman primates as the source of human food poisoning are very unlikely. The reverse, infection of monkey food by staphylococci originating from the oropharynx or wounds of caretakers may be more possible, since e.g. rhesus monkeys are quite susceptible to staphylococcal enterotoxins A and B (Sheahan et al., 1970; McGann et al., 1971; Liu et al., 1976a, b), but so far no such incidences have occurred. Staphylococcal food poisonings are therefore not considered in the present chapter.

Infections of man through nonhuman primates are probably uncommon as well, since phage-types highly pathogenic in man are rarely found in simian primates (Fiennes and Dzhikidze, 1972). Nevertheless the simian strains are potentially pathogenic to man and especially simians with signs of respiratory disease must be regarded as potentially dangerous, especially to children (Fiennes, 1967). Human infections by staphylococci of simian origin in the form of purulent wound infection are a potential threat particularly to animal handlers, veterinarians, technicians etc.

Clinical Symptoms

Nonhuman primates: The diseases in nonhuman primates resulting from *S. aureus* infections ranged from post-operative wound infections (Blouse et al., 1976) to pyodermia (Selbitz et al., 1980), omphalitis (Hänichen et al., 1983), pyometritis (Lang and Benjamin, 1969), suppurative bronchopneumonia (Kim and Wolf, 1980), polyarthritis and polyadenitis (Vickers, 1962), or septicemia (Vickers, 1962; Blouse et al., 1976; Griner, 1983).

Man: Staphylococci in man may be harmless commensals, but they can also cause life threatening diseases especially in children or debilitated patients. They also cause food poisonings. In children *S. aureus* causes purulent pneumonias, enterocolitis, dyspepsia, pyodermia, osteomyelitis, renal abscesses, conjunctivitis, otitis media, meningitis and septicemia (Kienitz, 1962, 1964). In patients of all ages wound infections, enterocolitis, pneumonia, endocarditis, meningitis, mastitis and endometritis have been associated with *S. aureus*-infections (Irmer, 1964; Knörr, 1964; Wellman and Senft, 1964; Cohen, 1980).

Pathology

Nonhuman primates: In an *E. patas* which died from a generalized *S. aureus*-infection multiple small, white abscesses were present in the lungs, liver, spleen and small intestines (Vickers, 1962). In *S. mystax* the *S. aureus* infections induced subacute suppurative bronchopneumonia with diffuse squamous metaplasia of the terminal bronchioles (Kim and Wolf, 1980). In an adult *C. aethiops* the staphylococci caused thrombotic endocarditis, lung abscesses, bronchopneumonia and renal infarctions (Wood et al., 1978).

Man: The lesions in human staphylococcal infections may range from epidermal abscesses as cutaneous pustules and furuncles to impetigo contagiosa, pneumonia leading to empyema, pulmonary abscessation or bronchiectasis.

The skin lesions range from superficial folliculitis to deep-seated carbuncles extending into the subcutaneous tissues. Impetigo is characterized by generalized or localized erythema, bullae and desquamation of the epidermis. Involvement of tissues underlying the skin lesions may lead to necrotizing fasciitis and myositis.

Staphylococcal pneumonia is commonly complicated by parenchymal cavitation or empyema of the thoracic cavity.

Endocarditis induced by staphylococcal bacteremia is often associated with metastatic abscess formation in the myocardium, lungs, kidneys, brain, bone, liver and spleen.

Haematogenous osteomyelitis, generally occurring in children or young adults affects most commonly the metaphyseal ends of long bones as necrosis of the marrow, bone death, subperiostal abscessation and occasionally pyogenic arthritis (Hawiger and Koenig, 1981).

Treatment

S. aureus is frequently resistant to penicillin (Vickers, 1962). Satisfactory results have been obtained using tetracycline, erythromycin, neomycin, chloramphenicol or sulfamethoxazoletrimethoprim (Vickers, 1962; Selbitz et al., 1980).

References

1. Baird-Parker, A.C.: Micrococcaceae; pp. 478–490 in "Bergey's Manual of Determinative Bacteriology", 8th ed. (Buchanan, R.E. and N.E. Gibbons, eds.) Williams
2. Bergdoll, M.S.: The nature of bacterial toxins. Clin. Toxicol. *5*, 441–451 (1972)

3. Blouse, L.E.; R.M. Brockett; P.J. Homme, and E.F. Jones: Epizootic staphylococcal infections in subhuman primates after surgical operation. Amer. J.vet. Res. *37*, 731-733 (1976)
4. Boncyk, L.H. and S.S. Kalter: Aerobic bacterial flora of chimpanzees - a four-year study. J.med. Primatol. *1*, 333-343 (1972)
5. Camporiondo, M.P. and P.de Mori: Study of the bacterial flora of a colony of "Macaca fuscata" living in the Zoo of Rome. Anthropologia Contemporanea *4*, 139-143 (1981)
6. Cohen, J.O.: Staphylococcal diseases. Pp.47-64 in "CRC Handbook Series in Zoonoses". (Steele, J.H., Series ed.), Section A: Bacterial, Rickettsial and Mycotic Diseases Vol. II (Stoenner, H., W.Kaplan and M.Torten, Section eds.), CRC Press, Boca Raton (1980)
7. Dzhikidze, E.K.; H.D. Schröder; Z.K. Stasilewitsch; W.Witte, and A.A. Trotz: Beitrag zu den Staphylokokken-Infektionen der Affen; pp.271-273, XXV. Internat. Symp. Erkr. Zootiere, Wien (1983)
8. Fiennes, R.: Zoonoses of Primates. The Epidemiology and Ecology of Simian Diseases in Relation to Man. Weidenfeld and Nicolson (1967)
9. Fiennes, R.N.T-W. and E.K. Dzhikidze: Respiratory pathogens and other organisms. Pp.227-282 in "Pathology of Simian Primates", Part II (Fiennes, R.N.T-W., ed.) Karger, Basel (1972)
10. Griner, L.A.: Pathology of Zoo Animals. Zool. Soc. San Diego (1983)
11. Grün, L.: Die Biologie der Staphylokokken; pp.1-27 "Staphylokokken in Klinik und Praxis" (Grün, L.ed.) Wissenschaftliche Verlagsgesellschaft, Stuttgart (1964)
12. Hänichen, T.; H.Wiesner und W.Rietschel: Beitrag zur physiologischen und pathologischen Gravidität und perinatale Probleme beim Orang-Utan (Pongo pygmaeus). Pp.357-362 in XXV. Intern. Symp. Erkr. Zootiere, Wien (1983)
13. Hawiger, J.and M.G. Koenig: Staphylococcal Diseases. Pp.177-184 in "Infectious Diseases" (Sanford, J.P. and J.P. Luby, eds.), Grune and Stratton, New York (1981)
14. Irmer, W.: Staphylokokkeninfektionen in der Chirurgie; pp. 187-208 in "Staphylokokken in Klinik und Praxis" (Grün, L., ed.), Wissenschaftliche Verlagsgesellschaft, Stuttgart (1964)
15. Kienitz, M.: Die enteralen Staphylokokken-Infektionen des Kindes. Ein Beitrag zu deren Klinik und Laboratoriumsdiagnostik. Bibliotheca Microbiologica, Fasc.2 (Grumbach, A., ed.), Karger, Basel (1962)
16. Kienitz, M.und W.Ritzerfeld: Staphylokokkentoxine und ihre klinische Bedeutung; pp.28-54 in "Staphylokokken in Klinik und Praxis" (Grün, L., ed.) Wissenschaftliche Verlagsgesellschaft, Stuttgart (1964)
17. Kim, J.C.S. and R.J. Wolf: Diseases of moustached marmosets; pp.431-435 in "The Comparative Pathology of Zoo Animals" (Montali, R.J. and G.Migaki, eds.) Smithsonian Institution Press, Washington (1980)
18. Knörr, K.: Staphylokokkeninfektionen in Geburtshilfe und Gynäkologie; pp.209-244 in "Staphylokokken in Klinik und Praxis" (Grün, L., ed.), Wissenschaftliche Verlagsgesellschaft, Stuttgart (1964)
19. Lang, C.M. and S.A. Benjamin: Acute pyometra in a rhesus monkey *(Macaca mulatta)*. J.Amer. vet. med. Ass. *155*, 1156-1157 (1969)
20. Liu, C.T.; M.J. Griffin, and R.T. Faulkner: Effect of staphylococcal enterotoxin B and body fluid compartments in conscious rhesus monkeys. J.med. Primatol. *5*, 36-344 (1976a)
21. Liu, C.T.; J.D. Helm III, and W.R. Beisel: Cardiovascular and vomiting responses to a lethal intravenous dose of staphyloenterotoxin A in rhesus monkeys. J.med. Primatol. *5*, 353-359 (1976b)
22. McGann, V.G.; J.Brent Rollins, and D.W. Mason: Evaluation of resistance to staphylococcal enterotoxin B: Naturally acquired antibodies of man and monkey. J.infect. Dis. *124*, 206-213 (1971)
23. Selbitz, H.-J.; K.Elze und S.Seifert: Staphylococcus aureus als Erreger einer Pyodermie beim Gorilla (Gorilla gorilla) im Zoologischen Garten Leipzig; pp.83-86, XXII. ISEZ, Arnhem (1980)
24. Sheahan, D.G.; H.R. Jervis; A.Takeuchi, and H.Sprintz: The effect of staphylococcal enterotoxin on the epithelial mucosubstance of the small intestine of rhesus monkeys. Amer. J.Path. *60*, 1-18 (1970)
25. Vickers, J.H.: Generalized staphylococcic infection in a monkey. J.Amer. vet. med. Ass. *141*, 256-257 (1962)

26. Wellman, W.E. and R.A. Senft: Bacterial meningitis. III. Infections caused by Staphylococcus aureus. Mayo Clin. Proc. *39*, 263-269 (1964)
27. Wood, L.L.; M.G. Bond, and B.C. Bullock: Bacterial endocarditis with obstruction of the right atrioventricular orifice and the pulmonary outflow tract in an African monkey *(Cercopithecus aethiops)*. Lab. Anim. Sci. *28*, 85-88 (1978)

3.10 Streptococci

Streptococci are gram-positive globoid to ovoid cocci appearing in chains of different length. They are non-motile. Their growth is usually poor on ordinary media, better growth is obtained on media containing blood or serum, on trypticase - soy - or on heart infusion agar.

Streptococci are facultatively aerobic. Carbohydrate fermentation results in dextrorotatory lactic acid as the mayor end product (in contrast to the anaerobically growing peptococci and peptostreptococci). Streptococci do not produce catalase or oxidase and they do not reduce nitrate.

A major criterium in the classification of streptococci is the production and type of haemolysis. The definitions are:
α-haemolysis: distinct zone of *partial* destruction of the erythrocytes around the colonies, often accompanied by a greenish or brownish discoloration of the blood agar;
β-haemolysis: a clear, colorless zone of *total* erythrocyte destruction around the colonies,
γ-haemolysis: no apparent haemolysis
α' (alpha prime) or wide zone alpha haemolysis: a small halo of intact or only partially lysed erythrocytes surrounded by a zone of complete haemolysis. At macroscopical examination of the agar plates this type is easily confused with ß-haemolysis (Facklam, 1974).

Streptococci are usually serogrouped according to Lancefield's scheme, which depends on the presence of carbohydrate substances, particularly substance C, a group-specific polysaccharide.

Several species of streptococci occur in both man and nonhuman primates:

3.10.1 Streptococcus Pyogenes (Group-A-Streptococci)

S.pyogenes, the only species in group A is predominantly a pathogen of man. It occasionally causes disease in mice and domestic animals. A few isolations have been made from nonhuman primates (Seegal et al., 1936; Krushak et al., 1970; Hänichen et al., 1983).

Agent

S.pyogenes organisms are of 0,6 to 1,0 μm diameter, but usually grow in pairs or short chains. It is β-haemolytic, ferments trehalose but not sorbitol or inulin. Ammonia is produced from arginine (Deibel and Seeley, 1975).

Epidemiology
Group-A-streptococcoses in human beings occur throughout the world, the streptococci dwelling primarily on pharyngeal tissues. Similarly the agent has been isolated from the pharyngeal tissues of nonhuman primates (Seegal et al., 1936). The contagion depends primarily upon direct contact, whereas transmission by inanimate objects is unlikely due to the vulnerability of *S. pyogenes* to environmental influences. Group-A-streptococci are usually man - man or man - animal transmitted, rarely from animals to man except in the case of raw milk from a bovine *S. pyogenes*-mastitis.

Clinical Symptoms
Nonhuman primates: Information on *S. pyogenes*-diseases in nonhuman primates is almost nonexistent. Hänichen et al. (1983) isolated the agent together with *S. aureus* from cases of omphalitis in orang-utan babies. Sauer and Fegley (1960) considered β-haemolytic streptococci as the causative agents of 11 out of 32 bacterial pneumonias in monkeys. Experimental pharyngeal infection of *M. mulatta* induced minor myocarditis, periarteritis and valvulitis in some experimental animals (Vanace, 1960).

Man: In man a wide variety of clinical manifestations is related to group-A-streptococci according to Erickson (1980) - Table 24.

3.10.2 Streptococcus Agalactiae (Group-B-Streptococci)

S. agalactiae, believed to be primarily a bovine pathogen, is the only member of group B. *S. agalactiae* has been isolated with increasing frequency from human beings (Erickson, 1980). It has occasionally been isolated from nonhuman primates (Karasek, 1969).

Table 24. Clinical Manifestation of Group-A-Streptococcal Infections in Man

Purulent:	Pharyngitis
	Peritonsillar abscesses
	Cervical adenitis
	Pneumonia
	Empyema
	Erysipelas
	Impetigo
	Wound infections
	Otitis media
	Pectoral abscesses
	Perianal cellulitis
	Puerperal fever
Secondary:	Arthralgia
	Chorea
	Erythema marginata
	Glomerulonephritis
	Rheumatic carditis

Agent

S. agalactiae organisms are spherical to ovoid, 0,6 to 1,2 μm in diameter. They usually form chains of more than 4 cells and frequently longer chains (Deibel and Seeley, 1974). On blood agar they may be β-haemolytic. *S. agalactiae* hydrolyses sodium hippurate, produces ammonia from arginine, and ferments glycerol aerobically.

Lancefield 's serological classification distinguishes 5 serotypes: Ia, Ib, Ic, II and III, with types II and III dominating in human infections (Larsen et al., 1978; Goroff and Kasper, 1980).

Epidemiology

S. agalactiae causes primarily several forms of mastitis in cattle and until recently cattle have been considered to be the only significant source of *S. agalactiae* (Erickson, 1980). The role of group B-streptococci in human health has only been realized during the last 15 years (Goroff and Kasper, 1980). Recent evidence has even suggested that man plays a carrier role with regard to this organism. Infection is by contact or by aerosol.

Clinical Symptoms

Nonhuman primates: B-streptococci have been isolated from stillborn baboons and a cotton top marmoset (Karasek, 1969). Experimental intracerebral infections of newborn rhesus monkeys by serotype Ic or III resulted in fatal neonatal meningitis without any associated pneumonia. Intravenous or intraamniotic inoculation resulted in variable clinical manifestations ranging from no apparent disease to fatal meningitis and pneumonia (Larsen et al., 1978) to intraamniotic infections, or placentitis (Larsen et al., 1981).

Man: *S. agalactiae* in man causes sepsis in neonates, puerperal infections and occasionally other syndromes such as wound infections, respiratory symptoms, osseous and urogenital tract infections. In neonates the acute stage is typically a respiratory distress syndrome, occasionally followed by meningitis (Erickson, 1980). Group B-streptococci are now the most frequent agent associated with serious infections during the first 3 months of life in the USA. Particularly the early onset type of the infection, affecting neonates 5 days of age or less has a case-fatality rate of sometimes more than 50% (Goroff and Kasper, 1980).

3.10.3 Group-D-Streptococci

Group D is comprised of *S. faecalis* including its subspecies, *S. faecium, S. durans, S. equinus* and *S. bovis.* The only species occurring in both man and nonhuman primates is *S. faecalis,* which is indigenous to the human and simian digestive tract. It also causes endocarditis lenta and occasionally caries in man.

Agents

The characteristics of group-D-streptococci are given in Table 25.

S. faecalis occurs in ovoid or elongated cells of 0,5 to 1,0 μm diameter, mostly growing in pairs or short chains. It grows in methylene blue milk, ferments sorbi-

Table 25. Characteristics of Group D-Streptococci

Species	Haemolysis	Test						
		Reduction of Potassium tellurite	Mannitol	Arginine dihydrolase	Lactose	Glyzerol (anaerobically)	Gelatin	Starch
Str. faecalis var. faecalis	γ	+	+	+	+	+	−	−
Str. faecalis var. liquefaciens	γ	+	+	+	+	+	+	−
Str. faecalis var. zymogenes	β	+	−	+	+	+	−	+
Str. faecium	α, Vergr.	−	+	+	+	−	−	−
Str. durans	α, β, α	−	−	+	+	−	−	−
Str. equinus	α, γ	−	−	−	−	−	−	−
Str. bovis	α, γ	−	−	−	+	−	−	+

tol, glucose, citrate, serine or malate but not arabinose. Pyruvate and arginine are utilized. *S. faecalis liquefaciens* liquefies gelatin.

On blood agar *S. faecalis* grows usually with γ-, occasionally with α-haemolysis.

Clinical Symptoms

Nonhuman primates: in callitrichids *S. faecalis liquefaciens* was associated with diarrhea and death within 2 days (own results, unpublished).

Man: Group-D-streptococci (enterococci) are a cause of bacterial endocarditis (endocarditis lenta), characterized by a slowly developing, moderate fever, with occasional painful splenic infarctions, as well as cutaneous and mucosal haemorrhages.

Pathology

Nonhuman primates: In the generalized *S. faecalis liquefaciens* infection in marmosets acute catarrhal enteritis is the leading lesion, accompanied by signs of septicemia.

Man: Endocarditis lenta or subacute bacterial endocarditis is a proliferating and ulcerating endocarditis of pre-damaged valves (Doerr, 1974).

3.10.4 Non-Grouped-Streptococci

S. Viridans (S. Mitior)

S. viridans is characterized by a green discoloration of blood agar. Its taxonomic position is debated. As with D-streptococci it causes endocarditis lenta in man, it has also been incriminated as a cariogenic agent, as it is present in the oral cavity of virtually everyone (Köhler und Mochmann, 1975).

Agent

S. viridans produces greenish discoloration of sheep blood agar when grown in superficial colonies, whereas colonies growing under the surface of the agar produce α-haemolysis.

Clinical Symptoms

Nonhuman primates: α-haemolytic *S. viridans* have been associated with unusually high numbers of abortions and stillbirths in a breeding colony of rhesus monkeys (Swindle et al., 1982).

Man: Viridans-streptococci are the most frequent cause of subacute bacterial endocarditis following dental or surgical procedures in patients with a previously damaged endocardium (Finegold and Martin, 1982).

Pathology

Nonhuman primates: The isolations of *S. viridans* in aborted and stillborn rhesus monkeys were associated with an acute, sometimes necrotic placentitis, placental infarctions or fetal bronchopneumonia (Swindle et al., 1982).

Man: *S. viridans* is considered as one causative agent of endocarditis lenta and of caries (Köhler and Mochmann, 1975).

S. Mutans

S. mutans as a member of the viridans-group is, like *S. viridans*, usually regarded as a commensal organism of the human oral cavity and is believed to cause dental caries (Walker et al., 1982). It has also been isolated from the gastrointestinal tract of monkeys, particularly from dental plaques, the caecum and colon (Walker et al., 1982). Transmission from nursery attendants to *M. mulatta* infants delivered by cesarian section was reported by Jordan et al. (1985).

Agent

S. mutans is either β- or γ-haemolytic. The barely marginal role of *S. mutans* as a zoonotic agent does not require further description. More details are given in Finegold and Martin (1982).

3.10.5 Diplococcus (Streptococcus) Pneumoniae

D. pneumoniae is a well known respiratory pathogen in man which also causes respiratory diseases and generalized infections in ruminants and guinea pigs. Pneumococci are also a common cause of disease in nonhuman primates (Sauer and Fegley, 1960; Kaufmann and Quist, 1969; Henderson et al., 1970; Vickers, 1971; Cobb et al., 1976; Watson et al., 1983; Solleveld et al., 1984).

Agent

D.pneumoniae is a gram-positive, lanceolate coccobacterium which typically grows in pairs or occasionally in short chains. In fresh smears of sputum or exudates the organisms are frequently surrounded by a capsule.

D.pneumoniae requires enriched media such as trypticase or brain heart infusion agar enriched with 5% defibrinated blood for cultivation. 5 to 10% of strains require microaerophilic conditions for primary isolation, occasional strains even depend on anaerobic incubation. Typical *D.pneumoniae* colonies after 24 hrs of incubation are round, transparent, glistening, sometimes mucoid and of about 1 mm diameter. They are surrounded by an approximately 2 mm zone of α-haemolysis after aerobic incubation, anaerobically grown colonies produce β-haemolysis. The haemolytic zone stains greenish after one day due to methaemoglobin-formation. Young colonies are usually dome-shaped, but on aging they become flattened and umbilicated. *D.pneumoniae* is distinguished from other α-haemolytic streptococci by its optochin-susceptibility, its bile (sodium desoxycholate) solubility (except R-colonies) and a positive Neufeld reaction (Finegold and Martin, 1982).

Based on specific capsular polysaccharides, 83 capsular types have been recognized. For practical purposes usually only a selection of serotypes are tested for (Köhler und Mochmann, 1975). Types 1, 2, 3, 4, 7, 8, 12 and 14 are highly pathogenic in man (Jawetz et al., 1964; Austrian, 1974), whereas types 1, 9, 18, 19 and 33 have been reported in nonhuman primates (Kaufmann and Quist, 1969; Good and May, 1971; Keeling and McClure, 1974).

Epidemiology

A common inhabitant of the upper respiratory tract, the pneumococci can be isolated from the pharynx of up to 20% of normal human beings depending on the season of the year. It has also been found in the sinuses and nasal cavities of healthy rhesus monkeys (Shapiro et al., 1982). Human pneumococci can be transmitted to other people and to animals (Rolle and Mayr, 1978).

Pathogenesis

From the pharynx *D.pneumoniae* can invade the sinuses, the middle ear and the meninges or the lower respiratory tract. Failure of normal cellular and humoral mechanisms to check the spread of bacteria within the organism is followed by passage of the pneumococci into the systemic circulation, giving rise to bacteremia. The virulence of the bacteria depends on the possession of the capsule which partially protects the bacteria from phagocytosis. The capsular polysaccharides of the pneumococci are, however, nontoxic to man. In pneumococcal pneumonia a sensibilisation of the lung tissues by previous subclinical infections has been suggested (Köhler und Mochmann, 1975).

Clinical Symptoms

Nonhuman primates: In nonhuman primates the pneumococci, besides causing bronchopneumonias (Fegley and Sauer, 1960; Sauer and Fegley, 1960; Henderson et al., 1970; Vickers, 1971) are a leading cause of bacterial meningitis or meningoencephalitis (Kaufmann and Quist, 1969; Henderson et al., 1970; Fox and Soave,

1971; Fox and Wikse, 1971; Keeling et al., 1971; Manning, 1971; Keeling and McClure, 1974; Cobb et al., 1976; Solleveld et al., 1982, 1984; Watson et al., 1983).

Signs of pneumonia include respiratory distress, inappetence, shivering and rough coat, often the animals are found dead. Monkeys or apes suffering from pneumococcal meningitis/meningoencephalitis are similarly frequently found dead without clinical signs. Other animals show anorexia, listlessness or lethargy, neuromuscular incoordination, paralysis, tilting or pressing of the head, nystagmus, pupillary constriction, muscular tremors, clonic seizures, blindness etc. Most cases of pneumococcal meningitis are associated with pneumococcal pneumonia.

Pneumococcal peritonitis has been observed in a male chimpanzee (Gunders et al., 1957) and a male rhesus monkey (Kaufmann and Quist, 1969).

On one occasion, pneumococcal septicemia resulted in the development of bilateral panophtalmitis in a rhesus monkey (Herman and Fox, 1971) with blepharospasm, haemorrhagic conjunctivitis, keratitis, hypopyos, nystagmus and blindness.

Man: The typical diseases caused by *D. pneumoniae* in man are lobar croupos pneumonias and bronchopneumonias. They are most commonly observed during the winter season and are probably secondary to virus infections, colds etc.

Pneumococcal pneumonia begins abruptly with fever, shaking chills, cough, chest pain, rales, bronchophony, friction rubs and often dyspnea.

In some surveys pneumococcal meningitis is the most common type of bacterial meningitis in adults, which may begin abruptly with a fulminant clinical course or may have a more protracted development extending over days or weeks. The signs of pneumococcal meningitis are stiff neck, fever, headache, lethargy, pupillary abnormalities, seizures and coma (Bryant, 1981).

Otitis media caused by type 3 pneumococci (mucoid strains) is a very serious disease preceeding meningitis or leading to mastoitis.

Other, occasional pneumococcal diseases are a peritonitis of an increasing clinical incidence seen in children with nephrosis or patients with cirrhosis (Bryant, 1981), and *ulcus serpens corneae*, usually caused by types 1 to 3. *Ulcus corneae* frequently follows trauma to the eye and does not result from exogenous infections. *Ulcus serpens corneae* leads to the destruction of corneal tissues within a few hours (Köhler and Mochmann, 1975).

Pathology

Nonhuman primates: Grossly visible lesions of pneumococcal pneumonia in nonhuman primates are either those of bronchopneumonia or of lobar pneumonia. In bronchopneumonia, the entire lobes are edematous and congested (Kaufmann and Quist, 1969), in lobar pneumonia the affected lobes show hepatization and on the cut surface may appear jelly-like. The alveoli, bronchioli and bronchi are filled with a dense exudate of polymorphonuclear leucocytes. Perivascular tissues around the larger vessels are edematous and infiltrated by polymorphonuclear leucocytes, large mononuclear cells and lymphocytes; the lymph vessels are similarly distended. The pneumonic areas are surrounded by alveolar edema.

Pneumococcal pleuritis and pericarditis consist of a substrate of fibrin and polymorphonuclear granulocytes, a similar exudate is found in the arachnoidal space in cases of pneumococcal meningitis. A necrotizing vasculitis is common in both the grey and white substance (Carlton and Hunt, 1978), whilst focal or areal necrosis can appear in the parenchymas of the cerebrum and cerebellum (Kaufmann and Quist, 1969; Fox and Soave, 1971).

Man: Pneumococcal pneumonia in its initial stages is characterized by heavy alveolar edema, followed by stages of red and grey hepatization. In red hepatization the alveolar spaces become filled with erythrocytes, in grey hepatization, they are filled with fibrin and polymorphonuclear granulocytes. The pulmonary arterioles may become thrombosed. On recovery the inflammatory cells are replaced by macrophages (Spencer, 1977).

In *ulcus serpens* the corneal epithelium and the stroma are necrotic and infiltrated by polymorphonuclear granulocytes (Hinzpeter and Naumann, 1980).

Treatment

The drug of choice remains penicillin, in patients or species allergic to penicillin, erythromycin can be used instead unless the actual strains are resistant (Bryant, 1981).

References

1. Austrian, R.: *Streptococcus pneumoniae* (Pneumococcus); pp. 109-115 in "Manual of Clinical Microbiology", 2nd ed. (Lennette, E.H., E.H. Spaulding, and J.P. Truant, eds.) Amer. Soc. Microbiol. Washington, D.C. (1974)
2. Bryant, R.E.: Pneumococcal infection. Pp. 202-208 in "Infectious Diseases" (Sanford, J.P. and J.P. Luby, eds.) Grune and Stratton, New York (1981)
3. Carlton, W.W. and R.D. Hunt: Bacterial diseases; pp. 1368-1480 in "Pathology of Laboratory Animals" (Benirschke, K., F.M. Garner, and T.C. Jones, eds.) Springer, New York (1978)
4. Cobb, L.M.; P.L. Hepworth, and R. Heywood: Pneumococcal leptomeningitis in cynomolgus monkeys *(Macaca fascicularis)*. Vet. Rec. *99*, 84 (1976)
5. Deibel, R.H. and H.W. Seeley, jr.: Streptococcaceae *fam. nov.*; pp. 490-517 in "Bergey's Manual of Determinative Bacteriology" 8th ed. (Buchanan, R.E. and N.E. Gibbons, eds.) Williams and Wilkins, Baltimore (1974)
6. Doerr, W.: Herz und Gefäße; pp. 1.1-1.83 in "Organpathologie", Band I (Doerr, W., W. Giese, L.D. Leder und E. Remmele, eds.) Georg Thieme, Stuttgart (1974)
7. Erickson, E.D.: Streptococcosis; pp. 65-82 in "CRC Handbook Series in Zoonoses" (Steele, J.H., Series ed.) Section A: Bacterial, Rickettsial and Mycotic Diseases. Vol. II (Stoenner, H., W. Kaplan and M. Torten, Section eds.) CRC Press, Boca Raton (1980)
8. Facklam, R.R.: Streptococci, pp. 96-108 in "Manual of Clinical Microbiology" 2nd ed. (Lennette, E.H., E.H. Spaulding and J.P. Truant, eds.) Amer. Soc. Microbiol. Washington D.C. (1974)
9. Fegley, H.C. and R.M. Sauer: Mass treatment of monkeys with antibacterial substances. Ann. N.Y. Acad. Sci. *85*, 777-784 (1960)
10. Finegold, S.M. and W.J. Martin: Diagnostic Microbiology. 6th ed. C.V. Mosby, St. Louis (1982)
11. Fox, J.G. and O.A. Soave: Pneumococcic meningoencephalitis in a rhesus monkey. J. Amer. vet. med. Ass. *159*, 1595-1597 (1971)
12. Fox, J.G. and S.E. Wikse: Bacterial meningoencephalitis in rhesus monkeys: Clinical and pathological features. Lab. Anim. Sci. *21*, 558-563 (1971)
13. Good, R.C. and B.D. May: Respiratory pathogens in monkeys. Infect. Immun. *3*, 87-93 (1971)
14. Goroff, K.D. and D.L. Kasper: Current status of group B streptococcal infections in man; pp. 85-86 in "CRC Handbook Series in Zoonoses" (Steele, J.H., Series ed.) Section A: Bacte-

rial, Rickettsial and Mycotic Diseases. Vol. II. (Stoenner, H., W. Kaplan and M. Torten, Section eds.) CRC Press, Boca Raton (1980)
15. Gunders, A. E.; R. Lainson, and D. A. Bailey: Pneumococcal peritonitis in a male chimpanzee. J. trop. med. Hyg. *60*, 100-102 (1957)
16. Hänichen, T., H. Wiesner und W. Rietschel: Beitrag zur physiologischen und pathologischen Gravidität und perinatale Probleme beim Orang-Utan (Pongo pygmaeus). XXI. Internat. Symp. Erkr. Zootiere, Wien, pp. 357-362 (1983)
17. Henderson, J. D., jr.; W. S. Webster; B. C. Bullock; N. D. M. Lehner, and T. B. Clarkson: Naturally occurring lesions seen at necropsy in eight woolly monkeys (*Lagothrix* sp.). Lab. Anim. Care *20*, 1087-1097 (1970)
18. Herman, P. H. and J. G. Fox: Panophthalmitis associated with diplococcic septicemia in a rhesus monkey. J. Amer. vet. med. Ass. *159*, 560-562 (1971)
19. Hinzpeter, E. N. und G. O. H. Naumann: Hornhaut und Sklera; pp. 313-407 in "Spezielle pathologische Anatomie" (Doerr, W., G. Seiffert und E. Uehlinger, Series eds.), Bd. 12 "Pathologie des Auges" (Naumann, G. O. H., Vol. ed.) Springer, Berlin (1980)
20. Jawetz, E.; J. L. Melnick, and E. A. Addberg: Review of Medical Microbiology. Lange Medical Pub., Los Altos, CA (1964)
21. Jordan, H. V.; J. van Houte, and J. Russo: Transmission of the oral bacterium *Streptococcus mutans* to young *Macaca fascicularis* monkeys from human nursery attendants. Archs. oral Biol. *11/12*, 863-864 (1985)
22. Karasek, E.: Streptokokkeninfektionen bei Zootieren, pp. 81-84 in: XI. Internat. Symp. Erkr. Zootiere, Zagreb (1969)
23. Kaufmann, A. F. and K. D. Quist: Pneumococcal meningitis and peritonitis in rhesus monkeys. J. Amer. vet. med. Ass. *15*, 1158-1162 (1969)
24. Keeling, M. E.; W. B. Bonner, and H. M. McClure: Pneumococcal meningitis in a chimpanzee. Primate Zoonoses Surveil. Rep. *3*, 10-11 (1971)
25. Keeling, M. E. and H. M. McClure: Pneumococcal meningitis and fatal enterobiasis in a chimpanzee. Lab. Anim. Sci. *24*, 92-95 (1974)
26. Köhler, W. und H. Mochmann: Grundriß der Medizinischen Mikrobiologie. VEB Gustav Fischer, Jena (1975)
27. Krushak, D. H.; R. A. Zimmermann, and B. L. Murphy: Induced group B beta-hemolytic streptococcic infection in chimpanzees. J. Amer. vet. med. Ass. *157*, 742-744 (1970)
28. Larsen, J. W.; W. T. London; A. E. Palmer; J. W. Tossell; R. A. Bronsteen; M. Daniels; B. L. Curfman, and J. L. Sever: Experimental group B streptococcal infection in the rhesus monkey. Amer. J. Obstet. Gynec. *132*, 686-690 (1978)
29. Larsen, J. W.; W. T. London; C. J. Baker; B. L. Curfman, and J. L. Sever: Intraamniotic infection due to group B *Streptococcus*: Treatment and antibody response. Amer. J. Obstet. Gynec. *58*, 222-226 (1981)
30. Manning, P. J.: Diplococcal leptomeningitis in a rhesus monkey. Lab. Anim. Dig. *7*, 52-53 (1971)
31. Rolle, M. und A. Mayr: Mikrobiologie, Infektions- und Seuchenlehre für Tierärzte, Biologen und Agrarwissenschaftler. Enke, Stuttgart (1978)
32. Sauer, R. M. and H. C. Fegley: The roles of infectious and non-infectious diseases in monkey health. Ann. N. Y. Acad. Sci. *85*, 866-888 (1960)
33. Seegal, B. C.; G. Heller, and J. Jablonowitz: Incidence of hemolytic streptococci and pneumococci in the pharyngeal flora of normal rhesus monkeys. Proc. Soc. exp. Biol. *34*, 812-816 (1936)
34. Shapiro, E. D.; E. R. Wald; W. Doyle, and D. Rohn: Bacteriology of the maxillary sinus of rhesus monkeys. Ann. Otol. Rhinol. Laryngol. *91*, 150-151 (1982)
35. Solleveld, H. A.; P. J. Heidt; P. M. C. A. van Eerd, and M. J. van Zwieten: Meningitis in chimpanzees *(Pan troglodytes)*. A clinicopathological study of six cases. Lab. Anim. Sci. *32*, 425 (1982)
36. Solleveld, H. A.; M. J. van Zwieten; P. J. Heidt, and P. M. C. A. van Eerd: Clinicopathologic study of six cases of meningitis and meningoencephalitis in chimpanzees *(Pan troglodytes)*. Lab. Anim. Sci. *34*, 86-90 (1984)
37. Spencer, H.: Pathology of the Lung (Excluding Pulmonary Tuberculosis). Vol. I, 3rd ed., Pergamon Press, Oxford (1977)

38. Swindle, M.M.; C.F. Craft; B.M. Marriott; J.D. Strandberg, and M. Luzarraga: Ascending intrauterine infections in rhesus monkeys. J. Amer. vet. med. Ass. *181*, 1367-1370 (1982)
39. Vanace, P.W.: Experimental streptococcal infection in the rhesus monkey. Ann. N.Y. Acad. Sci. *85*, 910-930 (1960)
40. Vickers, J.H.: Respiratory diseases of primates. Curr. vet. ther. *4*, 412-414 (1971)
41. Walker, J.; H. Hayday, and D. Beighton: The distribution of the cariogenic bacterium *Streptococcus mutans* in the gastrointestinal tract of monkeys *(Macaca fascicularis)*. Arch. oral Biol. *27*, 697-700 (1982)
42. Watson, E.A.; L.N. Martin, and G.B. Baskin: An outbreak of *Streptococcus pneumoniae* in a large outdoor rhesus monkey colony. Lab. Anim. Sci. *33*, 494 (1983)

3.11 Clostridium

Clostridia are obligate anaerobic to aerotolerant, grampositive, sporulating rods. Some clostridia are pathogenic to both man and nonhuman primates. Clostridia are grouped according to their sporulation and gelatin hydrolization.

I. Spores subterminal
 A. gelatin not hydrolyzed - Group I
 B. gelatin hydrolyzed - Group II
II. Spores terminal
 A. gelatin not hydrolized - Group III
 B. gelatin hydrolized - Group IV
III. Species with special growth requirements.

3.11.1 C. tetani

C. tetani causes lockjaw (Tetanus), a disease known since Hippocrates (McComb, 1980) in man and a number of animal species with horses being most susceptible. Cases of tetanus have been reported in nonhuman primates without specification (Goss, 1942; Fukuda and Asakura, 1960; Kronberger, 1966; Kaufmann, 1972), in "macaques" (Hill, 1939), in free-ranging *M. mulatta* (DiGiacomo and Missakian, 1972; Kessler et al., 1982; Rawlins and Kessler, 1982a, b), *M. mulatta* (Sakurai, 1967 - cit. from McComb, 1980), *S. sciureus* (Kessler and Brown, 1979) and a *Callicebus moloch* (Anon, 1971). The agent has also been isolated from *P. cynocephalus* (Boncyk and Kalter, 1980).

Agent
C. tetani is a slender, straight rod of 0,5-0,7 × 2,5-4,0 μm size, that is actively motile using peritrichous flagella and forms oval to rounded terminal spores giving a characteristic drumstick-like appearance. *C. tetani* grows well at 37 °C in cooked meat broth. On agar which contains substances to create a low redox-potential, flat translucent, grey, dull, circular superficial colonies of 4 to 6 mm diameter develop if grown anaerobically. The margins of the colonies are irregular. On moist agar even swarming can be observed. On blood agar the colo-

nies are surrounded by a haemolytic halo. *C. tetani* usually does not ferment sugars, there are, however, exceptions with glucose-positive strains occasionally occurring. It liquefies gelatin slowly usually with H_2S formation, milk is slowly clotted or precipitated (Smith and Hobbs, 1975).

All *C. tetani* possess a common 0-antigen, additional type-specific H-antigens distinguish 10 different serotypes. The serotypes II, IV, V, and IX are known to elaborate a powerful exotoxin of protein nature (Köhler and Mochmann, 1975), consisting of 3 components:

1. tetanospasmin, the neurotoxin responsible for muscle spasms;
2. a peripherally active nonspasmogenic toxin, and
3. tetanolysin producing haemolysis on blood agar plates.

The exotoxin is extremely stable when dried while in solution it is quite labile.

Vegetative *C. tetani* are readily killed by heating but spores are highly resistant to dry heat and ordinary boiling.

Epidemiology

C. tetani is an inhabitant of the digestive tract of homoiothermic animals and man. It is also commonly found in the soil parallel to the extend of fertilization or grassing. Transmission to man and animals is by contamination of wounds, especially those with enough tissue necrosis to lower the pH sufficiently and provide a low redox-potential. Therefore any bite wound by any animal is of potential danger. Even laceration may be the port of entry since concurrently growing aerobic bacteria can provide the low redox potential below the superficial crust within a short period.

Pathogenesis

C. tetani elaborates its toxins after multiplying at the site of infection. The toxin spreads via blood or along neurons to the central nervous system where it acts in several ways. Chief among those is the inhibition of the normal function of the interneuronal Renshaw cells in the anterior horn of the spinal cord (Meadow, 1978; Smith, 1981) (Fig. 9).

The level of action is the presynaptic cholinergic synapse with inhibition of the release of acetylcholine (Blobel and Schliesser, 1980). Renshaw cells normally limit the duration and intensity of α-motoneurons impulses, therefore inhibition

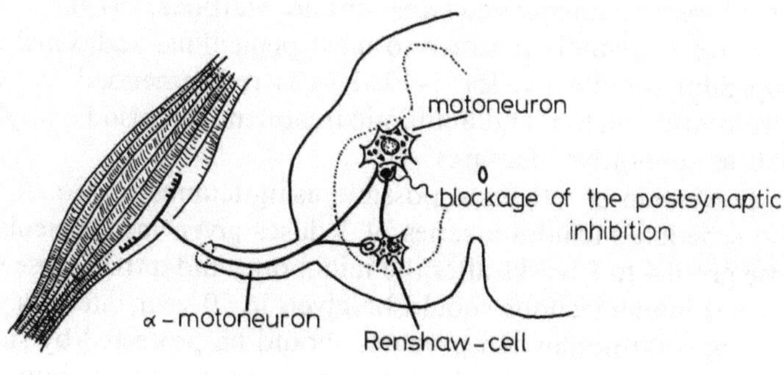

Fig. 9. Action of tetanus toxin

of the Renshaw cell activity results in unopposed continuous stimulation of the muscles.

Tetanospasmin also binds to the gangliosides of the central nervous system (van Heyningen, 1969).

Clinical Symptoms

Nonhuman primates: Tetanus has been reported as a major cause of mortality in the Cayo Santiago *M. mulatta* colony (Kessler et al., 1982; Rawlins and Kessler, 1982a, b). The mean age of the affected animals was around 7 years with both sexes being equally affected. The peak incidence for both sexes occurred during the mating season. Neonatal tetanus has not been observed.

The clinical signs in the monkeys were: torpor, inability to prehend food, difficulties in swallowing, progressive stiffening and adduction of the pectoral limbs, bipedal running, "kangaroo hopping", toppling over, piloerection and with progressing disease the classical signs of tetanus: trismus, opisthotonus, extensor rigidity, status epilepticus and death from respiratory paralysis 24 hrs to 10 days after onset of the symptoms. The case-fatality rate was 85% (DiGiacomo and Missakiani, 1972; Rawlins and Kessler, 1982a). Tetanic monkeys also suffer from harassment by the other group members (Kessler et al., 1982).

Man: After an incubation period varying from 2 to 14 days depending on the site of infection and the patient's age, the first symptom is often a burning sensation or muscular spasms in the region of the local infection. This is followed by trismus, risus sardonicus, opisthotonus and sensitivity to sounds and light. The patient remains completely conscious throughout the painful disease. Death may occur from laryngeal spasm, from spasms of the respiratory muscles, or from gastrointestinal bleeding. The case-fatality rate in man is still over 50% (Smith, 1981).

Pathology

Tetanus is not associated with any characteristic lesion, although lesions might, however, be related to secondary complications.

Treatment

Treatment should be directed at neutralization of the toxin by hyperimmune globulin given intramuscularly. The dosage is 3.000 to 5.000 units in adult human beings (Smith, 1981). In rhesus monkeys 15.000 units in addition to 0,5 ml tetanus toxoid were recommended (Kessler and Martinez, 1979).

Since *C. tetani* is sensitive to most penicillins, additional application of 1 gm ampicillin sodium i.v. for 5–10 days is recommended by Kessler et al. (1982). Barbiturate-sedation and antibiotic treatment can also be applied in man and animals as supportive measures.

Prevention of tetanus is possible using tetanus toxoid. A complete immunization schedule includes a series of 3 doses given intramuscularly with the second dose given 4 to 8 weeks after the initial one, and a third dose 6 to 12 months later. Booster immunizations should be given in 10 year intervals. All animal handlers working with nonhuman primates should be protected by such an immunization program to minimize the dangers resulting from bite wounds.

3.11.2 Clostridium Perfringens (C. Welchii)

The only other pathogenic *Clostridium* species shared by man and nonhuman primates is *C. perfringens*, which causes gas gangrene, enterotoxemia and enteritis necroticans in man (Haagsma, 1979), and gastric dilatation in nonhuman primates (Newton et al., 1971; Bennett et al., 1980; Christie and King, 1981; Stein et al., 1981).

Agent

C. perfringens is an immobile, plump rod of variable length (4–8 × 0,8–1,0 μm) with rounded ends, which is encapsulated in the host and in cultures grown in serum-containing media. Sporulation is usually poor, with the spores being oval, medial or subterminal.

C. perfringens grows well at 37 °C under anaerobic conditions with β-haemolysis on blood agar (sheep or human blood). The colonies are rounded, almost half-spherical, yellowish-grey, opaque with a glistening surface. R-forms are common in some strains.

C. perfringens liquefies gelatin, produces fibrinolysin, hyaluronidase, collagenase, haemolytic and lethal lecithinase, desoxyribonuclease, haemolysin and haemagglutinin. It ferments glucose, maltose and sucrose with gas formation. Litmus milk is acidified with gas formation and fast clotting. *C. perfringens* is indol negative and H_2S positive.

Serotyping distinguishes the groups A–F. Infections in nonhuman primates as far as they were serotyped, all belonged to group A, the serotype causing gas gangrene and food poisoning in man due to exotoxin production. Type A *C. perfringens* produces an enterotoxin during sporulation which causes secretion of fluid, sodium, and chloride into the ileum and shloughing of epithelial cells in mice and rats (McDonel et al., 1975; Yamamoto et al., 1979). Enteritis necroticans in comparison is mostly caused by type C (Blobel and Schliesser, 1980, Shandera et al., 1983). *C. perfringens* is quite resistant to heat.

Epidemiology

C. perfringens is not directly transmitted from animals to man or reverse, but infected animals may contaminate the environment including food.

Clinical Symptoms

Nonhuman primates: *C. perfringens* infections have most frequently been associated with acute gastric dilatation (bloat syndrome) particularly in macaques, although *C. perfringens*-infections may not be the only cause of that condition (Brack, 1982). Acute gastric dilatation of whatever cause is a very dramatic disease of extremely high case-fatality rate in nonhuman primates. It is characterized by rapid development of gastric gas and fluid distention, collapse and death within hours. Subcutaneous emphysema may be observed especially over the abdominal wall and the hind limbs. The disease is obviously extremely painful judging from the fact that the animals are frequently found with their teeth clenched hard enough as to break the teeth. In most cases the animals are found dead in an upright sitting position or lying flat on their belly.

In experimental intoxications of adult *M. fascicularis* by purified *C. perfringens* type A enterotoxin, emesis and diarrhea ensued usually 30 to 120 min after challenge (Hauschild et al., 1971). The diarrhea lasted up to 19 hr (Uemura et al., 1975).

Man: Signs of enterotoxemia develop within a few hours after ingestion of contaminated food as nausea, abdominal pain, and diarrhea without accompanying fever and usually without vomitus. The symptoms disappear usually within a few hours.

Gas gangrene is a clostridial myonecrosis, often following contussion or deep puncture wounds. Sudden severe pain is an early symptom. The local lesions are characterized by sudden onset within hours after the incident. The skin overlying the wounds becomes bronzed, vesicles or haemorrhagic bullae appear near the wound, a brownish fluid of sweet odor exudes from the wound, crepitus in surrounding tissues may be noted (Mader, 1981). Anorexia, high fever, elevated pulse rate and signs of intoxication such as vomitus, watery or bloody diarrhea, profound shock etc. accompany the severe disease.

Pathology

Nonhuman primates: The lesions of *C. perfringens* infections in monkeys are essentially those of the bloat syndrome. They are dominated by extreme gastric dilatation by gaseous (CO_2 and ethanol) and semiliquid, brownish contents of sweetish odor (Smith et al., 1969; Turner and Cowey, 1971). Similar dilatation may occur in the upper small intestines. A variable amount of subcutaneous gas and gelatinous edema may be present particularly in the flank region (Bennett et al., 1980; Brack, 1982). Other secondary lesions are lung edema, petechial haemorrhages in the intestinal mucosa, congestion of the portal system caused by the distended and rotated stomach. The stomach may even be ruptured prior to or shortly after death. Serofibrinous peritonitis may be present (Newton et al., 1971; Bennett et al., 1980; Christie and King, 1981).

Man: Enterotoxemia is rarely fatal, and if so, is related to other, preexisting diseases.

In Gas gangrene the lesions are characterized by swelling, edema and the disruption of the tissues by gas bubbles.

Treatment

The treatment of choice in gas gangrene is penicillin G, up to 20 million units/day intravenously (Mader, 1981).

For clostridial food poisoning as the cause of enterotoxemia prevention by proper food handling is essential, while treatment of patients suffering from enterotoxemia is usually limited to supporting measures and symptomatic.

References

1. Anon: Surveillance of disease in nonhuman primates. Primate Zoonoses Surveil. Rep. *3*, 1-26 (1971)
2. Bennett, B.T.; L. Cuasay; T.J. Welsh; F.Z. Beluhan, and L. Schofield: Acute gastric dilatation

in monkeys: A microbiologic study of gastric contents, blood and feed. Lab. Anim. Sci. *30*, 241-244 (1980)
3. Blobel, H. und T. Schliesser: Handbuch der bakteriellen Infektionen bei Tieren. Bd. II, Gustav Fischer, Stuttgart - New York (1980)
4. Boncyk, L.H. and S.S. Kalter: Bacteriologic findings in a nonhuman primate colony. Develop. biol. Standard. *45*, 23-28 (1980)
5. Brack, M.: Erkrankungen nichthumaner Primaten: Akute Magendilatation. Kleintierpraxis *27*, 313-317 (1982)
6. Christie, R.J. and R.E. King: Acute gastric dilatation and rupture in *Macaca arctoides* associated with *Clostridium perfringens.* J. med. Primatol. *10*, 263-264 (1981)
7. DiGiacomo, R.F. and E.A. Missakian: Tetanus in a free-ranging colony of *Macaca mulatta*: A clinical and epizootiologic study. Lab. Anim. Sci. *22*, 378-383 (1972)
8. Fukuda, N. and S. Asakura: Diseases of monkeys in the Ueno Zoo. II. Internat. Symp. Zoo Animals Pathology (2. Sympzium Patologii Zwierzat Zoo) Warszaw, pp. 191-194 (1960)
9. Goss, L.J.: Diagnosis and treatment of disease of wild animals in captivity. Cornell Vet. *32*, 155-151 (1942)
10. Haagsma, J.: Clostridial diseases in Europe; pp. 225-236 in "CRC Handbook Series in Zoonoses" (Steele, J.H., Series ed.) Section A: Bacterial, Rickettsial, and Mycotic Diseases, Vol. 1 (Stoenner, H., W. Kaplan, and M. Torten, Section eds.) CRC Press, Boca Raton (1979)
11. Hauschild, A.H.W.; M.J. Walcroft, and W. Campbell: Emesis and diarrhea induced by enterotoxin of *Clostridium perfringens* type A in monkeys. Canad. J. Microbiol. *17*, 1141-1143 (1971)
12. Kaufmann, A.F.: The effects of spontaneous disease during quarantine, conditioning, and production of nonhuman primates for scientific use following importation. Pan Amer. Hlth Org. Publ. No. *235*, 39-48 (1972)
13. Kessler, M.J. and R.J. Brown: A clinical description of tetanus in squirrel monkeys *(Saimiri sciureus)*. Lab. Anim. Sci. *29*, 240-242 (1979)
14. Kessler, M.J. and H.S. Martinez: Treatment of tetanus in the rhesus monkey *(Macaca mulatta)*. J. Zoo Anim. Med. *10*, 119-122 (1979)
15. Kessler, M.J.; R.G. Rawlins, and W.T. London: Tetanus in monkeys: Symptomatology, epizootiology, treatment and prophylaxis. Amer. Ass. Zoo Vets. Ann. Proc. 1982, 37-39 (1982)
16. Köhler, W. und H. Mochmann: Grundriß der Medizinischen Mikrobiologie. VEB Gustav Fischer Verlag, Jena (1975)
17. Kronberger, H.: Ergebnisse der Sektion von 890 Säugetieren aus Zoologischen Gärten. VIII. Internat. Symp. Erkr. Zootiere, Leipzig; pp. 259-268 (1966)
18. Mader, J.T.: Clostridial myonecrosis (Gas gangrene); pp. 310-312 in "Infectious Diseases" (Sanford, J.P. and J.P. Luby, eds.), Grune and Stratton, New York (1981)
19. McComb, J.A: Tetanus (Lockjaw); pp. 89-105 in "CRC Handbook Series in Zoonoses" (Steele, J.H., Series ed.), Section A: Bacterial, Rickettsial, and Mycotic Diseases, Vol. II (Stoenner, H.; W. Kaplan, and M. Torten, Section eds.), CRC Press, Boca Raton (1980)
20. McDonel, J.L. and C.L. Duncan: Effects of *Clostridium perfringens* enterotoxin on metabolic indexes of everted rat ileal sacs. Infect. Immun. *12*, 274-280 (1975)
21. Meadow, M.: Medical and surgical management of canine tetanus. Gaines Progress *8*, 4-5 (1978)
22. Newton, W.M.; P.D. Beamer, and H.E. Rhoades: Acute bloat syndrome in stumptailed macaques *(Macaca arctoides)*: A report of four cases. Lab. Anim. Sci. *21*, 193-196 (1971)
23. Osman Hill, W.C.: Notes on malaria and tetanus in monkeys. J. comp. Path. *49*, 274-278 (1939)
24. Rawlins, R.G. and M.J. Kessler: A five-year study of tetanus in the Cayo Santiago rhesus monkey colony: Behavioral description and epizootiology. Amer. J. Primatol. *3*, 23-39 (1982a)
25. Rawlins, R.G. and M.J. Kessler: Behavioral description and epidemiology of tetanus in the rhesus monkeys of Cayo Santiago. Int. J. Primatol. *3*, 326 (1982b)
26. Shandera, W.X.; C.O. Tacket, and P.A. Blake: Food poisoning due to *Clostridium perfringens* in the United States. J. infect. Dis. *147*, 167-170 (1983)
27. Smith, A.W.; H.W. Casey; J.T. LaCroix, and D.K. Johnson: Acute bloat syndrome (Gastric dilatation) in *Macaca mulatta.* J. Amer. vet. med. Ass. *155*, 1241-1244 (1969)

28. Smith, J.W.: Tetanus; pp.312-314 in "Infectious Diseases" (Sanford, J.P. and J.P. Luby, eds.) Grune and Stratton, New York (1981)
29. Smith, L.D.S. and G.Hobbs: Clostridium Prazmowski 1880, 23; pp.551-572 in "Bergey's Manual of Determinative Bacteriology", 8th ed. (Buchanan, R.E. and N.E. Gibbons, eds.) Williams and Wilkins, Baltimore (1975)
30. Stein, F.J.; D.H. Lewis; G.G. Stott, and R.F. Sis: Acute gastric dilatation in common marmosets *(Callithrix jacchus).* Lab. Anim. Sci. *31*, 522-523 (1981)
31. Turner, A.E. and A.Cowey: Bloat syndrome in *Macaca mulatta*: Report on twelve cases. Lab. Primate Newsl. *10* (3), 1-6 (1971)
32. Uemura, T.; G.Sakaguchi; T.Itoh; K.Okazawa, and S.Sakai: Experimental diarrhea in cynomolgus monkeys by oral administration with *Clostridium perfringens* type A viable cells or enterotoxin. Japan J.Med. Sci. Biol. *28*, 165-177 (1975)
33. Van Heyningen, W.E.: The fixation of tetanus toxin and strychnine by gangliosides. J.gen. Microbiol. *31*, 375-385 (1969)
34. Yamamoto, K.; I.Ohishi, and G.Sakaguchi: Fluid accumulation in mouse ligated intestine inoculated with *Clostridium perfringens* enterotoxin. Appl. Environ. Microbiol. *37*, 181-186 (1979)

3.12 Listeria Monocytogenes

Listeriosis is a world-wide bacterial zoonosis affecting a wide variety of domestic and wild animals as well as man. Nonhuman primates are not a major host of these organisms but nevertheless are occasionally infected. Infections have been reported in *C.jacchus* (Zwart and Donker-Voet, 1959), *G.demidovii* (Fiennes, 1968), *Papio* sp. (Pinkerton, 1972), *C.mona* (Vetesi et al., 1973), *M.nigra* (McClure and Strozier, 1975), *M.mulatta* (Anon, 1980), *C.cephus* (Jentzsch et al., 1969), *P.troglodytes* (Heldstab and Rüedi, 1982) and *M.fascicularis* (Chalifoux and Hajema, 1981; Tribe, 1982; Heldstab and Thommen, 1984).

Agent

L.monocytogenes is a gram-positive, motile, nonsporing, non-acid-fast rod of 0,7 to 0,8 × 0,9 to 4,0 μm size with rounded ends. The organisms might be slightly curved or, more frequently, can be coccoid or filamentous. In older cultures the bacteria may be gram-negative. The coccoid forms are easily mistaken as cocci and the gram-negative ones as unimportant contaminants. *L.monocytogenes* grows in aerobic to microaerophilic conditions. When it is grown at 20 to 25 °C, it develops peritrichous flagella, whereas at 37 °C only 1 to 4 polar flagella are formed. The motility is quite characteristic as an end-over-end tumbling in liquid cultures grown at 20 °C.

Growth can be obtained on blood agar or on selective media such as trypaflavin-nalidixic-acid-serum-agar (TNSA) or acridine-nalidixic-acid-agar (Welshimer, 1981). The colonies are slightly raised, of 0,2 to 1,5 mm diameter after 24 hs incubation. They are stained blue green on clear, solid media when viewed by oblique light (45 angle), and on sheep blood agar are surrounded by a narrow zone of β-haemolysis.

For enrichment, storage at 4 °C for periods of a few days to 3 months before successful isolation is recommended (Killinger, 1974).

The biochemical reactions of *L. monocytogenes* are (Hartwigk, 1958; Seeliger and Welshimer, 1974; Welshimer, 1981):

Table 26. Biochemical Characteristics of *L. monocytogenes*

Catalase	+	Esculin	+
Oxidase	∅	Mannitol	∅
Glucose	+ (no gas)	Salicin	+
Lactose	V	Raffinose	∅
L-Arabinose	∅	Dulcitol	∅
D-Galactose	V	Sorbitol	∅
Melibiose	∅	Methyl red	+
Rhamnose	V		
Sucrose	V		
Xylose	∅		
Trehalose	+		

Table 27. Seeliger-Donker-Voet Scheme for *L. monocytogenes* serovars

Serovar	O-Antigens													H-Antigens
1/2a	I	II	(III)											AB
1/2b	I	II	(III)											ABC
1/2c	I	II	(III)											BD
3a		II	(III)	IV										AB
3b		II	(III)	IV										ABC
3c		II	(III)	IV										BD
4a			(III)		(V)		VII		IX					ABC
4ab			(III)		V	VI	VII		IX					ABC
4b			(III)		V	VI								ABC
4c			(III)		V		VII							ABC
4d			(III)		(V)	VI		VIII						ABC
4e			(III)		V	VI		(VIII)	(IX)					ABC
4f			(III)		V							XV		ABC
4g			(III)		V	VI	VII			X,	XI			ABC
5			(III)		(V)	VI		VIII		X				ABC
6			(III)				VII	VIII			XI			ABC
7			(III)									XII	XIII	ABC

L. monocytogenes is quite resistant to environmental factors, remaining viable in dust and dirt even after prolonged exposure to sunlight or for several months in damp, fertile soil (Bowmer et al., 1979).

Several serovars of *L. monocytogenes* can be distinguished (Table 27).

Epidemiology

L. monocytogenes has been isolated from healthy sheep, cattle, and man as well as from pathological materials. The organism has also been recovered from a wide variety of mammals, birds, reptiles, fish, crustaceans, molluscs and terrestrial arthropods as well as from mud, sewage, dust etc.

The modes of transmission to animals and man may be complicated (Tab. 28).

Table 28. Transmission of *Listeria monocytogenes*

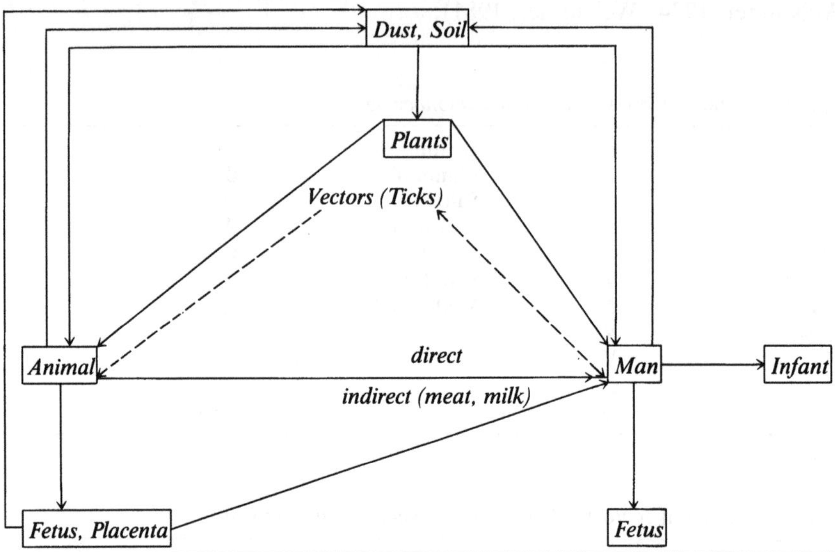

The intestinal tract of man and animals may be a reservoir for *L. monocytogenes* with animal manure and human waste possibly being the source of soil infection (Bowmer et al., 1979). Infected soil or dust may be directly infectious or can transmit the agents via plants. Infection of pregnant females can result in fetal or placental infections in both animals and man. Listeriosis in man is therefore an occupational hazard with farmers, veterinarians or animal caretakers or people who handle meat or poultry being at greatest risk.

Pathogenesis

It is suggested that *L. monocytogenes*, when entering the oral cavity contaminates the nasopharynx where the infected droplets possibly gain access to the ethmoturbinate region, thus permitting the bacteria to enter the brain via the olfactory tract (Bowmer et al., 1979).

When invading the body via the alimentary tract, *L. monocytogenes* tends to localize in the intestinal wall, especially in the Peyer's patches, from where it may pass to the liver and other organs to produce micro-abscesses.

Clinical Symptoms

Nonhuman primates: Listeriosis of nonhuman primates has been associated with uterine infections in at least five cases. Tribe (1982) recovered serotype I from two *M. fascicularis*-fetuses aborted at about 4 gestational months, Vetesi et al. (1973) isolated serotype 4 b from aborted *C. mona* fetuses. Chalifoux and Hajema (1981) recovered serotype 4 from a neonatal *M. fascicularis*, Heldstab and Thommen (1984) also isolated serotype 4b from 3 *M. fascicularis* which died from neonatal septicemia.

In an anonymous news item (1980) the isolation of *L. monocytogenes* from the placenta of an aborted rhesus monkey was reported.

Other infections seen were listeric meningoencephalitis in a chimpanzee (Heldstab and Rüedi, 1982), which presents itself as apathy, anorexia, nuchal rigidity and unilateral paralysis, or by hepatitis in a Demidov's galago.

Pinkerton (1972) isolated two strains of Listeria from blood cultures of 100 apparently healthy wild baboons, and lesions were missed as well in an infected *C.jacchus* (Zwart and Donker-Voet, 1959). Experimental aerogenous infections in *M.mulatta* produced only a short-lasting rise of temperature without further sequelae (Kautter et al., 1963).

Man: Listeriosis in man may induce quite different symptoms. The most frequently observed syndrome is acute meningoencephalitis with or without septicemia. Other syndromes are septicemia with pharyngitis and mononucleosis, oculoglandular forms, cervicoglandular forms, granulomatosis, typhoidal-pneumonic forms, listeriosis of the newborn, listeriosis in pregnancy, cutaneous listeriosis and other rare forms (Seeliger, 1961).

Listeric meningoencephalitis is characterized by headache, myalgia, fever, chills, nausea, vomiting, photophobia and meningismus, it is also often associated with pharyngitis, otitis media, gastrointestinal or respiratory illness.

Neonatal listeriosis appears as disseminated granulomatosis, aspiration pneumonia, and meningitis.

Listeriosis during pregnancy is essentially an infection of the uterine contents with minimal involvement of maternal tissues and thus few, if any, maternal flu-like symptoms. The resultant infant may be well, acutely ill or stillborn, but abortions are only infrequently observed (Bowmer et al., 1979).

Pathology

Nonhuman primates: Necrotic foci in the liver and focal necrotic placentitis were the most prominent lesions in the aborted *C.mona* fetus (Vetesi et al., 1973), the stillborn *M.niger* (McClure and Strozier, 1975) and the neonatal *M.fascicularis* (Chalifoux and Hajema, 1981). Other lesions in the *M.fascicularis* were adrenal abscesses, purulent pyelitis and proctitis (Chalifoux and Hajema, 1981). The listeric meningoencephalitis in the chimpanzee and the neonatal *M.fascicularis* was of a fibrinopurulent nature (Chalifoux and Hajema, 1981; Heldstab and Rüedi, 1982), whilst the hepatitis of the dwarf-galago consisted of multiple miliary abscesses (Fiennes, 1968).

Man: The pathologic hallmark of listeriosis is the miliary abscess in the different organs.

In listeric meningoencephalitis the leptomeninges contain miliary granulomata, whilst the subarachnoidal space might be filled with pus and granulomata can be observed in the pons, medulla, cerebellum and cerebral cortex. The granulomata consist of a necrotic center and are surrounded by inflammatory cells with prominent mononuclear cells (Siegel and Nelson, 1981). Lymphonodular (cervicoglandular) listeriosis may lead to purulent liquefaction of the affected lymph nodes with fistula formation and purulent drainage.

In listeriosis of pregnancy the placenta may show macroabscesses, focal villositis or chorioamnionitis (Siegel and Nelson, 1981). In the stillborn foetuses focal hepatic necroses or necroses in other organs may be found (Bowmer et al., 1981).

Treatment

L. monocytogenes is susceptible *in vitro* to sulfonamides, ampicillin, penicillin, tetracycline, chloramphenicol, erythromycin and novobiocin. Ampicillin is the drug of choice because of its penetration of the blood-brain barrier. Cephalosporins and streptomycin are not consistently active against Listeria.

References

1. Anon: News items. Primate Supply *5* (1), 23 (1980)
2. Bowmer, E.J.; R.H. Conklin, and J.H. Steele: Listeriosis; pp. 423-443 in "CRC Handbook Series in Zoonoses" (Steele, J.H., Series ed.), Section A: Bacterial, Rickettsial and Mycotic Diseases, Vol. I (Stoenner, H.W. Kaplan and M.Torten, Section eds.), CRC Press, Boca Raton (1979)
3. Chalifoux, L.V. and E.M. Hajema: Septicemia and meningoencephalitis caused by *Listeria monocytogenes* in a neonatal *Macaca fascicularis.* J.med. Primatol. *10*, 336-339 (1981)
4. Fiennes, R.N.T-W.: The Zoological Society of London. Report of the Society's Pathologist for the year 1965. J.Zool., Lond. *154*, 173-191 (1968)
5. Hartwigk, H.: Morphologische und kulturelle Eigenschaften von Listeria monocytogenes (L.m.); pp.5-16 in "Listeriosen" (Roots, E.und D.Strauch, eds.) Parey, Berlin (1958)
6. Heldstab, A.and D.Rüedi: Listeriosis in an adult female chimpanzee *(Pan troglodytes).* J.comp. Path. *92*, 609-612 (1982)
7. Heldstab, A.and D.R. Thomen: Neonatale Mortalität bei Javaneraffen (Macaca irus) im Zoologischen Garten Basel. XXVI. Internat. Symp. Erkr. Zootiere, Brno, pp.103-110 (1984)
8. Jentzsch, K.D.; H.D. Schroeder und E.Karasek: Beitrag zum Vorkommen von Zooanthroponosen bei Zootieren. XI. Internat. Symp. Erkr. Zootiere, Zagreb, pp.7-17 (1969)
9. Kautter, D.A.; S.J. Silverman; W.G. Roessler, and J.F. Drawdy: Virulence of *Listeria monocytogenes* for experimental animals. J.Infect. Dis. *112*, 167-180 (1963)
10. Killinger, A.H.: *Listeria monocytogenes*; pp.135-139 in "Manual of Clinical Bacteriology" (Lennette, E.H., E.H. Spaulding, and J.Truant, eds.) Amer. Soc. Microbiol., Washington D.C. (1974)
11. McClure, H.M. and L.M. Strozier: Perinatal listeric septicemia in a Celebes black ape. J.Amer. vet. med. Ass. *167*, 637-638 (1975)
12. Pinkerton, M.: Miscellaneous organisms; pp.283-313 in "Pathology of Simian Primates", Part II (Fiennes, R.N.T-W, ed.) Karger, Basel (1972)
13. Seeliger, H.P.R.: Listeriosis, Hafner Press, New York (1961)
14. Seeliger, H.P.R. and H.J. Welshimer: Genera of uncertain affiliation. Genus Listeria Pirie 1940, 383 Nom. cons. Opin.12.Jud. Comm.1954, 151; pp.593-596 in "Bergey 's Manual of Determinative Bacteriology", 8th ed. (Buchanan, R.E. and N.E. Gibbons, eds.) Williams and Wilkins, Baltimore (1974)
15. Siegel, J.D. and J.D. Nelson: Listeriosis, pp.222-225 in "Infectious Diseases" (Sanford, J.P. and J.P. Luby, eds.) Grune and Stratton, New York (1981)
16. Tribe, G.W.: Listeria monocytogenes associated with abortion in cynomolgus monkeys. Primate Supply *7* (2), 9-13 (1982)
17. Vetesi, F.; A.Balsai, and F.Kemenes: Abortion in Gray's monkey *(Cercopithecus mona)* associated with Listeria monocytogenes. Acta microbiol. Acad. Sci. hung. *19*, 441-443 (1973)
18. Welshimer, H.J.: The genus *Listeria* and related organisms; pp.1680-1687 in "The Prokaryotes. A Handbook on Habitats, Isolation and Identification of Bacteria." (Starr, M.P., H.Stolp, H.G. Trüper, A.Balows and H.G. Schlegel, eds.) Springer, Berlin (1981)
19. Zwart, P.and J.Donker-Voet: Listeriosis bij in gevangenschap gehouden dieren. T.Diergeneesk. *84*, 712-716 (1959)

3.13 Erysipelothrix Insidiosa (Rhusiopathiae)

E. insidiosa, the causative agent of erysipelas in pigs and of mouse septicemia occasionally infects man, causing erysipeloid or more serious cutaneous lesions. *E. insidiosa* has also been isolated occasionally from nonhuman primates such as *S. oedipus* (Schröder, 1966; Jentzsch et al., 1969), *S. nigricollis* (Hirsh et al., 1975), *C. aethiops* (Barnes, 1970) or *C. diana* (Wallach, 1977).

Agent

E. insidiosa was first isolated by Koch (1878) from mice and 1886 by Loeffler from pigs that died of erysipelas. *E. insidiosa* is a gram-positive rod of 1 to 2 μm size, that is easily decolorized, thus appearing gram-negative. Cells from rough colonies are somewhat larger and form long filaments or chains resembling actinomycetes except for the lack of branching. Metachromatic granules are often present. The bacteria are non-encapsulated, nonmotile, and are not acid-fast. *E. insidiosa* is a facultative aerobe, and grows at a wide range of temperatures on most nutrient media. The best growth is obtained in 1% glucose broth, in tryptase broth, beef infusion broth or in liquid media with added serum or Tween 80.

The colonies appearing on the agar-surface in initial isolations after 24 hs of incubation are small (0.1 to 0.5 mm in diameter), glistening, clear, convex and circular, which become opaque and bluish with age. On blood agar they are usually surrounded by a small zone of alpha-haemolysis.

In subcultures a transition to rough (R) forms is quite common, but the R-forms may also occasionally occur in initial isolations. R-colonies are slightly larger (0.5 to 1 mm in diameter), flat and more irregular.

In TSI-agar the butt and the slant are acidified with H_2S production and not gas. *E. insidiosa* is catalase-negative and is inhibited by potassium tellurite in contrast to the *Corynebacterium* spp. and to *Listeria monocytogenes*. Other biochemical characteristics are (Conklin and Steele, 1979; Ewald, 1981):

Table 29. Biochemical Characteristics of *E. insidiosa*

Glucose	+ (no gas)	Sorbitol	Ø
Lactose	+	Adonitol	Ø
Galactose	+	Indole	Ø
Maltose	d	Salicin	Ø
Fructose	+	Esculin	Ø
Mannose	d or Ø	Hippurate	Ø
Trehalose	Ø or d	Nitrate	Ø
Rhamnose	d or Ø	Voges Proskauer	Ø
Xylose	d or Ø	Methyl red	Ø
Melibiose	Ø		
Cellobiose	Ø		

(For carbohydrate fermentation tests, Andrades base with 10% serum should be used.)

E. insidiosa is considered quite resistant to moderate temperatures, and to phenol, salt, sunlight etc.

Epidemiology

E. insidiosa infects pigs, mice, birds, fish etc. under natural conditions. It is usually transmitted by direct contact or via a contaminated environment. Indirect transmission by insect vectors may also play a certain role.

E. insidiosa infections of nonhuman primates have been related to avian species (Schröder, 1966; Hirsh et al., 1975) or to free ranging mice (Wallach, 1977), while human infections have been frequently traced back to fish.

Clinical Symptoms

Nonhuman primates: The *E. insidiosa* infected monkeys showed a wide range of nonspecific symptoms, such as depression, or respiratory symptoms (Schröder, 1966; Hirsh et al., 1975; Wallach, 1977).

Man: *E. insidiosa* infections of man can result in

1. erysipeloid,
2. generalized cutaneous infections, or
3. septicemia.

Ad 1. Erysipeloid is a localized cutaneous infection usually on the fingers or hands. It is characterized by a well-demarcated, focal, violaceous erythema, that spreads peripherally and fades in the centre. The peripheral rim is slightly elevated and progresses slowly. The pain associated with those local, slightly swollen lesions seems to be out of proportion to their size. Lymphangitis or lymphadenitis are usually absent.

Ad 2. Generalized cutaneous lesions may progress from a local lesion. The majority of patients affected by generalized cutaneous lesions have fever and articular pain. The clinical course is usually much more protracted than that of the localized form, with recurrences being not uncommon.

Ad 3. Septicemia may occur with or without skin lesions, and frequently is associated with endocarditis. Septicemia is very rare in non-immunosuppressed patients (Conklin and Steele, 1979).

Pathology

Nonhuman primates: The morphological lesions of simian *E. insidiosa*-infections were as nonspecific and variable as the clinical signs. They were reported as cachexia, interstitial pneumonia (Schröder, 1966), multifocal neutrophilic myocarditis and encephalitis (Barnes, 1970) or petechial haemorrhages, congestion, hepatomegaly and icterus (Wallach, 1977).

Treatment

Although penicillin is the drug of choice in *E. insidiosa*-infections, in the callitrichids erythromycin is more advisable.

References

1. Barnes, D.M.: *Erysipelothrix insidiosa* infection in an African green monkey. Primate Zoonoses Surveil. Rep. *2*, 7 (1970)
2. Conklin, R.H. and J.H. Steele: *Erysipelothrix* infections; pp. 327-336 in "CRC Handbook

Series in Zoonoses" (Steele, J.H., Series ed.), Section A: Bacterial, Rickettsial and Mycotic Diseases. Vol. I (Stoenner, H., W. Kaplan and M. Torten, Section eds.) CRC Press, Boca Raton (1979)
3. Ewald, F.W.: The genus *Erysipelothrix*; pp. 1688-1700 in "The Prokaryotes. A Handbook on Habitats, Isolation and Identification of Bacteria", Vol. II (Starr, M.P., H. Stolp, H.G. Trüper, A. Balows and H.G. Schlegel, eds.) Springer, Berlin (1981)
4. Hirsh, D.C.; G.A. Boorman, and S.S. Jang: Erysipelas in a black and red tamarin. J. Amer. vet. med. Ass. *167*, 646-647 (1975)
5. Jentzsch, K.D.; H.D. Schröder und E. Karasek: Beitrag zum Vorkommen von Zooanthroponosen bei Zootieren. XI. Internat. Symp. Erkr. Zootiere, Zagreb, pp. 7-17 (1969)
6. Koch, R.: Untersuchungen über die Ätiologie der Wundinfektionskrankheiten. Vogel, Leipzig (1878)
7. Loeffler, F.: Experimentelle Untersuchungen über Schweinerotlauf. Arbeiten aus dem Kaiserlichen Gesundheitsamt. *1*, 46-55 (1886)
8. Schröder, H.D.: Erysipelothrix insidiosa-Infection bei Pincheäffchen (Leontocebus oedipus) - zugleich ein Beitrag zur seuchenhygienischen Überwachung von frischimportierten Zootieren. VIII. Internat. Symp. Erkr. Zootiere, Leipzig, pp. 49-57 (1966)
9. Wallach, J.D.: Erysipelas in two captive diana monkeys. J. Amer. vet. med. Ass. *171*, 679-980 (1977)

3.14 Corynebacteriaceae

Corynebacteriaceae are small, non-sporulating, non-motile (except *L. monocytogenes*), gram-positive rods, some of which having a distinctive pleomorphism.

3.14.1 Corynebacterium Ulcerans

C. ulcerans was first reported as a pathogen in man in 1927 by Gilbert and Stewart. It has also been isolated in a few instances from rhesus monkeys (May, 1972) and a bonnet macaque (Fox and Frost, 1974).

Agent
C. ulcerans is a non-motile, small, plump rod measuring 0,6 × 0,1 µm. In older cultures the bacteria may be longer and more club shaped with metachromatic granulation (Babes - Ernst bodies) or appear more cocco-bacillary resembling staphylococci. These metachromatic granules in the corynebacteriaceae consist of long-chain polyphosphate. *C. ulcerans* grows in a 10% CO_2 containing atmosphere on sheep blood agar with a fine beta-haemolytic line around the colonies. After 18 to 24 hrs white and mucoid colonies appear, which become yellowish after being exposed to air and on further incubation appear like typical diphtheroidal colonies (May, 1972).

The biochemical reactions of *C. ulcerans* are: (see Tabl. 30)

Most *C. ulcerans* strains produce two toxins, one immunologically identical to diphtheria toxin and the other related to *C. pseudotuberculosis*-toxin (Knight, 1979).

Table 30. Biochemical Characteristics of *C. ulcerans*

Glucose	+	Trehalose	(+)
Lactose	Ø	Arginine dihydrolase	Ø
Sucrose	+ (72 hs)	Nitrate	Ø
Levulose	+	Catalase	+
Galactose	+	Urease	+
Maltose	+	Gelatine	+ (5 days)
		DNAase	+
		Starch	+
		Glycerol	+

Epidemiology

C. ulcerans has been reported as a cause of episodes of mastitis in cattle (Jayne-Williams and Skerman, 1966; Higgs et al., 1967) and has also been isolated from the human throat (Jebb, 1948). Most human *C. ulcerans* infections are probably endogenous, but carriers working in food processing can be the source of contamination. Outbreaks of *C. ulcerans* pharyngitis attributed to the ingestion of contaminated raw milk have occurred (Knight, 1979).

C. ulcerans has also been isolated from nasal swabs, lung abscesses and bite wounds in nonhuman primates, suggesting a transmission between animals and possibly to man via the contaminated oral flora. Even in the case of *C. ulcerans* mastitis of a *M. radiata* (Fox and Frost, 1974) trauma to the teats from the nursing infant has been suggested to be the port of entry (Knight, 1979).

In endogenous infections clinical disease may be induced by damage of the competitive resident flora, by lowering of the hosts resistance or by changes in toxigenicity.

Clinical Symptoms

Nonhuman primates: *C. ulcerans* caused chronic subcutaneous, maxillary or lung abscesses and pneumonia in *M. mulatta* and, in one case, necrotizing mastitis in a bonnet monkey.

Man: *C. ulcerans* has been the cause of sore throat, or infrequently of severe tonsillitis, often in people working with cattle (Jebb, 1948; Tomlinson, 1966; Saxholm, 1951).

Pathology

Nonhuman primates: The lung lesions varied from pneumonia to focal abscesses or necrosis filled with a yellowish, mucoid pus (May, 1972).

Treatment

C. ulcerans is sensitive to chloramphenicol, cephalothin, streptomycin, erythromycin, novobiocin, penicillin and tetracycline, it also responded well to 30 mg kanamycin i.m. twice daily for 7 consecutive days (Fox and Frost, 1974). It is, however, resistant to furazolidone and polymyxin.

The action of the diphtheria-like toxin might be counteracted by diphtheria antitoxin in severe cases.

3.14.2 C. Pseudotuberculosis

C.pseudotuberculosis is the etiologic agent of caseous lymphadenitis in sheep, of ulcerative lymphangitis in horses and of some pyogenic infections in cattle. Single cases have been found in a variety of exotic animal species including one case in a chimpanzee (Kim, 1976). According to Knight, 1979, four reports of the infection have been in human beings.

Agent
C.pseudotuberculosis does not possess a capsule and shows the typical Chinese-letter configurations on direct smears. It grows best on blood or serum containing media in an atmosphere with an increased partial pressure of CO_2. Like most corynebacteria it grows readily on tellurite blood agar, forming whitish-grey, friable colonies. Most strains produce haemolysin on blood agar.

The biochemical characteristics are:

Table 31. Biochemical Characteristics of *C. pseudotuberculosis*

Catalase	+	Starch	Ø
Glucose	+	Gelatin	V
Maltose	+	Urea	+
Sucrose	V	Nitrate (equine strains:	V
Lactose	V	ovine or caprine strains)	Ø

C.pseudotuberculosis produces an exotoxin characterized as a phospholipase D, which is lethal to guinea pigs, rabbits, mice and sheep.

Epidemiology
Contamination of wounds or abrasions seems to be the primary means of infection in both animals and humans. The organism is not commonly isolated from soil unless heavily contaminated by clinically ill sheep. In some areas transmissions by biting flies are suspected.

Direct transmission from animals to man would seem a strong possibility for persons handling animals or animal products (Knight, 1979).

Clinical Symptoms
Nonhuman primates: The one case in a chimpanzee was marked by a brownish diarrhea not affected by furoxone-treatment (Kim, 1976).

Man: In three of the four registered human cases the isolants were working with sheep or other animals. They all developed axillar or inguinal lymphadenitis following wounds or probable insect bites. Systemic signs, if at all present, consisted of fatigue, malaise and mild temperature rise.

Pathology
Nonhuman primates: The most prominent lesion in the chimpanzee were a subacute purulent interstitial nephritis and chronic, serous pleuritis (Kim, 1976).

Treatment
As with all diseases that form external abscesses, incisions and drainage is the principal method of treatment. Antibiotic treatment usually failed.

3.14.3 Corynebacterium Equi

C. equi typically infects 2 to 3 months-old foals causing pneumonia, arthritis, ulcerative lymphangitis and internal abscesses. It rarely affects other mammalian species, 1 case has been reported in a cottontop marmoset (Stein and Stott, 1979), 5 others in man (Knight, 1979).

Agent
C. equi is a heavily encapsulated corynebacterium of sometimes spheroidal appearance and partially acid-fast staining properties. It grows readily on ordinary culture media forming moist, pale pink colonies without haemolysis on blood agar.

The biochemical reactions of *C. equi* are:

Table 32. The biochemical reactions of *C. equi* are

Catalase	+	Starch	ø
Glucose	ø	Gelatin	ø
Maltose	ø	Urea	V
Sucrose	ø	Nitrate	+
Lactose	ø		

C. equi apparently does not produce any potent exotoxin.

Epidemiology
The habitat of *C. equi* is probably the soil, with inhalation of the bacteria being suggested as the major route of infection in horses (and probably in other species as well). Three of the five human patients had been exposed to animals, all five were immunosuppressed to a certain degree. Infection of man probably occurs from contaminated environment rather than by direct animal to man contact.

Clinical Symptoms
Nonhuman primates: In the one *S. oedipus* loss of appetite, labored respiration and general unthriftiness were observed. The animal died after three days of illness (Stein and Stott, 1979).

Man: Four of the five patients developed pulmonary infiltrates or pneumonia additionally to their initial disease treated by immunosuppressants (Knight, 1979).

Pathology
Nonhuman primates: The primary lesions of the marmoset were a pyogranulomatous bronchopneumonia, a necrotic pleuritis, fibrinopurulent epicarditis and pyogranulomatous serositis (Stein and Stott, 1979).

Treatment

C. equi is susceptible to tetracycline, erythromycin and chloramphenicol, but can develop antibiotic-resistance. Penicillin seems to be ineffective (Knight, 1979).

References

1. Fox, J.G. and W.W. Frost: *Corynebacterium ulcerans* mastitis in a bonnet macaque (*Macaca radiata*). Lab. Anim. Sci. *24*, 820-822 (1974)
2. Gilbert, R. and F.C. Stewart: *Corynebacterium ulcerans*: A pathogenic micro-organism resembling *C. diphtheria*. J. Lab. Clin. Med. *12*, 756-771 (1927)
3. Higgs, T.M.; A. Smith; L.M. Cleverly, and F.K. Neave: *Corynebacterium ulcerans* infections in a dairy herd. Vet. Rec. *81*, 34-35 (1967)
4. Jayne-Williams, D.J. and T.M. Skerman: Comparative studies on coryneform bacteria from milk and dairy sources. J. appl. Bact. *29*, 72-92 (1966)
5. Jebb, W.H.H.: Starch-fermenting gelatin liquefying *Corynebacterium* isolated from the human nose and throat. J. Path. Bact. *60*, 403-412 (1948)
6. Kim, J.C.S.: *Corynebacterium pseudotuberculosis* as a cause of subacute interstitial nephritis in a chimpanzee. Vet. Med. small Anim. Clin. 1976, 1093-1095 (1976)
7. Knight, H.D.: Corynebacterial diseases;pp. 311-326 in "CRC Handbook Series in Zoonoses" (Steele, J.H., Series ed.) Section A: Bacterial, Rickettsial and Mycotic Diseases, Vol. I (Stoenner, H., W. Kaplan and M. Torten, Series eds.) CRC Press, Boca Raton (1979)
8. May, B.D.: *Corynebacterium ulcerans* infections in monkeys. Lab. Anim. Sci. *22*, 509-513 (1972)
9. Saxholm, R.: Toxin-producing diphtheriae-like organisms isolated from cases of sore throat. J. Path. Bact. *63*, 303-311 (1951)
10. Stein, F.J. and G. Stott: *Corynebacterium equi* in the cottontop marmoset (*Saguinus oedipus*): A case report. Lab. Anim. Sci. *29*, 519-520 (1979)
11. Tomlinson, A.J.H.: Human pathogenic coryneform bacteria: Their differentiation and significance in public health today. J. appl. Bact. *29*, 131-137 (1966)

3.15 Actinomycetales

Actinomycetales are bacteria that produce branching filamentous hyphae, thus superficially resembling fungi. Their taxonomy is somewhat confused, three genera are of medical importance: Actinomyces, Nocardia and Dermatophilus. All three genera occur rarely in nonhuman primates.

3.15.1 Actinomyces Spp.

Actinomyces spp. produce diseases particularly in cattle, swine and dogs, 4 species *(A. israeli, A. naeslundi, A. viscosus, Arachnia proprionica)* are human pathogens (Pier, 1979). Single *Actinomyces* infections, probably *A. israelii*, have been described in *Cebus capucinus, Ateles geoffroyi* (Weidman, 1923, 1927, 1935 - all cit. from Al-Doory, 1972), *Hylobates moloch, M. maura* (Fiennes, 1967) and a *Mandrillus leucophaeus* (Altman and Small, 1973).

Agent
Actinomyces spp. are microaerophilic to anaerobic, gram-positive, nonmotile and non-spore forming, branching filamentous hyphal elements that fragment into short rods or coccoid forms. They are indole- and urease-negative, whilst carbohydrates are fermented without gas production.

Cultivation is possible on thioglycollate media enriched with serum or on brain heart infusion agar.

Life Cycle and Epidemiology
Actinomycosis is not ordinarily considered as a transmissible disease but rather as an endogenous infection (Pier, 1979). *Actinomyces* spp. are residents of the oral mucosal membranes, and once they have established the infection of the tissues they spread within the individual by erosion of blood vessels, organs etc.

Geographical Distribution
A. israeli and probably also the other *Actinomyces* spp. are distributed worldwide.

Clinical Symptoms
Nonhuman primates: The isolations were made from the discharge of nasal abscesses (Weidman, 1923, 1927, 1935 - all cit. from Al-Doory, 1972), from acute enteritis (Fiennes, 1967) or from a nonspecific illness characterized by anorexia and facial edema (Altman and Small, 1973).

Man: Human actinomycosis occurs either as cervico-facial, thoracic or abdominal types with the formation of typical actinomycotic mycetomas: swelling, erosion of bones, fistula formation and discharge of granules (Sulphur granules) (Pier, 1979).

Pathology
Nonhuman primates: The types of lesions recorded were acute haemorrhagic enteritis, caseous necroses in liver and pancreas (Fiennes, 1967), subcutaneous edema, lingual and oral erosions and ulcers, or peritonitis with multiple "sulfur granules", containing abscesses on the surface of most abdominal organs (Altman and Small, 1973).

Man: The lesions vary from typical actinomycotic mycetomas to those of a protean progressive suppurative process (Pier, 1979).

3.15.2 Nocardia Spp.

Nocardiae are aerobic actinomycetes which may produce disease in man and animals, especially of the respiratory tract. Two species *(N. asteroides* and *N. brasiliensis)* are of medical importance in man (Pier, 1979), although a few cases of human nocardiosis have been caused by *N. caviae* (for reference see Boncyk et al., 1975).

Nocardiosis has also been reported in *Macaca mulatta* (Jonas and Wyand, 1966; Liebenberg and Giddens, 1985), *M. nemestrina* (Liebenberg and Giddens, 1985), *M. fascicularis* (Sakakibara et al., 1984), *Cercopithecus aethiops* (Al-Doory et al., 1969), *Papio cynocephalus* (Boncyk et al., 1975) and *Pongo pygmaeus* (McClure et al., 1976).

Agents

Nocardiae are aerobic, partially acid fast, branching filamentous bacteria that fragment into rods and coccoid forms. They grow readily on nonselective media sometimes with production of karotenoid pigments. The colonies are raised, irregular, opaque and adhere to the underlying media. Most strains produce aerial hyphae.

N. brasiliensis and *N. asteroides* reduce nitrates to nitrites, but they do not hydrolyse xanthine. *N. brasiliensis* hydrolyses gelatine, whereas *N. asteroides* is gelatine negative.

Life Cycle

Nocardiosis is not considered a true zoonosis but is acquired from a contaminated environment. Similarly to *Actinomyces* spp. infections the agent is spread by erosion of the vasculature or other anatomical barriers.

Geographical Distribution

N. asteroides occurs throughout the world, whereas *N. brasiliensis* is a tropical microorganism, being prevalent especially in the neotropics.

Clinical Symptoms

Nonhuman primates: Pulmonary infections are characterized by upper respiratory signs, severe dyspnea and lethargy or weakness, anorexia or coma (Al-Doory et al., 1969; McClure et al., 1976; Sakakibara et al., 1984). Infections of abdominal organs cause lethargy, depression, abdominal distention, intestinal destruction and diarrhea (Liebenberg and Giddens, 1985).

Man: Human nocardiosis may appear as a systemic infection with primary involvement of the lungs exhibiting cough, chest pain, weight loss and fever (*N. asteroides*) or as mycetomas on the extremities or trunk (*N. brasiliensis*) (Pier, 1979; Rhoades, 1981).

Pathology

Nonhuman primates: *N. asteroides* induced predominantly lung lesions as catarrhal-necrotic or granulomatous pneumonias (Jonas and Wyand, 1966; Al-Doory et al., 1969; McClure et al., 1976; Sakakibara et al., 1984; Liebenberg and Giddens, 1985). Dissemination to extrapulmonary organs involved the brain, kidneys, heart, liver, intestines, and peritoneum (Sakakibara et al., 1984; Liebenberg and Giddens, 1985).

Treatment

The therapy of nocardiosis is usually unsatisfactory (Rhoades, 1981), sulfonamides are the drugs of choice, but they have to be administered for several months. Trimethoprim-sulfamethoxazole has also been used with success.

3.15.3 Dermatophilus Sp.

Dermatophilus Congolensis

Dermatophilosis is an infectious dermatitis of terrestrial and aquatic mammals. The disease is most common in cattle, sheep and horses, with nonhuman primates and man being only occasionally infected. As far as nonhuman primates are concerned dermatophilosis has only been reported in New World monkeys, mostly in *Aotus trivirgatus* (King et al., 1971; McClure et al., 1971; Fox et al., 1973), with single cases being reported in a *Lagothrix lagotricha* (Fraser and Garcia, 1971), a *Callicebus moloch* (Migaki and Seibold, 1976) and (not confirmed by cultivation) in a *Saimiri sciureus*) (Kessler and Brown, 1981).

Agent

D. congolensis is an aerobic, gram-positive bacterium of variable morphology depending on the maturity of the culture (Fig. 10). The infectious form is coccoid, motile, and about 1 µm in diameter. In suitable environments long filamentous germ tubes of approximately 0,5 to 1,0 µm diameter are produced, which subsequently branch (Fig. 10). Septation begins at the original spore and extends down the filament with decreasing proximity of the septa. The filament widens, successive multiplanar divisions dissect it into pockets or rows of coccoid elements which can also be demonstrated in microscopical slides of the lesions. Upon moistening the coccoid spores are released, develop flagellae and become motile.

Epidemiology

Dermatophilosis is a true zoonosis, as human infections have been caused by direct contact with infected animals (Pier, 1979). *D. congolensis* is released from infected animals or man when the lesions become wet, transmission is by direct or indirect contact. Vectors may be flies, ticks, thorns, rubbing posts, clippers etc.

Geographical Distribution

Dermatophilosis is found in the tropical to temperate zones of Africa, Australia, South and North America, the Mediterranean Middle East and Europe (Pier, 1979).

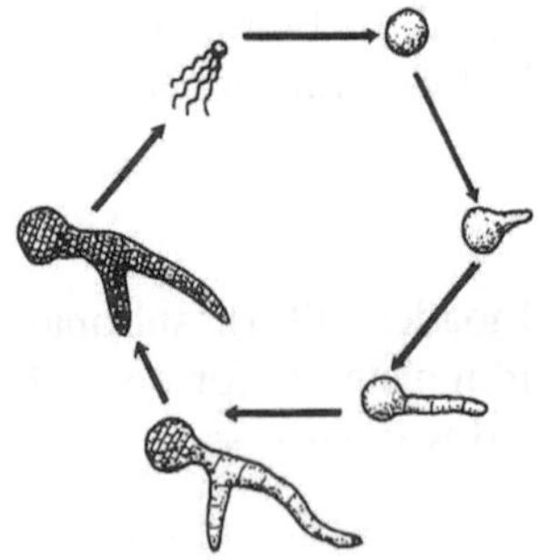

Fig. 10. Life cycle of *Dermatophilus congolensis*

Clinical Symptoms

Nonhuman primates: The lesions appear as small, scaly areas which enlarge to papillomatous wart-like, encrusted lesions in the skin of the head and extremities, or the entire body in severe infections.

Man: The exudative pustulous dermatitis is characterized by discrete cutaneous pustules usually on the hands and forearms.

Pathology

Nonhuman primates: The lesions are limited to the skin, particularly the epidermis. The epidermis is focally thickened due to hyperkeratosis, parakeratosis, acanthosis, necrosis and abscess formation. The underlying dermis shows minor inflammatory congestion and small accumulations of mononuclear leukocytes. Numerous basophilic filaments were present in the parakeratotic and suppurative foci (King et al., 1971; Fox et al., 1973; Migaki and Seibold, 1976).

Treatment

Dermatophilosis can be treated by many antibiotics, most commonly used is dihydrostreptomycin (Pier, 1979).

References

1. Al-Doory, Y.: Systemic mycoses; pp. 224-242 in "Pathology of Simian Primates", Part II (Fiennes, R. N. T-W, ed.) Karger, Basel (1972)
2. Al-Doory, Y.; M. E. Pinkerton; T. E. Vice, and V. Hutchinson: Pulmonary nocardiosis in a vervet monkey. J. Amer. vet. med. Ass. *155*, 1179-1180 (1969)
3. Altman, N. H. and J. D. Small: Actinomycosis in a primate confirmed by fluorescent antibody technics in formalin fixed tissues. Lab. Anim. Sci. *23*, 696-700 (1973)
4. Boncyk, L. H.; B. McCullough; D. D. Grotts, and S. S. Kalter: Localized nocardiosis due to *Nocardia caviae* in a baboon (*Papio cynocephalus*). Lab. Anim. Sci. *25*, 88-91 (1975)
5. Fiennes, R.: Zoonoses of Primates. The Epidemiology and Ecology of Simian Diseases in Relation to Man. Weidenfeld and Nicolson, London (1967)
6. Fox, J. G.; L. H. Campbell; C. Reed, and O. A. Soave: Dermatophilosis (Cutaneous streptothricosis) in owl monkeys. J. Amer. vet. med. Ass. *163*, 642-644 (1973)
7. Fraser, C. E. O. and F. G. Garcia: Cutaneous streptothricosis (dermatophilosis) in a woolly monkey. Primate Zoonoses Surveil. Rep. *5*, 7-8 (1971)
8. Jonas, A. M. and D. S. Wyand: Pulmonary nocardiosis in the rhesus monkey. Path. Vet. *3*, 588-600 (1966)
9. Kessler, M. J. and R. J. Brown: Mycetomas in a squirrel monkey (*Saimiri sciureus*). J. Zoo Anim. Med. *12*, 91-93 (1981)
10. King, N. W.; C. E. O. Fraser; F. G. Garcia; L. A. Wolf, and M. E. Williamson: Cutaneous streptothricosis (Dermatophiliasis) in owl monkeys. Lab. Anim. Sci. *21*, 67-73 (1971)
11. Liebenberg, S. P. and W. E. Giddens, jr.: Disseminated nocardiosis in three macaque monkeys. Lab. Anim. Sci. *35*, 162-166 (1985)
12. McClure, H. M.; J. Chang; W. Kaplan, and J. M. Brown: Pulmonary nocardiosis in an orangutan. J. Amer. vet. med. Ass. *169*, 943-945 (1976)
13. McClure, H. M.; W. Kaplan; W. B. Bonner, and M. E. Keeling: Dermatophilosis in owl monkeys. Sabouraudia *9*, 185-190 (1971)
14. Migaki, G. and H. R. Seibold: Dermatophilosis in a titi monkey (*Callicebus moloch*). Amer. J. Vet. Res. *37*, 1225-1226 (1976)
15. Pier, A. C.: Actinomycetes; pp. 17-30 in "CRC Handbook Series in Zoonoses" (Steele, J. H.,

Series ed.), Section A: Bacterial, Rickettsial, and Mycotic Diseases. Vol. I (Stoenner, H.; W. Kaplan, and M. Torten, Section eds.), CRC Press, Boca Raton (1979)
16. Rhoades, E.R.: Nocardiosis; pp.340–341 in "Infectious Diseases" (Sanford, J.P. and J.P. Luby, eds.), Grune and Stratton, New York (1981)
17. Sakakibara, I.; Y. Sugimoto; H. Minato; M. Takasaka, and S. Honjo: Spontaneous nocardiosis with brain abscess caused by *Nocardia asteroides* in a cynomolgus monkey. J. med. Primatol. *13*, 89–95 (1984)

3.16 Mycobacteriaceae - Tuberculosis Type

Mycobacteria are aerobic, acid- and alcohol fast, nonmotile rods, some of which are classical pathogens, or at least potential pathogens, whilst others are purely saprophytic. Their characteristic acid fastness depends on high molecular weight esterified mycolic acid residues, which are covalently linked to arabino-galactone in the cytoskeleton of the bacteria (Goren and Brennan, 1979; Draper, 1982). This has also therapeutic consequences.

Pathogenic and potentially pathogenic mycobacteria if growing in vitro at all, require special nutrient media, such as Löwenstein-Jensen agar, Ogawa-medium or other enriched media.

3.16.1 Mycobacterium Tuberculosis

M. tuberculosis is the classical agent of human tuberculosis. Human tuberculosis is a world wide disease of man, which is easily transmissible to Old World monkeys and great apes. The tuberculous monkey in turn is a health hazard especially to other monkeys, but retransmission of the infection to man has been reported as well (Anon, 1970; Duffy et al., 1970; Keeling et al., 1971; Hyatt et al., 1973; Mayhall et al., 1981).

Agent
M. tuberculosis is a slender, straight or slightly curved rod of 1,3 to 3,5 μm × 0,2 to 0,6 μm size. It grows eugonic on egg-containing media incubated at 35 °C with small, dry, friable colonies appearing after 2 to 3 weeks. The colonies are rough, warty, granular or bread-crump like and buff colored. After several weeks they increase in size up to 5 to 8 mm diameter. Typical colonies have a "cauliflower"-like center and flat, irregular margins. Colonies are easily detached from the medium's surface, but are difficult to emulsify. The biochemical characteristics of *M. tuberculosis* and of some other slowly growing medically important mycobacteria are given in Table 33, derived from Jenkins et al. (1982).

Mycobacteria possess protein groupings common to many or all members, but also species specific antigenic components (Affronti and Seibert, 1965).

Tubercle bacteria are highly resistant to drying. Cultures maintained at 37 °C have been found both viable and virulent after 12 years when protected from

Table 33. Identification of Slowly Growing Mycobacteria

	Pigment production			Temperature range (°C)			Resistance to														OAA
	Photochromic	Scotochromic	Nonchromic	33	37	42	INH	TCH	HA	PNB	Catalase > 45 mm	Niacin production	Nitrate reduction	Acid phosphatase	Tween 80 hydrolysis	β-galactosidase	Growth in 5% (w/v) NaCl	Urease	Nicotinamidase	Pyrazinamidase	Colonial morphology
M. tuberculosis	–	–	+	+	+	+	–	+	–	–	–	+	+	–	F	+	–	+	+	+	R
M. bovis BCG	–	–	+	+	+	–	–	–	–	–	–	F	–	+	–	–	–	+	–		R
M. africanum	–	–	+	+	+	F	–	–	–	–	–	F	–	–	–	–	–	+	F	F	R
M. bovis	–	–	+	+	+	+	–	–	–	–	–	–	–	–	–	–	–	+	–	–	R
M. ulcerans	–	F	M	M	F	–	M	+	F	–	–	–	–	–	–	–	–	F	F	F	R
M. kansasii	+	–	–	+	+	+	–	+	+	+	+	–	+	+	+	–	–	+	+	–	SmK
M. simiae	M	–	F	+	+	F	F	+	+	+	+	M	–	–	F	F	–	+	–	–	SmS
M. scrofulaceum	–	+	–	+	+	+	+	+	+	+	M	–	–	–	–	–	–	+	+	+	
M. avium	–	–	+	+	+	+	M	+	+	+	–	–	–	–	–	–	–	–	+	+	SmS/ SmT
M. intracellulare	F	–	M	+	+	M	–	+	+	M	F	–	–	–	–	–	–	F	+	+	SmS/ SmT
M. paratuberculosis	–	–	+	+	+	+	+	+	+	–	–	–	–	–	M	–	–	–	+	+	R

Key: + = >85% of strains positive; – = <15% of strains positive; *M*=50 to 85% of strains positive; *F*=15 to 49% of strains positive; Blank spaces=few or no data; *INH*=isoniazid (10 μg ml^{-1}); *TCH*=thiophene-2-carboxylic acid hydrazide (1 μg ml^{-1}); *HA*=hydroxylamine hydrochloride (250 μg ml^{-1}); *PNB*=p-nitrobenzoic acid (500 μg ml^{-1}); *OAA*=oleic acid albumin agar; *R*=Rough; *SmK*=Smooth Kansasii; *SmG*=Smooth Granular; *OC*=Opaque ciliate; *SmS*=Smooth scotochromogenic; *SmT*=Smooth transparent; *Xf*=Xenopi filamentous.

direct sunlight (Chadwick, 1982). UV- or sunlight will kill mycobacteria in culture within 2 hrs, but mycobacteria contained in sputum require an exposure time of 20 to 30 hrs. Mycobacteria are also killed by 60 °C within 30 min.

Epidemiology

Monkeys living in the jungle are considered to be free of tuberculosis (Francis, 1956; Mulder, 1976; Fourie and Kleeberg, 1978; Fourie and Odendahl, 1983). The animals are probably infected on first contact to man or the human environment immediately after capture, due to the high prevalence of tuberculosis in the human population in the countries of origin.

In most of these countries tuberculosis is the most important specific public health problem. In many areas 50 to 80% of adults are infected with active infections at rates of 0,5 to 1% (Kleeberg, 1975), with as many as 5% of tuberculin-negative children converting to positive each year (Bates and Stead, 1981). Especially on the Indian subcontinent the monkeys are held at the hunter's place for some time before being delivered to the exporter's compound or the laboratory. Of 1.100 young *M. mulatta* tested in New Delhi after being delivered by their trappers 2 were positive reactors and 3 were doubtful (Fremming et al., 1955 - cit. from Fiennes, 1972). Other human sources of infection later in captivity may be actively infected caretakers (Cappucchi et al., 1972; Lau et al., 1972). Once the infection is introduced into a monkey colony it may spread bushfire-like depending on the host species involved, and on the conditions of housing and management. Typical examples are 3 outbreaks which respectively involved 28% (Lindsey and Melby, 1966), 27% (Keeling et al., 1969), or 31% (Indzhita et al., 1977) of the animals before brought under control.

Based on the incidence following importation, the disease is common (1–2% incidence) in Asian monkey species, particularly those from the Indian subcontinent. Tuberculosis is less frequent in African species and rare in New World monkeys and prosimians before being infected in captivity (Kaufmann and Anderson, 1978 a, b). Rhesus monkeys are usually considered to be the most susceptible primate species (Habel, 1947; Benson et al., 1955; Keeling et al., 1969; Tribe, 1969; Fredrickson et al., 1971; Schmidt, 1972; Renquist and Whitney, 1978), at least most reported outbreaks of simian tuberculosis have occurred in rhesus monkeys (Fig. 11). Lapin and Yakovleva (1963), however, suggest a crucial influence of the indoor housing conditions rather than a particular susceptibility.

Tuberculosis due to *M. tuberculosis* has been less frequently reported in stumptailed macaques (*M. arctoides*) (Wolf et al., 1967; Hyatt et al., 1973; Indzhita et al., 1977; Rao et al., 1982), pigtailed macaques (*M. nemestrina*) (Doty and Greenstein, 1964; Jentzsch et al., 1969; Lau et al., 1972; Morton and Blakely, 1972), and Assamese macaques (*M. assamensis*) (Rao et al., 1982). Cynomolgus monkeys (*M. fascicularis*) are thought to be relatively resistant (Tribe, 1969) although several cases have been reported (Hamm, 1974; Renquist and Whitney, 1978; Welshman, 1981). In African species most *M. tuberculosis* infections have been reported in baboons (Hamner, 1965; Huber und Maran, 1968; Heywood and Hague, 1969; Heywood et al., 1970; Bartsch et al., 1972; Tribe and Welburn, 1976; Fourie and Owendahl,

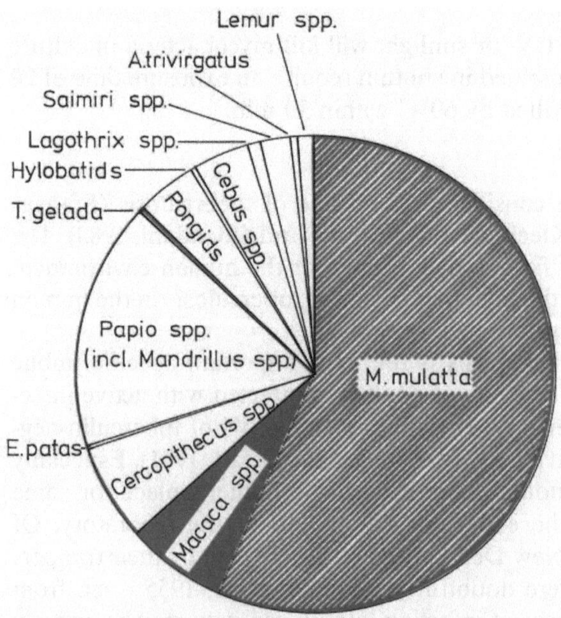

Fig. 11. Incidence of tuberculosis in nonhuman primates (literature review 1960–1977)

1983), with some cases having occurred in *Cercopithecus* spp. (Huber und Maran, 1968; Stones, 1969; Karesseva and Vichkanova, 1970), and *E. patas* (Migaki et al., 1971). The anthropoid apes are somewhat less susceptible, being comparable to the adult man, but nevertheless infections have been reported (Piwowarczyk, 1964; Fineg et al., 1966; Keeling et al., 1971; Zwart et al., 1980).

In New World monkeys, which generally are regarded as highly resistant, most of the cases have occurred in squirrel monkeys *(Saimiri sciureus)* (Fiennes, 1965; Chrisp et al., 1968; Hessler and Moreland, 1968; Huber und Maran, 1968; Duffy et al., 1970; Leathers and Hamm, 1976; Mayhall et al., 1981) occasionally in *Cebus* spp. (Piwowarczyk, 1964; Huber und Maran, 1968; Jentzsch et al., 1969; Leathers and Hamm, 1976) and owl monkeys *(Aotus trivirgatus)* (Snyder et al., 1970).

In the prosimians 23 cases have been reported by Fox (1923 - cit. from Knezevic and McNulty, 1967), 1 case in a *Lemur mongoz* by Knezevic and McNulty, 1967, and 1 case in a ringtailed lemur (Fredrickson et al., 1971).

The high susceptibility of rhesus monkeys and the greater resistance of New World monkeys has also been reflected by experimental results. Rhesus monkeys can be experimentally infected with 1 to 10 viable mycobacteria (Fiennes, 1972), whereas only 1 of 4 *A. trivirgatus* experimentally infected with 40.000 viable *M. tuberculosis* became fatally infected. The other 3 owl monkeys neither showed any clinical symptoms nor did they develop any macroscopical or histological lesions (Bone and Soave, 1970). In man 5 to 200 tubercle bacteria have to be inhaled to establish the infection (Dannenberg, 1978; Youmans, 1979b).

Transmission of *M. tuberculosis* (and *M. bovis*) in man and nonhuman primates is primarily aerogenous, by droplet infection with a droplet size of 1 to 5 microns. In experimentally infected monkeys the shedding of *M. tuberculosis* from lung lesions preceeded the development of a positive tuberculin reaction in the animals (Clark, 1968 - cit. from Kaufmann and Anderson, 1978).

Tuberculosis has also been transmitted via bite wounds (Lindsey and Melby, 1966) or infected tatoo needles (Allen and Kinard, 1958) between monkeys.

Pathogenesis

In aerogenous transmission only fine, suspendable particles up to 10 µm diameter containing 1 to 3 mycobacteria are infectious as they enter the alveolar spaces. Greater and heavier particles which impinge on the nasopharyngeal and bronchial mucosal membranes, are removed by the mucociliary stream into the oropharynx, and are eventually swallowed. In primary pulmonary tuberculosis 1 or more mycobacteria will lodge within an alveolus.

Owing to the type of delayed hypersensitivity to the bacterial antigens either proliferative or more exudative tuberculous lesions are elicited. In the *proliferative* type, typical tubercles are formed intra- or extravascularly. Within the liver lobule they always arise within the sinusoids (Youmans, 1979b). In the tubercles themselves, the mononuclear cells accumulate around the necrotic - caseous center and destroy the small numbers of bacteria released from the center. Epitheloid cells, which are altered mononuclear phagocytes, are common and by fusion of these cells Langhans' giant cells are formed. Argyrophilic reticulum fibers appear between the epitheloid cells, which, in man, are gradually transformed to collagen fibers. In nonhuman primates the macrophages around the necrotic (caseous) center cannot limit the growth of the ingested bacteria. They die and the caseous center enlarges. The mycobacteria ultimately enter the blood stream and are transported to other organs, resulting in generalized miliary tuberculosis. In monkeys also, the collagenization of the reticulum fibers of the tubercles is rather poor or altogether lacking. Lymphogenous dissemination results in early infection of the hilar lymphnodes. The resulting caseation enables the drainage of the bacteria into the vena cava, and thus into the circulation, they then return via pulmonary arteries back into the lung. The enlarging caseous lung lesions may also discharge their contents into the bronchial tree and so into the environment. The solid caseous material has an inhibitory effect on the growth of the tubercle bacteria, whereas after liquefaction, which is common in monkey tuberculosis, the mycobacteria often grow extracellularly in tremendous numbers. The bronchial wall erodes, releasing the bacteria, which overwhelmingly infect other areas of the lungs and other organs. If the amount of bacterial antigen is great in a host with a high level of hypersensitivity the *exudative* tuberculous bronchopneumonia results. In exudative lesions mononuclear cells accumulate in the alveolar spaces as a rather loose exudate, resulting in a localized bronchopneumonia with rapid caseation (Dannenberg, 1978).

Clinical Symptoms

Nonhuman primates: In Old World monkeys clinical signs are minimal or absent until the disease is far advanced (Mulder, 1976). Most strikingly observed are weakness with reluctance to move or even flaccid paralysis, loss of appetite, sitting in a crouched position and depression. Other monkeys, on the contrary, may be more aggressive (Heywood and Hague, 1969). The skin may become grey, powdery and folded, while the hair coat may become disheveled and fall out. The tuberculous monkey tires easily, it has dyspnea, loses weight, it may develop a chronic, dry, soft, intermittent cough, the subcutaneous lymphnodes may enlarge and cutaneous abscesses may develop (Innes, 1963; Vice et al., 1968; Tribe, 1969; Lau et al., 1972; Hyatt et al., 1973; Renquist and Whitney, 1978). Auscultation may reveal vesicular or bronchial sounds or even rales (Trolldenier, 1964). Most monkeys die within 6 to 9 months after tuberculin conversion (Habel, 1947).

In fulminating cases the hair coat often remains glossy, the animal is well nourished and may look perfectly healthy until it is found dead (Clarkson and Smith, 1969; Wolf and Seibold, 1971). On the other hand, as suggested by Lapin and Yakovleva (1963) even rhesus monkeys might be able to arrest the infection without any clinical signs (Houser et al., 1982) although that certainly is not the rule.

The symptoms of extrapulmonary tuberculosis are determined by the organs involved.

Primary and secondary cutaneous tuberculosis are marked by ulceration and enlargement of the subcutaneous lymphnodes (Allen and Kinard, 1958; Lindsey and Melby, 1966). Vertebral tuberculosis (Pott‘s disease) results in paraplegia (Tryphonas and Wong, 1982) or kyphosis of the vertebral column and paralysis (Fox et al., 1974). Cerebral tuberculosis causes epileptiform seizures, visual impairment and mydriasis (Machotka et al., 1975).

Man: The incubation period is 4 to 6 weeks (Bates and Stead, 1981). Usually neither radiographic changes nor any symptoms are associated with the initial stages of the infection which may therefore remain undetected for life unless discovered through tuberculin skin testing. Approximately 5% of initially infected people develop clinical tuberculosis within 2 years, especially when the infection occurs early in life.

If tubercle bacteria, after having remained dormant for an unknown period of time, begin to multiply rapidly due to other influences, then chronic tuberculosis results. The onset of active disease is usually insidious and may involve major portions of one or more lung lobes. The earliest symptoms are general malaise, night sweats due to unrecognized fever, weight loss, and cough, sometimes associated with production of a green or yellow sputum which may be streaked with blood. Chest pain may be present as well as headaches, depression and generalized intermittent abdominal pain. Uncommonly, an abrupt onset with high fever and acute tuberculous pneumonia may be seen after rupture of a caseous lymph node into a lobar bronchus. Anemia may develop in the advanced stages. Pulmonary function impairment, causing shortness of breath, is noted in patients with extensive acute tuberculous pneumonia. Haemoptysis is uncommon but may occasionally be fatal. Acute tuberculous pneumonia in patients with depressed resistance is characterized by high fever, chills, shortness of breath, weakness, and respiratory complaints of several weeks duration (Bates and Stead, 1981).

Diagnosis

In nonhuman primates the *in vivo* diagnosis of tuberculosis depends primarily on tuberculin tests, which were first introduced for use in monkeys by Schröder (1938). The tuberculin test in man and animals relies on the development of delayed hypersensitivity against a 5800 dalton molecular weight peptide contained in the mycobacterial cell (Nagai et al., 1980). The delayed hypersensitivity to tuberculin develops in nonhuman primates in about 3 to 4 weeks after infection, compared to 3 to 8 weeks in human beings (Clarke, 1968). The protein fraction contained in both the tubercle bacteria and the tuberculin is recognized by sensitized T-lymphocytes passing through the injection site. These cells release lymphokines which cause local vascular effluents and thus edema, and the accumulation of mononuclear cells, mainly sensitized lymphocytes, monocytes, and macrophages (Muscoplat, 1978; Dannenberg, 1978). The accurate response depends on a certain number of tubercle bacteria present in the host, on a sufficient number of circulating T-cells (Ott, 1979, 1980), and on an adequate amount of specific antigen contained in the tuberculin employed.

According to David and Selin (1976) a threshold amount of killed tubercle bacteria (10-4 mg/guinea pig) is required to induce hypersensitivity detectable by the tuberculin reaction. There is also an upper limit, because larger amounts, exceeding 1,0 mg/guinea pig, interfered with the immunological response in the same experiments.

Monkeys require 1.000 to 10.000 times more tuberculin to elicit the response than is necessary for Mantoux-testing in man. This may be due to a lower number of circulating sensitized T-lymphocytes, or to hypersensitivity to antigens other than those in the tuberculin (Ott, 1979, 1980).

Either Koch's Old Tuberculin (OT) or purified protein derivatives (PPD) are used. Old tuberculins are the unfractionated heated 10 : 1 concentrates of the culture filtrates of mycobacteria grown on liquid media. PPD 's are precipitated fractions of culture filtrates (Angus, 1978). Contract (USDA) OT and JenSal OT are made from *M. tuberculosis*, others are produced either from *M. tuberculosis* or *M. bovis*.

For use in nonhuman primates OT of higher potency is usually preferred over PPD because of its greater reactivity (Smith and Wolochow, 1973; McLaughlin and Marrs, 1978). Others (Acred et al., 1972; Fox et al., 1982) preferred PPD or mammalian tuberculin over OT in order to avoid false positive reactions. At least 1.500 tuberculin units (TU) or 15 mg OT are recommended in nonhuman primates (McLaughlin and Marrs, 1978; Ott, 1979, 1980), mostly used are 25 mg OT (Fox and Drake, 1974; Renquist and Whitney, 1978). The dosage in man is 5 TU PPD (Hyatt et al., 1973).

The tuberculin (0,1 ml) is injected intradermally - in monkeys usually into the upper eyelid. Care has to be taken to deposit the tuberculin close to the lid margin halfway between the medial and lateral canthi, and as superficially as possible to produce a well defined skin bleb. For proper dilution of the tuberculin distilled water is recommended over the use of buffered diluent, which has a significant tendency for false positive reactions (Fox and Drake, 1974).

Another site for intradermal tuberculin tests is the paraumbilical abdominal skin (Clarke, 1968), which, according to Schmidt (1968) has the advantages of easier intradermal application and, in low grade reactions, better readability. In questionable reactions the simultaneous use of both sites in repeat tests may be advisable.

In either site the tuberculin test should be read after 48 and 72 hs, in re-tests also after 2 to 8 hs, because anergic animals may demonstrate a "flash" reaction which would be missed at 24 hs (Sinski and Panier, 1966; Martin, 1978). Positive reactions in the palpebral tests are marked by local reddening and edematous swelling of the eyelid, and in heavy reactions even by complete closure of the eye, with necrosis and ulceration of the skin (Hartley, 1969).

During quarantine the test should be repeated at least two times at two weeks intervals and only if all members of the group or batch are clearly negative, can the group be released. Otherwise, the test has to be repeated until all remaining members of the group are negative in two consecutive tests. The tuberculin test, is, however, not totally reliable due to a certain number of false positive and false negative reactions.

False positive reactions can be caused by buffer diluents or impurities of the tuberculin (especially Old Tuberculins), by trauma, by previous BCG-vaccination or by cross reactions with atypical or saprophytic mycobacteria since the mycobacteria share some antigens with each other. Questionable tuberculin reactions should therefore be repeated and avian tuberculin be included (Ott, 1979, 1980). False negative results to tuberculin testing can be caused by

a) overwhelming pulmonary tuberculosis leading to anergic conditions;
b) testing early in infection before development of delayed hypersensitivity;
c) systemic fungal disease (Ganaway, 1974);
d) measles virus infections (Potkay et al., 1966) or injection of polio-, measles- or yellow fever virus vaccines (Ott, 1979);

e) chemoprophylaxis with INH, which suppresses a positive tuberculin reaction within 10 days after initiation of therapy (Schmidt, 1956; Burkholder et al., 1967; Gibson et al., 1971);
f) administration of immunosuppressive drugs for other reasons (Fox et al., 1978);
g) technical errors or failures such as suboptimal tuberculin doses, leaking out of the reagent, exposure of the tuberculin to light;
h) localized desensitization in repeat tests, if the injection site has not been changed.

Because of the false reactions of the tuberculin test, the diagnostic support by other methods is advisable.

X-ray tests may be of some help in certain stages of the disease in monkeys. Radiographs should be taken in lateral and dorsoventral projection. For dorsoventral radiographs the elbows and anterior extremities should be stretched in order to lift the scapulae off the thorax (Hime, 1969). Thoracic radiographs in nonhuman primates are generally more difficult to evaluate regarding tuberculosis as compared to man since tuberculous lesions in monkeys rarely calcify (Clarke, 1968). The usual miliary tubercle thus produces only small, discrete foci (Hime, 1969) which are easily missed, limiting the X-ray technique to the more advanced stages when the greatly enlarged mediastinal or hilar lymph nodes are visible as shades of greater radioopacity or cause elevation of the trachea (Lau et al., 1972; Ott, 1980). Tuberculous pneumonia can not be differentiated from other lobar pneumonias unless associated with gross cavitation which is uncommon (Hime, 1969).

Serotests demonstrate either cell mediated or humoral immune responses. The lymphocyte transformation test (LT), although depending on the same phenomena as the skin test is reported to be more sensitive and more objective than the former one (Chaparas et al., 1970, 1975; Condoulis et al., 1974). The LT-test becomes positive earlier in the disease and should remain positive even in anergic animals (Ott, 1980). The assay is based on the demonstration of increased tritiated thymidine incorporation by lymphocytes from tuberculin-sensitive animals. The limitations of the tests are its technical and safety requirements, which are beyond the capacities of many laboratories.

Humoral antibodies can be visualized by Soluble Antigen Fluorescent Antibody (SAFA) - or by enzyme immunoassay (ELISA) - tests, but humoral immunity is less sensitive after vaccination and/or challenge (Ott, 1979).

The SAFA-test, developed by Toussaint et al. (1969) demonstrates circulating humoral antibodies in the sera of infected animals and man, and, it has at least in monkeys proved to be more sensitive and more specific than the customary tuberculin tests. It becomes reactive 17 to 74 days before conversion of the skin sensitivity (Fife et al., 1970). Most experimentally infected monkeys become reactive 42 days after exposure whereas in the same animals the skin tests converted 42 to 134 days after exposure (Affronti et al., 1973). The SAFA-technique is an indirect fluorescent antibody procedure with fixation of soluble antigen on an artificial matrix. The antibodies are fluorescein-labelled and the tests are read on a fluorometer in the reflected beam. In man the usefulness of the SAFA test is questionable since it gives false negative results in extrapulmonary, and in disseminated tuberculosis (Bhardwai et al., 1981).

Another, but entirely nonspecific supporting test is the erythrocyte sedimentation rate (ESR) which is increased in tuberculosis (Welshman, 1981), but it has to be kept in mind that an increased ESR results in many chronic diseases. It may be useful in advanced anergic animals (Tribe and Welburn, 1976), but is unreliable in detecting early infections (Kaufmann and Anderson, 1978b).

Pathology

Nonhuman primates: Scott and Beattie (1928) distinguished lesions of:

1. pulmonary origin without alimentary infection
   a) early lesions;
   b) advanced lesions;
   c) advanced lesions with the beginning of dissemination by the blood stream;
   d) a generalized infection spread by the blood-vascular route;
2. pulmonary origin with secondary alimentary infection from swallowing sputum
   a) infection of lungs and intestines without generalisation;
   b) infection of lungs and intestines with generalisation by the blood stream;
3. alimentary origin
   a) early or primary intestinal lesions;
   b) later stages of such lesions;
4. alimentary origin with secondary pulmonary infection;
5. anomalous cases.

Basically the lesions of primary tuberculosis in nonhuman primates are firm, yellowish-white or greyish nodules ranging in size from pin point dimensions to several millimeters in diameter. The lung is without any doubt the most commonly affected organ (Benson et al., 1955; Francis, 1956; Wolf et al., 1967; Heywood and Hague, 1969; Vice et al., 1969; Snyder et al., 1970; Lau et al., 1972). The lung tubercles are filled with caseous material and are surrounded by areas of consolidation, which quite often progress to coalescing large, caseous tubercles or to tuberculous pneumonia affecting one or more lobes (Huber und Maran, 1964; Vice et al., 1969; Snyder et al., 1970; Lau et al., 1972; Renquist and Whitney, 1978). Microscopically the caseous centers are surrounded by epitheloid cells with usually only a few Langhans' giant cells contained within the lung tubercles (Hessler and Moreland, 1968; Renquist and Whitney, 1978). The mononuclear cells of Old World monkeys around the caseous primary focus cannot stop the growth of the ingested bacteria, they therefore die and the area of caseation enlarges (Dannenberg, 1978). Since around lung tubercles of nonhuman primates a fibrous capsule is usually not formed, the pulmonary primary lesions continue to enlarge and by confluence form large complexes of tubercles. Acid-fast bacteria are usually easily demonstrated in appreciable numbers in histological slides stained acid fast (Snyder et al., 1970; Lau et al., 1972). Calcification is usually rare or lacking as is the fibrous encapsulation (Heywood and Hague, 1969; Vice et al., 1969; Lau et al., 1972; Hamm, 1974).

Lung tuberculosis in monkeys is associated with similar lesions in the enlarged hilar or mediastinal lymph nodes as primary complex (Vice et al., 1969; Tribe, 1976). The lymph nodes may even appear as a large abscess filled with a viscous, yellowish fluid due to liquefaction (Lau et al., 1972). Once a primary focus is established spread is remarkably rapid, akin to the fulminating infections seen in children which rapidly result in miliary tuberculosis (Innes, 1963; Renquist and Whitney, 1978), with frequent involvement of the pleura and occasional extension to the spinal cord (Martin, 1968; Fox et al., 1974).

Cutaneous tuberculosis is often indicated by an ulcerating peripheral lymph node (Hamm, 1974). In the advanced stages of cutaneous tuberculosis, ulceration develops with the cavity being filled with white, creamy debris (Allen and Kinard, 1958; Lindsey and Melby, 1966).

Infections originating in the gastrointestinal tract, more commonly seen in *M. avium* infections, are characterized by the presence of large tubercles in the mesenteric lymph nodes (Renquist and Whitney, 1978). With progression of the infection the spleen, kidneys, liver, inguinal and axillary lymph nodes are often affected by miliary tuberculosis. In rare cases the myocardium is involved (Nair and Ray, 1955; Bartsch et al., 1972).

In tuberculous spondylitis (Potts' disease) portions of the spongy bone of the affected vertebrae undergo liquefactive necrosis with numerous Langhans' giant cells being present at the periphery of the tuberculous granulomas and bone sequestra. The vertebral lesions extend into the dura and compress the spinal cord, but do not invade the nervous tissues (Innes, 1963; Fox et al., 1974; Tryphonas and Wong, 1982). The lesions sometimes cause vertebral collapse (Martin et al., 1968; Fox et al., 1974).

Cerebral tuberculosis, following dissemination from the lung, is characterized by multiple yellow grey necrotic granulomas up to 1,0 cm in diameter with abundant Langhans giant cells, which are present in both the superficial and deep cerebral parenchymas (Machotka et al., 1975).

Great apes and New World monkeys are usually more resistant to tuberculosis, with the lesions being more of a proliferative type with less extensive caseation (Dannenberg, 1978).

The gross lesions of tuberculosis in nonhuman primates may be confused with lung mite lesions, pseudotuberculosis, nocardiosis or deep mycoses.

Lung mite lesions consist of a thick walled capsule with scattered inflammatory cells of several types, whilst the parasites or remnants of the mites are often visible in the center (Fig. 12). The origin of the lesions from terminal bronchi can usually be recognized.

Pseudotuberculomas are small abscesses consisting of a necrotic center without caseation, surrounded and infiltrated by large numbers of neutrophilic granulocytes.

Nocardiosis and deep mycoses are uncommon in nonhuman primates. They usually produce abscesses in which, in the case of mycoses, the fungi can be demonstrated by the appropriate staining methods (Al-Doory et al., 1969).

Man: In proliferative lung tuberculosis, epitheloid cellular granulations of 1 to 2 mm diameter are characteristic. The granuloma consists of central caseous necrosis (often present), surrounded by epitheloid cells and Langhans' giant cells and a peripheral wall of lymphocytes. The granulomas may heal, hyalinize or, in progressive disease undergo complete caseation.

Exudative tuberculosis in a sensibilized organism in its earliest stage may result in a sero-fibrinous exudate, which can be reabsorbed. If no reabsorption occurs,

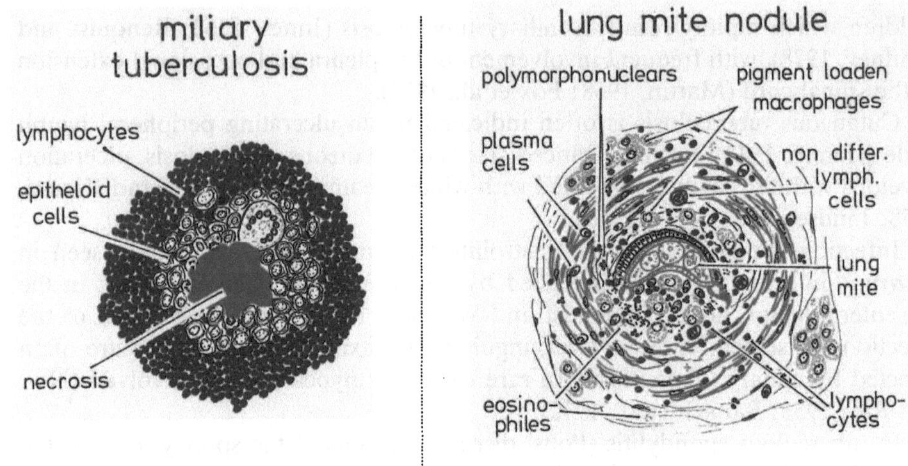

Fig. 12. Comparison of miliary tuberculosis and lung mite nodules

caseous necrosis follows. Caseation is a fine granular or homogenous coagulation of the exudate associated with lysis of all tissue components except the elastic fibers. In exudative tuberculomas the alveolar structures of the pulmonary framework remains preserved for quite some time, whereas in productive tubercles it is destroyed. Exudative tuberculosis tends to lead to rapid expansion resulting in caseous pneumonia, which can infiltrate entire lobes. If the exudative inflammation stops or regresses, caseous, encapsulated foci are formed. These consist of caseous centers surrounded by epitheloid cellular granulations containing Langhans' giant cells in the inner "specific" part of the capsule. The outer, nonspecific part of the capsule consists of collagen fibers formed by proliferating fibrocytes in the lymphocytic margin. If the tuberculous process comes to a halt the epitheloid inner part undergoes hyalinization, with the caseous necrosis in the center drying out and calcifying.

Calcification is common in primary tuberculosis and in childhood, but is rare in post primary stages and in older patients. Progression of the encapsulated foci in seemingly resting tuberculosis is heralded by a granular dissolution of the caseous necrosis leading to liquefaction. The liquefaction is often associated with rupture of the capsule and tremendous multiplication of the mycobacteria. If the liquefied necrosis drains into a bronchus, tuberculous cavities and bronchogenous spread of the mycobacteria (phthisis) result.

In primary infections usually a single exudative primary focus of pea to walnut size (Ghon's focus) results at the port of entry. The mycobacteria are soon transported from this site to the regional bronchopulmonary lymph nodes, where similar tubercles develop. In comparison in postprimary lung infections lymphonodular tubercles are lacking.

If the primary infection progresses beyond the infection of the bronchopulmonary (hilar) lymph nodes, the tubercle bacteria drain via the venous system into the circulation, resulting in an early lympho-haematogenous generalisation. A similar generalisation can also follow after a long post-primary interval as late generalisation, again usually originating from the hilar lymph nodes.

Haematogenous spread can result in miliary tuberculosis of virtually any organ, most commonly of the kidneys, lung, liver, and spleen, with great numbers of small, grey, firm nodules appearing in the affected organs (Giese, 1974).

Treatment

Tuberculosis in man can be treated by a number of synthetic compounds or antibiotics (Doub, 1979), whereas tuberculous nonhuman primates normally should be destroyed. Only in certain instances (extremely valuable breeder, long lasting experiments etc.) and only under strict quarantine, can treatment of tuberculous monkeys or apes be considered, using the same drugs as in man.

I. P-aminosalicylic acid (PAS): acting presumably by interference with the salicylate and benzoate utilization by virulent mycobacteria.

It is of only limited usefulness, particularly in animals, as it is frequently refused because of its extremely vile taste and the large quantities required (12 to 16 gr daily in man) which cause dyspepsia and diarrhea.

II. Thiacetazone: Thiacetazone suppresses the growth of *M. tuberculosis* by an unknown action. Its therapeutical usefulness is questionable, whilst its toxicity is moderate.

III. Pyrazinamide: presumably affecting the nicotinamide or nicotinic acid metabolism as a cofactor in dehydrogenases.

IV. Isoniazid (Isonicotinic acid hydrochloride - INH): Several different modes of action of INH have been proposed, the primary one being the inhibition of mycolic acid synthesis. It is neurotoxic due to its interference of pyridoxine metabolism, and is also hepatotoxic (Hinshaw, 1979). The pyridoxine-antagonism also results in suppression of immune mechanisms, notably the tuberculin reaction (Axelrod and Trakatellis, 1964; Burkholder et al., 1967; Gibson et al., 1971). Other disadvantages are its inactivation in stored, moist food and the development of INH-resistant mycobacteria strains (Anon, 1966).

V. Ethionamide: Ethionamide is bactericidal and causes loss of acid fastness also by inhibition of mycolic acid synthesis. Its use is limited by a distressing anorexia, metallic taste, nausea, vomiting, neuro- and liver toxicity (Hinshaw, 1979).

VI. Ethambutol: It is primarily bacteriostatic or weakly bactericidal in an unknown way. It is readily absorbed from the gastrointestinal tract, but is also neurotoxic especially to the optic nerve (Hinshaw, 1979).

VII. Streptomycin and dihydrostreptomycin probably act by inhibition of the initiation of polypeptide synthesis in the ribosome after binding to the ribosome (Hinshaw, 1979). Its drawbacks are i.e. acquired resistance in some strains of *M. tuberculosis* or its neurotoxicity especially towards the eighth cranial nerve causing impairment of hearing.

IIX. Viomycin: Viomycin is a cyclic peptide antibiotic. It inhibits protein synthesis by binding to ribosomes as do aminoglycosides. Its toxic action is also similar to that of streptomycin (Hinshaw, 1979).

IX. Cycloserine: By its structural resemblance to D-alanine cycloserine, it is an inhibitor of enzyme reactions involving D-alanine. It is neurotoxic to the motor cortex and frontal lobe centers, thus causing grand mal seizures (Hinshaw, 1979).

X. Kanamycin: The mode of action is probably similar to that of streptomycin. It is also ototoxic, has curare-like action and hypersensitivity reactions may be induced (Hinshaw, 1979).

XI. Rifampin: It inactivates the DNA-dependent RNA polymerase and thus inhibits the RNA-synthesis. Hepatitis similar to that induced by INH is an adverse side-effect of the drug (Hinshaw, 1979).

For the treatment of tuberculous nonhuman primates, INH is usually given in a dosage of 40,6 to 50 mg/kg body weight over several months (Heymann, 1959; Doty and Greenstein, 1964; Fineg et al., 1966; Allgood and Price, 1972). Burkholder et al. (1967) found viable *M. tuberculosis* still after 10 weeks treatment with INH and Gibson et al. (1971) demonstrated typical tubercles containing acid fast bacteria after 3 1/2 months treatment with 20 mg/kg INH. Schmidt (1972) suggests the existance of two populations of tubercle bacteria, one being readily destroyed within a matter of a few weeks INH, and a minor one that can survive for months.

Streptomycin has also occasionally been used in nonhuman primates. In combination with other drugs it was reported as a successful treatment in *M. arctoides* (INDZHIITA et al., 1977; Ward et al., 1985).

BCG vaccination, known in children, is also used in nonhuman primates especially in European zoo collections (Klöppel, 1964; Dannenbaum, 1971). It can be given intracutaneously, intravenously or by aerosol. The recommended doses are 7.3 × 10 7 vu in intracutaneous vaccination (Good and Mc Carroll, 1978) 1.6 × 10 6 vu in aerosols (Good and McCarroll, 1978) 1 mg in intravenous application (Barclay et al., 1970). In intracutaneous vaccination, the amount of vaccine given should be 0,1 ml in most cercopithecids and 0,05 ml in marmosets and tamarins (Klöppel, 1964). Intravenous or aerogenous vaccination induced a much greater degree of protection than intracutaneous vaccination (Barclay et al., 1973).

A major drawback of its usefulness is the induction of tuberculin hypersensitivity, eliminating the further diagnostic use of the tuberculin test (Kaufmann and Anderson, 1978b). The tuberculin hypersensitivity is much lower after aerogenous or intravenous vaccination than after intracutaneous application (Barclay et al., 1973). Another problem is the induction of sterile abscesses in some BCG batches. Finally, the resistance achieved by BCG might be overwhelmed by the massive numbers of bacteria shed by sick animals (Renquist and Whitney, 1978).

3.16.2 Mycobacterium Bovis

M. bovis is an important pathogen for human beings in certain regions, producing many forms of tuberculosis (Konyha et al., 1980); although after World War II its general importance has declined parallel to the eradication programs in cattle.

In nonhuman primates it can cause severe losses similar to *M. tuberculosis* (Francis, 1958; Fiennes, 1966; Fölsch, 1971; Bartholomew et al., 1972; Renner and Bartholomew, 1974; Kaufmann et al., 1975; Thoen et al., 1977; McLaughlin, 1978; Silberman, 1980; Thoen and Himes, 1980; Zumpe et al., 1980; Himes et al., 1982; Wilson et al., 1984; Tarara et al., 1985). It is assumed that up to 10% of simian tuberculosis is caused by *M. bovis* (McLaughlin, 1978). Other authors have computed 36% of simian tuberculosis as caused by *M. bovis* (Thoen and Himes, 1980).

Agent

M. bovis morphologically is usually plumper and shorter than *M. tuberculosis*. Its primary isolation is more difficult and often requires the addition of sodium pyruvate (4,5 g/L^{-1}) to the medium (Jenkins et al., 1982). Its growth is slow and dysgonic, forming small, colorless, smooth, flat, whitish colonies on Löwenstein-Jensen-medium (Chadwick, 1982). *M. bovis* will not grow on media containing more than 1% glycerol. Its other distinguishing features are given in the table in the preceding chapter.

Epidemiology

The epidemiology of *M. bovis* with respect to man and nonhuman primates is very similar to that of *M. tuberculosis*. Both man and nonhuman primates are primarily infected by droplet infection, with rhesus monkeys usually acquiring the infection in their country of origin. Also two wild olive baboons *(P. cynocephalus anubis)* from the Masai Mara Game Reserve in Kenya suffered from *M. bovis*-tuberculosis acquired as natural infections (Tarara et al., 1985). The frequency of bovine tuberculosis in cattle roaming free in towns and villages in India is described as "high" (Kleeberg, 1975). The danger of *M. bovis*-infected monkeys to other nonhuman primates is identical to that of *M. tuberculosis*-infected monkeys, whilst the risk to human health might even be greater. The U.S. Public Health Service has reported that the conversion rate of tuberculosis tests is 27 times higher for laboratory workers exposed to monkeys than for the general population (Konyha et al., 1980).

In the past, human infections also took a digestive route by consumption of raw dairy products.

Reversely *M. bovis* infected human beings may be a danger to animal health, as demonstrated in cattle in Germany (Schliesser, 1974) and Switzerland (own experiences). *M. bovis* frequently induces renal tuberculosis in man with silent urinary shedding of the bacteria over considerable periods of time.

Clinical Symptoms

Nonhuman primates: Evidence suggest a slower progression of *M. bovis* infections in nonhuman primates as compared to simian *M. tuberculosis* infections (Kaufmann and Anderson, 1978; McLaughlin, 1978; Zumpe et al., 1980). Most simian *M. bovis*-infections were clinically silent and incidentally discovered by routine tuberculin tests (Renner et al., 1972; Renner and Bartholomew, 1974; Muscoplat et al., 1975; Thoen et al., 1977; Silberman, 1980). Other affected animals suffered from pneumonia-like symptoms (Fiennes, 1966; Zumpe et al., 1980; Wilson et al., 1984), whilst in one case conjunctivitis, episcleritis and cloudiness of one eye were the only visible symptoms (West et al., 1981).

Man: *M. bovis* can cause progressive lung tuberculosis indistinguishable from the disease caused by *M. tuberculosis*. Extensive lung lesions are usually accompanied by dyspnea and coughing with associated fever, anorexia, and nausea.

The symptoms of extrapulmonary *M. bovis* infections depend on the organs involved. The most commonly affected extrapulmonary organ is the kidney (Eknoyan and Olivero, 1981). Renal tuberculosis, in general, remains unrecognized for prolonged periods, its presence may be suspected from the finding of sterile pyuria in an acid urine (Eknoyan and Olivero, 1981).

Pathology

Nonhuman primates: The lesions of bovine tuberculosis in nonhuman primates are essentially very similar to those caused by *M. tuberculosis* with the primary complex located in the respiratory system. *M. bovis* infections have, however, a more distinctive tendency to generalize as compared to *M. tuberculosis* infections, with the spleen, liver and mesenteric lymph nodes being the most commonly affected extrathoracic organs (Fiennes, 1966; Renner and Bartholomew, 1974; Thoen et al., 1977; Costallat et al., 1978; McLaughlin, 1978; Renner et al., 1979; Silberman, 1980; Zumpe et al., 1980; West et al., 1981).

Man: Most primary complexes of bovine tuberculosis are located in the lung and are essentially of the same structures as *M. tuberculosis* infections. In children the primary focus may be located in the pharynx, causing *lymphadenitis tuberculosa cervicola* and scrofuloderma in the cervical region.

The most common extrapulmonary sites of tubercular infections in man are the kidneys. The lesions begin as miliary tuberculosis with the initial localization in the glomeruli and shedding of the tubercle bacteria with the urine. Once within the tubules, further dissemination of the organisms to the renal medulla and pyramids can take place with further progression to caseation and cavitation, predominantly seen in the medulla (Eknoyan and Olivero, 1981). These lead again to discharge of tubercle bacteria in the urine (Asscher, 1980).

Treatment

For treatment see *M. tuberculosis.*

3.16.3 Mycobacterium Africanum

M. africanum, first isolated from a tuberculous patient in Senegal (Astets et al., 1968) is considered to be the causative agent of human tuberculosis in Africa (Thorel, 1980). It is estimated that 60% of tuberculosis patients in the Cameroons are infected with *M. africanum* rather than classical *M. tuberculosis* (McClure and Silcox, 1970). *M. africanum* has also been isolated from 3 chimpanzees and one cercopithecid, all of which but one chimpanzee originating from African countries (Thorel, 1980). The strain isolated from a rhesus monkey (McClure and Silcox, 1970) subsequently has been identified as an atypical *M. bovis* (Silcox and McClure, 1971).

Agent

M. africanum has properties intermediate between *M. bovis* and *M. tuberculosis.* Whether it is a separate species or a subspecies of *M. bovis* remains as yet unsettled (Goodfellow and Wayne, 1982). Colonies grown 3 to 4 weeks on egg medium are flat, dull and rough. Growth in Lebek agar goes 15 mm deep below the surface (Runyon et al., 1975; Thorel, 1980). It differs from *M. bovis* by its susceptibility to

50 mg/ml pyrizinamide. *M. africanum* is virulent for guinea pigs but not rabbits (Sommers, 1979).

Clinical Symptoms

Nonhuman primates: Clinical symptoms other than positive tuberculin tests have not been reported (Thorel, 1980).
Man: The symptoms are those of typical tuberculosis.

Pathology

Nonhuman primates: The lesions were described as broncho-pneumonia; hypertrophy of lymph nodes and spleen, and as tubercles in lungs and kidneys.

3.16.4 Mycobacterium Avium - Intracellulare Complex

M. avium is the classical causative agent of tuberculosis in fowl. *M. intracellulare*, originally referred to as "Battey-bacillus" was first isolated from patients at the Battey State Hospital in Rome/Georgia. Both species have been united into a common group. They are pathogenic in man (Engbaek, 1964; Kubin et al., 1966a, b; Chapman, 1977) and nonhuman primates (Kronberger, 1960; Sedgwick et al., 1970; Smith et al., 1973; Weiszfeiler and Karczag, 1973; Sesline et al., 1975; Kalter et al., 1978; Sesline, 1978; Fleischman et al., 1982; Holmberg et al., 1982; Lowenstine et al., 1983).

Agents

M. avium and *M. intracellulare* are closely related members of Runyon's group III nonphotochromogenic mycobacteria. The similarity of both bacteria and the lack of easily determined taxonomic characteristics to distinguish between them led to the formation of a common group *M. avium-intracellulare* complex. The complex contains at least 20 serotypes and also *M. xenopis*, a species not isolated from nonhuman primates so far (Goodfellow and Wayne, 1982).

Group III nonchromogenic mycobacteria are slow-growing, non pigmented or only faintly coloured without influence by light. The colonies of *M. avium-intracellulare* are thin, transparent, smooth, although particularly *M. avium* has a tendency to grow in rough cultures.

The metabolic properties of *M. avium-intracellulare* and of other mycobateria are summarized in table 33, derived from Runyon et al. (1974).

Epidemiology

Because of the similarity of *M. avium* strains highly pathogenic to birds with *M. intracellulare*, which are moderately pathogenic to man and with other strains that seem nonpathogenic for any species, accurate epidemiology is poorly developed (Chapman, 1979).

Reservoir hosts of avian tuberculosis are avian species and possibly pigs, but the microorganisms can also contaminate the environment. Environmental contamination seems to be even more common with *M. intracellulare*. The mycobacte-

Table 34. *M. avium* - intracellulare Infections in Nonhuman Primates

Host species	Agent	References
M. mulatta	M. avium	Kronberger, 1960
M. nemestrina	M. avium	Sedgwick et al., 1970
M. mulatta M. nemestrina M. speciosa M. fascicularis	M. avium serotype 1 M. intracellulare serotype Davis	Smith et al., 1973
M. mulatta	M. avium serotype 1, 4, 8, 18	Sesline et al., 1975
Papio cynocephalus Pan troglodytes	M. intracellulare	Kalter et al., 1978
M. mulatta	M. intracellulare serotype 10	Fleischmann et al., 1982
M. mulatta	M. avium - intracellulare serotypes 1, 2, 4, 8, 18	Holmberg et al., 1982
M. arctoides	M. avium - intracellulare	Lowenstine et al., 1983

ria are excreted in large numbers with the feces of sick animals. Infection occurs by ingestion or inhalation (Chapman, 1979).

As fas as nonhuman primates are concerned, the great majority of cases have been reported in Asian macaques, simulating the situation in *M. tuberculosis* infections (Table 34).

Clinical Symptoms

Nonhuman primates: The symptoms of avian tuberculosis in nonhuman primates are those of a chronic enteric infection with continuous or intermittent diarrhea unresponsive to therapy, weight loss, debilitation, sometimes with peripheral lymphadenopathy and splenomegaly (Smith et al., 1973; Sesline et al., 1975; Sesline, 1978; Fleischman et al., 1982; Lowenstine et al., 1983). Pulmonary signs, such as dyspnea are rarely observed.

Man: *M. avium* infection in children produces cervical lymphadenitis (Dragsted, 1949; Grigelova et al., 1967) or pneumonia (Feldman et al., 1949). The infection, although usually of a benign nature, can cause fatalities in children (Engbaek, 1964).

In adults the symptoms of avian tuberculosis are generally mild, often unremarkable as a slight temperature elevation with few physical findings which are frequently clouded by the presence of other pulmonary diseases (Chapman, 1977). Cervical lymphadenopathy of adults, attributed to *M. avium* has been described in Europe, especially in Germany (Kubin et al., 1966).

Pathology

Nonhuman primates: Avian tuberculosis of nonhuman primates is primarily an intestinal infection. Gross lesions occur particularly in the lower small intestine and the colon. They are characterized by irregular thickening and yellowish-white coloration of the intestinal mucosa due to a diffuse infiltration with large, foamy, eosinophilic epitheloid cells or macrophages. Plasma cells and lymphocytes may occasionally be associated with the cellular infiltrations but do not form tubercle-like granulomas. The chronic inflammatory response extends focally into all layers of the intestinal wall.

The intestinal lesions are sometimes accompanied by prominent lymphatic channels in the mesentery and enlargement of the mesenteric lymph nodes. The latter is caused by similar cellular accumulations in the subcapsular sinuses as seen in the intestinal wall. Epitheloid cells were also found in other organs such as the liver, lung, kidney, spleen, heart, and bone marrow.

Necrosis, calcification, fibrosis or signs of organisation are usually lacking. Multinucleated giant cells are usually rare, but in 25% of cases amyloidosis is present.

Acid fast organisms can usually be demonstrated in large numbers within the cytoplasm of the epitheloid cells (Smith et al., 1973; Sesline et al., 1975; Sesline, 1978; Holmberg et al., 1982; Lowenstine et al., 1983).

Man: Gross pathology and histological characteristics in *M. avium-intracellulare* infections of human beings, does not differ much from that produced by *M. tuberculosis* (Corpe, 1964; Smyth et al., 1964; Reid and Wolinsky, 1969).

Treatment

Infected nonhuman primates usually have to be destroyed because of epidemiological reasons and of the interference with the tuberculin tests.

M. avium-intracellulare and atypical mycobacteria are resistant to INH and PAS. They rapidly acquire almost total resistance to streptomycin, although many strains have been found susceptible to this drug (Chapman, 1977). *M. avium* is usually sensitive to cycloserine. Rifampin resistant strains are common in Japan, but rarely found in the USA. The majority of strains are susceptible to Kanamycin and sulfadiazine.

Atypical Mycobacteria Other Than M. Avium-Intracellulare

Mycobacteria occurring in man and lower animals which are neither *M. tuberculosis* nor *M. bovis* are called atypical mycobacteria. They lack pathogenicity for rabbits and guinea pigs. Some are facultatively pathogenic to man and some have also been found in nonhuman primates.

With reference to their ability to produce pigments during in vitro growth they are grouped according to Timpe and Runyon (1954) and Runyon (1959) into:

1. Group I (photochromogenic) mycobacteria: growth in 3 or 4 weeks with pigment production only after exposure to light;
2. Group II (scotochromogenic) mycobacteria: growth in 10 to 14 days with production of a strong yellow pigment that deepens somewhat after exposure to light;
3. Group III (nonphotochromogenic) mycobacteria: growth in 3 to 4 weeks usually in small dysgonic colonies exhibiting either no or only faint yellowish pigment;
4. Group IV (rapid growing) mycobacteria: growth in 4 to 6 days with a wide variation in pigmentation and colonial morphology.

Other, biochemical variations are shown in table 33 of the previous chapter.

3.16.5 M. Kansasii

Human *M. kansasii* infections have occurred throughout the world with particular geographical concentration in the midwestern United States (Sommers, 1979) or in the heavily populated industrial areas of Germany (Krebs and Käppler, 1973).

In nonhuman primates an outbreak in a *M. mulatta* colony was reported by Valerio et al. (1979) involving a total of 74 animals.

Agent

M. kansasii is effective only against urea and nicotinamide (Bönicke, 1960), other characteristics are given in table 33.

Antigenically *M. kansasii* is positioned closer to *M. tuberculosis* than any other atypical mycobacteria (Lind, 1961). The heteropolysaccharides from the cell walls of *M. tuberculosis* and *M. kansasii* are identical (Birnbaum and Affronti, 1968), and in consequence, heterologous reactions to tuberculins (and PPD) are common (Pfuetze et al., 1965).

Epidemiology

The source or reservoir of *M. kansasii* is not clear (Sommers, 1979). *M. kansasii* has been recovered from tuberculin-positive cattle (Worthington and Kleeberg, 1964; Siam, 1965) and pigs (Tison et al., 1966) it was also isolated from tap- and discharge waters of hospitals (Tison et al., 1967; Bailey et al., 1970). Contrary to *M. tuberculosis* it is rarely man - man transmitted (Sommers, 1979).

Clinical Symptoms

Nonhuman primates: The signs of infection were limited to conversions of tuberculin tests using Kochs OT.

Man: *M. kansasii* infections resemble mild tuberculosis (Sommers, 1979) with slight fever, weight loss, cough, expectoration and malaise (Pfuetze et al., 1965; Chapman, 1977, 1979). Only a very small proportion of patients experience an abrupt onset with high, spiking fever and severe cough (Chapman, 1979). In rare cases fatal infections have occurred in children (Wood et al., 1956).

Lymphonodular infections involve most frequently the lymphnodes that drain the oro- or nasopharynx, as a painless swelling. *M. kansasii* was also recovered from osteomyelitic lesions.

Pathology

Nonhuman primates: At necropsy following euthanasia in the majority of animals minor tuberculous lesions were localized in the lungs and/or mediastinal lymph nodes, liver or spleen.

Pulmonary nodules were often confluent, containing dry, yellowish-green caseous material. If held for prolonged periods after tuberculin conversion, the gross lesions were undistinguishable from those produced by *M. tuberculosis*, with cavitation being common in such animals.

Histologically the lesions consisted of small foci of large macrophages resembling primitive reticulum cells or epitheloid cells in the early stages. In more advanced cases the lesions appeared as typical tubercles with central caseation and occasional multinucleated giant cells (Valerio et al., 1979).

Man: The most significant change in resected lungs were tuberculous endobronchitis with a paucity of giant cells. Of the giant cells present many were of foreign-body type rather than the Langhans' type (Takimura and Thompson, 1965).

3.16.6 Mycobacterium Simiae

M. simiae was first isolated in 1964 from 33 *M. mulatta* and *C. aethiops* soon after importation (Karesseva et al., 1965; Weiszfeiler et al., 1970). In rare cases it also infects man (Yue and Cohen, 1966; Valdivia et al., 1971; Krasnow and Gross, 1975; Donovan et al., 1976; Lavy and Yoshpe-Purer, 1982).

Agent

M. simiae differs from *M. kansasii* by its ability to synthesize niacin (Karesseva et al., 1965; Meissner and Schroeder, 1975; Weiszfeiler et al., 1981). The only other niacin-positive atypical mycobacterium is *M. habana*, which is considered to be synonymous to *M. simiae* because of identical antigens and biochemical reactions (Meissner and Schroeder, 1975; Weiszfeiler and Karczag, 1976).

Epidemiology

The information on the epidemiology of *M. simiae* is rather scarce. The agent, so far, has only been recovered from monkeys and man. In an abstracted paper by Donovan et al. (1976) the patient had handled monkeys and monkey tissues for several years.

Clinical Symptoms

Nonhuman primates: The isolations have apparently been made from clinically healthy animals.

Man: Diseases in man occurred in either immunosuppressed (Yue and Cohen, 1966) or cancer patients (Donovan et al., 1976) or secondary to tuberculosis (Lavy and Yoshpe-Purer, 1982) and were determined by the original disease.

Pathology

Man: In a few patients the infection has been associated with a granulomatous tissue reaction (Sommers, 1979). In a patient receiving generous cortisone-treatment, dense obliteration of the pleural and pericardial cavities, with numerous multilocular lung cavities were seen (Yue and Cohen, 1966). The inflammatory response consisted of unusual epitheloid cell accumulations without the formation of the typical tuberculoid granulomas, although Langhans' giant cells were visible in some sections. This abnormal response might have been caused by the extensive cortisone-treatment, since in another case reported by Donovan et al. (1976) the more characteristic tuberculoid granulomas and caseation were demonstrated.

3.16.7 Mycobacterium Scrofulaceum

M. scrofulaceum is the etiological agent of a form of lymphadenitis in children (Prissick and Masson, 1956; Weed et al., 1956). It has also been isolated from 2 *Erythrocebus patas* in one colony (Renquist and Potkay, 1979).

Agent
M. scrofulaceum grows in compact, domed colonies with a strong yellow-orange color, which further deepens on exposure to light. Flecks of red carotenoid pigment appear in the yellow stained colonies. The biochemical properties are shown in table 33.

Epidemiology
Scotochromogenic bacteria, usually without species differentiation have been isolated from soil, water and raw milk (Chapman, 1977). In children the portal of entry seems to be the oropharynx. The distribution of *M. scrofulaceum* and its pathogenicity in man vary widely.

Clinical Symptoms
Nonhuman primates: One of the two *E. patas* infected with *M. scrofulaceum* developed gradual weight loss and lack of appetite several weeks after positive tuberculin tests following importation (Renquist and Potkay, 1979). Radiographs showed extensive opacity in the lungs.

Man: Lymphadenitis is usually confined to submandibular and/or preauricular nodes, or occasionally those of the posterior cervical triangle. In rare cases primary infections of the lung have been reported, these resembled primary tuberculosis. Lymphonodular abscessation often leads to skin lesions, whilst the disseminated disease may be associated with osteomyelitis (Chapman, 1977).

Pathology
Nonhuman primates: At necropsy numerous caseous white nodules of 0.5–2.0 cm diameter were noticed in the liver, spleen, omentum, lungs and lymphnodes. Histologically the lesions consisted of a granulomatous pneumonia, with granulomas containing giant cells in the liver and spleen (Renquist and Potkay, 1979).

Man: There is little difference between the lesions caused by *M. tuberculosis* and those caused by atypical mycobacteria including *M. scrofulaceum.*

3.16.8 Mycobacterium Chelonei (Abscessus)

The species *M. chelonei*, belonging to the *M. fortuitum-chelonei* complex, consists of two subspecies *M. chelonei* ss. *borstelense* and *M. chelonei* ss. *abscessus (M. runyoni)* (Sommers, 1979). *M. chelonei abscessus* is considered a significant but rare pathogen in man (Finegold and Martin, 1982). It has been isolated secondary to trauma (Sommers, 1979) or after accidental injection (Inman et al., 1969). Occasional *M. chelonei/M. abscessus* infections have been reported in a *Papio anubis* (Quimby et al., 1979), an *Aotus trivirgatus* (Karlson et al., 1970) or in *M. mulatta* (Karasseva and Weiszfeiler, 1969; Weiszfeiler et al., 1969). Nonspecified Runyon group III mycobacteria have been isolated from a tuberculous rhesus monkey (Latt, 1975).

Agent

Bacterial growth is obtained *in vitro* in 4 or 5 days. The colonies are smooth, white or pink (Weiszfeiler et al., 1969), the biochemical reactions are noted in table 33.

Clinical Symptoms

Nonhuman primates: A baboon developed an ulcerated abscess, progressive lymphopenia and progressive weight loss several weeks after injection of human IgG and killed *M. smegmatis* (Quimby et al., 1979). The owl monkey reported by Karlson et al. (1970) had been splenectomized and died during an experiment with "Delta" herpesvirus without showing signs of infection.

Man: Moore and Frerichs (1953 - cit. from Weiszfeiler et al., 1969) isolated the agent from a patient suffering from arthritis with abscessation and named it *M. abscessus.* Other observations were made after implantation of accidentally contaminated porcine heart valves, leading to sternal infections, or as in one patient, an abscess at the base of the aorta (Anon, 1977). Injection of contaminated histamine resulted in a subcutaneous, tender, warm and firm mass after a few days in another patient. Upon incision a persistent sinus formed (Inman et al., 1969).

Pathology

Nonhuman primates: At necropsy the baboon had multiple large pulmonary lesions containing viable *M. chelonei* (Quimby et al., 1979). In the owl monkey, the axillary lymph nodes were prominent and haemorrhagic. Several small dark areas in its lungs proved to be small granulomas of histiocytes and lymphocytes, some showing necrosis or degeneration (Karlson et al., 1970). The lesions caused by Runyon group III atypical mycobacteria in the rhesus monkey consisted of ulceration of the cheek pouch membranes and enlargement of the regional lymph nodes (Latt, 1975).

Man: The lesions in man vary from caseating granulomas resembling tuberculosis to simple suppuration, necrosis and nonspecific granulation (Inman et al., 1969).

References

1. Acred, P.; P.A. Hunter; J.E.C. Bywater, and J. James: Diagnosis of tuberculosis in experimentally infected rhesus monkeys (*Macaca mulatta*); pp. 2-12 in Medical Primatology 1972, Proc. 3rd Conf. exp. Med. Surg. Primates, Lyon 1972, part III, Karger, Basel (1972)
2. Affronti, L.F.; E.H. Fife, and L. Grow: Serodiagnosis test for tuberculosis. Amer. Rev. resp. Dis. *107*, 822-825 (1973)
3. Affronti, L.F. and F.C. Seibert: Some early investigations of *Mycobacterium tuberculosis.* Amer. Rev. resp. Dis. *92*, 1-8 (1965)

4. Al-Doory, Y.; M.E. Pinkerton; T.E. Vice, and V. Hutchinson: Pulmonary nocardiosis in a vervet monkey. J. Amer. vet. med. Ass. *155*, 1179-1180 (1969)
5. Allen, A.M. and R.F. Kinard: Primary cutaneous inoculation tuberculosis in the Macaca mulatta monkey. Amer. J. Path. *34*, 337-347 (1958)
6. Allgood, M.A. and G.T. Price: Isoniazid therapy of tuberculosis in baboons. Primates *12*, 81-90 (1971)
7. Angus, R.D.: Tuberculins for use in animals: pp. 109-114 in "Mycobacterial Infections of Zoo Animals" (Montali, R.J., ed.), Smithson. Instit. Press, Washington, D.C. (1978)
8. Anon: Report on simian tuberculosis seminar. Lab. Primate Newsl. *5* (1), 17-18 (1966)
9. Anon: Squirrel monkey suspected source of tuberculosis in a human. Lab. Primate Newsl. *9* (3), 14 (1970)
10. Anon: Isolation of mycobacterial species from porcine heart valve prothesis. CDC Morb. Mortal. Weekly Rep. *26*, 42 (1977)
11. Asscher, A.W.: The Challenge of Urinary Tract Infections. Academic Press, London (1980)
12. Astets, M.; H. Boisvert; F. Grumbach; M. Brunel, et N. Rist: Les bacilles tuberculeux de type Africain. Note preliminaire. Rev. Tuberc. Pneumol. *32*, 179-184 (1968)
13. Axelrod, A.R. and A.C. Trakatellis: Relationship of pyridoxine to immunological phenomena. Vitam. and Horm. *22*, 591-607 (1964)
14. Bailey, R.K.; S. Wyles; M. Dingley; F. Hesse, and G.W. Kent: The isolation of high catalase *Mycobacterium kansasii* from tap water. Amer. Rev. resp. Dis. *101*, 430-431 (1970)
15. Barclay, W.R.; R.L. Anacker; W. Brehmer; W. Leif, and E. Ribi: Aerosol-induced tuberculosis in subhuman primates and the course of the disease after intravenous BCG-vaccination. Infect. Immun. *2*, 574-582 (1970)
16. Barclay, W.R.; W.M. Busey; D.W. Dalgard; R.C. Good; B.W. Janicki; J.E. Kasik; E. Ribi; C.E. Ulrich, and E. Wolinsky: Protection of monkeys against airborne tuberculosis by aerosol vaccination with Bacillus Calmette-Guerin. Amer. Rev. resp. Dis. *107*, 351-358 (1973)
17. Bartholomew, W.R.; T. Grafton; S. Al-Nakeeb, and G.W. McConnell: *Mycobacterium bovis* infections in recently imported rhesus monkeys. Primate Zoonoses Surveil. Rep. *9*, 4-5 (1972)
18. Bartsch, R.C.; B.C. Wessels, and E.E. McConnell: Myocardial tuberculoma in a chacma baboon (*Papio ursinus* Kerr, 1792). Lab. Anim. *6*, 41-47 (1972)
19. Bates, J.H. and W.W. Stead: Tuberculosis; pp. 320-328 in "Infectious Diseases" (Sanford, J.P. and J.P. Luby, eds.), Grune & Stratton, New York (1981)
20. Benson, R.E.; B.G. Fremming, and R.J. Young: A tuberculosis outbreak in a Macaca mulatta colony. Amer. Rev. Tuberc. *72*, 204-209 (1955)
21. Bhardwaj, O.P.; V.K. Srivastava, and K. Balakrishnan: Soluble antigen fluorescent antibody (SAFA) test in serodiagnosis of disseminated and extrapulmonary tuberculosis. Indian J. Med. Res. *73*, 150-156 (1981)
22. Birnbaum, S.E. and L.F. Affronti: Chemical and serological relationship between the heteropolysaccharides of *Mycobacterium tuberculosis* and *Mycobacterium kansasii*. J. Bact. *95*, 559-564 (1968)
23. Bönicke, R.: Über das Vorkommen von Acylamidasen in Mykobakterien. V. Die Identifizierung von "atypischen" Mykobakterien der photochromogenen Gruppe mit Hilfe der "Amid Reihe". Zbl. Bakt., I. Abt. Orig. *179*, 223-230 (1960)
24. Bone, J.F. and O.A. Soave: Experimental tuberculosis in owl monkeys (*Aotus trivirgatus*). Lab. Anim. Care *20*, 946-948 (1970)
25. Burkholder, C.R.; D.C. Hirsh; R.L. Hickman, and O.A. Soave: Influence of isoniazid therapy on the course of tuberculosis in a rhesus monkey. J. Amer. vet. med. Ass. *151*, 918-919 (1967)
26. Cappucci, D.T. jr.; J.L. O 'Shea, and G.D. Smith: An epidemiologic account of tuberculosis transmitted from man to monkey. Amer. Rev. resp. Dis. *106*, 819-823 (1972)
27. Chadwick, M.V.: Mycobacteria. Wright and Sons, Bristol (1982)
28. Chaparas, S.D.; R.C. Good, and B.W. Janicki: Tuberculin-induced lymphocyte transformation and skin reactivity in monkeys vaccinated or not vaccinated with Bacille Calmette-Guerin, then challenged with virulent *Mycobacterium tuberculosis*. Amer. Rev. resp. Dis. *112*, 43-47 (1975)
29. Chaparas, S.D.; S.R. Hedrick; R.G. Clark, and R. Garman: Comparison of the lymphocyte

transformation test with the tuberculin test in rhesus monkeys and chimpanzees. Amer. J. vet. Res. *31*, 1437-1441 (1970)

30. Chapman, J.S.: The Atypical Mycobacteria and Human Mycobacteriosis. Plenum Med. Book 6. New York (1977)
31. Chapman, J.S.: Atypical mycobacteria; pp.473-488 in "CRC Handbook Series in Zoonoses" (Steele, J.H., Series ed.), Section A: Bacterial, Rickettsial, and Mycotic Diseases, Vol. I (Stoenner, H., W. Kaplan, and M. Torten, Section eds.), CRC Press, Boca Raton (1979)
32. Chrisp, L.E.; B.J. Cohen; D.H. Ringler, and G.D. Adams: Tuberculosis in a squirrel monkey (*Saimiri sciureus*). J. Amer. vet. med. Ass. *153*, 918-922 (1968)
33. Clarke, G.L.: The relationship of hypersensitivity of shedding of *Mycobacterium tuberculosis* in experimentally infected *Macaca mulatta*. Amer. Rev. resp. Dis. *98*, 416-423 (1968)
34. Clarkson, M.J. and M.W. Smith: Tuberculosis in rhesus monkeys. Lab. Anim. Handb. *4*, 51-55 (1969)
35. Condoulis, W.V.; L.M. Soltysik, and P. Baram: The *in vitro* assay of tuberculosis hypersensitivity in *Macaca mulatta*. II. A modified assay requiring 1 milliliter of blood. J. med. Primatol. *3*, 251-253 (1974)
36. Corpe, R.F.: Clinical aspects, medical and surgical, in the management of Battey-type pulmonary disease. Dis. Chest *45*, 380-382 (1964)
37. Costallat, L.F.; A.F. Pestana de Castro; F. Simon; L. de S.M. Diniz, and W. Giorgi: Mycobacteriosis in monkeys: a report of two cases. Vet. Med. small Anim. Clin. *73*, 651-655 (1978)
38. Dannenbaum, P.: BCG-Impfung bei Affen. Erfahrungen aus dem Zoologischen Garten Frankfurt. Med. Welt *22*, 672-674 (1971)
39. Dannenberg, A.M.: Pathogenesis of pulmonary tuberculosis in man and animals: Protection of personnel against tuberculosis; pp.65-75 in "Mycobacterial Infections of Zoo Animals" (Montali, R.J. ed.) Smithson. Inst. Press, Washington, D.C. (1978)
40. David, H.L. and M.J. Selin: Immune response to mycobacteria; pp.332-337 in "Manual of Clinical Immunology" (Rose, N.R. and H. Friedman, eds.), Amer. Soc. Microbiol., Washington, D.C. (1976)
41. Donovan, W.N.; I. Krasnow; E.M. Donowho, jr., and W.G. Johanson, jr.: *Mycobacterium simiae*. Amer. Rev. resp. Dis. *113* (4), part 2: 55 (abstract) (1976)
42. Doty, R.W. and E.T. Greenstein: Tuberculosis transmission in *Macaca nemestrina*. Lab. Primate Newsl. *3* (3), 7 (1964)
43. Doub, L.: The chemical structures, properties, and mechanisms of action of the antituberculous drugs; pp.435-456 in "Tuberculosis" (Youmans, G.P. ed.) Saunders, Philadelphia (1979)
44. Dragsted, I.: Avian tuberculosis in man. Lancet 1949 (II), 103-105 (1949)
45. Draper, P.: The anatomy of mycobacteria; pp.9-52 in "The Biology of the Mycobacteria", Vol. I. Physiology, Identification and Classification. (Ratledge, C. and J. Stanford, eds.) Academic Press (1982)
46. Duffy, M.; L. Frederickson, and R.D. Linnabary: Tuberculosis in a squirrel monkey with an associated tuberculin conversion in a handler. Primate Zoonoses Surveil. Rep. *1*, 9-10 (1970)
47. Eknoyan, G. and J. Oliveiro: The kidney in infectious disease; pp.541-566 in "The Kidney in Systemic Disease" 2nd ed. (Suki, W.M. and G. Eknoyan, eds.) Wiley & Sons, New York (1981)
48. Engbaek, H.C.: Three cases in the same family of fatal infection with *M. avium*. Acta Tuberc. Scand. *45*, 105-117 (1964)
49. Feldman, W.H.; D.W. Hutchinson; V.M. Schwarting, and A.G. Karlson: Juvenile tuberculous infection, probably of avian type. Amer. J. Path. *25*, 1183-1196 (1949)
50. Fiennes, R.N.T-W: Incidence of T.B. in squirrel monkeys and marmosets. Lab. Primate Newsl. *4*, 10 (1965)
51. Fiennes, R.N.T-W: The Zoological Society of London. Report of the society 's pathologist for the year 1963. J. Zool. *148*, 341-362 (1966)
52. Fiennes, R.N.T-W: Tuberculosis; pp.314-334 in "Pathology of Simian Primates". Part II (Fiennes, R.N.T-W, ed.), Karger, Basel (1972)
53. Fife, E.H.; R.H. Kruse; A.J. Toussaint, and E.V. Staab: Serodiagnosis of simian tuberculosis by soluble antigen fluorescent antibody (SAFA) tests. Lab. Anim. Care *20*, 969-978 (1970)
54. Fineg, J.; W.C. Hanly; J.R. Prine; D.C. van Riper, and W.P. Day: Isoniazid therapy in the chimpanzee. Lab. Anim. Care *16*, 436-446 (1966b)

55. Finegold, S.M. and W.J. Martin: Bailey and Scott 's Diagnostic Microbiology. 6th ed., Mosby, St. Louis (1982)
56. Fleischmann, R.W.; G.C. du Moulin; H.J. Esber; V. Ilievski, and A.E. Bogden: Nontuberculous mycobacterial infection attributable to *Mycobacterium intracellulàre* serotype 10 in two rhesus monkeys. J. Amer. vet. med. Ass. *181*, 1358-1362 (1982)
57. Fölsch, D.W.: Endemische Tuberkulose durch den Typus bovinus bei nichthumanen Primaten. Acta zool. path. antverp. *52*, 11-28 (1971)
58. Fourie, P.B. and H.H. Kleeberg: Tuberculosis in laboratory animals. J.S. Afr. vet. Ass. *49*, 219-221 (1978)
59. Fourie, P.B. and M.W. Odendaal: *Mycobacterium tuberculosis* in a closed colony of baboons. Lab. Anim. *17*, 125-128 (1983)
60. Fox, J.G.; L.H. Campbell; S.B. Snyder; C. Reed, and O.A. Soave: Tuberculosis spondylitis and Potts' paraplegia in a rhesus monkey, *Macaca mulatta*. Lab. Anim. Sci. *24*, 335-339 (1974)
61. Fox, J.G. and K.F. Drake: Tuberculin testing in *Macaca species*: Evaluation of two tuberculin diluents. Lab. Anim. Sci. *24*, 763-767 (1974)
62. Fox, J.G.; C.C. Murphy; R.E. Esser; F.L. Delmonico, and A.B. Cosimi: Tuberculosis outbreak in rhesus monkeys immunosuppressed with antithymocyte globulin. J. med. Primatol. *7*, 264-273 (1978)
63. Fox, J.G.; S.M. Niemi, and J.C. Murphy: A comparison of two tuberculins in nonsensitized macaques. J. med. Primatol. *11*, 380-388 (1982)
64. Francis, J.: Natural and experimental tuberculosis in monkeys. With observations on immunization and chemotherapy. J. comp. Path. *66*, 123-135 (1956)
65. Francis, J.: Tuberculosis in Animals and Man - A Study in Comparative Pathology. Cassell, London (1958)
66. Frederickson, L.E.; C.E. Barton; J.R. Ragan, and J.W. Roberts: An epizootic of tuberculosis in a municipal zoo: A public health problem. J. Amer. vet. med. Ass. *159*, 1474-1476 (1971)
67. Ganaway, J.R.: Zoonoses of laboratory animals - bacterial; pp. 245-257 in "CRC Handbook of Laboratory Animal Science", Vol. 2 (Melby, E.D. jr. and N.H. Altman, eds.) CRC Press, Cleveland, Ohio (1974)
68. Gibson, J.P.; M.W. Rohovsky, and J.W. Newberne: Modification of the tuberculin response of rhesus monkeys by isoniazid therapy. Lab. Anim. Sci. *21*, 62-66 (1971)
69. Giese, W.: Atemwege und Lungen; pp. 3.1-3.108 in "Organpathologie" Vol. 1 (Doerr, W., ed.), Thieme Stuttgart (1974)
70. Good, R.C. and N.E. McCarroll: BCG-vaccination in rhesus monkeys: Study of skin hypersensitivity and duration of protective immunity; pp. 115-121 in "Mycobacterial Infections of Zoo Animals" (Montali, R.J., ed.) Smithson. Inst. Press, Washington (1978)
71. Goodfellow, M. and L.G. Wayne: Taxonomy and nomenclature; pp. 471-521 in "The Biology of the Mycobacteria", Vol. I. Physiology, Identification and Classification (Ratledge, C. and J. Stanford, eds.), Academic Press, London (1982)
72. Goren, M.B. and P.J. Brennan: Mycobacterial lipids: Chemistry and biologic activities; pp. 63-193 in "Tuberculosis" (Youmans, G.P. ed.), W.B. Saunders, Philadelphia (1979)
73. Grigelova, R.; M. Turzova; V. Dornetzhuber, and R. Urbancik: A contribution to the problem of lymphatic system diseases provoked in children by *Mycobacterium avium*. Scand. J. resp. Dis. *48*, 71-75 (1967)
74. Habel, K.: Tuberculosis in a laboratory monkey colony; its spread and its control. Amer. Rev. Tuberc. *55*, 77-92 (1947)
75. Hamm, T.E.: Spontaneous tuberculosis in a cynomolgus monkey (*Macaca fascicularis*). Lab. Anim. Dig. *9* (2), 34-35 (1974)
76. Hamner, J.E. III: Pathological findings in a tuberculous baboon. Lab. Anim. Care *15*, 140-150 (1965)
77. Hartley, E.G.: Tuberculin testing of monkeys. Lab. Anim. Handb. *4*, 27-31 (1969)
78. Hessler, J.R. and A.F. Moreland: Pulmonary tuberculosis in a squirrel monkey (*Saimiri sciureus*). J. Amer. vet. med. Ass. *153*, 923-927 (1968)
79. Heyman, H.: INH-Anwendung bei Zootieren. Kleintierpraxis *4*, 123-126 (1959)
80. Heywood, R. and P.H. Hague: Tuberculosis in baboons. Lab. Anim. Handb. *4*, 43-50 (1969)

81. Heywood, R.; R.K. Medd, and A.E. Street: The early clinical diagnosis of tuberculosis in baboons. Brit. vet. J. *125*, 372-383 (1970)
82. Hime, J.M.: Radiological examination for simian pulmonary tuberculosis. Lab. Anim. Handb. *4*, 35-40 (1969)
83. Himes, E.M.; W.A. Wendt; D.W. Luchsinger, and J.L. Jarnagin: *Mycobacterium bovis* isolated from a dusky langur with granulomas in the intestine. J.Amer. vet. med. Ass. *181*, 1355-1357 (1982)
84. Hinshaw, H.C.: Treatment of tuberculosis; pp.457-488 in "Tuberculosis" (Youmans, G.P., ed.) Saunders, Philadelphia (1979)
85. Holmberg, C.A.; R.V. Henrickson; C.Malaga; R.Schneider, and D. Gribble: Nontuberculous mycobacterial disease in rhesus monkeys. Vet. Path. *19* (suppl.7), 9-16 (1982)
86. Houser, W.D.; H.Uno, and J.C. Gonder: Nonreactive and nonprogressive tuberculosis in a rhesus macaque. Lab. Anim. Sci. *32*, 425-426 (1982)
87. Huber, I.und B.Maran: Das Vorkommen von Tuberkulose bei Affen im Zoologischen Garten der Stadt Zagreb; X.Internat. Symp. Erkr. Zootiere, Salzburg, pp.63-64 (1968)
88. Hyatt, C.; H.Marshall, and H.F. Luke: Pet-associated tuberculosis - Washington. Primate Zoonoses Surveil. Rep. *11*, 7-8 (1973a)
89. Hyatt, C.; H.Marshall; H.F. Luke; R.B. Watkins; J.A. Beare; A.R. Hamlin; E.V. Bayer, and J.Chin: Pet monkey associated tuberculosis. CDC Morb. Mortal. Wkly. Rep. *22*(17), 151-152 (1973b)
90. Indzhita, L.V.; L.A. Yakovleva; V.G. Sunovonjan; E.K. Dzhikidze; V.I. Kovaljova, and V.N. Popova: The character and results of comparative experimental therapy in *Macaca arctoides* monkeys. Z.Versuchstierk. *19*, 13-25 (1977)
91. Inman, P.M.; A.Beck; A.E. Brown, and J.L. Stanford: Outbreak of injection abscesses due to *Mycobacterium abscessus*. Arch. Derm. *100*, 141-147 (1969)
92. Innes, J.R.M.: Tuberculosis of the central nervous system in *Macaca mulatta*. Three cases with vertebral and dural involvement. Brit. vet. J. *119*, 393-398 (1963)
93. Jenkins, P.A.; S.R. Pattyn, and F.Portaels: Diagnostic bacteriology; pp.441-470 in "The Biology of the Mycobacteria, Vol.1: Physiology, Identification and Classification" (Ratledge, C. and J.Stanford, eds.) Academic Press, London (1982)
94. Jentzsch, K.D.; H.D. Schröder und E.Karasek: Beitrag zum Vorkommen von Zooanthroponosen bei Zootieren. XI. Internat. Symp. Erkr. Zootiere, Zagreb, pp.7-17 (1969)
95. Kalter, S.S.; C.H. Millstein; L.H. Boncyk, and L.B. Cummins: Tuberculosis in nonhuman primates as a threat to humans. Develop. biol. Standard. *41*, 85-91 (1978)
96. Karasseva, V.and S.A. Vichkanova: Occurrence of mycobacteria in *Cercopithecus aethiops* monkeys. Proc. Microbiol. Res. Group Hung. Acad. Sci. *III*, 67-70 (1970)
97. Karasseva, V.and J.G. Weiszfeiler: Untersuchungen über die Virulenz und die Infektiosität von aus Affen angezüchteten Mykobakterien (M.simiae, M.abscessus). Zbl. Bakt., I.Abt. Orig. *210*, 524-530 (1969)
98. Karasseva, V.; J.G. Weiszfeiler, and E.Krasznay: Occurrence of atypical mycobacteria in *Macacus rhesus*. Acta microbiol. Acad. Sci. hung. *12*, 275-282 (1965)
99. Karlson, A.G.; H.R. Seibold, and R.H. Wolf: *Mycobacterium abscessus* infection in an owl monkey (*Aotus trivirgatus*). Pathol. vet. *7*, 448-454 (1970)
100. Kaufmann, A.F. and D.C. Anderson: Trends in the incidence of simian tuberculosis; pp.51-53 in "Mycobacterial Infections in Zoo Animals" (Montali, R.J. ed.), Smithson. Inst. Press, Washington D.C. (1978a)
101. Kaufmann, A.F. and D.C. Anderson: Tuberculosis control in nonhuman primate colonies; pp.227-234 in "Mycobacterial Infections of Zoo Animals" (Montali, R.J. ed.), Smithson. Inst. Press, Washington D.C. (1978b)
102. Kaufmann, A.F.; J.I. Moulthrop, and R.M. Moore, jr.: A perspective of simian tuberculosis in the United States - 1972.J. med. Primatol. *4*, 278-286 (1975)
103. Keeling, M.E.; W.B. Bommer, and H.M. McClure: Preliminary note on an epizootic of tuberculosis in a great ape population. Primate Zoonoses Surveil. Rep. *4*, 9-10 (1971)
104. Keeling, M.E.; R.E. Froehlich, and R.D. Ediger: An epizootic of tuberculosis in a rhesus monkey conditioning colony. Lab. Anim. Care *19*, 629-634 (1969)
105. Kleeberg, H.H.: Tuberculosis and other mycobacterioses; pp. 301-360 in "Diseases Transmitted from Animals to Man", 6th ed. (Hubbert, W.T.; W.F. McCulloch, and P.R. Schnurrenberger, eds.), C.C. Thomas, Springfield (1975)

106. Klöppel, G.: BCG-Impfung bei Affen. VI. Internat. Symp. Erkr. Zootiere, Wien, pp. 118-120 (1964)
107. Knezevic, A. L. and W. P. McNulty: Tuberculosis in *Lemur mongoz.* Folia primatol. *6,* 153-159 (1967)
108. Konyha, L. D.; E. M. Himes, and C. O. Thoen: Bovine tuberculosis; pp. 109-139 in "CRC Handbook Series in Zoonoses" (Steele, J. H., Series ed.), Section A: Bacterial, Rickettsial and Mycotic Diseases, Vol. II (Stoenner, H., W. Kaplan, and M. Torten, Section eds.) CRC Press, Boca Raton (1980)
109. Krasnow, I. and W. Gross: *Mycobacterium simiae* infection in the United States. A case report and discussion of the organism. Amer. Rev. resp. Dis. *111,* 351-360 (1975)
110. Krebs, A. and W. Käppler: Contribution to the epidemiology of pulmonary diseases caused by atypical mycobacteria; pp. 201-205 in "Atypical Mycobacteria" (Weiszfeiler, J. G. ed.), Akademiai Kiado, Budapest (1973)
111. Kronberger, H.: Befunde bei Affensektionen. 2. Miedzynarodowe sympozjum patologii zwierzat ogrodow zoologicznych. (2. Internat. Symp. Erkr. Zootiere, Warszawa, pp. 221-229; Panstowe Wydawnictwo Rolnicze/Lesvne (1960)
112. Kubin, M.; K. Dvorsky; R. Eisnerova; I. Mesensky; K. Franc, and M. Matejka: Pulmonary and nonpulmonary disease in humans due to avian mycobacteria. Amer. Rev. resp. Dis. *94,* 31-39 (1966 a)
113. Kubin, M.; J. Kruml; Z. Horak; J. Lukavsky, and C. Vanek: Pulmonary and nonpulmonary disease in humans due to avian mycobacteria. I. Clinical and epidemiological analysis of nine cases observed in Czechoslovakia. Amer. Rev. resp. Dis. *94,* 20-30 (1966 b)
114. Lapin, B. A. und L. A. Jakovleva: Comparative Pathology in Monkeys. C. C. Thomas, Springfield (1963)
115. Latt, R. H.: Runyon group III atypical mycobacteria as a cause of tuberculosis in a rhesus monkey. Lab. Anim. Sci. *25,* 206-209 (1975)
116. Lau, D. T.; J. M. Fuller, and P. E. Sumner: Tuberculosis in a pig-tailed macaque. J. Amer. vet. med. Ass. *161,* 696-699 (1972)
117. Lavy, A. and Y. Yoshpe-Purer: Isolation of *Mycobacterium simiae* from clinical specimens in Israel. Tubercle *63,* 279-285 (1982)
118. Leathers, C. W. and T. E. Hamm, jr.: Naturally occurring tuberculosis in a squirrel monkey and a cebus monkey. J. Amer. vet. med. Ass. *169,* 909-911 (1976)
119. Lind, A.: Serological studies of mycobacteria by means of diffusion - in - gel techniques. Inaug. Diss. Goteberg (1961)
120. Lindsey, J. R. and E. C. Melby: Naturally occurring primary cutaneous tuberculosis in the rhesus monkey. Lab. Anim. Care *16,* 369-385 (1966)
121. Lowenstine, L. J.; R. V. Henrickson, and J. H. Anderson: Diagnostic exercise. Lab. Anim. Sci. *33,* 245-246 (1983)
122. Machotka, S. V.; F. E. Chapple, and J. L. Stookey: Cerebral tuberculosis in a rhesus monkey. J. Amer. vet. med. Ass. *167,* 648-650 (1975)
123. Martin, D. P.: "Primates"; pp. 525-552 in "Zoo and Wild Animal Medicine" (Fowler, M. W., ed.) Saunders, Philadelphia (1978)
124. Martin, J. E.; W. C. Cole, and R. A. Whitney: Tuberculosis of the spine (Potts 's disease) in a rhesus monkey (*M. mulatta*). J. Amer. vet. med. Ass. *153,* 914-917 (1968)
125. Mayhall, C. G.; V. A. Lamb, and P. H. Coleman: Infection in rhesus (*Macaca mulatta*) and squirrel (*Saimiri sciureus*) monkeys due to *Mycobacterium tuberculosis* phage type B. J. med. Primatol. *10,* 302-311 (1981)
126. McClure, H. M. and V. A. Silcox: Possible *Mycobacterium africanum* infection in a rhesus monkey. Primate Zoonoses Surveill. Rep. *1,* 7-8 (1970)
127. McLaughlin, R. M.: *Mycobacterium bovis* in nonhuman primates; pp. 151-155 in "Mycobacterial Infections of Zoo Animals" (Montali, R. J., ed.), Smithson. Inst. Press, Washington D. C. (1978)
128. McLaughlin, R. M. and G. E. Marrs: Tuberculin testing in nonhuman primates: OT vs. PPD; pp. 123-128 in "Mycobacterial Infections of Zoo Animals" (Montali, R. J., ed.), Smithson. Inst. Press, Washington D. C. (1978)
129. Meissner, G. and K. H. Schröder: Relationship between *Mycobacterium simiae* and *Mycobacterium habana.* Amer. Rev. resp. Dis. *111,* 196-200 (1975)

130. Migaki, G.; H.R. Seibold; H.R. Wolf, and F.M. Garner: Pathologic conditions in the patas monkey. J. Amer. vet. med. Ass. *159*, 549-556 (1971)
131. Morton, W.R. and G.A. Blakely: A comparison of quarantine period in pigtail macaque monkeys (*Macaca nemestrina*) from Sumatra and Malaysia. Primate Zoonoses Surveill. Rep. *8*, 13-14 (1972)
132. Mulder, J.B.: Tuberculosis in non-human primates. Vet. Med. small Anim. Clin. *71*, 1286 (1976)
133. Muscoplat, C.C.: Mechanisms of cell-mediated immunity, tolerance and anergy; pp. 71-84 in "Mycobacterial Infections in Zoo Animals" (Montali, R.J., ed.) Smithson. Inst. Press, Washington, D.C. (1978)
134. Muscoplat, C.C.; C.O. Thoen; R.M. McLaughlin; J.R. Thoenig A.W. Chen, and D.W. Johnson: Comparison of lymphocyte stimulation and tuberculin skin reactivity in *Mycobacterium bovis*-infected *Macaca mulatta*. Amer. J. vet. Res. *36*, 699-701 (1975)
135. Nagai, S.; T. Nagasuga, and J. Matsumoto: Tuberculin peptide from culture filtrate of *Mycobacterium tuberculosis*. Amer. Rev. resp. Dis. *121*, 551-557 (1980)
136. Nair, C.P. and A.P. Ray: Observations on the incidence and types of tuberculosis in rhesus monkeys on autopsy studies. Indian J. Malariology *9*, 185-189 (1955)
137. Ott, J.E.: Tuberculin testing in primates. Amer. Ass. Zoo Vets. Ann. Proc. 1979, 75-83 (1979)
138. Ott, J.E.: Tb testing in primates; XXII. Internat. Symp. Erkr. Zootiere, Arnhem, pp. 93-103 (1980)
139. Pfuetze, K.H.; L.V. Vo; A.F. Reimann; G.S. Berg, and W. Lester: Photochromogenic mycobacterial pulmonary disease. Amer. Rev. resp. Dis. *62*, 470-475 (1965)
140. Piwowarczyk, S.: Die Typen von Tuberkelbakterien, die bei den Tieren im Zoologischen Garten in Warszawa auftreten. VI. Internat. Symp. Erkr. Zootiere, Wien; pp. 102-108 (1964)
141. Potkay, S.; J.R. Ganaway; N.G. Rogers, and R. Kinard: An epizootic of measles in a colony of rhesus monkeys (Macaca mulatta). Amer. J. vet. Res. *27*, 331-334 (1966)
142. Prissick, F.H. and A.M. Masson: Cervical lymphadenitis in children caused by chromogenic mycobacteria. Canad. Med. Ass. J. *75*, 798-802 (1956)
143. Quimby, F.W.; K. Miller, and C.R. Valeri: Atypical mycobacteriosis in a baboon. J. Amer. vet. med. Ass. *174*, 954 (1979)
144. Rao, A.T.; L.N. Acharjyo, and B.C. Nayak: Tuberculosis in some ungulates and primates at Nandankanan biological park. Indian J. Pathol. Microbiol. *25*, 199-202 (1982)
145. Reid, J.D. and E. Wolinsky: Histopathology of lymphadenitis caused by atypical mycobacteria. Amer. Rev. resp. Dis. *99*, 8-12 (1969)
146. Renner, M. and W.R. Bartholomew: Mycobacteriologic data from two outbreaks of bovine tuberculosis in nonhuman primates. Amer. Rev. resp. Dis. *109*, 11-16 (1974)
147. Renner, M.; W.R. Bartholomew, and K. Wicher: Monkey colonies as a potential reservoir for bovine type tuberculosis. Abstr. Ann. Meet. Amer. Soc. Microbiol. *72*, 118 (1972)
148. Renquist, D.M. and S. Potkay: *Mycobacterium scrofulaceum* infection in *Erythrocebus patas* monkeys. Lab. Anim. Sci. *29*, 97-101 (1979)
149. Renquist, D.M. and R.A. Whitney: Tuberculosis in nonhuman primates - an overview; pp. 9-16 in "Mycobacterial Infections of Zoo Animals" (Montali, R.J., ed.) Smithson. Inst. Press, Washington D.C. (1978)
150. Runyon, E.H.: Anonymous mycobacteria in pulmonary disease. Med. Clin. N. Amer. *43*, 273-291 (1959)
151. Runyon, E.H.; A.G. Karlson; G.P. Kubica, and L.G. Wayne: Mycobacterium; pp. 148-174 in "Manual of Clinical Microbiology" 2nd ed. (Lennette, E.H.; E.H. Spaulding, and J.P. Truant, eds.), Amer. Soc. Microbiol., Washington D.C. (1974)
152. Runyon, E.H.; L.G. Wayne, and G.P. Kubica: Mycobacteriaceae Chester 1897; pp. 681-701 in "Bergey 's Manual of Determinative Bacteriology." 8th ed. (Buchanan, R.E. and N.E. Gibbons, eds.), Williams and Wilkins, Baltimore (1975)
153. Schliesser, T.: Tuberkulose bei Haus- und Wildtieren. Prax. Pneumol. *28*, 511-516 (1974)
154. Schmidt, L.H.: Some observations on the utility of simian pulmonary tuberculosis in defining the therapeutic potentialities of isoniazid. Amer. Rev. Tuberculosis *74*, 138-153 (1956)
155. Schmidt, L.H.: Further on Koch's O.T. tuberculin versus PPD for tuberculin testing. Lab. Primate Newsl. *7* (4), 28 (1968)
156. Schmidt, L.H.: The John Barnwall lecture: Improving existing methods of control of tuberculosis: A prime challenge to the experimentalist. Amer. Rev. resp. Dis. *105*, 183-205 (1972)

157. Schroeder, C. R.: Acquired tuberculosis in the primate in laboratories and zoological collections. Amer. J. Publ. Hlth Nat. Hlth *28*, 469-475 (1938)
158. Scott, H. E. and J. Beattie: The distribution of tuberculous lesions in man and other primates with an account of the lymphatic glands and vessels of the thorax and upper abdomen. J. Path. Bact. *31*, 49 (1928)
159. Sedgwick, C.; J. Parcher, and R. Durham: Atypical mycobacterial infection in the pig-tailed macaque (*Macaca nemestrina*). J. Amer. vet. med. Ass. *157*, 724-725 (1970)
160. Sesline, D.H.: *Mycobacterium avium* enteritis in nonhuman primates; pp. 157-159 in "Mycobacterial Infections of Zoo Animals" (Montali, R. J. ed.), Smithson. Inst. Press, Washington (1978)
161. Sesline, D.H.; L.W. Schwartz; B.I. Osburn; C.O. Thoen; T. Terrell; C. Holmberg; J.H. Anderson, and R. V. Hendrickson: *Mycobacterium avium* infection in three rhesus monkeys. J. Amer. vet. med. Ass. *167*, 639-645 (1975)
162. Siam, M.A.: Differenzierungen von sog. atypischen Mycobakterien menschlicher und tierischer Herkunft. Inaug. Diss. Gießen (1965)
163. Silberman, M.S.: Tuberculosis in a band of imported hamadryas baboons; pp. 161-162 in "The Comparative Pathology of Zoo Animals" (Montali, R. J. and G. Migaki, eds.), Smithson. Inst. Press, Washington D. C. (1980)
164. Silcox, V.A. and H.M. McClure: Possible *Mycobacterium africanum* infection in a rhesus monkey (con't.). Primate Zoonoses Surveil. Rep. *3*, 11 (1971)
165. Smith, A.W. and H. Wolochow: Comparison of old tuberculin and purified protein derivative in *Macaca mulatta*. Lab. Anim. Sci. *23*, 373-376 (1973)
166. Smith, E.K.; R.D. Hunt; F.G. Garcia; C.E.O. Fraser; R.S. Merkal, and A.G. Karlson: Avium tuberculosis in monkeys. Amer. Rev. resp. Dis. *107*, 469-471 (1973)
167. Smyth, J.T.; N. Kovacs, and W.P. Harris: Pulmonary disease due to unclassified mycobacteria (Battey-type): Report of 14 cases with histological confirmation. Tubercle *45*, 223-228 (1964)
168. Snyder, S.; T. Peace; O. Soave, and J. Lund: Tuberculosis in an owl monkey (*Aotus trivirgatus*). J. Amer. vet. med. Ass. *157*, 712-713 (1970)
169. Sommers, H.M.: Disease due to mycobacteria other than *Mycobacterium tuberculosis*, pp. 386-403 in "Tuberculosis" (Youmans, G. P., ed.), Saunders, Philadelphia (1979)
170. Stones, P.B.: Incidence of tuberculosis in *Macaca mulatta* and *Cercopithecus aethiops* monkeys with special reference to the tuberculin test. Lab. Anim. Handb. *4*, 11-17 (1969)
171. Takimura, Y. and J.R. Thompson: Photochromogenic mycobacterial pulmonary infection in a group of hospitalized patients in Chicago. I. Laboratory studies. Amer. Rev. resp. Dis. *91*, 583-591 (1965)
172. Tarara, R.; M.A. Suleman; R. Sapolsky; M.J. Wabomba, and J.G. Else: Tuberculosis in wild olive baboons, *Papio cynocephalus anubis* (Lesson), in Kenya. J. Wildl. Dis. *21*, 137-140 (1985)
173. Thoen, C.D.; F.Z. Beluhan; E.M. Himes; V. Capek, and B.T. Bennett: *Mycobacterium bovis* infection in baboons (*Papio papio*). Arch. Path. Lab. Med. *101*, 291-293 (1977)
174. Thoen, C.D. and E.M. Himes: Mycobacterial infections in exotic animals; pp. 241-245 in "The Comparative Pathology of Zoo Animals". (Montali, R.J. and G. Migaki, eds.), Smithson. Inst. Press, Washington D.C. (1980)
175. Thorel, M.F.: Isolation of *Mycobacterium africanum* from monkeys. Tubercle (Edinb.) *61*, 101-104 (1980)
176. Timpe, A., and E.H. Runyon: Relationship of "atypical" acid-fast bacilli to human disease: Preliminary report. J. Lab. Clin. Med. *44*, 202-209 (1954)
177. Tison, F.; B. Devulder; P. Roos, and A. Tacquet: Techniques et résultats de la recherche des mycobactéries dans les viandes. Ann. Inst. Pasteur *17*, 156-160 (1966)
178. Tison, F.; A. Tacquet, and B. Devulder: Recherche des mycobactéries dans les eaux des piscines et les eaux usées de la région du nord. Ann. Inst. Pasteur *18*, 167-176 (1967)
179. Toussaint, A.J.; E.H. Fife; R.C. Parlett; L.F. Affronti; G.L. Wright; M. Reich, and W.C. Morse: A soluble antigen fluorescent antibody test for the serodiagnosis of *Mycobacterium tuberculosis* infections. Amer. J. clin. Path. *52*, 708-713 (1969)
180. Tribe, G.W.: Clinical aspects of the detection and control of tuberculosis in newly imported monkeys. Lab. Anim. Handb. *4*, 19-23 (1969)

181. Tribe, G.W.: Tuberculosis in the baboon (Papio anubis). Primate Supply *1*(1), 13-14 (1976)
182. Tribe, G.W. and A.E. Welburn: Value of combining the erythrocyte sedimentation rate test with tuberculin testing in the control of tuberculosis in baboons. Lab. Anim. *10*, 39-46 (1976)
183. Trolldenier, H.: Zur Klinik, Diagnostik und Therapie der Tuberkulose der Affen. Zbl. Vet.-Med. B *11*, 337-364 (1964)
184. Tryphonas, L.and J.Wong: Pott 's paraplegia in a tuberculous rhesus monkey, *Macaca mulatta*. J.med. Primatol. *11*, 211-220 (1982)
185. Valdivia, A.J. and M.R. Suarez: *Mycobacterium habana*: probable nueva especie dentro de las microbacterias no clasificadas. Bol. Hig. Epid. (La Habana) *9*, 65-73 (1971)
186. Valerio, D.A.; D.W. Dalgard; R.W. Voelker; N.E. McCarrol, and R.C. Good: *Mycobacterium kansasii* infection in rhesus monkeys; pp.145-150 in "Mycobacterial Infections of Zoo Animals" (Montali, R.J., ed.), Smithson. Inst. Press, Washington D.C. (1979)
187. Vice, T.E.; M.E. Pinkerton; F.A. Fear, and S.S. Kalter: An outbreak of tuberculosis in a group of experimental baboons. Primates *9*, 105-122 (1968)
188. Ward, G.S.; M.R. Elwell; M.Tingpalapong, and J.Pomsdhit: Use of streptomycin and isoniazid during a tuberculous epizootic in a rhesus and cynomolgus breeding colony. Lab. Anim. Sci. *35*, 395-399 (1985)
189. Weed, L.A.; H.M. Keith, and G.M. Needham: Non-tuberculous acid-fast cervical adenitis in children. Proc. Staff Meetings Mayo Clin. *31*, 259-163 (1956)
190. Weiszfeiler, J.G.; V.Karasseva, and E.Karczag: Comparative studies on the taxonomic relationship between Mycobacterium abscessus and Mycobacterium borstelense. Incidence of Mycobacterium abscessus in monkeys. Acta microbiol. Acad. Sci. hung. *16*, 371-379 (1969)
191. Weiszfeiler, J.G.; V.Karasseva, and E.Karczag: Studies on *Mycobacterium simiae* and some other atypical mycobacteria isolated from monkeys. Proc. Microbiol. Res-Grp Hung. Acad. Sci. *III*, 31-39 (1970)
192. Weiszfeiler, J.G.; V.Karasseva, and E.Karczag: *Mycobacterium simiae* and related mycobacteria. Rev. Infect. Dis. *3*, 1040-1045 (1981)
193. Weiszfeiler, J.G. and E.Karczag: Study of mycobacteria strains belonging to the avian-intracellular group isolated from monkeys. Ann. Soc. belge Med. trop. *53*, 315-320 (1973)
194. Weiszfeiler, J.G. and E.Karczag: Synonymy of *Mycobacterium simiae* Karasseva et al.1965 and *Mycobacterium habana* Valdivia et al.1971. Int. J.System. Bacteriol. *26*, 474-477 (1976)
195. Welshman, M.D.: An outbreak of tuberculosis in Macaca fascicularis monkeys. Primate Supply *6*(1), 3-5 (1981)
196. West, C.S.; S.J. Vainisi; C.M. Vygantas, and F.Z. Beluhan: Intraocular granulomas associated with tuberculosis in primates. J. Amer. vet. med. Ass. *179*, 1240-1244 (1981)
197. Wilson, P.; E.Weavers; B.West; M.Taylor; J.Kavanagh, and P. Jones: *Mycobacterium bovis* infection in primates in Dublin Zoo: epidemiological aspects and implications for management. Lab. Anim. *18*, 383-387 (1984)
198. Wolf, R.H.; B.C. Bullock, and T.B. Clarkson: Tuberculosis in the stumptailed macaque (*Macaca speciosa*). J.Amer. vet. med. Ass. *151*, 914-917 (1967)
199. Wolf, R.and H.R. Seibold: An unusual outbreak of simian tuberculosis. Primate Zoonoses Surveil. Rep. *4*, 10-11 (1971)
200. Wood, L.E.; V.B. Buhler, and A.Pollak: Human infection with the "yellow" acid-fast bacillus. A report of fifteen additional cases. Amer. Rev. Tuberc. Pulm. Dis. *73*, 917-929 (1956)
201. Worthington, R.W. and H.H. Kleeberg: Isolation of *Mycobacterium kansasii* from bovines. S.Afr. Vet. Med. Ass. *35*, 29-33 (1964)
202. Youmans, G.P.: Biologic activities of mycobacterial cells and cell components; pp.46-62 in "Tuberculosis" (Youmans, G.P., ed.), W.B. Saunders, Philadelphia (1979a)
203. Youmans, G.P.: Epidemiology of tuberculosis; pp.356-369 in "Tuberculosis" (Youmans, G.P., ed.), W.B. Saunders, Philadelphia (1979b)
204. Yue, W.Y. and S.S. Cohen: Pulmonary infection caused by niacin-positive *Mycobacterium kansasii*. Amer. Rev. resp. Dis. *94*, 447-449 (1966)
205. Zumpe, D.; M.S. Silbermann, and R.P. Michael: Unusual outbreak of tuberculosis due to *Mycobacterium bovis* in a closed colony of rhesus monkeys (*Macaca mulatta*). Lab. Anim. Sci. *30*, 237-240 (1980)
206. Zwart, P.; C.Blok, and S.J. van den Anker: Tuberkulose und Nebennierenatrophie bei einem Schimpansen (Pan paniscus Sch.). XXII. Internat. Symp. Erkr. Zootiere, Arnhem, pp.87-91 (1980)

3.17 Mycobacterium Leprae

M. leprae is the causative agent of leprosy (Lepra, Aussatz), one of the oldest infections of man.

Leprosy has long been regarded as a strictly human disease, but recently it became known, that natural infections occur in nine-banded armadillos *(Dasypus novemcinctus)* of the southern USA (Stoors, 1971; Walsh et al., 1975, 1981; Binford et al., 1977; Meyers et al., 1977, 1980a; Smith et al., 1978; Marchiondo et al., 1980). The disease has also been reported in a chimpanzee (Donham and Leininger, 1977; Leininger et al., 1978, 1980a, b) and a mangabey *(Cercocebus atys)* (Meyers et al., 1980b, 1981).

Agent

M. leprae is an obligate intracellular parasite. It survives without multiplication in cultures of human cells, but it can not be cultivated on artificial media. After experimental infection *M. leprae* multiplies in mouse foot pads (Shepard, 1960; Shepard and Chang, 1962); in nine-banded armadillos (but only poorly in eight-banded armadillos *(D. sabanicola)* (Convit, 1977). It multiplies also in experimentally infected rhesus monkeys *(Macaca mulatta)*, sooty mangabeys *(Cercocebus atys)* and African green monkeys *(Cercopithecus aethiops)* (Meyers et al., 1984; Baskin et al., 1985).

Within the cells the bacteria grow in characteristic parallel bundles ("globi"). The single organism measures approximately 0.3–0.5 × 1.0–8.0 μm in size and appears as a straight rod with rounded ends. *M. leprae* usually stains heavily acid fast but may be beaded. For demonstration of *M. leprae* in paraffin embedded tissues the Fite-Faraco-staining method (Fite et al., 1947) seems to work better than the Ziehl-Neelsen method (Walsh et al., 1981). *M. leprae* can further be identified using oxidase-tests (Prabhakaran and Kirchheimer, 1966), or pyridine extraction tests with pyridine removing the acid fast staining properties (Convit and Pinardi, 1972).

Epidemiology

Although spontaneous *M. leprae*-infections have been demonstrated in armadillos and nonhuman primates the source of human infections seems to be the human patient with lepromatous leprosy. Only in one case transmission from armadillos to man has been suspected in a Texan who had killed and dressed many armadillos (Freiberger and Fudenberg, 1981 - cit. from Walsh et al., 1981).

Transmission requires the extended and close contact with infectious patients - patients suffering from lepromatous leprosy. Those patients excrete large numbers of *M. leprae* within their nasal secretions (de Souza-Araujo, 1950; Davey and Rees, 1972; Köhler and Mochmann, 1975). Leprosy bacteria remain viable for up to seven days in the dried nasal mucus from lepromatous patients (Davey and Rees, 1972).

Non-experimental infections of animals probably require human contacts. In the case of the armadillos, contamination of the environment by human lepers is suspected (Walsh et al., 1981). Nonhuman primates could be infected through close contact to lepromatous patients after being captured in highly endemic areas such as Sierra Leone (Walsh et al., 1981). However, rhesus monkeys kept as pets in contact to patients even with lepromatous leprosy did not show signs of infection (Hagsted, 1983).

Clinical Symptoms

Nonhuman primates: Both cases observed in nonhuman primates have to be classified as lepromatous leprosy. The disease started as a crusted maculopapular rash involving the entire trunk and limbs, followed by nodular lesions on the lips, nostrils, nasal septum, eyebrows, ears, carpus, forearm and scrotum. Periodic exacerbations were characterized by increased swelling and reddening of the nodules with ulceration of the metacarpeal nodules. In the sooty mangabey paralysis of the intrinsic muscles of the foot and hand, as well as of the extensor and peroneal muscles of the foot developed (Donham and Leininger, 1977; Meyers et al., 1980, 1981; Walsh et al., 1981).

In experimentally infected rhesus monkeys the nodules appeared at 10 days p.i. (de Souza-Araujo, 1950). These lesions then enlarged and extended to uninoculated areas within 20 months postinoculation and subsequently regressed (Martin et al., 1984). In African green monkeys the lesions were noted on the ears at 3 to 25 months post inoculation (Meyers et al., 1984). Experimentally infected mangabeys developed lesions after 4 to 6 months at the inoculations sites on the ears (Wolf et al., 1985).

Man: In human leprosy the bacteria first become apparent within the Schwann's cells of peripheral nerves, then they spread into the surrounding skin. In man several clinical types of leprosy are distinguished (Köhler and Mochmann, 1975; Drutz, 1981):

1. Lepromatous type
2. tuberculoid type
3. indeterminate type
4. borderline type

Ad 1. The lepromatous type is of a progressive, malignant nature with characteristic cutaneous infiltrates (Fig. 13) (facies leoninina), involvement of the peripheral nerves, mucosal membranes and even glandular tissues. Trophical disturbances result in atrophy and ulceration. Both trophical breakdown and neglect of the denervated tissues result in severe mutilation and automutilation. The lepromatous form tends to appear later than tuberculoid leprosy, and reflects a low resistance of the patient. Skin, nerves, nasal mucous membranes, blood and reticuloendothelial cells are crowded by the mycobacteria, with damage to the nerves and skin being widespread. Lepromintests are usually negative.

Ad 2. Tuberculoid leprosy is of a more benign character, occurring in the relatively immune host. This type is characterized by sharply demarcated superficial foci of different appearance. The typical tuberculoid lesion is a large erythematous plaque with sharply elevated outer margins fading centrally to a flattened clear zone that is rough, anhidrotic, hairless, hypopigmented and anesthetic (Fig. 14). Acid-fast bacteria are usually not seen in the skin lesions, but they may however be demonstrated in distant, otherwise normal peripheral nerves. The lepromintest is usually positive.

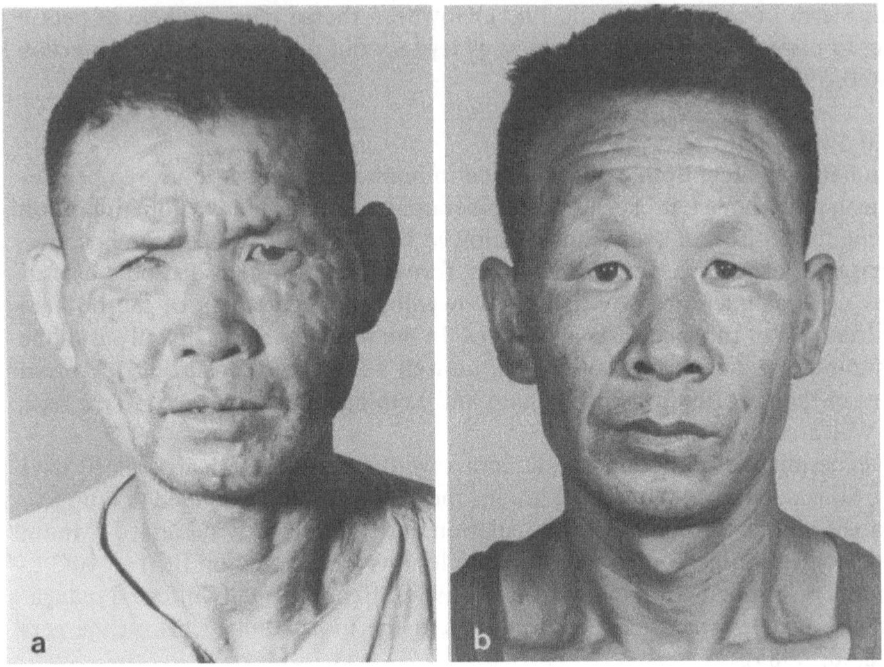

Fig. 13a, b. Lepromatous Leprosy (Drutz, 1981)

Ad 3. In the indeterminate group the leprosy-lesions remain macular for a long period, and can be possibly followed by either the lepromatous or tuberculoid form. The indeterminate form is of a relatively few hypopigmented or erythematous macules in the skin, which consist of small round cells around the skin appendages.

Ad 4. Borderline type: Between the relatively stable lepromatous and tuberculoid types there are several unstable borderline cases. As the degree of resistance in tuberculoid leprosy decreases, the number of lesions increase with simultaneous regression of the demarcation. If the tendency of changing is toward the tuberculoid type it is called a "reversal reaction". More commonly the progression is to lepromatous leprosy - the "downgrading" reaction".

Pathology

Nonhuman primates: The nodular lesions in both the chimpanzee and the mangabey were characterized by a diffuse dermal infiltration by pale, foamy histiocytes, lymphocytes, plasma cells and neutrophils; with loss of dermal collagen, dermal papillae and of the rete pegs. With the progressive disease only the histiocytes were left. Large vacuoles were present in the histiocytic cytoplasm, which contained tangled masses of fine, weakly eosinophilic staining reticular structures. The histiocytes were otherwise filled with acid fast organisms in Fite-Faraco stained sections (Leininger et al., 1978, 1980; Meyers et al., 1980). Neural involvement consisted of focal invasion of the perineurium by histiocytes.

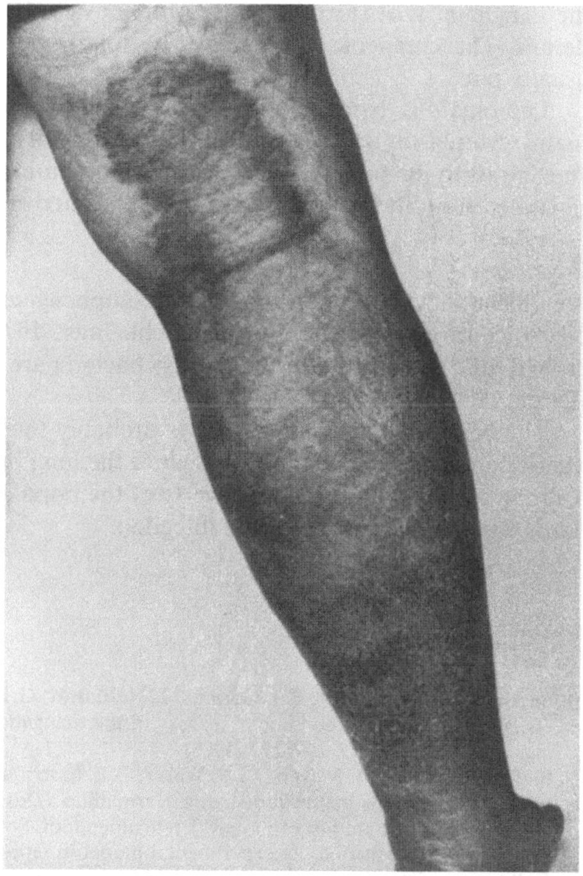

Fig. 14. Tuberculoid Leprosy (Drutz, 1981)

33 months after their first appearance, the lesions in the chimpanzee consisted of marked athrophy of the skeletal muscles, and alopecia. Diffuse thickening of the hands, feet and digits were present whereas the nodules on ears and lips diminished in size. The axillary and inguinal lymph nodes were large and yellow-tan coloured. The soft palate, pharyngeal tissues and nasal passages were similarly stained and thicked (Leininger et al., 1980).

The lesions consisted at that time mainly of diffuse infiltrations with large histiocytes. Only in the nasal mucosa did neutrophils and lymphocytes participate in the infiltrates. The peripheral nerves in the forearm showed segmental invasion by histiocytes, lymphocytes and acid fast bacteria. In an intravenously and intracutaneously inoculated *C. atys* which died 46 months after inoculation the lesions consisted of severe lepromatous infiltrates in the skin, nasal mucosa, peripheral nerves, and testicles. Internal organs were only minimally involved (Baskin et al., 1985).

Man: In indeterminate leprosy only small lymphocyte type cells are dispersed around the skin appendages. Acid-fast bacteria are usually few but may be found in the cutaneous nerves.

In tuberculoid leprosy the lesions consist of epitheloid cells and giant cells of the Langhans' type surrounded by lymphocytes. These lesions extend into the epidermis. The cutaneous nerves may be totally destroyed. Acid fast bacteria are usually rare.

Lepromatous leprosy is characterized by accumulations of histiocytes with foamy cytoplasm, which are stuffed with myriads of leprosy bacteria (lepra cells). In contrast to the tuberculoid type the lepromatous dermal infiltrations are always separated from the epidermis by an uninfiltrated zone (Kirchheimer, 1979).

Treatment

Treatment with dapsone (DDS, diaminodiphenylsulfone) is presently the method of choice in established leprosy (Kirchheimer, 1979). It seems to inhibit the microbial folic acid synthesis and leprosy bacteria are extremely sensitive to the drug (Drutz, 1981).

The best prophylactic measure is probably the improvement of the socioeconomic conditions in endemic areas since the long history of the disease has shown a strong relationship between poverty of the population at risk, low hygienic standards and the occurrence of the infection.

References

1. Baskin, G.B.; R.H. Wolf; B.J. Gormus; L.N. Martin; G.P. Walsh; C.H. Binford; W.M. Meyers, and R. Malaty: Experimental leprosy in the mangabey *(Cercocebus atys)*: Necropsy findings. Int. J. Leprosy *53*, 269-277 (1985)
2. Binford, C.H.; W.M. Meyers; G.P. Walsh; E.E. Storrs, and H.L. Brown: Naturally acquired leprosy-like disease in the nine-banded armadillo *(Dasypus novemcinctus)*: Histopathologic and microbiologic studies of tissues. J. reticuloendoth. Soc. *22*, 377-388 (1977)
3. Convit, J.: The armadillo as experimental model in leprosy research; pp. 53-56 in "Leprosy: Cultivation of the Etiologic Agent, Immunology, Animal Models." Proc. Workshop on Future Problems in the Microbiology of *M. leprae*. Sci. Publ. No. 342 Pan Amer. Hlth Org. Washington, D.C. (1977)
4. Convit, J. and M.E. Pinardi: A simple method for the differentiation of *Mycobacterium leprae* from other mycobacteria through routine staining techniques. Int. J. Leprosy *40*, 130-132 (1972)
5. Davey, T.F. and R.J.W. Rees: The nasal discharge in leprosy. Clinical and bacteriological aspects. Leprosy Rev. *45*, 121-134 (1972)
6. de Souza-Araujo, H.C.: Case of acute malignant leprosy, with infection of the consort within three months of matrimonial life. Isolations from a cutaneous lesion of the same patient on an acid-alcohol fast bacillus (Chromogenic culture) pathogenic for rats, mice, *Macacus rhesus* and man. Mem. Inst. Osw. Cruz *48*, 76-99 (1950)
7. Donham, K.J. and J.R. Leininger: Spontaneous leprosy-like disease in a chimpanzee. J. Infect. Dis. *136*, 132-136 (1977)
8. Drutz, D.J.: Leprosy; pp. 334-340 in "Infectious Diseases" (Sanford, J.P. and J.P. Luby, eds.), Grune and Stratton, New York (1981)
9. Fite, G.L.; P.J. Cambre, and H.M. Turner: Procedure for demonstrating lepra bacilli in paraffin sections. Arch. Path. *43*, 624-625 (1947)
10. Hagstad, H.V.: Leprosy in sub-human primates: Potential risk for transfer of *M. leprae* to humans. Int. J. Zoonoses 127-131 (1983)
11. Kirchheimer, W.F.: Leprosy (Hansen's disease); pp. 447-463 in "CRC Handbook Series in Zoonoses" (Steele, J.H., Series ed.), Section A: Bacterial, Rickettsial, and Mycotic Diseases, Vol. 1 (Stoenner, H., W. Kaplan, and M. Torten, Section ed.), CRC Press, Boca Raton (1979)

12. Köhler, W. and H. Mochmann: Grundriß der Medizinischen Mikrobiologie. VEB Gustav Fischer, Jena (1975)
13. Leininger, J.R.; K.J. Donham, and M.J. Rubino: Leprosy in a chimpanzee. Morphology of the skin lesions and characterization of the organism. Vet. Path. *15*, 339-346 (1978)
14. Leininger, J.R.; K.J. Donham; M.J. Rubino, and W.M. Meyers: Spontaneous leprosy in a chimpanzee. Int. J. Leprosy *47*, 342-343 (1979)
15. Leininger, J.R.; K.J. Donham, and W.M. Meyers: Leprosy in a chimpanzee. Postmortem lesions. Int. J. Leprosy *48*, 414-421 (1980 a)
16. Leininger, J.R.; K.J. Donham; M.J. Rubino, and W.M. Meyers: Naturally acquired leprosy in a chimpanzee: Necropsy findings and experimental transmission to other animals. Lab. Invest. *42*, 132 (1980 b)
17. Marchiondo, A.A.; J.H. Smith, and S.K. File: Naturally-occurring leprosy-like disease of wild armadillos: Ultrastructure of lepromatous lesions. J. reticuloendoth. Soc. *27*, 311-325 (1980)
18. Martin, L.N.; B.J. Gormus; R.H. Wolf; G.P. Walsh; W.M. Myers, and C.H. Binford: Experimental leprosy in nonhuman primates. Advanc. Vet. Sci. Comp. Med. *28*, 201-236 (1984)
19. Meyers, W.M.; C.H. Binford; H.L. Brown, and G.P. Walsh: Leprosy in wild armadillos. Proc. Symp. Path. Zoo Anim., National Zoological Park, Smithsonian Inst. Press, pp. 543-549 (1980 a)
20. Meyers, W.M.; G.P. Walsh; H.C. Binford; R.H. Wolf; B.J. Gormus; L.N. Martin; P.J. Gerone, and G.B. Baskin: Experimental leprosy in three species of monkeys. 37rd Ann. Meeting Internat. Acad. Pathology, San Francisco (1984)
21. Meyers, W.M.; G.P. Walsh; H.L. Brown; Y. Fukunishi; C.H. Bonford; P.J. Gerone, and R.H. Wolf: Naturally acquired leprosy in a mangabey monkey (*Cercocebus* sp.). Int. J. Leprosy *48*, 495-496 (1980 b)
22. Meyers, W.M.; G.P. Walsh; H.L. Brown; Y. Fukunishi; C.H. Binford; P.J. Gerone, and R.H. Wolf: Naturally acquired leprosy in a mangabey monkey (*Cercocebus* sp.). Lab. Invest. *44*, 44 A (1981)
23. Meyers, W.M.; G.P. Walsh; H.L. Brown; R.J.W. Rees, and J. Convit: Naturally-acquired leprosy-like disease in the nine-banded armadillo *(Dasypus novemcinctus)*: Reactions in leprosy patients to lepromins prepared from naturally infected armadillos. J. reticuloendoth. Soc. *22*, 369-375 (1977)
24. Prabhakaran, K. and W.F. Kirchheimer: Use of 3,4-dihydroxyphenylalanine oxidation in the identification of *Mycobacterium leprae.* J. Bact. *92*, 1267-1268 (1966)
25. Shepard, C.C.: The experimental disease that follows the injection of human leprosy bacilli into food-pads of mice. J. exp. Med. *112*, 445-454 (1960)
26. Shepard, C.C. and Y.T. Chang: Effect of several anti-leprosy drugs on multiplication of human leprosy bacilli in foot-pads of mice. Proc. Soc. exp. Biol. *109*, 636-638 (1962)
27. Smith, J.H.; S.K. File; B.A. Nagy; D.S. Folse; J.A. Buckner; L.J. Webb, and A.M. Beverding: Leprosy-like disease of wild armadillos in French Acadiana, Louisiana. J. reticuloendoth. Soc. *24*, 705-719 (1978)
28. Storrs, E.E.: The nine-banded armadillo: A model for leprosy and other biomedical research. Int. J. Leprosy *39*, 703-714 (1971)
29. Walsh, G.P.; W.M. Meyers; C.H. Binford; P.J. Gerone; R.H. Wolf, and J.R. Leininger: Leprosy - a zoonosis. Leprosy Rev. *52* (Suppl. 1), 77-83 (1981)
30. Walsh, G.P.; E.E. Storrs; H.P. Burchfield; E. Cottrell; M.F. Vidrine, and C.H. Binford: Leprosy-like disease occurring naturally in armadillos. J. reticuloendoth. Soc. *18*, 347-351 (1975)
31. Wolf, R.H.; B.J. Gormus; L.N. Martin; G.B. Baskin; G.P. Walsh; W.M. Meyers, and C.H. Binford: Experimental leprosy in three species of monkeys. Science *227*, 529-531 (1985)

4 Fungi

4.1 Superficial Mycoses

Superficial mycoses are fungal diseases confined to the outermost layer of the skin and the hair (Kaplan, 1980).

4.1.1 Piedraia Hortae (P. Hortai)

P. hortae is the causative agent of black piedra in man and nonhuman primates (Kaplan, 1959). *P. hortae* has been isolated from *Lemur fulvus* (Kaplan, 1959); *Pan troglodytes, Pan paniscus* (Lochte, 1937; Takashio et de Vroey, 1975); *Ateles* sp., *Cebus* sp., *Saimiri* sp., and *Callithrix* sp. (Kaufmann et al., 1970).

Agent
P. hortae is a member of the Ascomycotina (Myriangiales, Loculoascomycetes) forming asci and ascospores of 25 to 55 μm size if growing around hairs (Fig. 15). It forms hard, gritty nodules around the hairs, which gave the condition its name (piedra in spanish means stone). The nodules are composed of cemented mycelium surrounding the hair shafts.

For *in vitro* isolation Sabouraud agar is recommended. Addition of thiamine supports growth, whilst media containing cycloheximide inhibit the growth of *P. hortae.* The extremely slowly growing colonies on Sabouraud agar are greenish-black to black or dark-brown, some strains may be cream colored. They are heaped, glabrous or covered by a fine downy mycelium giving a velvety or waxy appearance.

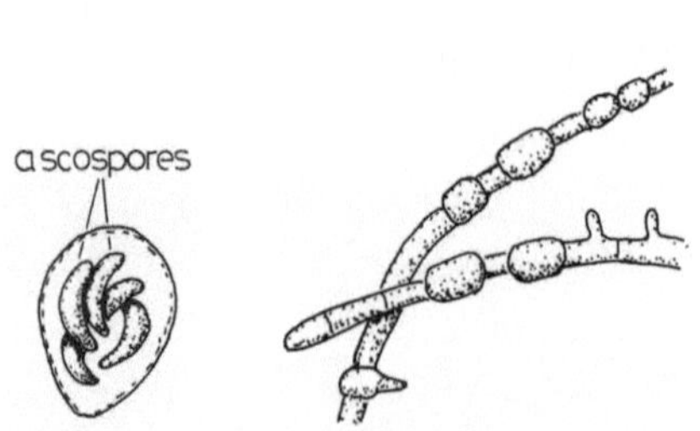

Fig. 15. *Piedraia hortae*

The colonies consist of dark, thick walled, closely septate hyphae with numerous swollen and irregularly shaped cells (Fig. 15), with asci and ascospores being rare in cultures (Ajello and Padhye, 1974; Campbell and Stewart, 1980). Identification depends more or less on the demonstration of tissue forms by examination of suspicious hairs in wet mounts of 10% KOH. After gentle heating the presence of septate, dematiaceous mycelium can be demonstrated on the surface of the hair filaments. Mature nodules, when crushed, reveal the presence of oval asci containing 8 aseptate, curved, spindle-shaped ascospores that bear a filament at each pole. The asci of Loculoascomycetes are double-walled (Gedek, 1980).

Epidemiology
Black piedra is limited to tropical and subtropical areas (Africa, Asia, Americas). *P. hortae* presumably occurs as a saprophyte in nature, probably in soil, but it has not yet been isolated from a nonliving source (Ajello and Padhye, 1974; Gedek, 1980). Man-man or animal-man transmission is not known.

Pathogenesis
The molds first penetrate into the hair shaft, multiply, and subsequently break out to the outside to form fructification mycelia which form the nodules around the hair. In other cases, the fungus invades the hair shaft and causes extensive damage to the hair by destruction of the cortex and medulla (Ajello and Padhye, 1974).

Clinical Symptoms
Nonhuman primates: *P. hortae* has been found particularly in the anogenital region (Takashio et de Vroey, 1975).

Man: Black piedra affects mainly the scalp hair, but only rarely the axillary and pubic hair. The hair filaments above the follicular orifice are overgrown by the mycelium, forming the pathognomonic hard, gritty nodules.

Pathology
No morphological lesions other than those of the hair are described.

4.1.2 Trichosporon Cutaneum (T. Beigelii)

T. cutaneum causes "White piedra" in man and horses. It has also been reported in a black spider monkey (Kaplan, 1959 - cit. from Kaplan, 1980) and from captive baboon skin (Al-Doory, 1967). Nonspecified *Trichospores* were reported from wild

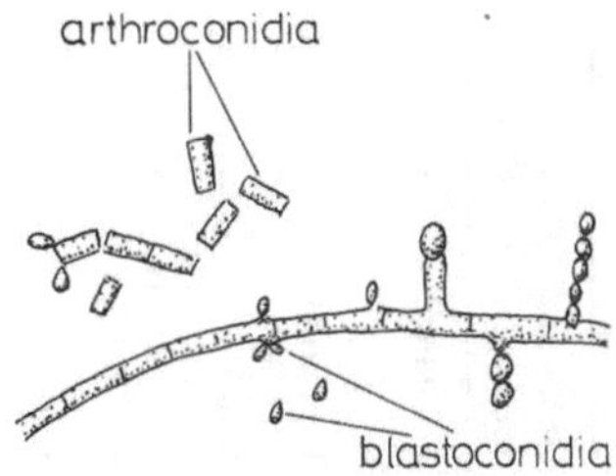

Fig. 16. *Trichosporon cutaneum*

Table 35. Biochemical Characteristics of *Trichosporon* Species and *Geotrichum candidum*

Genus and Species	Dex-trose	Mal-tose	Su-crose	Lac-tose	Galac-tose	Meli-biose	Cello-biose	Inosi-tol	Xy-lose	Raffi-nose	Treha-lose	Dulci-tol	Urease	KNO_3	Pseudo-hyphae	Growth at 37 °C	Growth in malt extract broth
T. cutaneum	+	+*	+*	+	+	+*	+	+*	+	+*	+*	+*	+*	–	+	+*	Pellicle or ring blastospores seen
T. pullulans	+	+	+	+	+	+*	+	+*	+	+*	+	–	+	+	+	+	Pellicle or rings with islets
T. penicillatum	+	–	–	–	+	–	–	–	+	–	–	–	–	–	+	+	White wrinkled pellicle
T. capitatum	+	–	–	–	+	–	–	–	–	–	–	–	–	–	+	+	Wrinkled pellicle, may be submerged
T. inkin	+	+	+	+	+	–	+	+	+	–	+	–	–	–	+	+	Pellicle formed
G. candidum	+	–	–	–	+	–	–	–	+	–	–	–	–	–	–	–	Pellicle or white islets

* Strain variation

living baboons (Al-Doory, 1969). White piedra is more sporadic and less common in occurrence and is not strictly limited to tropical areas as black piedra is (Kaplan, 1980).

Agent

T. cutaneum grows rapidly on Sabouraud agar or other artificial media not containing cycloheximide. The colonies growing within a few days are dull white or cream-colored, translucent, yeast-like, with age they become yellowish-gray and wrinkled.

Microscopically the colonies consist of hyaline hyphae which break up into arthroconidia, arthrospores and some characteristic blastospores (Fig. 16). The arthroconidia are of 2–4 × 3.5–9 μm size (Campbell and Stewart, 1980; Kaplan, 1980). *T. cutaneum (beigelii)* assimilates dextrose, lactose and xylose, whilst other carbon compounds are only occasionally assimilated (Ajello and Padhye, 1974; Silva-Hutner and Cooper, 1974; Koneman et al., 1978). For details see table 35.

Epidemiology

The infection is probably acquired by exposure to sources in the environment. Direct contact transmitted the infection from the beard to pubic hair or reverse between a patient and his two female companions (Grainger, 1986).

Clinical Symptoms

Nonhuman primates: No description of any lesions of white piedra in nonhuman primates could be found.

Man: White piedra is characterized by the development of soft, white to light-brown nodules on and within the hairshaft that may appear as a sheath surrounding the shaft. White piedra generally affects the hair of the beard and mustache, occasionally also the axillary, genital and scalp regions.

References

1. Ajello, L. and A. A. Padhye: Dermatophytes and the agents of superficial mycoses; pp. 469–481 in "Manual of Clinical Microbiology" 2nd ed. (Lennette, E. H., E. H. Spaulding and J. P. Truant, eds.). Amer. Soc. Microbiol. Washington, D. C. (1974)
2. Al-Doory, Y.: The mycoflora of the subhuman primate. III. The flora of the skin of the baboon, vervet, and gelada in captivity. Folia primatol. *7*, 292–298 (1967)
3. Al-Doory, Y.: The mycology of the freeliving baboon (Papio sp.). Mycopath. Mycol. Appl. (Den Haag) *38*, 7–15 (1969)
4. Campbell, M. C. and J. L. Stewart: The Medical Mycology Handbook. John Wiley and Sons, New York (1980)
5. Gedek, B.: Kompendium der medizinischen Mykologie. Pareys Studientexte 25. Parey, Berlin (1980)
6. Grainger, C. R.: White piedra: a case with evidence of spread by contact. Trans. roy. Soc. Trop. Med. Hyg. *80*, 87 (1986)
7. Kaplan, W.: The occurrence of black piedra in primate pelts. Trop. geogr. Med. *11*, 115–126 (1959)
8. Kaplan, W.: Superficial mycoses; pp. 437–440 in "CRC Handbook Series in Zoonoses" (Steele, J. H., Series ed.); Section A: Bacterial, Rickettsial, and Mycotic Diseases, Vol. II (Stoenner, H., W. Kaplan, and M. Torten, Section ed.), CRC Press, Boca Raton (1980)

9. Kaufmann, A.F.; G. Morris; J.H. Richardson; G. Healy, and W. Kaplan: A survey of newly arrived South American monkeys for potential human pathogens. Primate Zoonoses Surveil. Rep. *1*, 11-14 (1970)
10. Koneman, E.W.; G.D. Roberts, and S.F. Wright: Practical Laboratory Mycology, 2nd ed., Williams and Wilkins (1978)
11. Lochte, T.: Über das Vorkommen der Piedra beim Schimpansen und über die Beziehungen der tierischen Piedra zur menschlichen. Arch. Derm. Syph. *175*, 107-113 (1937)
12. Silva-Hutner, M. and B.H. Cooper: Medically important yeasts; pp. 491-507 in "Manual of Clinical Microbiology", 2nd ed. (Lennette, E.H., E.H. Spaulding, and J.P. Truant, eds.), Amer. Soc. Microbiol., Washington D.C. (1974)
13. Takashio, M. et C. de Vroey: Piedra noire chez des chimpanzés du Zaire. Sabouraudia *13*, 58-62 (1975)

4.2 Cutaneous Mycoses

Cutaneous mycoses are caused by specialized fungi able to penetrate and parasitize all fully keratinized tissues, but are unable to invade the subcutaneous or deeper layers of the skin. The most frequently encountered are caused by *Trichophyton* and *Microsporon* species.

4.2.1 Trichophyton Spp.

Trichophyton without further specification have been reported in captive baboons and vervets (Al-Doory, 1967), and *Hapalemur griseus* and *Phaner furcifer* (Niaussat et al., 1977).

4.2.1.1 Trichophyton Mentagrophytes (Perfect Stage: Arthroderma Benhamiae)

T. mentagrophytes ssp. induce ringworm or tinea in man and a variety of animals (Gedek, 1980; Padhye, 1980). The spectrum of animal hosts showing clinical signs of the infection include several species of nonhuman primates, such as *Macaca mulatta* (Böhm, 1968; Gugnani, 1971), *Cercopithecus ascanius* (Hauck and Klehr, 1977) and *Hylobates concolor* (Gugnani, 1971). Transmission of *T. mentagrophytes* from monkey to man has been established in at least two cases (Gugnani, 1971; Hauck and Klehr, 1977).

Agent
Most *Trichophyton* spp. are characterized by their club-shaped, smooth, thin-walled macroconidia (macroaleuriospores) with 8 to 10 septa ranging in size from 8 × 4 μm to 15 × 8 μm. They are positioned singly at the terminal ends of the hyphae or on short branches (Fig. 17). In *T. mentagrophytes* the macroconidia are usually abundant in zoophilic strains, but are lacking in anthropophilic strains

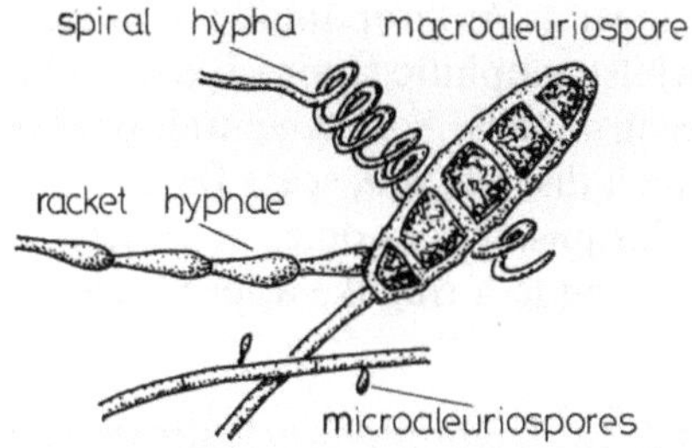

Fig. 17. *Trichophyton mentagrophytes*

(Campbell and Stewart, 1980). Microconidia (microaleuriospores) are generally very numerous, being spherical or clavate and 2 to 4 μm in size. They are formed singly along the septate hyphae (en threses) or in grape-like clusters (en grappe) (Koneman et al., 1978; Finegold and Martin, 1982). A characteristic of *T. mentagrophytes* are its spiral hyphae (Campbell and Stewart, 1980).

T. mentagrophytes grows rapidly and abundantly on Sabouraud-dextrose agar. It appears either as a "downy" variety (commonly in human tinea pedis) with flat, cream- or peach colored, coarsely granulated colonies or as a "granular" variety (common in zoophilic strains) with white, cottony colonies, some of which becoming distinctly folded. The reverse side of the colonies is rose-brown to orange or even deep-red (Finegold and Martin, 1982). *T. mentagrophytes* is urease-positive (Koneman et al., 1978). On potato dextrose agar it grows in stellate colonies (Campbell and Stewart, 1980).

Epidemiology

T. mentagrophytes is a dermatophyte with world wide distribution. Infections are usually contracted through direct contact with infected human beings or animals, especially wild rodents. The saprophytic life of the dermatophytes outside the host organisms with its survival for a long time in shedded hairs and crusts enables indirect transmission by contaminated equipment.

Pathogenesis

T. mentagrophytes invades the skin and nails. Invasion of the hair shaft is ectothric, first producing characteristic conical shaped points of entry.

Zoophilic strains cause an extreme inflammatory response (Campbell and Stewart, 1980).

Clinical Symptoms

Nonhuman primates: The lesions in rhesus monkeys were described as alopecia, scaly erythematous patches or scaly dermatitis, appearing as small foci or as an asbestos-like eczema (Böhm, 1968; Gugnani, 1971; Hauck and Klehr, 1977). Pruritus was reported only in a case with granulomatous lesions on the head (Gugnani, 1971).

The head seems to be the most commonly affected area of the body surface, but infections of the entire body or the flank region only, have been reported as well.

Infected hairs do not fluoresce under a Wood's lamp.

Man: Anthropophilic *T. mentagrophytes* cause several forms of tinea in man, whilst zoophilic strains induce a favus-like syndrome. The local infection spreads centrifugally from a central, primary focus simultaneously in all directions, forming a discoid, red, scaly focus whose border is usually walled and frequently contains pustules, nodules or vesicles. The healing stages also begin in the central part leading to a ringlike appearance (Gedek, 1980).

4.2.1.2 Trichophyton Rubrum (Perfect Stage Unknown)

T. rubrum is one of the main causative agents of tinea pedum and of onychomycosis in man (Gedek, 1980). *T. rubrum* or a closely related fungus was also isolated from two separate chimpanzee colonies (Otcenasek et al., 1967; Mortelmans, 1980).

Agent
T. rubrum is slowly growing, producing a flat or heaped-up colony with a white to reddish cottony or velvety surface and at first yellow, subsequently cherry red color on the reverse side. In older strains rings of deep red develop, and rugal folds are common.

Microconidia are rare in fluffy strains, but are more common in the velvety or granular strains. They grow in clusters of globose to clavate spores. Macroconidia are generally present, although in the granular strains they are not common. If formed, they appear as thin-walled, sausage-like, 3 to 8 septa containing cells with blunted ends (Koneman et al., 1978; Campbell and Stewart, 1980; Finegold and Martin, 1980).

T. rubrum does not penetrate the hair, urease production is slow or negative.

Clinical Symptoms
Nonhuman primates: Early lesions occur in different parts of the body in the chimpanzee and are covered with scales, with the hairs being lost in these affected areas (Otcenasek et al., 1967; Mortelmans, 1980).

Man: *T. rubrum* is the most common agent of tinea corporis, tinea pedis and onychomycosis; it rarely causes tinea capitis. A rare subcutaneous infection (Majocchis granuloma) is also caused by *T. rubrum.*

4.2.1.3 Trichophyton Simii (Perfect Stage: Arthroderma Simii)

A. simii, the perfect stage of *T. simii* (Stockdale et al., 1965) causes ringworm in man (Gugnani et al., 1967; Klokke and Durairaj, 1967; Rippon et al., 1968; Tewari, 1969), nonhuman primates (Stockdale et al., 1965; Okoshi et al., 1966) and domestic animals (Gupta et al., 1968; Tewari, 1969; Gugnani, 1971).

Agent
A. simii forms finely powdered colonies on Sabouraud-dextrose agar and malt extract agar, with a characteristic vinaceous pigment being produced on malt extract agar.

Clavate to pyriform microconidia are abundant. Macroconidia are numerous, cylindrical to fusiform and rapidly degenerating into chlamydospores, a feature typical and diagnostic of *A. simii* (Gugnani, 1971).

Clinical Symptoms
A. simii causes the typical ringworm lesions in both man and animals.

Treatment
The drug of choice in all *Trichophyton* spp.-infections is griseofulvin (Allen and Taplin, 1981), supported by topical tolnaftate or similar drugs.

4.2.2 Microsporum (Microsporon) Spp.

The genus *Microsporum* is characterized by large (8–15 × 35–150 μm), thick-walled, spindle-shaped, rough or spiny macroconidia containing 4 to 5 septa. The microconidia are clubshaped, borne at the hyphae either sessile or connected to it by short sterigmata.

4.2.2.1 Microsporum Canis (Perfect: Nannizzia Otae)

M. canis is primarily a pathogen of animals, being the principal agent of canine and feline ringworm. Man acquires the infection through contact with infected animals. Infections of nonhuman primates with *M. canis* were reported in a rhesus monkey (Baker et al., 1971), *Colobus polykomos* (Dougpagne, 1960 - cit. from Böhm, 1966), *Ateles* sp. (Kaufmann et al., 1970), gibbons (Kaben, 1963; Seeliger et al., 1963; Taylor et al., 1973), chimpanzees (Klokke and de Vries, 1963), orang-utans and gorillas (Böhm, 1966).

Agent
M. canis macroconidia are usually numerous, spindle-shaped, with some having a slight curve to their tips. The walls of the macroconidia become roughened first at the tips and subsequently all over.

M. canis is fast growing on Sabouraud dextrose agar. Within one week, the colonies have a flat, thin to velvety white surface and a radiating border. The reverse side of the colonies is pale to deep yellow (Koneman et al., 1978; Campbell and Stewart, 1980).

Epidemiology
Cats and dogs are the principal and reservoir hosts of the fungus. Spread of the fungus between canids and from dogs to man is well known. Man-man transmissions also occur, but after several human passages the fungus loses its infectivity for man and dies (Padhye, 1980).

Transmissions of *M. canis* from nonhuman primates to man (usually caretakers) have been reported or suspected several times (Böhm, 1966; Taylor et al., 1973).

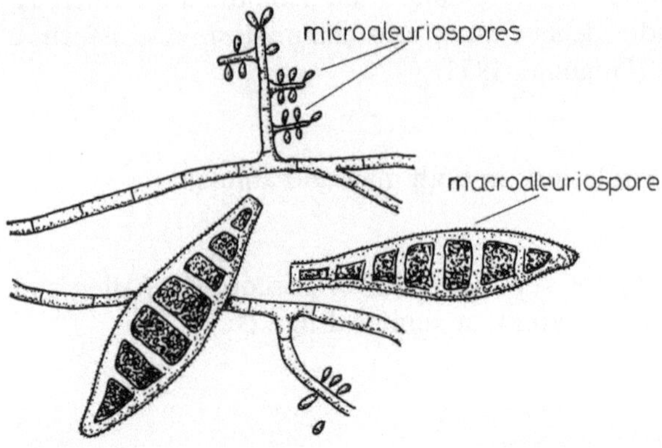

Fig. 18. *Microsporum canis*

Clinical Symptoms

Nonhuman primates: The signs associated with *M. canis*-infections are focal to generalized alopecia and circumscribed raised plaques on the skin without particular localization. The plaques are covered by yellowish-grey scales or white crusts, and are frequently surrounded by a narrow rim of erythema. The hairs within the plaques are broken after ectothrix invasion. Itching is not necessarily associated with the skin disease, fluorescence of the hairs under a Wood's lamp is usually lacking.

Man: *M. canis* causes tinea capitis or tinea corporis in man. Tinea capitis is characterized by round, scaly patches that have become bald by breakage of the infected hairs. Formation of purulent, boggy, granulomatous masses (kerions) is believed to be a hypersensitivity reaction (Allen and Taplin, 1981).

Pathology

In skin biopsies taken from plaques, marked thickening of the epidermis and dermis, hyperkeratosis and acanthosis, and inflammatory cells concentrated around the hair follicles were reported. Some follicles were completely replaced by polymorphonuclear leukocytes forming dermal abscesses (Baker et al., 1971).

4.2.2.2 Microsporum Gypseum (Perfect: Nannizzia Gypsea, Nannizzia Incurvata)

M. gypseum is a free-living geophilic saprophyte, that only rarely causes human or animal infections (Finegold and Martin, 1982). Accordingly the fungus has rarely been reported in nonhuman primates *(M. mulatta)* (Jänisch und Koch, 1964; Gugnani et al., 1971; Smith and Mehren, 1985).

Agent

M. gypseum grows readily on Sabouraud-dextrose agar forming flat, irregularly fringed colonies with a coarse powdery, fawn to buff or cinnamon-brown surface and an orange to brownish stained reverse side.

Macroconidia are usually present in large numbers and are characteristically large (25-60 × 7.5-16 µm), rounded - ellipsoidal, multiseptate (3 to 9 cells) spindles with echinulate surfaces (Koneman et al., 1978; Finegold and Martin, 1982). Microconidia are also seen (Campbell and Stewart, 1980).

Epidemiology
M. gypseum is common in soil. The most commonly infected animals are dogs and horses (Ajello and Padhye, 1974).

Clinical Symptoms
Nonhuman primates: The lesions correspond to those described with *M. canis.* They were limited to the head in *M. mulatta* (Jänisch und Koch, 1964), whereas in barbary apes *(Macaca sylvanus)* they consisted of hairless rings and patchy scaly dermatitis on the abdomen, head and arms (Göldenboth and Klös, 1971 - cit. from Smith and Mehren, 1985). In an orangutan crusty lesions involved one nostril as deep cracks associated with mild haemorrhage (Smith and Mehren, 1985).

Man: *M. gypseum* is another causative agent of tinea capitis, tinea corpis and tinea favosa (Campbell and Stewart, 1980).

4.2.2.3 Microsporum Fulvum (Perfect: Nannizzia Fulva)

M. fulvum, which is very similar to *M. gypseum,* has been isolated from *M. fascicularis* by Hasegawa and Yamamoto (1963 - cit. from Jänisch und Koch, 1964).

Agent
Growth, epidemiology and pathogenicity are very similar to that of *M. gypseum* (Ajello and Padhye 1974; Campbell and Stewart, 1980).

Treatment
The treatment of *Microsporum* spp.-infections is identical to that of *Trichophyton*-infections.

References

1. Ajello, L. and A.A. Padhye: Dermatophytes and the agents of superficial mycoses; pp. 469-481 in "Manual of Clinical Microbiology", 2nd ed. (Lennette, E.H., E.H. Spaulding, and J.P. Truant, eds.), Amer. Soc. Microbiol., Washington, D.C. (1974)
2. Al-Doory, Y.: The mycoflora of the subhuman primates. III. The flora of the skin of the baboon, vervet, and gelada in captivity. Folia primatol. *7,* 292-298 (1967)
3. Allen, A.M. and D. Taplin: "Superficial" fungal and bacterial skin infections; pp. 358-365 in "Infectious Diseases" (Sanford, J.P. and J.P. Luby, eds.); Grune and Stratton, New York (1981)
4. Baker, H.J.; L.G. Bradford, and L.F. Montes: Dermatophytosis due to *Microsporum canis* in a rhesus monkey. J. Amer. vet. med. Ass. *159,* 1607-1611 (1971)
5. Böhm, K.: Dermatomykosen bei Zootieren. Kleintierpraxis *13,* 139-141 (1968)
6. Böhm, K.H.: Eine Mikrosporie bei Orang-Utans und Gorillas. Kleintierpraxis *11,* 185-189 (1966)

7. Campbell, M.C. and J.L. Stewart: The Medical Mycology Handbook. John Wiley and Sons, New York (1980)
8. Finegold, S.M. and W.J. Martin: Bailey and Scott's Diagnostic Microbiology. 6th ed., C.V. Mosby, St. Louis (1982)
9. Gedek, B.: Kompendium der medizinischen Mykologie. Pareys Studientexte 25. Parey, Berlin (1980)
10. Gugnani, H.C.: *Trichophyton mentagrophytes* infection in monkeys and its transmission to man. Hindustan Antibiot. Bull. *14*, 11-13 (1971)
11. Gugnani, H.C.: *Arthoderma simii* infection in dogs. Hindustan Antibiot. Bull. *14*, 83-85 (1971)
12. Gugnani, H.C.; D.N. Mulay, and D.K. Murphy: Fungus flora of dermatophytosis and *Trichophyton simii* infection in North India. Indian J. Derm. Vener. *33*, 73-82 (1967)
13. Gugnani, H.C.; H.S. Randhawa, and J.B. Srivastav: Isolation of dermatophytes and other keratinophilic fungi from apparently healthy skin coats of domestic animals. Indian J. med. Res. *59*, 1699-1702 (1971)
14. Gupta, P.K.; R.P. Singh, and I.P. Singh: Dermatophytes from man, dogs and pigs with special reference to *Trichophyton simii* and *Microsporum nanum*. Indian J. Anim. Hlth *7*, 247-253 (1968)
15. Hauck, H. und N. Klehr: "Meerkatzenfavus" als Ursache für die Pilzinfektion eines Menschen. Z. Allgemeinmed. *53*, 331-332 (1977)
16. Jänisch, W. und H.A. Koch: Dermatomykosen bei Rhesusaffen. Kleintierpraxis *9*, 200-202 (1964)
17. Kaben, U.: Griseofulvintherapie bei einem an Mikrosporie erkrankten Weißhandgibbon. Mh. Vet. Med. *18*, 667-670 (1963)
18. Kaufman, A.F.; G. Morris; J.H. Richardson; G. Healy, and W. Kaplan: A survey of newly arrived South American monkeys for potential human pathogens. Primate Zoonoses Surveil. Rep. *1*, 11-14 (1970)
19. Klokke, A.H. and G.A. de Vries: Tinea capitis in chimpanzees caused by Microsporum canis Bodin 1902 resembling *M. obesum* Conant 1937. Sabouraudia *2*, 268-270 (1963)
20. Klokke, A.H. and P. Durairaj: The causal agents of superficial mycoses from rural areas of South India. Sabouraudia *5*, 153-158 (1967)
21. Koneman, E.W.; G.D. Roberts, and S.F. Wright: Practical Laboratory Mycology. 2nd. ed., Williams and Wilkins, Baltimore (1978)
22. Mortelmans, J.: Mycotic infections in captive wild mammals and birds: Some considerations on epizootiology, pathology, and prophylaxis; pp. 277-281 in "The Comparative Pathology of Zoo Animals" (Montali, R.J. and G. Migaki, eds.), Smithson. Inst. Press, Washington (1980)
23. Niaussat, M.M.; T.L. N'Guyen, et D. Mollin: Mycose à trichophyton chez des Lemurien. Sci. Tech. Anim. Lab. *2*, 51-54 (1977)
24. Okoshi, S.; M. Takashio, and A. Hasegawa: Ringworm caused by *Trichophyton simii* in a captive chimpanzee. Jap. J. med. Mycol. *7*, 204-207 (1966)
25. Otcenasek, M.; J. Dvorak, and K. Ladzianska: Trichophyton rubrum-like dermatophyte as a causative agent of dermatophytosis in chimpanzees. Mycopath. Mycol. appl. (Den Haag) *31*, 33-37 (1967)
26. Padhye, A.A.: Cutaneous mycoses; pp. 441-458 in "CRC Handbook Series in Zoonoses" (Steele, J.H., Series ed.), Section A: Bacterial, Rickettsial and Mycotic Diseases, Vol. II (Stoenner, H., W. Kaplan, and M. Torten, Section eds.), CRC Press, Boca Raton (1980)
27. Rippon, J.W.; A. Eng, and F.D. Malkinson: *Trichophyton simii* infection in United States. Arch. Derm. *98*, 615-619 (1968)
28. Seeliger, H.P.R.; W. Bisping und H.P. Brandt: Über eine Microsporum-Enzootie bei Kappen-Gibbons *(Hylobates lar)* verursacht durch eine Variante von *Microsporum* canis. Mykosen *6*, 61-68 (1963)
29. Smith, D.A. and K.G. Mehren: Dermatophytosis due to *Microsporum gypseum* in an orangutan *(Pongo pygmaeus)*. J. Zoo Anim. Med. *16*, 3-6 (1985)
30. Stockdale, P.M.; D.W.R. Mackenzie, and P.K.C. Austwick: *Arthroderma simii* sp. nov., the perfect state of *Trichophyton simii* (Pinoy) comb. nov. Sabouraudia *4*, 112-123 (1965)
31. Taylor, R.L.; F.C. Cadigan, jr. and V. Chaicumpa: Infections among Thai gibbons and humans caused by atypical *Microsporum canis*. Lab. Anim. Sci. *23*, 226-231 (1973)
32. Tewari, R.P.: *Trichophyton simii* infections in chicken, dog and man in India. Mycopath. Mycol. appl. (Den Haag) *39*, 293-298 (1969)

4.3 Subcutaneous Mycoses

4.3.1 Sporotrichosis

4.3.1.1 Sporothrix Schenkii (Sporotrichum Schenkii)

Sporotrichosis is a subacute to chronic fungal skin infection in man and animals, particularly of the deep subcutaneous tissues and lymphatics. Spread to other organs, although occurring, is generally rare. The most frequently affected animals are horses, dogs and cats. A natural infection has been reported in a chimpanzee (Jungerman and Schwartzman, 1972 - cit. from Kaplan and Ajello, 1980). Experimental infection of a finger of a rhesus monkey resulted in generalized lymphadenopathy, especially on the infected side (Benham and Keston, 1932).

Agent
S. schenkii is a dimorphic hyphomycete growing as a mold at room temperature (22 °C), and as a yeast at 37 °C.

The mycelium growing in colonies on standard fungus media does not look like that of a mold. On Sabouraud agar at room temperature grow small, white to cream-colored colonies within 3 to 5 days, the colonies become membranous, wrinkled and dark brown or black (Finegold and Martin, 1982). Microscopically the branched hyphae bear bouquet-like clusters of egg- or pear-shaped microconidia (2 to 6 µm), which are born directly from the hyphae or from the tips of small conidiophores. Older cultures may produce larger, thick-walled chlamydospores.

The yeast phase is characterized by small spindle- to cigar-shaped, round or oval budding cells of 1 to 4 µm size growing in smooth, tan colonies within 3 to 5 days.

The tissue form appears as small, oval, budding, yeastlike cells which can be demonstrated by Gomori- or Gridley-staining techniques.

Epidemiology
Sporotrichosis is a cosmopolitic, but sporadic disease favored by high environmental temperatures and humidity. *S. schenkii* is found in soil and on decaying plant material. Infections usually require traumatic puncture wounds by thorns, splinters or possibly insects. Pulmonary infections, however, may result from inhalation of the fungus. Direct transmission from host to host is unusual, although human infections by dogs or cats, probably via scratches or bites, have been reported (Gedek, 1980).

Clinical Symptoms
Man: The most common form is cutaneous sporotrichosis, which is associated with regional lymphangitis and lymphadenitis of the draining lymph nodes. Several days or weeks following infection, the lymphatics become corded and strings of nodules are formed radiating from the port of entry along the regional lymphatics. The nodules eventually ulcerate and discharge a liquid pus. The infec-

tion may become generalized with infection of the bones, joints, muscles, testicles and internal organs.

Pathology
Man: The nodules consist of centrally necrotic granulation tissue containing epitheloid cells and giant cells which is surrounded by a peripheral zone of connective tissue. The epidermis may be thickened due to infiltration by lymphocytes and occasional giant cells.

Treatment
As potassium iodide is specific for sporotrichosis, the ulcers and wound should therefore be treated accordingly (Smith, 1981).

4.4 Systemic Mycoses

4.4.1 Zygomycosis

Zygomycosis, formerly termed phycomycosis, hyphomycosis etc. is a general term that includes diseases caused by fungi of the class Zygomycetes (formerly Phycomycetes). The genera causing zygomycosis belong to the orders Entomophthorales (Basidiobolus and Conidiobolus) and Mucorales (Absidia, Cunninghamella, Mucor, Rhizomucor, Rhizopus, Saksenea, Mortiella).

4.4.1.1 Entomophthorales

Entomophthorales are primarily parasites of insects with hyphal bodies being formed in the infected insects. Multiplication is by budding or by zygospores after copulation of hyphal bodies. Conidia are developed on the hyphae which grow through the stigmata of the insects to the outside (Fromentin, 1975; Gedek, 1980).

4.4.1.1.1 Entomophthora (Delacroixia) Coronata

E. coronata causes a disease known as *Rhinophycomycosis entomophthorae* in man and horses (Emmons and Bridge, 1961; Fromentin, 1975, 1976; Gedek, 1980). Most cases have been reported as occurring in Nigeria (Campbell and Stewart, 1980). *Rhinophycomycosis entomophthorae* also occurred in a chimpanzee in the wild in East Africa (Roy and Cameron, 1972; Roy, 1974) and probably in a mandrill of a zoo as judged from the tissue section (Migaki et al., 1982).

Agent
The fungus develops on Sabouraud agar as a flat, gray colony with usually satellite colonies around it, due to its heavy sporulation (Greer, 1974). Microscopically the hyphae are not or only poorly septated and produce many chlamydospores.

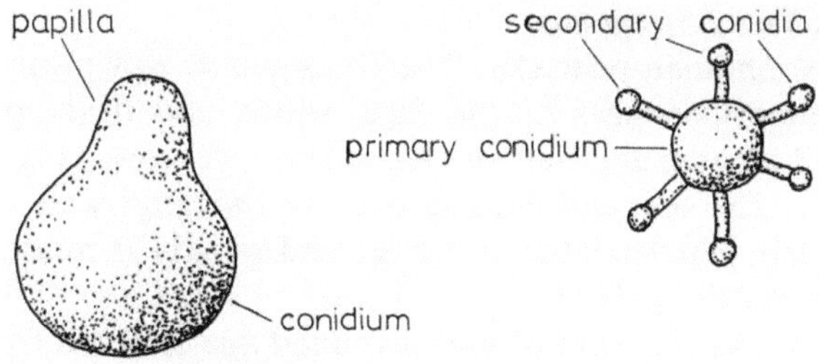

Fig. 19. *Entomophthora coronata*

Fig. 20. Chimpanzee 8 months after biopsy and commencement of treatment, showing enlargement and extension of the lesions to involve the tissues surrounding the right eye. (Roy and Cameron 1972)

Numerous globose conidia are born singly on short, slender conidiophores and are forcibly expelled on maturity without an attached hyphal fragment. The conidia have a prominent papilla (Fig. 19) and produce multiple secondary conidia in the form of a "corona" (Fig. 19). Aging conidia develop characteristic hairlike appendages.

E. coronata produces proteases *in vitro* (Fromentin, 1975, 1976).

Epidemiology

The exact epidemiology of the fungal disease is unknown. Microtraumas e.g. by insects (Gedek, 1980) or exposure to soil (Campbell and Stewart, 1980) are considered to be important factors.

Clinical Symptoms

Nonhuman primates: The chimpanzee was presented with a subcutaneous hard, tumorlike mass of the face, which had developed gradually over a period of 18 months and had spread to both eyebrows (Fig. 20).

The skin and mucosa over the swellings were intact (Roy and Cameron, 1972). One year later the swellings had greatly increased and the animal was virtually blind (Roy, 1974).

The mandrill of undetermined age suffered from a nonspecific illness, characterized by severe depression, weakness, excessive salivation and vague respiratory distress (Migaki et al., 1982). The animal died seven days after the initial signs of illness.

Man: The disease so far has occurred only in adults from 20 to 40 years of age, being more frequent in males than in females. The disease starts in the nasal tissues, where a painless, swollen mass develops, which may disfigure the entire face (Campbell and Stewart, 1980; Gedek, 1980).

Pathology

Nonhuman primates: The tumor-like tissues consisted of fibromuscular tissue with focal inflammatory reactions containing eosinophils, plasma cells, lymphocytes, and multinucleated giant cells (Fig. 21). The fungi were demonstrable as broad, septate, eosinophilic, PAS-positive hyphae. The hyphae frequently elicited Hoeppli's phenomena (Roy and Cameron, 1972; Migaki et al., 1982).

4.4.1.2 Mucorales

Of the Mucorales only the genus Mucor is a pathogen in both man and nonhuman primates.

4.4.1.2.1 Mucor Sp.

Mucorales are saprophytes common throughout the world which in rare instances become pathogens in man and animals. Infections by *Mucor* sp. in man are of a more acute nature and occur particularly in immunocompromised hosts (diabetics, leukemia patients, severe burns, cortisone treatment). Similarly to man the mucormycoses of nonhuman primates appeared as acute diseases in contrast to the more chronic nature of mucormycosis in domestic animals. Although most cases of mucormycosis in nonhuman primates occurred in Old World monkeys – *M. mulatta* (Gisler and Pitcock, 1963; Hessler et al., 1967; Martin et al., 1969), *Cercocebus galeritus chrysogaster* (Kageruka and de Vroey, 1972) and *Mandrillus sphinx* (Lucke and Linton, 1965; Lucke and Migaki, 1979), at least one case was reported in a South American spider monkey *(Ateles belzebuth hybridus)* kept at the London Zoo (Fiennes, 1967).

Agent

Mucorales are characterized by broad, nonseptated hyphae and the formation of sporangia filled by sporangiospores. Within the mucorales the genera are distin-

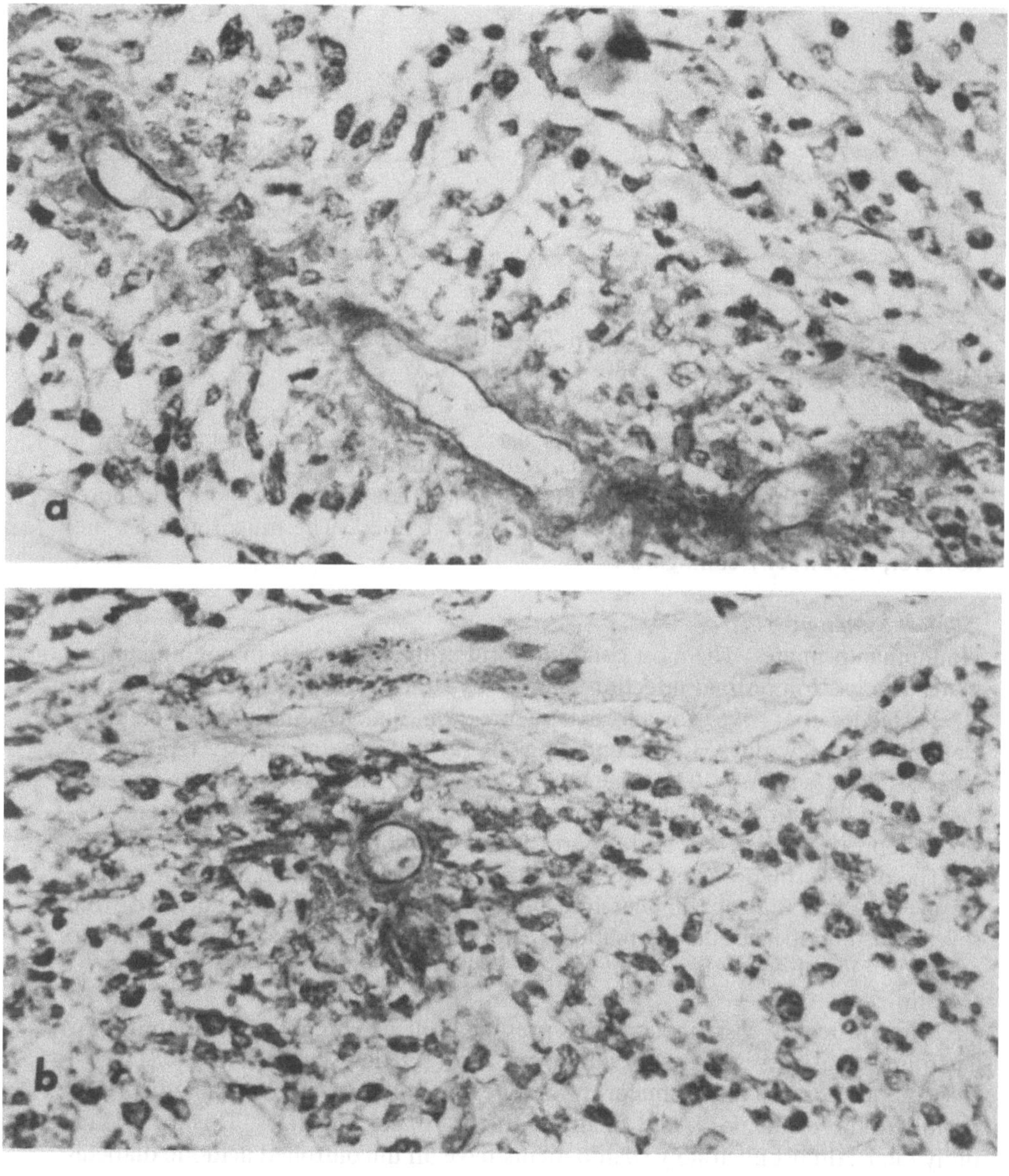

Fig. 21a, b. Fungal hyphae cut longitudinally. Note surrounding zone of "fibrinoid" material (PAS; ×700; Roy and Cameron 1972)

guished according to their fructification organs and the existence of mycoids. In the *Mucor* sp. the sporangia are always supported by a columella-bearing sporangiophore (Fig. 22).

Rhizoids are lacking in contrast to the closely related *Rhizopus* and *Absidia* spp.

Mucor grows rapidly on almost all nutrient media not containing cyclohexi-

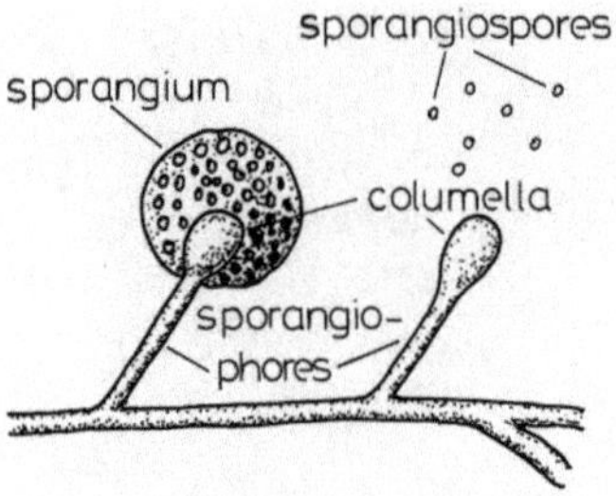

Fig. 22. *Mucor* sp.

mide. A Petri-dish is entirely filled in 2 to 5 days with a loose, gray, woolly mycelium (Campbell and Stewart, 1980; Gedek, 1980).

Epidemiology
The mucorales are common in soil, particularly if fertilized, on organic matter and other substrates high in carbohydrate contents. They liberate sporangiospores which can be transmitted by air or contaminated food. The mucormycoses are, however, not directly contagious.

Clinical Symptoms
Nonhuman primates: The most common form of mucormycosis in nonhuman primates is a gastrointestinal infection marked by sudden death (Gisler and Pitcock, 1962; Hessler et al., 1967) or anorexia, depression, diarrhea and progressive emaciation over a few days (Lucke and Migaki, 1979). In the mangabey reported by Kageruka and de Vroy (1971) a generalized infection was associated with klebsiellosis. As in gastrointestinal mucormycosis the only symptoms observed were anorexia, depression, diarrhea, and progressive emaciation.

The other form, observed once in a rhesus monkey suspected of having diabetes mellitus was that of rhinocerebral phycomycosis. The monkey suffered from severe pustular periorbital dermatitis, a tumorlike fistulous mass progressing to extensive ulceration of the orbital and paranasal skin and of mucous membranes (Martin et al., 1969).

Man: Mucormycosis in man can present itself as rhinocerebral, pulmonary, gastrointestinal, disseminated or cutaneous form (Gedek, 1980; Mohr, 1981).

Rhinocerebral (cranial) mucormycosis begins in the nasal mucosa and spreads rapidly to the palate, face, and finally to the brain and meninges. This form is rapidly fatal. Almost invariably such patients have an uncontrolled acidotic diabetes (Ajello and Kaplan, 1980).

Pulmonary mucormycosis involves the bronchi and the pulmonary arteries, producing pulmonary infarction. Pulmonary mucormycosis may start gradually or suddenly with chest pain, fever, haemoptysis and friction rub.

Gastrointestinal forms are usually associated with malnutrition or amebic colitis. The symptoms include abdominal pain, diarrhea and bloody stools.

Disseminated mucormycosis almost always develops in patients with severe underlying diseases. The symptoms are dominated by the primary disease.

Subcutaneous forms are relatively common in tropical Africa and are characterized by hard, freely movable lumps in the thighs, forearms and buttocks.

Pathology

Nonhuman primates: Gastrointestinal mucormycosis is characterized by severe haemorrhagic-ulcerative gastroenteritis (Hessler et al., 1967; Lucke and Migaki, 1979) or a typhlo-colitis of the same kind (Gisler and Pitcock, 1962).

In disseminated mucormycosis tumorlike, haemorrhagic-necrotic nodules and ulcers were distributed throughout the gastrointestinal tract, liver, spleen, kidneys, myo- and endocardium, lungs and muscles (Kageruka and de Vroey, 1972).

The rhinocerebral form in the rhesus monkey observed by Martin et al. (1969) involved the dermis of the frontal region, which was markedly thickened and contained numerous abscesses and ulcers; as well as the frontal bone, the frontal and paranasal sinuses, the turbinates and the periorbital connective tissues.

Nonseptate, branched hyphae can usually be demonstrated in the granulomas by conventional histologic staining methods.

Man: Gastrointestinal mucormycosis produces ulcers and gangrene in parts of the gastrointestinal tract. Dissemination to other organs also occurs.

In pulmonary forms the fungus grows mainly in bronchi, but it can also invade the pulmonary arteries, inducing pulmonary infarction and haematogenous spread to other organs including the brain.

Rhinocerebral mucormycosis, which begins in the nasal mucosa, spreads rapidly by vascular thrombosis and infarction into the nasal mucosa and may reach the CNS.

Treatment

Treatment of nonhuman primates is usually impossible due to the diagnostic imponderabilities. In man amphothericin B or 5-fluorcystin are recommended, but more important seems to be the treatment of the underlying diseases, particularly the diabetes control (Gedek, 1980).

References

1. Ajello, L. and W. Kaplan: Systemic mycoses; pp. 469-485 in "CRC Handbook Series in Zoonoses" (Steele, J.H., Series ed.), Section A: Bacterial, Rickettsial, and Mycotic Diseases, Vol. II (Stoenner, H.; W. Kaplan, and M. Torten, Section eds.) CRC Press, Boca Raton (1980)
2. Benham, R.W. and B. Keston: Sporotrichosis: Its transmission to plants and animals. J. Infect. Dis. *50*, 437-458 (1932)
3. Campbell, M.C. and J.L. Stewart: The Medical Mycology Handbook. John Wiley & Son, New York (1980)
4. Emmons, C.W. and C.H. Bridges: *Entomophthora coronata*, the etiologic agent of phycomycosis of horses. Mycologia (N.Y.) *53*, 307-312 (1961)
5. Fiennes, R.: Zoonoses of Primates. The Epidemiology and Ecology of Simian Diseases in Relation to Man. Weidenfeld and Nicolson, London (1967)
6. Finegold, S.M. and W.J. Martin: Bailey and Scott's Diagnostic Microbiology. C.V. Mosby, St. Louis (1982)
7. Fromentin, H.: Résultats préliminaires sur les protéases *d'Entomophthora coronata* pathogène de mammifères. Bull. Soc. Mycol. Med. *4*, 215-220 (1975)
8. Fromentin, H.: Activité caséinolytique d'Entomophthora coronata au cours de sa croissance en culture stable. Mycopathologia *59*, 43-45 (1976)
9. Gedek, B.: Kompendium der medizinischen Mykologie. Pareys Studientexte 25. Parey, Berlin (1980)

10. Gisler, D.B. and J.A. Pitcock: Intestinal mucormycosis in the monkey (Macaca mulatta). Amer. J.vet. Res. *23*, 365-366 (1962)
11. Greer, D.L.: Fungi of phycomycosis; pp.541-549 in "Manual of Clinical Microbiology", 2nd ed. (Lennette, E.H., E.H. Spaulding, and J.P. Truant, eds.) Amer. Soc. Microbiol., Washington, D.C. (1974)
12. Hessler, J.R.; J.C. Woodard; R.J. Beattie, and A.F. Moreland: Mucormycosis in a rhesus monkey. J.Amer. vet. med. Ass. *151*, 909-913 (1967)
13. Kageruka, P. and C. de Vroey: Generalized mucormycosis in the golden-bellied mangabey (Cercocebus galeritus chrysogaster, Lydekker). Acta zool. path. antverp. *55*, 19-28 (1972)
14. Kaplan, W. and L. Ajello: The subcutaneous mycoses; pp.459-468 in "CRC Handbook Series in Zoonoses" (Steele, J.H., Series ed.), Section A: Bacterial, Rickettsial, and Mycotic Diseases, Vol. II (Stoenner, H., W. Kaplan, and M. Torten, Section eds.), CRC Press, Boca Raton (1980)
15. Lucke, V.M. and A.H. Linton: Phycomycosis in a mandrill *(Mandrillus sphinx)*. Vet. Rec. *77*, 1306-1309 (1965)
16. Lucke, V.M. and G. Migaki: Diagnostic exercise. Lab. Anim. Sci. *29*, 601-602 (1979)
17. Martin, J.E.; D.J. Kroe; R.E. Bostrom; D.J. Johnson, and R.A. Whitney, jr.: Rhino-orbital phycomycosis in a rhesus monkey *(Macaca mulatta)*. J.Amer. vet. med. Ass. *155*, 1253-1257 (1969)
18. Migaki, G.; J.D. Toft II, and R.E. Schmidt: Disseminated entomophthoromycosis in a mandrill *(Mandrillus sphinx)*. Vet. Path. *19*, 551-554 (1982)
19. Mohr, J.A.: Mucormycosis; pp.355-356 in "Infectious Diseases" (Sanford, J.P. and J.P. Luby, eds.), Grune and Stratton, New York (1981)
20. Roy, A.D.: Rhinophycomycosis entomophthorae occurring in a chimpanzee in the wild in East Africa. Further report. Amer. J.trop. Med. Hyg. *21*, 935 (1974)
21. Roy, A.D. and H.M. Cameron: Rhinophycomycosis entomophthorae occurring in a chimpanzee in the wild in East Africa. Amer. J.trop. Med. Hyg. *21*, 234-237 (1972)
22. Smith, J.W.: Sporotrichosis; pp.356-357 in "Infectious Diseases" (Sanford, J.P. and J.P. Luby, eds.), Grune and Stratton, New York (1981)

4.5 Histoplasmosis

The *Histoplasma*-species are dimorphic hyphomycetes, which exist in the environment as molds and in avian and mammalian organisms as yeasts.

4.5.1 Histoplasma Capsulatum

Human histoplasmosis due to *H. capsulatum* is an acute to chronic respiratory disease. The infection, although worldwide occurring, is concentrated to the Mississippi-area of the USA and parts of Latin America (Mexico, Guatemala, Venezuela, Peru). Besides man dogs are the most frequent victims (Ajello and Kaplan, 1980). *H. capsulatum*-histoplasmosis has also been reported in a squirrel monkey *(Saimiri sciureus)* from a Minneapolis area pet shop (Bergeland et al., 1970). Experimental infections of baboons *(Papio cynocephalus)* resulted in seroconversion (CF-tests, agar-gel diffusion tests), in positive skin tests, cultural recovery of the agent and histopathological lesions when sacrified 6 to 24 months later (Al-Doory et al., 1969). Similar infection of *M. fascicularis* induced only seroconversion in CF-tests and positive skin tests after 2 months but no disease or morphical lesions (Hill and Marcus, 1957).

Agent
Cultivation of *H. capsulatum* in its mycelial (mold) phase is readily achieved on both blood-containing or blood-free nutrient media incubated at room temperature (25 °C). After 7 to 10 days first moist and waxy, later white, cottony colonies appear on Sabouraud agar. The colonies subsequently become yellowish, followed by a brown to black color. Finally the colonies become powder-like. On blood-containing media the fungus grows as small, white, adherent colonies embedded into the agar. The colonies reach a diameter of 6 to 8 mm after several weeks and become wrinkled in the centre (Hartung and Salfelder, 1969). *H. capsulatum* is urease positive (Gedek, 1980).

Microscopically the mold phase consists of small, branched, septate hyphae. Small round to teardrop microconidia (microaleuriospores) develop on short lateral branches along the hyphae, but these are generally rare in cultures. The diagnostic forms are the large, spherical to pyriform, tuberculated (finger-like projections) macroconidia (macroleuriospores) on narrow conidiophores (Fig. 23).

Conversion to the yeast form is sometimes difficult. It develops as such at 37 °C on a rich medium such as brain heart infusion agar with 5% blood. The colonies of the yeast-form are soft, cream-colored, and pasty in texture. Microscopically, these are seen as small, oval budding cells of 2 to 4 μm size (Fig. 23), which in earlier stages may resemble arthrospores (Koneman et al., 1978; Newberry, 1981).

Within infected human or animal tissues only the yeast phase develops. The budding yeasts (1 to 3 μm in diameter) are found either in the cytoplasm of large macrophages or lying free within the tissue (Koneman et al., 1978). They appear to be surrounded by a capsule or halo thought to be a shrinking artifact caused by the fixation of the tissue (Fig. 23).

Epidemiology
Histoplasma spp. do not spread by contagion but by inhalation of the airborne spores. Gregarious birds (blackbirds, pigeons, chickens) are considered to be the main reservoir hosts. Their droppings infect the soil, chicken coops, bell- and church-towers, attics and caves (Al-Doory and Rhoades, 1968; Gedek, 1980). Disturbance of such sites may result in the release of large amounts of spore-containing dusts, which are disseminated by air currents (Ajello and Kaplan, 1980).

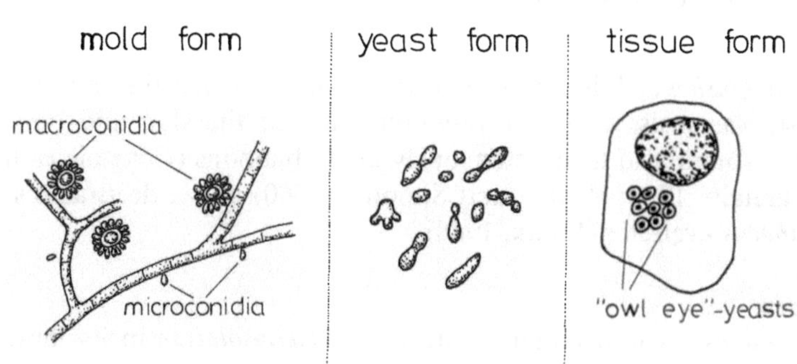

Fig. 23. *Histoplasma capsulatum*

Clinical Symptoms

Nonhuman primates: The squirrel monkey was noticed to be anorectic, depressed, and having respiratory difficulties. It died after an illness of 3 days duration.

Man: Most infections are subclinical, but sometimes they produce mild diseases consisting of headache, fever, chills, myalgia, erythema multiforma or nodosa, nausea and vomiting. Nonproductive cough, dyspnea and pleuritic chest pains may follow and persist for 2 to 3 months.

Chronic histoplasmosis, which occurs mostly in middle-aged or elderly white men simulates chronic tuberculosis. Symptoms include fever, malaise, anorexia, fatigue, chills, night sweats, chest pain, dyspnea, haemoptisis and a productive (purulent) cough.

Disseminated disease occurs most commonly in very young, very old or immunocompromised patients. Untreated disseminated histoplasmosis may be fatal (Newberry, 1981).

Pathology

Nonhuman primates: The disease is characterized by a pyogranulomatous pneumonia. White, irregular nodules are distributed throughout the entire lung, with the pleural cavity being filled by a serosanguinous exudate. Additional microgranulomas were present in the spleen and liver (Bergeland et al., 1970).

Man: The lesions of primary histoplasmosis consist of histiocyte-rich pyogranulomatas of varying size particularly in the lower parts of the lung. Confluence of the primary granulomas may result in large, centrally caseous, sometimes calcified masses which also involve the regional lymphnodes (Hartung and Salfelder, 1969). The primary infection and reinfection might also lead to extensive bronchopneumonia, which can be associated with the formation of caverns.

Dissemination causes similar metastases particularly in the spleen, but also in other organs including the brain (Hartung und Salfelder, 1969).

Treatment

No satisfactory treatment of clinical histoplasmosis is known, amphothericin B or saramycetin may be tried (Gedek, 1980).

4.5.2 Histoplasma Duboisii

H. duboisii which is very similar to *H. capsulatum,* is the causative agent of African histoplasmosis, a disease predominantly of the skin and bones. *H. duboisii* has been found also to spontaneously infect baboons (Courtois et al., 1955; Mariat et Segretain, 1956; Walker and Spooner, 1960) and a de Brazza's monkey - *Cercopithecus neglectus* (Frank, 1968).

Agent

H. duboisii is undistinguishable from *H. capsulatum* in its mold form. The yeast form, however, consists of larger (7 to 20 µm), round or oval, uninucleated cells with 1,5 µm thick walls. Buds are attached by a narrow neck (Campbell and Stewart, 1980). *H. duboisii* is an intracellular parasite like *H. capsulatum*. In cultures it is urease negative.

Epidemiology

The epidemiology of *H. duboisii* is probably identical to that of *H. capsulatum*. The fungus has so far only been isolated in Africa, it was recovered in its mold phase from soil in Kenya (Al-Doory and Kalter, 1967).

Clinical Symptoms

Nonhuman primates: The baboons suffered from cutaneous histoplasmosis with the lesions appearing on the tail, buttocks and fingers (Courtois et al., 1955; Walker and Spooner, 1960). The only sign in the de Brazza's monkey was weight loss which ultimately led to the sacrifice of the animal (Frank, 1968).

Man: African histoplasmosis produces granulomatous lesions particularly in the skin and skeletal system.

Pathology

Nonhuman primates: The cutaneous histoplasmosis of the baboons appeared as abscesses and ulcers in the affected parts of the skin (Courtois et al., 1955; Walker and Spooner, 1960). In the systemic histoplasmosis of the de Brazza's monkey the most severe lesions were noticed in both kidneys. These were greatly enlarged due to a massive, uniform infiltration by macrophages containing up to 3 yeasts in their cytoplasm. Minor lesions were also present in the liver as moderate periportal lymphocytic-histiocytic infiltration, with the fungi being present in some histiocytes. Fungi were also demonstrated in lung macrophages without inflammatory response.

Man: The skin is the principal site of infection, the normal tissues being replaced by a granulomatous tissue usually replete with giant cells (Al-Doory, 1972). The granulomatous tissues may erode the epidermis causing ulceration, or it may extend deep into the thickened subcutis.

Generalized forms spread to the lymph nodes, spleen, liver, lung and bone, with oesteolysis being induced in the latter.

The fungus, like *H. capsulatum* grows predominantly in large giant cells, which may break down, thereby liberating solid masses of extracellular *Histoplasma* yeast cells (Al-Doory, 1972).

References

1. Ajello, L. and W. Kaplan: Systemic mycoses; pp. 469-485 in "CRC Handbook Series in Zoonoses" (Steele, J. H., Series ed.); Section A: Bacterial, Rickettsial, and Mycotic Diseases, Vol. II (Stoenner, H.; W. Kaplan, and M. Torten, Section eds.), CRC Press, Boca Raton (1980)
2. Al-Doory, Y.: Systemic mycoses; pp. 224-242 in "Pathology of Simian Primates." Part II (Fiennes, R. N. T-W, ed.), Karger, Basel (1972)
3. AL-Doory, Y. and S. S. Kalter: The isolation of *Histoplasma duboisii* and keratinophilic fungi from soils of East Africa. Mycopath. Mycol. appl. (Den Haag) *31*, 289-295 (1967)
4. Al-Doory, Y. and E. R. Rhoades: Isolation of Histoplasma capsulatum from a Texas cave. Mycopath. Mycol. appl. (Den Haag) *35*, 201-207 (1968)

5. Al-Doory, Y.; T.E. Vice, and S.S. Kalter: Experimental histoplasmosis in the baboon (Papio sp.). Mycopath. Mycol. appl. (Den Haag) *38*, 83-92 (1969)
6. Bergeland, M.E.; D.M. Barnes, and W. Kaplan: Spontaneous histoplasmosis in a squirrel monkey. Primate Zoonoses Surveil. Rep. *1*, 10-11 (1970)
7. Campbell, M.C. and J.L. Stewart: The Medical Mycology Handbook. John Wiley and Sons, New York (1980)
8. Courtois, G.; G. Segretain; F. Mariot, and J.C. Levaditi: Mycose cutanée à corps leveriformes observée chez des singes africains en captivité. Ann. Inst. Pasteur *89*, 124-127 (1955)
9. Frank, H.: Systemische Histoplasmose bei einem afrikanischen Affen. Dtsch. tierärztl. Wschr. *75*, 371-374 (1968)
10. Gedek, B.: Kompendium der medizinischen Mykologie. Pareys Studientexte 25. Parey, Berlin (1980)
11. Hartung, M. und K. Salfelder: Die Histoplasmose; pp. 79-127 in "Infektionskrankheiten, Band III: Mykosen, Aktinomykosen und Nocardiosen, Pneumokokken- und Klebsiellenerkrankungen" (Gsell, O. und W. Mohr, eds.), Springer, Berlin (1969)
12. Hill, G.A. and S. Marcus: Challenge of Macaca irus with Histoplasma capsulatum. Amer. Rev. Tuberc. Pulmon. Dis. *75*, 849-851 (1957)
13. Koneman, E.W.; G.D. Roberts, and S.F. Wright: Practical Laboratory Mycology. 2nd ed. William + Wilkins, Baltimore (1978)
14. Mariat, E. and G. Segretain: Etude mycologique d'une histoplasmose spontanée du singe africain (Cynocephalus babuin). Ann. Inst. Pasteur *91*, 874-891 (1956)
15. Newberry, W.M.: Histoplasmosis; pp. 345-346 in "Infectious Diseases" (Sanford, J.P. and J.P. Luby, eds.), Grune and Stratton, New York (1981)
16. Walker, J. and E.T.C. Spooner: Natural infection of the African baboon *(Papio papio)* with the large-cell form of *Histoplasma*. J. Path. Bact. *80*, 436-438 (1960)

4.6 Candidiasis

Members of the genus *Candida* are yeast-like fungi. They are frequently present as part of the normal flora of the skin and mucous membranes, which, under certain circumstances may become true pathogens in man and animals causing moniliasis (thrush) or even systemic disease.

4.6.1 Candida Albicans

The most frequently encountered species in man and nonhuman primates is *C. albicans*, which in the latter causes predominantly lesions in the upper alimentary tract (Migaki et al., 1982). Candidiasis in nonhuman primates does not show any host-species preferences, although sometimes a higher incidence in New World species has been suggested (Nelson et al., 1966). The infection has been reported in orangutans (Schneider et al., 1983), gorillas (Fiennes, 1967; Schneider et al., 1983); chimpanzees (Schmidt and Butler, 1970); gibbons (Saez, 1975); siamangs (Schneider et al., 1983); baboons (Al-Doory, 1967a, 1969; Saez, 1969a, 1970a; Saez et al., 1978; Saez and Rinjard, 1978), mandrills (Schneider et al., 1983), macaques (Kerber et al., 1968; Kaufmann and Quist, 1969; Saez, 1969b, 1970b; Saez and Rinjard, 1970), *Cercopithecus petaurista* (Schneider et al., 1983), spider monkeys (Patterson et al., 1974; Schneider et al., 1983), capuchin monkeys (Wikse et al., 1970; McCullough et al., 1977) and marmosets (Nelson et al., 1966; own results - unpublished).

Agent

C. albicans grows within 3 days on most nutrient media, but Tween 80-cornmeal-agar is recommended for better demonstration of the characteristic features. The colonies growing on Sabouraud dextrose agar are white, creamy, shiny and wrinkled, whilst on EMB-agar incubated at 37 °C in 5 to 10% CO_2 feathering (Fig. 24) may be seen. On bromcresol green agar at room temperature, the colonies appear white to pale green or pale yellow (Koneman et al., 1978; Campbell and Stewart, 1980).

Microscopic examination of the colonies reveals yeast-like budding cells, pseudohyphae and hyphae. Those grown on cornmeal agar additionally produce spherical, thickwalled chlamydospores of 8 to 10 µm diameter. Tight clusters of budding cells (blastoconidia) are also seen along the hyphae (Fig. 25).

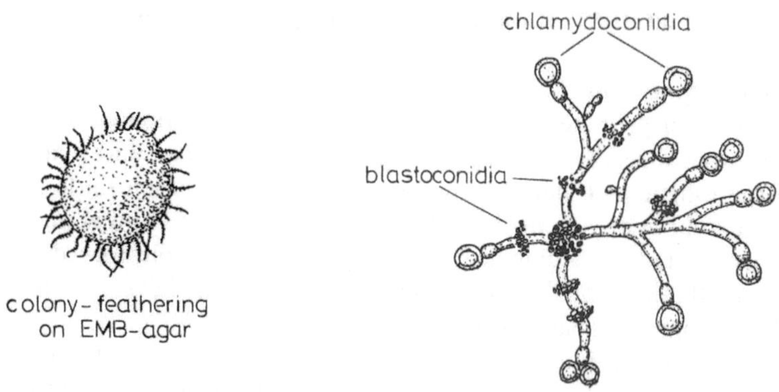

Fig. 24. *Candida albicans*

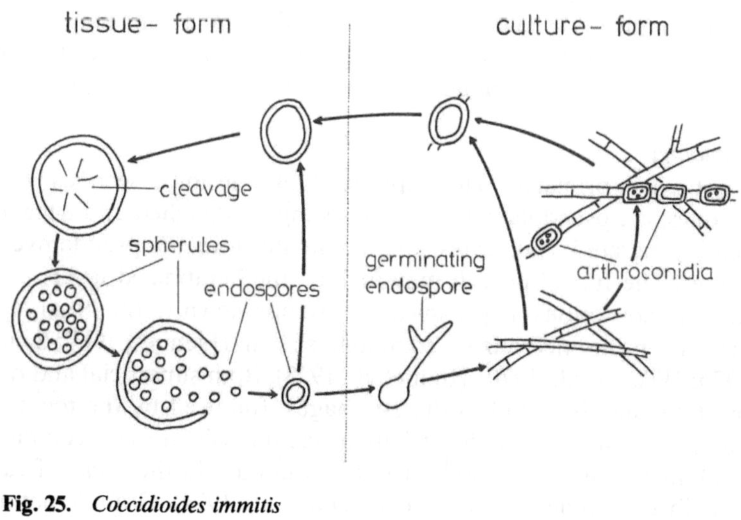

Fig. 25. *Coccidioides immitis*

Tissue forms are present as hyphae, pseudohyphae and budding cells. These can be better demonstrated by PAS-, Grocott- or Gridley-staining methods rather than by H. and E.-staining.

C. albicans assimilates glucose, maltose, sucrose, galactose, xylose and trehalose, but not lactose, melibiose, cellobiose, inositol, raffinose and dulcitol (Silva-Hutner and Cooper, 1974).

Epidemiology

C. albicans is part of the normal microflora of man, birds, canids and to some degree of other mammals. Infection occurs at an early age from the mother to the offspring. Isolation from the environment is commonly associated with fecal material. *C. albicans* is an inhibitant of the upper alimentary tract and of the lower female genital tract. It is normally held in check by the competitive bacterial flora. The yeasts become infectious after structural changes of the mucous membranes (contraceptives, gestation); after damage of the competitive bacteria (antibiotic treatment); or after immunosuppression (Gedek, 1980). Consequently, candidiasis frequently follows antibiotic treatment of other diseases in both man and animals (Young et al., 1957; Kaufmann and Quist, 1969; Schmidt and Butler, 1970).

Clinical Symptoms

Nonhuman primates: Candidiasis in nonhuman primates is mostly a disease of the upper digestive tract appearing as whitish streaks and plaques or even ulcers of the tongue or other oral mucous membranes. These oral lesions are often tolerated without any clinical signs (Nelson et al., 1966; Hunt et al., 1978) or they may be associated with diarrhea, dysphagia or even respiratory difficulties (Wikse et al., 1970). Enterocolitis candidosa in spider monkeys resulted in catarrhal to bloody diarrhea (Patterson et al., 1974).

Man: The primary infections of infants usually lead to stomatitis candidosa. Otherwise candidiasis in man may appear as an acute or chronic disease of the mucous membranes, skin, nails or internal organs, the clinical signs varying accordingly; i.e. the disease may appear as oral thrush, intertrigo, paronychia, onychia, vulovaginitis or, in males, balanitis. Intestinal infections may occur as catarrhal to necrotizing enterocolitis. Systemic forms include endocarditis, meningitis, septicemia or pulmonary candidiasis (Ajello and Kaplan, 1980; Gedek, 1980).

Pathology

Nonhuman primates: The superficial oral lesions, seen as whitish streaks or plaques, are pseudomembranes consisting of sloughed and degenerated epithelial cells, emigrated leukocytes and abundant *Candida*-pseudomycelia and blastospores. The fungal growth may include the keratinized layers of the epithelia in these minor lesions. If the invasion extends down to the basal layer and the propria, ulcers and necrotizing granulomas result (Fiennes, 1967; Schmidt and Butler, 1970; Wikse et al., 1970; Hunt et al., 1978). Both superficial and deeper lesions are most commonly found in the esophagus, followed by the tongue and other oral mucous membranes. Other forms of candidiasis are less commonly observed in nonhuman primates. Enterocolitis without further classification has been described in spider monkeys (Patterson et al., 1974). Granulomatous inflammation of the small intestine occurred in a *Cebus apella* suffering from nasal, pharyngeal and lymphonodular candidiasis (McCullough et al., 1977). Candida onychomycosis and balanitis was observed in a rhesus monkey (Wikse et al., 1970).

Man: *Stomatitis candidosa* quite often appears simply as swelling and reddening of the affected mucous membranes. *Esophagitis candidosa* can occur as yellowish-whitish pseudomembranes or as haemorrhagic-necrotizing inflammation of all layers with vascular invasion resulting in thrombosis. Intestinal candidiasis can be distinguished into:

1. catarrhal enterocolitica with small, yellowish pseudomembranes on the enteric surfaces.
2. haemorrhagic-necrotizing enterocolitis, affecting particularly the small intestines.
3. pseudomembranous-necrotizing enterocolitis, which is characterized by longitudinal or transversal ulcers especially of the small intestines.
4. follicular-ulcerating and abscessating enterocolitis originating from the Peyer's patches and solitary follicles.
5. enteritis gravis of the haemorrhagic-necrotizing type, which frequently penetrates the entire enteric wall.

In endocardial candidiasis ordinarily the aortic or mitral valve is involved, with the formation of large vegatations and frequent embolization.

Cutaneous candidiasis ranges from marked nail and limited skin involvement to massive disfiguring skin granulomas (Gedek, 1980; Louria, 1981).

Treatment

Amphotericin B is the most effective drug. Recently, ketoconazole has been found effective in the therapy of mucocutaneous candidiasis (Louria, 1981).

4.6.2 Candida Guillermondii (Pichia Guillermondii)

C. guillermondii, a possibly opportunistic yeast of man (Campbell and Stewart, 1980) has been isolated from the skin, mouth and genital tissues of free-living and captive baboons, particularly from females (Al-Doory, 1967b, c, 1969a; Al-Doory et al., 1967).

4.6.3 Candida Tropicalis

C. tropicalis, frequently implicated as a cause of endocarditis in man (Campbell and Stewart, 1980) has also been found on baboon skin, aural and nasal tissues (Al-Doory, 1967c, 1969a, b) and in the intestinal tract of *Macaca speciosa* (Saez, 1970b) and *Saguinus oedipus* (own results, unpublished).

4.6.4 Candida Krusei

C. krusei may invade burn-lesions or cause infant diarrhea in man (Campbell and Stewart, 1980). Occasionally it has been demonstrated on baboon skin, oral, rectal or vaginal tissues (Al-Doory, 1967b, c; Al-Doory et al., 1967).

4.6.5 Candida Pseudotropicalis

C. pseudotropicalis, one of the barely pathogenic species in man (Campbell and Stewart, 1980) has also been isolated from baboon large intestines (Al-Doory, 1969a, b).

References

1. Ajello, L. and W. Kaplan: Systemic mycoses; pp. 469-485 in "CRC Handbook Series in Zoonoses" (Steele, J. H., Series ed.), Section A: Bacterial, Rickettsial, and Mycotic Diseases, Vol. II (Stoenner, H. W. Kaplan, and M. Torten, Sect. eds.), CRC Press, Boca Raton (1980)
2. Al-Doory, Y.; S. S. Kalter, and M. Frederickson: The mycoflora of the subhuman primates. II. The flora of the rectum and vagina of the baboon in captivity. Mycopath. Mycol. appl. (Den Haag) *31*, 332-336 (1967)
3. Al-Doory, Y.: Studies on the pathogenicity of *Candida albicans* isolated from baboons. Canad. J. Microbiol. *14*, 443-448 (1967a)
4. Al-Doory, Y.: The mycoflora of the subhuman primates. I. The flora of the oral cavity of the baboon in captivity. Mycopath. Mycol. appl. (Den Haag) *31*, 43-48 (1967b)
5. Al-Doory, Y.: The mycoflora of the nonhuman primates. III. The flora of the skin of the baboon, vervet, and gelada in captivity. Folia primatol. *7*, 292-298 (1967c)
6. Al-Doory, Y.: The mycology of the freeliving baboon (Papio sp.). Mycopath. Mycol. appl. (Den Haag) *38*, 7-15 (1969a)
7. Al-Doory, Y.: The mycoflora of the subhuman primates. IV. The flora of the ear, nose, and intestinal contents of the baboon (Papio sp.). Mycopath. Mycol. appl. (Den Haag) *37*, 215-220 (1969b)
8. Campbell, M. C. and J. L. Stewart: The Medical Mycology Handbook. John Wiley and Sons, New York (1980)
9. Fiennes, R.: Zoonoses of Primates. The Epidemiology and Ecology of Simian Diseases in Relation to Man. Weidenfels and Nicolson, London (1967)
10. Gedek, B.: Kompendium der medizinischen Mykologie. Pareys Studientexte 25. Parey, Berlin (1980)
11. Hunt, R. D.; M. P. Anderson, and L. V. Chalifoux: Spontaneous infectious diseases of marmosets. Primates in Med. *10*, 239-253 (1978)
12. Kaufmann, A. F. and K. D. Quist: Thrush in a rhesus monkey: Report of a case. Lab. Anim. Care *19*, 526-527 (1969)
13. Kerber, W. T.; W. H. Reese, and J. van Natta: Balanitis, paronychia, and onychia in a rhesus monkey. Lab. Anim. Care *18*, 506-507 (1968)
14. Koneman, E. W.; G. D. Roberts, and S. F. Wright: Practical Laboratory Mycology. 2nd. ed. Williams and Wilkins, Baltimore (1978)
15. Louria, D. B.: Candidiasis; pp. 352-354 in "Infectious Diseases" (Sanford, J. P. and J. P. Luby, eds.), Grune and Stratton, New York (1981)
16. McCullough, B.; J. Moore, and R. E. Kuntz: Multifocal candidiasis in a capuchin monkey *(Cebus apella).* J. med. Primatol. *6*, 186-191 (1977)

17. Migaki, G.; R.E. Schmidt; J.D. Toft II, and A. Kaufmann: Mycotic infections of the alimentary tract of nonhuman primates: A review. Vet. Path. *19* (Suppl. 7), 93-103 (1982)
18. Nelson, B.; G.E. Cosgrove, and N. Gengozian: Diseases of an imported primate *Tamarinus nigricollis*. Lab. Anim. Care *16*, 255-257 (1966)
19. Patterson, D.R.; J.E. Wagner; D.R. Owens; N.C. Ronald, and C.S. Frisk: *Candida albicans* infections associated with antibiotic and corticosteroid therapy in spider monkeys. J. Amer. vet. med. Ass. *164*, 721-722 (1974)
20. Saez, H.: Stomatite myco-infectieuse du babouin *Papio papio* (Desm.). Ann. Med. Vet. *5*, 309-314 (1969a)
21. Saez, H. et J. Rinjard: Candidose et oedeme aigu du poumon chez un primate. Le Poumon et le Cœur *26*, 701-709 (1970)
22. Saez, H.: Levures de la cavité buccale du babouin, *Papio papio* (Desm.). Zbl. Vet.-Med. B *17*, 381-388 (1970a)
23. Saez, H.: Micromycetes hébergés par des macaques à face rouge "Macaca speciosa". F. Cuvier Experimentation Animal *3*, 183-189 (1970b)
24. Saez, H.: Candidose de l'estomac chez un gibbon à favoris blancs - Hylobates concolor leucogenys - âgé de quelques heures. Mykosen *18*, 519-522 (1975)
25. Saez, H.; J. Coudert et J. Rinjard: Candidose et tumeur obstructive intestinales chez un babouin captif, "Papio papio" (Desm.). Mycopathologia *64*, 173-177 (1978)
26. Saez, H. et J. Rinjard: Candidose buccale et invagination intestinale chez le babouin en captivité *Papio papio*. Rev. Elev. Méd. vét. Pays trop. *31*, 39-43 (1978)
27. Schmidt, R.E. and T.M. Butler: Esophagal candidiasis in a chimpanzee. J. Amer. vet. med. Ass. *157*, 722-723 (1970)
28. Schneider, H.-E.; R. Blaschke-Hellmessen; G.M. Klemm; G. Thess und V. Pohle: Zur Candidaproblematik im Primatenbestand des Zoologischen Gartens Dresden; pp. 299-308, 25. Int. Symp. Erkr. Zootiere, Wien (1983)
29. Silva-Hutner, M. and B.H. Cooper: Medically important yeasts; pp. 491-507 in "Manual of Clinical Microbiology", 2nd. ed. (Lennette, E.H., E.H. Spaulding, and J.P. Truant, eds.), Amer. Soc. Microbiol., Washington, D.C. (1974)
30. Wikse, S.E.; J.G. Fox, and R.M. Kovatch: Candidiasis in simian primates. Lab. Anim. Care *20*, 957-963 (1970)
31. Young, R.J.; B.D. Fremming; R.E. Benson, and M.D. Harris: Care and management of a *Macaca mulatta* monkey colony. Proc. Anim. Care Panel *7*, 67-83 (1957)

4.7 Coccidioides Immitis

Coccidioidomycosis is an acute or progressive fungal disease confined to the New World, where it is endemic in the semi-arid southwestern USA and parts of Mexico, with the highest infection rates in the San Joaquin Valley and a seasonal peak in the dry, dusty summer and autumn months (Smith, 1981).

A large variety of wild and captive animals besides man are susceptible to this disease. Natural infections have been reported in gorillas (McKenney et al., 1944; Bush et al., 1971; Benirschke et al., 1981), chimpanzees (Strutton, 1964; Ingram, 1975), baboons (Rosenberg et al., 1984), *Theropithecus gelada* (Rapley and Long, 1974), rhesus monkeys (Beaman et al., 1980; Castleman et al., 1980), and a bonnet monkey *(M. radiata)* (Beaman et al., 1980) all of which had been kept in the endemic area.

One report in a baboon in the UK, recently imported from Senegal (Welshman et al., 1981) raises several questions, especially with regard to the epidemiology of the infection in that case. It might possibly have been an *Emmonsia*-infection rather than a coccidioidomycosis.

Agent
C. immitis is a dimorphic fungus which occurs in a mold and a spherule-form (Fig. 25). It grows rapidly on Sabouraud agar when incubated at room temperature (25 °C) as floccose-cottony, white colonies, frequently with depressed centers. With age, the colonies darken. The colonies consist of septate hyphae of 2 to 4 μm in diameter, which rapidly produce barrel-shaped thick-walled arthrospores (2 to 4 × 3 to 6 μm) alternating with empty spaces.

Care: Mycelial arthrospores of C. immitis are very infectious in the laboratory, working in a safety hood is strongly recommended. The spherules in tissues are not infectious.

In the host organism, *C. immitis* is no longer filamentous, but develops as single spherules. The spherules are thick-walled, smooth and range in size from 20 to 200 μm in diameter. The internal cytoplasm matures by a process of progressive cleavage into small spherical endospores of 2 to 5 μm in diameter. After rupture of the wall of the spherule, the endospores are released and are both haematogenously and lymphogenously disseminated throughout the body. Each endospore may enter a new cycle and become a spherule that again produces and discharges endospores (Ajello and Kaplan, 1980); or if they are exposed to the nonliving environment or when cultivated on artificial media, they are retransformed to the mycelial form.

Epidemiology
C. immitis is known to exist in certain desert-like areas of the USA (Arizona, California, Nevada, New Mexico, Texas, Utah), Central- (Mexico, Guatemala, Honduras) and South America (Columbia, Venezuela, Bolivia, Paraquay and Argentina) (Ajello and Kaplan, 1980). Isolated cases in the Old World are believed to either have acquired the infection in the New World or to be misdiagnosed infections with *Emmonsia* spp., which are very similar to *C. immitis,* but are not geographically limited (Gedek, 1980).

C. immitis is a soil-borne fungus adapted to desert climates with the ecologic conditions around rodent burrows being important factors in the life cycle of *C. immitis* (Selby, 1975). Fomites created by dust storms favor the distribution and inhalation of arthrospores (Beaman et al., 1980; Castleman et al., 1980). Laboratory infections are easily acquired, because *C. immitis* cultures readily disintegrate leaving large numbers of arthrospores. In experimental aerosol infections of *M. mulatta* 50 to 10.000 arthrospores produced 100% infectivity (Converse et al., 1962). In another experiment spherules and microscopic sized lesions appeared in the lungs of rhesus monkeys on the fifth day after aerosol infection with arthrospores. On the seventh day the lesions became centrally necrotic, whilst 15 days p.i. similar lesions occurred in other organs (Blundell et al., 1969).

Transplacental infection, suspected to occur in man (Selby, 1975) has proven in a rhesus monkey after experimental infection of the mother (Castleberry et al., 1963).

Clinical Symptoms

Nonhuman primates: The symptoms of spontaneous coccidioidomycosis in nonhuman primates fall into two categories. The most commonly observed being respiratory symptoms such as coughs or pleural exudates (McKenney et al., 1944; Rapley and Long, 1974; Ingram, 1975; Benirschke et al., 1981). Almost as frequent were locomotory disorders such as lameness or paralysis (Ingram, 1975; Beaman et al., 1980; Castleman et al., 1980; Rosenberg et al., 1984). In the one questionable case reported by Welshman et al. (1981) the signs were limited to skin ulceration at the muzzle and tail.

Man: Human infections are asymptomatic in 60% of all cases. If disease results, coccidioidomycosis occurs in three basic forms: 1. pulmonary, 2. disseminated, and 3. residual pulmonary coccidioidomycosis (Ajello and Kaplan, 1980). Early symptoms in symptomatic pulmonary coccidioidomycosis include a flu-like cough and pleuritic pain. Patients with disseminated coccidioidomycosis may, in addition, develop cutaneous and subcutaneous abscesses; bone lesions as well as acute meningitis. Residual pulmonary coccidioidomycosis is diagnosed by coin-sized, thin-walled cavities in radiographs, which usually occur at the periphery of a lower lobe (Koneman et al., 1978). In man primary cutaneous coccidioidomycosis occurs (Gedek, 1980).

Pathology

Nonhuman primates: The principal types of lesions in nonhuman primates are granulomas or cavities in the lung. The granulomas appear as white nodules, which may progress to abscesses or to large granulomatous masses involving entire lobes. Within the granulomas large numbers of spherules are usually visible. In experimental aerosol infections of *M. mulatta* occasional hyphal elements were noticed around the residual lung cavities (Converse et al., 1962). In all spontaneous infections, the infection disseminates with development of similar granulomas in the enlarged thoracic lymph nodes, kidneys, liver and spleen.

In the paralysed animals, large granulomas destroyed parts of the vertebral bones by osteonecrosis and osteolysis. The granulomas also extended into the pachymeninx of the spinal cord or into the skeletal muscles.

The *C. immitis*-granulomas always contained numerous giant cells of the Langhans'-type in addition to histiocytes, lymphocytes and polymorphonuclear granulocytes.

Man: The lesions in the most frequently affected lungs are usually of granulomatous type with formation of multinucleated giant cells (Koneman et al., 1978).

Treatment

In the majority of patients no specific therapy is necessary, but patients with progressive pulmonary diseases or signs of dissemination require treatment with amphotericin B (Smith, 1981).

References

1. Ajello, L. and W. Kaplan: Systemic mycoses; pp. 469-485 in "CRC Handbook Series in Zoonoses" (Steele, J. H., Series ed.), Section A: Bacterial, Rickettsial, and Mycotic Diseases, Vol. II (Stoenner, H.; W. Kaplan, and M. Torten, Section eds.), CRC Press, Boca Raton (1980)
2. Al-Doory, Y.: Systemic mycoses; pp. 224-242 in "Pathology of Simian Primates", Part II (Fiennes, R. N. T-W., ed.), Karger, Basel (1972)
3. Beaman, L.; C. Holmberg; R. Henrickson, and B. Osburn: The incidence of coccidioidomycosis among nonhuman primates housed indoors at the California Primate Research Center. J. med. Primatol. *9*, 254-261 (1980)
4. Benirschke, K.; M. P. Anderson; J. E. Osterhuis, and L. S. Nelson: Coccidioidomycosis in gorillas and attempted prophylactic vaccination, 23. Internat. Symp. Erkr. Zootiere, Halle, pp. 119-122 (1981)
5. Blundell, G. P.; M. W. Castleberry; E. P. Lowe, and J. L. Converse: The pathology of *Coccidioides immitis* in the *Macaca mulatta*. Amer. J. Path. *39*, 613-630 (1969)
6. Bush, M.; J. A. Moore, and L. M. Neeley: Sedation for transportation of a lowland gorilla. J. Amer. vet. med. Ass. *159*, 546-548 (1971)
7. Castleberry, M.; W. Converse, and J. E. Favero: Coccidioidomycosis transmission to infant monkey from its mother. A case report. Arch. Path. *75*, 459-461 (1963)
8. Castleman, W. L.; J. Anderson, and C. A. Holmberg: Posterior paralysis and spinal osteomyelitis in a rhesus monkey with coccidioidomycosis. J. Amer. vet. med. Ass. *177*, 933-934 (1980)
9. Converse, J. L.; E. P. Lowe; M. W. Castleberry; G. P. Blundell, and A. R. Besemer: Pathogenesis of *Coccidioides immitis* in monkeys. J. Bact. *83*, 871-878 (1962)
10. Gedek, B.: Kompendium der medizinischen Mykologie. Pareys Studientexte 25. Parey, Berlin (1980)
11. Ingram, K. S.: Coccidioidomycosis in a colony of chimpanzees. Amer. Assoc. Zoo Vet. Ann. Proc. 1975, 127-131 (1975)
12. Koneman, E. W.; G. D. Roberts, and S. F. Wright: Practical Laboratory Mycology. 2nd ed. William and Wilkins, Philadelphia (1978)
13. McKenney, F. D.; J. Traum, and A. E. Bonestell: Acute coccidioidomycosis in a mountain gorilla *(Gorilla beringei)* with anatomical nodes. J. Amer. vet. med. Ass. *104*, 136-140 (1944)
14. Ophuls, W. and H. C. Moffit: A new pathogenic mould: (formerly described as a protozoan: *Coccidioides immitis* pyogenes) preliminary report. Philadelphia Med. J. *5*, 1471-1472 (1900)
15. Rapley, W. A. and J. R. Long: Coccidioidomycosis in baboon recently imported from California. Canad. Vet. J. *15*, 39-41 (1974)
16. Rixford, E. and T. C. Gilchrist: Two cases of protozoan (coccidioidal) infection of the skin and other organs. Johns Hopkins Hosp. Rep. *1*, 209-268 (1896)
17. Rosenberg, D. P.; C. A. Gleiser, and K. D. Carey: Spinal coccidioidomycosis in a baboon. J. Amer. vet. med. Ass. *185*, 1379-1381 (1984)
18. Selby, L. A.: Coccidioidomycosis; pp. 517-528 in "Diseases Transmitted From Animals to Man". 6th ed. (Hubbert, W. T.; W. F. McCulloch, and P. R. Schnurrenberger, eds.), C. C. Thomas, Springfield (1975)
19. Smith, J. W.: Coccidioidomycosis; pp. 348-349 in "Infectious Diseases" (Sanford, J. P. and J. P. Luby, eds.), Grune and Stratton, New York (1981)
20. Strutton, W. H.: A case report of *Coccidioides immitis* in an immature chimpanzee. 6571st Aeromedical Res. Lab. ARL-TR-64-16 (1964)
21. Welshman, M. D.; M. P. Fleming, and G. W. Tribe: Coccidiomycosis in a baboon. Vet. Rec. *108*, 62 (1981)

4.8 Paracoccidioides Brasiliensis

Paracoccidioidomycosis is a common systemic fungal infection in tropical and subtropical South America, especially in Columbia and Brazil (Smith, 1981). The disease in man is characterized by granulomatous inflammations containing macrophages and giant cells, which are most frequently present in the buccal mucosa and lymphoid tissues. In disseminated infections these lesions are scattered throughout the body (Johnson and Lang, 1977).

One suggested case of paracoccidioidomycosis, unfortunately not confirmed by fungal cultivation was reported in a squirrel monkey with gastric ulceration and a large granuloma in the liver (Johnson and Lang, 1977).

References

1. Johnson, W.D. and C.M. Lang: Paracoccidioidomycosis (South American blastomycosis) in a squirrel monkey *(Saimiri sciureus)*. Vet. Path. *1*, 368-371 (1977)
2. Smith, J.W.: South American blastomycosis; pp.351-352 in "Infectious Diseases" (Sanford, J.P. and J.P. Luby, eds.), Grune and Stratton, New York (1981)

4.9 Cryptococcus Neoformans (Perfect: Filobasidiella Neoformans)

Cryptococcosis is an acute, subacute or chronic pulmonary systemic or meningeal mycosis of man and a variety of animals with a worldwide distribution, although the largest number of patients is reported from the USA (Anderson, 1975; Muchmore, 1981). It is estimated that 5.000 to 15.000 cases of nonclinical and clinical human pulmonary cryptococcosis occur annually in New York City, with 200 to 300 cases of cryptococcal meningitis occurring each year in the USA (Ajello and Kaplan, 1980).

In nonhuman primates spontaneous infections have been described in a rhesus monkey and a Formosan macaque *(M. cylopis)* in Washington D.C. (Garner et al., 1969); a patas monkey *(Erythrocebus patas)* in Rockville/MD (Sly et al., 1977); a de Brazza's monkey *(Cercopithecus neglectus)* of the London Zoo (Fiennes, 1967); a sooty mangabey of the Washington Zoo (Littman and Zimmermann, 1956 - cit. from Al-Doory, 1972), and a *Callithrix geoffroyi* in Panama (Takos and Elton, 1953).

Agent
C. neoformans is a yeast which normally does not grow in its mycelial phase. The yeast cells are surrounded by characteristic mucopolysaccharide (D-xylose, D-glucuronic acid, D-mannositol) capsules. These capsules are generally narrow (1 to 4 μm) in cultures, but repeated subcultivation on brain-heart infusion agar at 37 °C

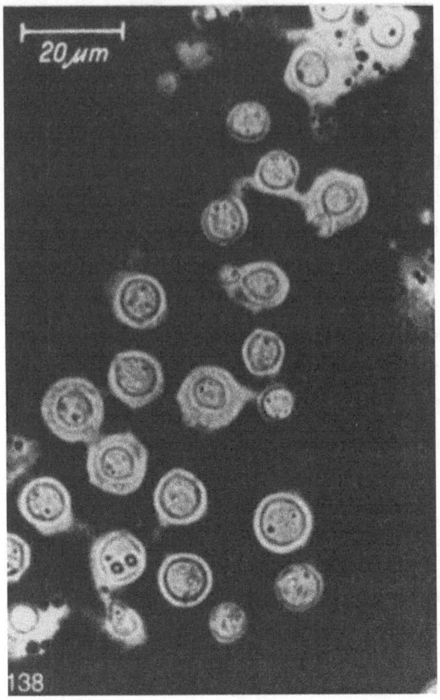

Fig. 26. *C. neoformans* culture - indian ink preparation (Gedek, 1980)

increases their thickness. In comparison in tissues the capsules may be twice the radius of the cell. The capsules are best demonstrated by indian ink staining techniques (Fig. 26). The blastospores are spherical, 5 to 30 µm in diameter and usually occur singly or in pairs.

The fungus grows well on Sabouraud-dextrose-agar or other nutrient media not containing cycloheximide. The colonies are usually cream-colored to tan, but if the fungus is grown on seed agar or on caffeic acid medium with added ferric citrate, brown colonies are produced in six hours (Campbell and Stewart, 1980).

C. neoformans is urea and inosit-positive in contrast to other yeasts (Gedek, 1980). Four serotypes (A–D) with somewhat different geographic distribution can be distinguished (Finegold and Martin, 1982).

The perfect form has been induced by Kwon-Chung by the mating of two compatible isolates with production of basidia, basidiospores and clamp-connections in the fertilized hyphae (Campbell and Stewart, 1980).

Epidemiology

C. neoformans, unlike other yeasts, is not a natural inhabitant of the mammalian organism. Cryptococcosis is not contagious but is an airborne disease acquired from aspiration of blastospore-containing dusts. The reservoirs of the fungus are pigeon-, chicken- or bird habitats, since certain N-components in the bird-droppings favor the growth of the fungus. The blastospores are highly resistant to drying and thus remain infectious for a long time. They enter the human or animal host via the respiratory or digestive tract or through skin lesions (Gedek, 1980). In man a strong association between clinical cryptococcosis and other preexisting debilitating diseases can be observed (Anderson, 1975; Ajello and Kaplan, 1980; Finegold and Martin, 1982).

Clinical Symptoms

Nonhuman primates: The symptoms vary widely. In the slender lori the fungus was isolated from skin and hair without disease being apparent. The de Brazza's monkey died from shock (Fiennes, 1967). The sooty mangabey had enlarged lymph-nodes (Al-Doory, 1972) and the rhesus monkey became anorectic and reluctant to move. This was followed by emaciation, weakness, and semicoma (Garner et al., 1969). At least two animals suffered from CNS-symptoms: both the Formosan macaque and the patas monkey experienced epileptiform seizures (Garner et al., 1969; Sly et al., 1977), which were accompanied by blindness and ataxias in the Formosan macaque.

Man: The symptoms of cryptococcosis are non-characteristic and frequently complicated by other underlying diseases. The primary infection of the lung is commonly asymptomatic or may produce slight cough and occasionally even thoracic pain (Gedek, 1980).

Cryptococcal meningitis, which is the most dramatic form of dissemination, usually begins with frontal or temporal headache, dizziness, vertigo, nausea or vomitus, but nuchal rigidity is mostly less pronounced. The well developed cryptococcal meningitis may simulate tuberculous meningitis and, if untreated, is highly fatal within days or weeks (Gedek, 1980; Muchmore, 1981).

Pathology

Nonhuman primates: The cryptococcal lesions in the reported cases were concentrated in the lungs and CNS. In the lungs unencapsulated cysts and focal inflammation as infiltrations of epitheloid cells, lymphocytes and polymorphonuclear cells were noticed (Takos and Elton, 1953; Garner et al., 1969; Sly et al., 1977). The cysts were frequently filled by gelatinous material. The CNS lesions consisted of granulomas or gelatinous cysts in the brain parenchyma, and of a granulomatous meningitis (Garner et al., 1969; Sly et al., 1977). Gelatinous splenomegaly and enlargement of the mesenteric lymph nodes were found in the Geoffroy's marmoset (Takos and Elton, 1953).

Man: Cerebral cryptococcosis may appear as slight cloudiness of the leptomeninges due to scattered histiocytic infiltration; as tubercle-like structures along the blood vessels consisting of epitheloid cell granulomas containing giant cells of the foreign-body type, plasma cells and lymphocytes; as thick exudates of similar composition or as cystic spaces on the cut surface of the brain parenchyma. The cryptococci may be present as pure cultures within these spaces. They are histolytic and may induce a chronic inflammation with macrophages, lymphocytes, giant cells and eosinophilia. For demonstration of the fungi in tissue section the normal H- and E-staining is insufficient, better results are obtained with Gram's-, PAS-, Giemsa-, or Grocott-staining.

Lung lesions may be solid, rounded, sometimes multiple, nodules of chronic inflammatory tubercle-like nature. Similar lesions may occur in bones as osteomyelitis; in the skin and subcutaneous tissues as tumorlike structures; or in lymph-nodes, where they may resemble Hodgkin's disease (Al-Doory, 1972; Gedek, 1980).

Treatment
Amphotericin B is the preferred drug in dosages up to 1 mg/kg/day. 5-fluorocytosine (Acontil/Roche), another useful drug, is less toxic than amphotericin B, but *C. neoformans* may rapidly develop resistance to it, thereby limiting its usefulness (Gedek, 1980; Muchmore, 1981).

References

1. Ajello, L. and W. Kaplan: Systemic mycoses; pp. 469-485 in "CRC Handbook Series in Zoonoses" (Steele, J. H., Series ed.), Section A: Bacterial, Rickettsial, and Mycotic Diseases, Vol. II (Stoenner, H.; W. Kaplan, and M. Torten, eds.), CRC Press, Boca Raton (1980)
2. Al-Doory, Y.: Systemic mycoses; pp. 224-242 in "Pathology of Simian Primates", Part II. (Fiennes, R. N. T-W., ed.), Karger, Basel (1972)
3. Anderson, K. L.: Pathogenic yeasts; pp. 537-542 in "Diseases Transmitted From Animals to Man", 6th. ed. (Hubbert, W. T.; W. F. McCulloch, and P. R. Schnurrenberger, eds.), C. C. Thomas, Springfield (1975)
4. Campbell, M. C. and J. L. Stewart: The Medical Mycology Handbook. John Wiley and Sons, New York (1980)
5. Fiennes, R.: Zoonoses of Primates. The Epidemiology and Ecology of Simian Diseases in Relation to Man. Weidenfels and Nicolson, London (1967)
6. Finegold, S. M. and W. J. Martin: Diagnostic Microbiology. 6th ed. Mosby Co., St. Louis (1982)
7. Garner, F. M.; D. F. Ford, and M. A. Ross: Systemic cryptococcosis in 2 monkeys. J. Amer. vet. med. Ass. *155*, 1163-1168 (1969)
8. Gedek, B.: Kompendium der medizinischen Mykologie. Pareys Studientexte 25. Paul Parey, Berlin (1980)
9. Muchmore, H. G.: Cryptococcosis; pp. 349-351 in "Infectious Diseases" (Sanford, J. P. and J. P. Luby, eds.), Grune and Stratton, New York (1981)
10. Sly, D. L.; W. T. London; A. E. Palmer, and J. M. Rice: Disseminated cryptococcosis in a patas monkey *(Erythrocebus patas)*. Lab. Anim. Sci. *27*, 694-699 (1977)
11. Takos, M. J. and J. W. Elton: Spontaneous cryptococcosis of marmoset monkeys in Panama. Arch. Path. *55*, 403-407 (1953)

4.10 Geotrichum Candidum

Geotrichosis is a rather rare abnormal pseudoinfection of the mucous membranes of man and animals by Geotrichum spp., most commonly by *G. candidum*. Geotrichum spp. do not invade the tissues but multiply by arthrospores in the desquamated cells of the mucous membranes, most commonly of the respiratory- and gastrointestinal tract. The fungal hyphae may produce whitish plaques similar to stomatomycosis of the oral cavity (Gedek, 1980; Finegold and Martin, 1982).

Gastrointestinal geotrichosis has been reported in six adult gorillas of the New York Zoo which suffered from watery diarrhea over a period of 4 days (Dolensek et al., 1977).

References

1. Dolensek, E.P.; R.L. Napolitano, and J.Kazimiroff: Gastrointestinal geotrichosis in six adult gorillas. J.Amer. vet. med. Ass. *171*, 975–976 (1977)
2. Finegold, S.M. and W.J. Martin: Diagnostic Microbiology. 6th ed. C.V. Mosby, St. Louis (1982)
3. Gedek, B.: Kompendium der medizinischen Mykologie. Pareys Studientexte 25. Paul Parey, Berlin (1980)

5 Pneumocystis Carinii

P. carinii is a microorganism of uncertain taxonomic position which infects a wide variety of domestic and wild animals. It is regarded as a protophyte or yeast by some or as protozoan by others. It was first isolated from guinea pigs and a *Callithrix penicillata* (Chagas, 1909 - cit. from Poelma, 1975). *P. carinii* infections of nonhuman primates were also described in *C. aurita*, *Aotus trivirgatus*, *Lagothrix lagotricha*, *Ateles belzebuth*, *Alouatta fusca*, *Pan troglodytes*, *Galago senegalensis*, *Galagoides demidovii*, and *Tupaia glis* (McClure and Keeling, 1971; Long et al., 1975; Poelma, 1975; Chandler et al., 1976).

In man it causes a special form of diffuse pneumonia with apparently increasing frequency (Minielly et al., 1970). Cases have been concentrated in medical centers that care for immunosuppressed patients (Krogstad, 1981), and recently pneumocystosis has been associated with AIDS (Letvin and King, 1984; Popovic et al., 1984).

Agent

P. carinii is an extracellular organism appearing as cysts which measure between 3,5 and 5 μm in diameter. The cysts are surrounded by a thick, rigid wall, and their cytoplasm typically contains several intracystic bodies. The organisms are poorly stained with H. and E. stain, better results are obtained using PAS- or Gomori's methenamine silver stains.

P. carinii can not be cultivated on artificial media, therefore information on its development and life cycle is lacking.

Geographical Distribution

P. carinii occurs world-wide.

Clinical Symptoms

Nonhuman primates: *P. carinii* has been observed in the lungs of apparently "healthy" monkeys as well as from animals which suffered from severe pneumonia as indicated by dyspnea, anorexia and weight loss (McClure and Keeling, 1971; Poelma, 1975; Chandler et al., 1976).

Man: *P. carinii* causes two types of disease:

a) Endemic pneumocystosis (infant pneumocystosis) in premature or debilitated infants. The disease begins with anorexia, weight loss and diarrhea, followed by tachypnea, dyspnea, and cyanosis without remarkable fever.
b) Pneumocystosis in the immunocompromised child or adult. It is a sporadic disease of usually abrupt onset and a more rapid clinical course than the one described in the infants.

Pathology

Nonhuman primates: In animals which died from pneumocystosis the lungs were diffusely or multifocally consolidated with petechial and ecchymotic haemorrhages spread over the pulmonary pleura. The histological lesions were characterized by extensive mononuclear or plasma cellular interstitial pneumonia (McClure and Keeling, 1971). Within the alveolar spaces a foamy, reticular, honeycomb-like material was accumulated, which contained small basophilic structures.

Man: The lungs do not collapse at necropsy in both forms of pneumocystosis. They contain pale grey and firm areas, alternating with emphysematous areas. Histopathologically the lesions are described as extensive interstitial plasma cellular to lymphohistiocytic interstitial pneumonia with "honeycombed" foamy material in the alveolar spaces.

Treatment

Trimethoprim-sulfamethoxazole (20 mg trimethoprim, 100 mg sulfamethoxazole) can be given for 2 weeks. It is less toxic than the previously used pentamidine.

References

1. Chandler, F.W.; H.M. McClure; W.G. Campbell, jr., and J.C. Watts: Pulmonary pneumocystis in nonhuman primates. Arch. Pathol. Lab. Med. *100*, 163-167 (1976)
2. Krogstad, D.J.: Pneumocystis; pp.383-384 in "Infectious Diseases" (Sanford, J.P. and J.P. Luby, eds.). Grune and Stratton, New York (1981)
3. Letvin, N.L. and N.W. King: Comparative pathology of acquired immunodeficiency syndrome (AIDS). Comp. Path. Bull. *10*, 1 and 5 (1984)
4. Long, G.G.; J.D. White, and J.L. Stookey: *Pneumocystis carinii* infection in splenectomized owl monkey. J. Amer. Vet. Med. Ass. *167*, 651-654 (1975)
5. McClure, H.M. and M.E. Keeling: Pneumocystis carinii in chimpanzees. Primate Zoonoses Surveill. Rep. (5), 5-7 (1971)
6. Minielly, J.A.; F.C. McDuffie, and K.E. Holley: Immunofluorescent identifications of *Pneumocystis carinii*. Arch. Path. *90*, 561-566 (1970)
7. Poelma, F.G.: *Pneumocystis carinii* infections in zoo animals. Z. Parasitenk. *64*, 61-68 (1975)
8. Popovic, M.; M.G. Sarngadharan; E.Read, and R.C. Gallo: Detection, isolation, and continuous production of cytopathic retroviruses (HTLV-III) from patients with AIDS and pre-AIDS. Science *224*, 497-500 (1984)

6 Protozoa

6.1 Sarcodina

6.1.1 Entamoeba Histolytica

E. histolytica causes intestinal and extraintestinal amebiasis in man and a variety of nonhuman primates after invading the tissues. *E. histolytica* infections of man are endemic almost worldwide but in most cases are restricted to the silent carrier state. Despite its cosmopolitan distribution the highest incidence of clinical amebiasis apparently occurs in tropical or subtropical countries (Seehy, 1981).

Nonhuman primates are natural hosts to several species of amebas, among those *E. histolytica.* In Asian macaques most infections are asymptomatic, whereas great apes, colobines and New World monkeys more commonly suffer from severe amebiasis.

E. histolytica infections have been seen in chimpanzees (Porter, 1953-54; Benson et al., 1954; Fremming et al., 1955; Edsall et al., 1960; Fineg et al., 1961; van Riper et al., 1966; Miller and Bray, 1966; Reardon and Rininger, 1968), gorillas (Linke, 1967); orangutans (Herman and Schroeder, 1939; Patten, 1939); langurs (Geisel et al., 1975; Palmieri et al., 1984); proboscis monkeys (Müller and Rüedi, 1981); *Colobus* spp. (Frank, 1982; Loomis et al., 1983); baboons (Kuntz and Myers, 1966; Myers et al., 1971; Kuntz and Moore, 1973); macaques (Johnson, 1941; Rowland and Vandenbergh, 1965; Kuntz et al., 1968; Sano et al., 1980; Sakakibara et al., 1982; Kessler et al., 1984); spider monkeys (Johnson, 1941; Schiefer and Loew, 1978), and *Lagothrix lagotricha* (own findings).

Agent

E. histolytica is a protozoan parasite with two stages in its life cycle. It exists either as a motile, feeding, ameboid trophozoite in tissues or as a nonmotile, nonfeeding cyst form in the intestinal luminal contents and feces.

The trophozoites are of 10 to 60 μm diameter. The smaller (10 to 15 μm) "minuta" race, now designated as *E. hartmanni* (Sheehy, 1981) are considered to be nonpathogenic, whilst the larger (more than 20 μm diameter) "magna" forms being pathogenic (Toft, 1982). The electrophoretic isoenzyme pattern of *E. histolytica* shows 11 cymodemes, which appear to be identical in simian and human strains (Sargeaunt et al., 1982), but different in pathogenic and apathogenic strains (Sargeaunt et al., 1978). Smith and Meerovitch (1985) found only zymodeme III and VIII in captive nonhuman primates, zymodemes characteristic of non-pathogenic strains in man. Potentially pathogenic strains of *E. histolytica* are also characterized by a higher rate of erythrophagocytosis; a greater propensity to agglutinate with concavalin A; a lack of overall surface charge; a more potent cytopathogenicity *in vitro*, and the ability to grow in axenic media (Davis, A. and Pawlowski, 1985).

The excentric, globoid nucleus is barely visible and the cytoplasm appears homogenous in unstained trophozoites. After iodine staining the cytoplasm is greenish and divided into a broad, clear, refractil, hyaline ectoplasm with extending pseudopodia, and a granular endoplasm, which, in pathogenic strains, contains erythrocytes and other phagocytized materials.

This is because the non-pathogenic trophozoites feed on the intestinal contents, whilst the pathogenic forms feed also on tissues and blood. The trophozoites are very delicate and easily killed (Lushbaugh and Pittman, 1982), they are not infectious.

The trophozoites develop to a smaller, round or ovoid precystic stage without cystwall, and then finally to globoid or slightly asymmetric cysts. Precysts contain one large nucleus, while mature infective cysts up to four nuclei. The cytoplasm of young cysts contains glycogen-vacuoles and one or more dark, sausage- to cigar-shaped chromatoid bodies, which are believed to be masses of ribosomes involved in protein synthesis (Ravdin and Guerrant, 1982).

The cysts are 5 to 10 μm in diameter in the "minute" form *(E. hartmanni)*, and 10 to 20 μm in the "magna" form. The cysts are very resistant to low temperatures. They can also survive outside the host for weeks to months in a moist environment (Ravdin and Guerrant, 1982) and they are remarkably resistant to chemical disinfectants (Davis and Pawlowski, 1985). Like trophozoites they are very susceptible to drying, and they are very vulnerable to radiation including UV-light. They are also destroyed by 5% acetic acid, chlorine, iodine and formalin. For disinfection heating to 56 °C or simple drying is sufficient.

The excystation of ingested cysts begins after passage through the stomach. It requires anaerobic conditions and bacteria, and, aided by the alkaline environment usually occurs in either the human small or large bowel. The nuclei divide forming a transient metacystic 8 nuclei-containing stage. This is followed by cytoplasmic division and the emergence of eight trophozoites (Ravdin and Guerrant, 1982). The trophozoites themselves reproduce by binary fission (Trissl, 1982).

Axenic growth can be obtained on a number of artificial media, such as TYI-S-33 (trypticase, yeast extract, iron, serum) or B 1-S-33 media (Diamond et al., 1978; Gold et al., 1982). Clonal growth of *E. histolytica* was recently achieved in semi-solid agar (Gillin and Diamond, 1978).

Epidemiology

Only mature cysts are infective, as immature cysts and trophozoites are destroyed during the gastric passage. The infective cysts passed in the stools do not require an extrinsic inoculation period (Lushbaugh and Pittman, 1982). After excystation the trophozoites migrate in man and most nonhuman primates to the large intestines from where they can be excreted immediately or they may feed on the enteric bacteria and debris without becoming pathogenic; or they may invade the intestinal mucosa and induce clinical disease. The chances of establishing the infection parallel the number of ingested cysts. This is also influenced by the nutritional status and immunological competence of the host and by the presence (or shortage) of growth factors essential for the ameba, such as B-complex vitamins.

Transmission is by direct fecal-oral transmission or via fecal contamination of water or vegetables. Captive nonhuman primates may be infected by passage of infective cysts among the animals (Sakakibara et al., 1982) or they may become infected directly from man. Transmission (or retransmission) of the amebae from nonhuman primates to man, although generally of little concern (Lushbaugh and Pittman, 1982) may be of more importance in animal handlers taking care of, or in children playing with nonhuman primates.

Pathogenesis

Commensalism which occurs in both "minuta" and "magna" forms is usually not associated with any recognizable clinical or morphological abnormalities in the host. The stimuli responsible for conversion from commensalism to invasiveness in the "magna" forms are only partially known. Pathogenicity depends, as far as known, on

1. The virulence of the *Entamoeba* strains. Virulence depends on the strain-characteristic invasiveness and tissue damaging potentials (Lopez-Revilla and Said-Fernandez, 1980). Virulence is also enhanced by repeated host passages: Amebae require direct association with viable bacteria for virulence (Wittner and Rosenbaum, 1970), as bacteria that contain mannose-binding sites in their cellular wall adhere to and are ingested by amebae (Bracha et al., 1982; Bracha and Mirelman, 1984). Virulent *E. histolytica*-strains secrete a polypeptide of 13.000 to 14.000 dalton (amoebapore) which is incorporated into lipid bilayers, liposomes and cells, leading to progressive and irreversible changes in the ion conductance of the target membranes (Lynch et al., 1982; Gitler et al., 1984). They also produce 25.000 to 35.000 dalton cytotoxic proteins of enterotoxin activity (Lushbaugh et al., 1979; Ravdin and Guerrant, 1982). Virulent *E. histolytica* contain serotonin or other neurohumoral substances that cause intestinal secretion (Mc Gowan et al., 1983) and have a haemolytic effect (Lopez-Revilla and Said-Fernandez, 1980). Virulent amebas have a higher proteolytic activity than nonvirulent ones (Gadasi and Kobiler, 1983).
2. The number of trophozoites attached to the mucosal lining.
3. The hypermotility of the intestine.
4. The creation of a low redox-potential (350 to 425 mv) by oxygen-consuming bacteria, because *E. histolytica* grows best at reduced oxygen tension. Even total anaerobiosis is tolerated in the presence of a well mixed bacterial flora and sufficient amounts of starch in the intestinal contents. The multiplication of the amebas is preceded by an increase in the bacterial population.
5. The host species involved. New World monkeys and great apes are more prone to clinical amebiasis than Old World monkeys (except colobines) (Hegner et al., 1932; Fremming et al., 1955; Belding, 1965; Amyx et al., 1978; own experiences - unpublished). Among the Old World monkeys the colobines and langurs are more susceptible to clinical amebiasis due to their special gastric morphology and physiology. The stomach in colobines and langurs consists of: 1. A greatly

distended presacculate and sacculate portion lined by an unilayered, mucinous, columnar epithelium containing sparse, short glands.
2. A tubular tubus gastricus with typical fundic glands containing chief- and parietal cells in its distal portion and 3. A pyloric segment with typical pyloric glands (Kuhn, 1964). Within the presaccular and saccular portions the pH is almost neutral (5,0 to 6,7), enabling the excystation of the amebas and invasion of the gastric wall (Frank, 1982; Loomis et al., 1983; Palmieri et al., 1984). Invasive trophozoites first attach to the intestinal (gastric) epithelial cells, possibly mediated by a lectin-like substance on the surface of the trophozoites (Kobiler and Mirelman, 1980).

Immediately before or after the trophozoites establish contact to the mucosal membrane, severe cytopathic alterations become apparent. As the trophozoites approach the brush border the microvilli of the enterocytes below shorten and become irregular. Dense amorphous material then accumulates between the ameba and the microvilli. The microvilli disappear, polymorphonuclear leukocytes emigrate from the capillaries and progress into the intercellular spaces of the epithelium. Further advancement of the ameba results in direct contact with the apical cytoplasm of the enterocytes, whose mitochondria and endoplasmic reticulum dilate due to an increase in lipid droplets. The pseudopodia of the ameba penetrate into the epithelium with dilatation of the intercellular bridges. The cytoplasm of the enterocytes degenerates and the polymorphonuclears in the vicinity also degenerate (Takeuchi and Phillipps, 1975). Two modes of toxic activity have been postulated, one acting through soluble compounds from the distance which are secreted by the trophozoites (Lushbaugh et al., 1979; Bos, 1979), the other acting via direct contact with the target cells (Ravdin et al., 1980). Lytic tissue destruction ensues within a few hours (Ravdin et al., 1980; Ravdin and Guerrant, 1982) with an amorphous eosinophilic material surrounding the immigrated trophozoites. Phagocytosis of tissue compounds by the amebae begins almost simultaneously with the CPE exerted by amebic "toxins", associated with lectin-like activities (Mattern et al., 1980; Kobiler et al., 1981). Inhibition of monocyte chemotaxis by *E. histolytica*, reduced phagocytic activity of peripheral polymorphonuclear leukocytes and destruction of leukocytes (Guerrant et al., 1981) hamper the inflammatory responses, further contributing to the extension of the tissue damage. After invasion of the propria mucosae the amebas may

a) remain in the superficial mucosa with only minor tissue damage, producing shallow ulcers which may be self-healing;
b) penetrate the muscularis mucosae and progress to the submucosa, resulting in deep ulcers. Interference with the mucosal circulation induces larger necrotic areas, which in man and great apes may appear as flask-like ulcers undermining the mucosa. In the lower monkeys the circulatory brake down induces superficial necrosis and haemorrhagic-fibrinous exudates rather than deep-reaching ulcers. There are also differences in the invasive pathways of the amebas between hominoids and lower monkeys. The invasion in man and the great apes occurs via the surface epithelium, whilst in macaques penetration near the crypt-openings is suggested with subsequent invasion of interglandular papillae (Bond et al., 1946 - cit. from Belding, 1965).

c) On reaching the intestinal vascular bed the amebas may invade the mucosal blood circulation and lymph vessels with subsequent transport via the portal vein system to other sites, particularly the liver, and subsequently, the lung (Aikat et al., 1979). In the liver the trophozoites induce focal necrosis or necrotic abscesses, possibly due to ischemia from amebic obstruction of portal vessels (Aikat et al., 1979). In the early stages of amebic liver abscesses an intense inflammatory response consisting of polymorphonuclear leukocytes is likely to kill the trophozoites (Guerrant et al., 1981; Ravdin and Guerrant, 1982).

Clinical Symptoms

Nonhuman primates: In most *E. histolytica* infections of nonhuman primates clinical symptoms are completely lacking. Clinical disease, however, can be observed particularly in the great apes and in New World monkeys (Ruch, 1959; Vickers, 1969), or in malnourished Old World monkeys (McCarrison, 1920). Symptoms are generally more common in infants, whereas adults tend to enter silent carrier stages (Johnson, 1941; Miller, 1952).

Clinical symptoms are also influenced by the host species. In members of the subfamily Colobinae *(Presbytis, Pygathrix, Nasalis, Colobus)* which suffer from *gastric* (and hepatic) amebiasis the symptoms varied from suddenly increased respiratory rate and abnormally bent posture (Müller and Rüedi, 1981) to diarrhea, anorexia, and weight loss (Palmieri et al., 1984). In massive hepatic amebiasis, icterus may be seen (Geisel et al., 1975). In other nonhuman primates affected primarily with colonic amebiasis, the following symptoms have been observed: apathy, muscular weakness, loss of appetite, bloody-mucoid diarrhea, vomitus, and general malaise (von Provazek, 1912; Eichhorn and Gallagher, 1916; Weidman, 1923; Soldey, 1924; Ratcliffe, 1931; Herman and Schroeder, 1939; Patten, 1939; Hill, 1953; Fremming et al., 1955; Levine, 1961; Miller and Bray, 1966; Linke, 1967; Vickers, 1968, 1969; Peters et al., 1976). Chronic infection may cause mild, chronic colitis and obstipation.

Man: The symptoms of amebiasis can be extremely variable. They are usually absent in cyst-passers with only luminal infections. Invasive amebiasis may result in (Sheehy, 1980):

a) amebic dysentery with gradual onset. The passage of 5 to 15 stools daily, associated with abdominal cramps, or colic and flatulence. The foul-smelling semiformed to liquid stools in amebic dysentery contain variable amounts of blood or bloodtinged mucus, trophozoites and sometimes cysts.
b) amebic colitis, a fulminant disease of abrupt onset. Diarrhea is profuse with 20 to 30 bloody, mucoid stools per day, which is accompanied by tenesmus, severe abdominal cramping, anorexia, nausea and vomitus.
c) amebic hepatitis, which occurs in 10 to 15 percent of dysenteric patients as a nonspecific response consisting of hepatomegaly, hepatic tenderness, fever and leukocytosis.
d) extraintestinal amebiasis, usually as hepatic abscesses. Amebic abscesses occur particularly in males (male to female ratio = 7 : 1) between 20 and 50 years of age. 80 percent of the abscesses are located in the right lobe, probably due to the greater width and straighter course of the right portal vein branch, or the

greater volume of the right lobe. Hepatic abscesses are associated with rapid physical detoriation, hepatomegaly, liver tenderness, distention, malaise and fever. The local symptoms depend on the localization of the abscesses.

e) other extraintestinal sites are the lungs and pleura, these occur usually secondary to liver abscesses. The symptoms are those found with pleural effusion, empyema, pulmonary consolidation or bronchopleural fistula.
Cerebral abscesses are rare complications and are almost always fatal. Cutaneous amebiasis may involve perianal or genital areas, in rare cases facial cutaneous amebiasis was reported (Mendoza and Barba, 1986).

Pathology

Nonhuman primates: The morphological lesions depend to a certain degree on the host species involved. Macaques are mostly in an equilibrium with the amebas, tissue invasion is rare, and usually only microscopically visible (Johnson, 1941 - cit. from Belding, 1965). Invasion of the intestinal wall, which is more common in the hominoids and New World monkeys takes place primarily in the caecum, but may spread to the colon, in severe cases even to the rectum. The invading amebas penetrate to, but not through the muscularis mucosae in monkeys inducing large areas of necrosis with depressed centers, that are covered by fibrinous exudates. Histologically the colonic mucosa is necrotic and ulcerated down to the muscularis mucosae with extensive areas of haemorrhage and neutrophilic and mononuclear cellular infiltration (King, 1976). Chronic colitis is associated with petechial haemorrhages and ulcers (Bond et al., 1946; Bostrom et al., 1968; Fremming et al., 1955; Miller and Bray, 1966; Vickers, 1969).

In the colobines the primary lesions of amebiasis are found in the gastric presaccus and saccus as an ulcerative gastritis (Fig. 27), which may lead to gastric perforation (Müller and Rüedi, 1981; Frank, 1982; Loomis et al., 1983; Palmieri et al., 1984). Gastric amebiasis of colobines and intestinal amebiasis of New World monkeys frequently cause hepatic abscesses (Patten, 1939; Fremming et al., 1955; Belding, 1965; Miller and Bray, 1966; Geisel et al., 1975; Amyx et al., 1978).

Pulmonary amebic abscesses have been described in a chimpanzee (Miller and Bray, 1966).

Man: The first lesions are tiny, focal necroses of the superficial mucosa connected by small openings to small cavities filled with yellowish-brown, necrotic contents. These small necrotic foci are characteristically in the sigmoid, the distal ileum, caecum and descending colon. They usually progress to superficial ulcers.

In dysentery the ulcers progress rapidly both laterally and in depth, producing large irregular ulcers or necrosis of larger parts of the mucosal membranes. Invasion of the submucosa induces typical flask-shaped ulcers. Larger blood vessels may become eroded with resulting massive haemorrhages or may be blocked with subsequent gangrene of the intestinal wall. The ulcers even may perforate the intestinal wall and cause perforation peritonitis. In the vertical portions of the colon, the caecum and the rectum pseudotumoral lesions (amebomas) may develop (Davis and Pawlowski, 1985).

Hepatic amebiasis involves particularly the portal areas as lymphocytic-plasmacellular infiltrates, with fibroblast proliferation. Amebic liver abscesses follow thrombosis of the small portal veins or of the small interlobular veins, amebas

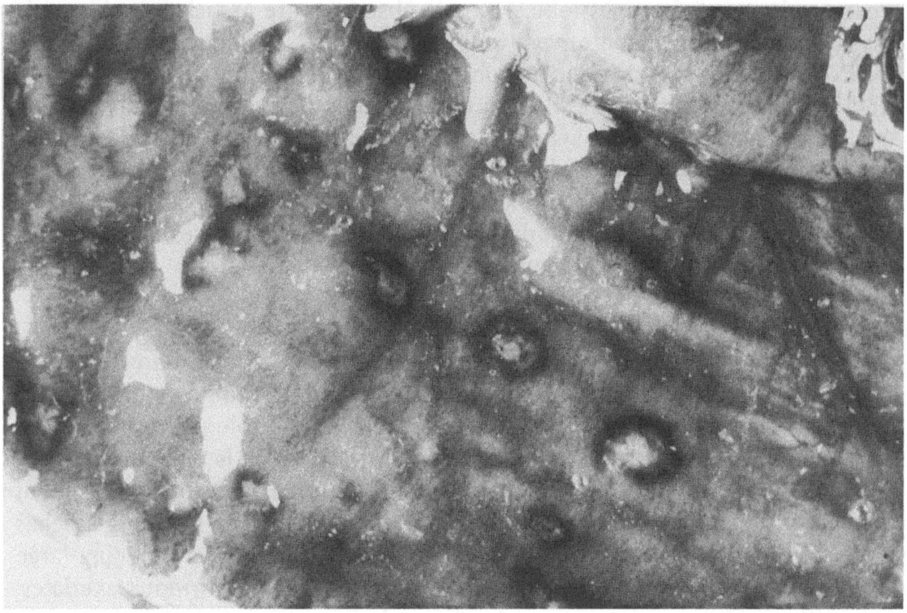

Fig. 27. *Procolobus Kirkii:* amebic gastritis

included in the thrombi cause necrosis and lysis of the surrounding parenchyma. The abscesses are surrounded by a minor inflammatory response and with time a fibrous capsule is formed. Small abscesses may become organized, leaving small, fibrous scars, whilst larger abscesses may calcify or they may progress and finally result in peritonitis, fistulation or, after further dissemination of the trophozoites, lead to respiratory infection.

Pulmonary amebiasis follows haematogenous spread of trophozoites or direct infection through the diaphragm. Particularly in haematogenous infections the lesions may simulate miliary tuberculosis or bronchopneumonic consolidation of pulmonary tissues.

Treatment

Today the imidazole derivatives are considered the drugs of choice for treatment of symptomatic intestinal amebiasis and have replaced most of the other drugs.

Metronidazole is recommended in man in dosages of 500 to 750 mg 3 times daily for 5 days orally (Sheehy, 1981); and in orangutans at 20 mg/kg/day for 5 days orally (Peters et al., 1976). In experimental infections of golden hamsters 40 mg/kg given 4 times were successful (Raether and Seidenath, 1984). Other imidazole-derivatives recommended for treatment of invasive amebiasis are tinidazole, ornidazole and nimorazole (WHO Scientific Working Group, 1980).

Asymptomatic luminal infections are best treated with amide-compounds in man (clefamide, etofamide, faramide, teclozan).

6.1.2 Entamoeba Chattoni (E. Polecki)

E. chattoni, like *E. histolytica* is a cosmopolite. Although usually a commensal of the large intestines, *E. chattoni* has been associated with intestinal disorders in man (Burrows and Klink, 1955; Salaki et al., 1979; Chacin-Bonilla, 1983).

E. chattoni has also been found in "monkeys" (Mudrow-Reichenow, 1956; Levine, 1961), *Macaca mulatta* (Kessel and Johnstone, 1949; Kessel and Kaplan, 1949; Mandal and Choudhury, 1984), and *Papio cynocephalus* (Myers et al., 1971).

Agent
E. chattoni is very easily confused with *E. histolytica.* Its trophozoites are 12 to 20 μm diameter. The coarsely granular endoplasm generally contains bacteria, yeasts and other intestinal contents. The nucleus in stained preparations contains delicately beaded peripheral chromatin and a small centric or subcentric karyosome. The cysts usually are 10 to 17 μm in diameter, spherical or subspherical, and generally contain a number of small glycogen vacuoles. They usually contain only one, rarely two nuclei. The chromatoidal bodies are small and numerous.

Epidemiology
The primary hosts of *E. chattoni* are pigs (Lushbaugh and Pittman, 1982). Transmission to other animals or man is probably direct, similar to that of *E. histolytica.* Most human infections occur in New Guinea and it is occasionally seen in Latin America (Chacin-Bonilla, 1983).

Clinical Symptoms
Most infections in nonhuman primates and man are subclinical (Lawless, 1954; Gay et al., 1985). Clinical symptoms characterized by abdominal cramps, diarrhea, nausea, and malaise have been rarely observed in man (Salaki et al., 1979).

Treatment
In man metronidazole: 750 mg 3 times daily for 10 days has proven effective (Salaki et al., 1979; Chacin-Bonilla, 1983).

6.2 Ciliata

6.2.1 Balantidium Coli

B. coli is an inhabitant of the large intestines of man and a variety of animals. It is distributed world wide, but is nevertheless an uncommon infection in man and sometimes has been regarded as little more than a medical curiosity. *B. coli* is, however, a potential pathogen in man and nonhuman primates, especially if other host factors favor the development and invasiveness of the balantidia.

Old World species are more frequently reported as infected than New World monkeys. The infection so far has not been found in prosimians. *B. coli* has been reported in gorillas (Teare and Loomis, 1982); orangutans (Peters et al., 1973, 1976); chimpanzees (Christeller, 1922; Benson et al., 1955; Schmitt, 1964; van Riper et al., 1966; McClure and Guilloud, 1971; Kim et al., 1978); macaques (Noc, 1908; Poindexter, 1942; Habermann and Williams, 1957; Young et al., 1957; Gisler et al., 1960; Ford and Speltie, 1972; Sano et al., 1980); African green monkeys (Reardon and Reininger, 1968); baboons (Kuntz and Myers, 1967; Myers and Kuntz, 1968; Kuntz and Moore, 1973; Movcan and Tscherner, 1982); mangabeys (Joyeux, 1913) and howler monkeys (Benton, 1976).

Agent

B. coli occurs in its trophozoite form in the mammalian host and as a cyst form in the environment.

The trophozoites are egg-shaped, of 35-90 × 20-50 µm size, with a pointed posterior end. Their pellicle is covered by spiral-longitudinal rows of short cilia, which originate from basal granules in the ectoplasm. A ciliated cystostome at the anterior end leads into the gullet. An excretion pore, the cytopyge, is situated at the posterior end. One contractile vacuole is located near the anterior end, whilst another one along with the food vacuoles is positioned near to the posterior end. Unstained trophozoites are grey-greenish. In stained preparations the macronucleus appears as an ovoid to sausage or even bar-shaped body. The globoid micronucleus immediately adjacent to the macronucleus is sometimes invisible.

The cysts are spherical or subspherical, 35 to 65 µm in diameter and surrounded by a thick, double wall. The macro- and micronucleus appear as in the trophozoites, but the vacuoles and other structures disappear after a while. The latter are present in the young cysts and after encystation the parasites still keep moving within the cyst for some time before losing the cilia which renders them immobile. Unstained cysts are greenish as are the trophozoites. The cysts can survive in moist environment for several weeks at 22 °C but are quite sensitive to drying and sunlight.

B. coli is a facultative anaerobe. It feeds on intestinal contents, other protozoa and bacteria, but also on tissues, erythrocytes etc. when given a chance. It produces a hyaluronidase-like substance which may aid its spread in the intestinal mucosa once infection has been established (Tempelis and Lysenko, 1957 - cit. from Walzer and Healy, 1982).

Reproduction occurs only in the trophozoite stage, mainly by binary fission.

Epidemiology

Balantidia are residents in the digestive tract of over 50 vertebrate and invertebrate hosts (Levine, 1973; Walzer and Healy, 1982). Organisms morphologically indistinguishable from human *B. coli* have been found in a variety of mammals, with pigs being frequently suggested as the reservoir hosts (Walzer and Healy, 1982). However, human and porcine *B. coli* showed differences in their temperature sensitivity and antigenic properties (Walzer and Healy, 1982). According to Fiennes (1967) *B. coli* can pass also from nonhuman primates to man and vice versa. Infection of man occurs by ingestion of infective cysts in either drinking water or food.

Pathogenesis

B. coli is ordinarily a harmless commensal in man and nonhuman primates. The trophozoites do not invade the healthy intact intestinal mucosa, but depend on another organism or situation to interrupt the integrity of the normal tissue to gain entrance into the mucosal membrane and thus cause disease (Phillips, 1980). Protein-deficient and carbohydrate-rich diets as well as bacterial infections support the activities of *B. coli* trophozoites.

Although *B. coli* can invade anywhere along the colon, in man this most commonly occurs in the rectosigmoid area. Invasion is most probably mechanical but is aided by cytolytic enzymes. The contacted epithelial cells shrink up, their cytoplasm becoming eosinophilic. The trophozoites then invade the propria mucosa, and even the submucosa as well as the local lymphatic vessels. They induce a plasmacellular, lymphocytic and eosinophilic inflammatory response. Secondary bacterial infections of the parasitic lesions induce an influx of polymorphonuclear leukocytes and the formation of submucosal abscesses.

Extraintestinal balantidiasis is rare and presumably follows perforation of the colon (Walzer and Healy, 1982).

Clinical Symptoms

Nonhuman primates: *B. coli* infections rarely cause clinical symptoms in the lower primates unless otherwise complicated. In own observations in *Macaca fascicularis* with concurrent shigellosis, the leading sign was a profuse watery-haemorrhagic diarrhea. Usually monkeys are asymptomatic shedders, occasionally soft stools may be noticed (Hegner, 1934; Moorman, 1941; Cockburn, 1948; Schmitt, 1964; McClure and Guilloud, 1971).

In the great apes the signs are more distinctive, consisting of watery-mucoid diarrhea, tenesmus, muscular weakness, loss of appetite, vomitus, lethargy, and in some cases death (Schmitt, 1964; Mortelmans et al., 1971; Peters et al., 1976; Kim et al., 1978; Teare and Loomis, 1982).

Man: Human *B. coli* infections may be

a) asymptomatic, or
b) chronic with recurring diarrhea, or
c) occur as a mild to profuse dysentery with bloody-mucoid or purulent stools. Headache, insomnia, nausea, vomitus, cachexia, weight-loss, pallor, weakness, anemia, dehydration, malaise and meteorism have been associated with balantidial infections (Neafie, 1976).

Pathology

Nonhuman primates: *B. coli* infections may completely lack any morphological lesions (McClure and Guilloud, 1971), but in other cases, particularly in hominoids, the parasite may cause severe ulcerative colitis with the ulcers extending down into the muscularis mucosae, with accompanying lymphocytic infiltration, coagulation, necrosis and haemorrhage (Phillips, 1980; Toft, 1982; own observations).

Man: The lesions of *B. coli* infections in man may vary as described in nonhuman primates from complete lack to catarrhal inflammation to ulcerative colitis. The ulcers are often flask-shaped, with undetermined edges. A zone of coagulation necrosis forms the base of the ulcers, which is further surrounded by edema and chronic inflammation (Neafie, 1976).

Treatment

The drugs of choice are imidazole-derivates. Metronidazole given at 20 mg/kg/day orally for 5 days (Peters et al., 1976) or a 2 months regimen of ½ tablet twice daily in tea mixture (Benton, 1976) was successful in hominoids.

In man metronidazole (750 mg in children or 1,25 gr in adults for 10 days) failed to cure several patients during an epidemic in Truk (Walzer and Healy, 1982).

References

1. Benson, R.E.; B.D. Fremming, and R.D. Young: Care and management of chimpanzees at the Radiobiological Laboratory of the University of Texas and the United States Air Force. School. of Aviat. Med. U.S. Air Force Rep. 55-58 (1955)
2. Benton, L.jr.: The establishment and husbandry of a black howler *Alouatta caraya* colony at Columbia Zoo. Internat. Zoo Yearbook *16*, 149-152 (1976)
3. Christeller, E.: Über die Balantidienruhr bei den Schimpansen des Berliner Zoologischen Gartens. Virchows Arch. *238*, 396-422 (1922)
4. Cockburn, T.A.: *Balantidium* infection associated with diarrhea in primates. Trans. roy. Soc. trop. Med. Hyg. *42*, 291-293 (1948)
5. Fiennes, R.: Zoonoses of Primates. The Epidemiology and Ecology of Simian Diseases in Relation to Man. Weidenfeld and Nicolson, London (1967)
6. Ford, A.C. and T.M. Speltie: Some aspects of the parasitology of nonhuman primates. Ann. Rep. TNO 1972, 160-162 (1972)
7. Gay, J.D.; T.L. Abell; J.H. Thompson, and V. Loth: *Entamoeba polecki* infection in Southeast Asian refugees: Multiple cases of a rarely reported parasite. Mayo Clin. Proc. *60*, 523-530 (1985)
8. Gisler, D.B.; R.E. Benson, and R.J. Young: Colony husbandry of research monkeys. Ann. N.Y. Acad. Sci. *85*, 758-768 (1960)
9. Habermann, R.T. and F.P. Williams, jr.: Diseases seen at necropsy of 708 *Macaca mulatta* (rhesus monkey) and *Macaca philippinensis* (cynomolgus monkey). Amer. J. vet. Res. *18*, 419-426 (1957)
10. Hegner, R.: Intestinal protozoa of chimpanzees. Amer. J. Hyg. *19*, 480-501 (1934)
11. Joyeux, C.: Note sur quelques protozoaires sanguicoles et intestinaux observés en Guinee française. Bull. Soc. Path. exot. *6*, 612-615 (1913)
12. Kim, J.C.S.; C.R. Abee, and R.H. Wolf: Balantidiosis in a chimpanzee *(Pan troglodytes)*. Lab. Anim. *12*, 231-233 (1978)
13. Kobiler, D.; D. Mirelman, and C.F. Mattern: Lectin and toxin-like activities of *Entamoeba histolytica*: Comparison of properties. Amer. J. trop. Med. Hyg. *30*, 955-959 (1981)
14. Kuntz, R.E. and J.A. Moore: Commensals and parasites of African baboons (*Papio cynocephalus* L. 1766) captured in Rift Valley province of central Kenya. J. med. Primatol. *2*, 236-241 (1973)
15. Kuntz, R.E. and B.J. Myers: Microbiological parameters of the baboon (Papio sp.): Parasitology, pp. 741-755 in "The Baboon in Medical Research", Vol. 2 (Vagtborg, H., ed.), Univ. Texas Press, Austin (1967)
16. Levine, N.D.: Protozoan Parasites of Domestic Animals and Man. 2nd ed., Burgess Publ. Co., Minneapolis (1973)
17. McClure, H.M. and N.B. Guilloud: Comparative pathology of the chimpanzee; pp. 103-272 in "The Chimpanzee", Vol. 4 (Bourne, G.H., ed.), Karger, Basel (1971)
18. Mendoza, J.B. and E.J.R. Barba: Cutaneous amebiasis of the face: a case report. Amer. J. trop. Med. Hyg. *35*, 69-71 (1986)

19. Moorman, A. E.: *Balantidium coli* and pinworms in a chimpanzee. J. Parasitol. *27*, 366 (1941)
20. Mortelmans, J.; J. Vercruysse, and P. Kageruka: Three pathogenic intestinal protozoa of anthropoid apes: *Entamoeba histolytica*, *Balantidium coli*, and *Troglodytella abrassarti*; pp. 187-191. Proc. 3rd. Internat. Congr. Primatol. Zürich, Vol. 2 (Biegert, J. and W. Leuteneger, eds.), Karger, Basel (1971)
21. Movcan, A.T. und W. Tscherner: Parasitenbefall bei Mantelpavianen (Papio hamadryas), pp. 337-339, 24. Internat. Symp. Erkr. Zootiere, Veszprem (1982)
22. Myers, B.J. and R.E. Kuntz: Intestinal protozoa of the baboon *Papio doguera* Pucheran 1856. J. Protozool. *15*, 363-365 (1968)
23. Neafie, R.C.: Balantidiasis, pp. 325-327 in "Pathology of Tropical and Extraordinary Diseases. An Atlas. Vol. 1" (Binford, C.H. and D.H. Connor, eds.) Armed Forces Institute of Pathology, Washington, D.C. (1976)
24. Noc, F.: Un cas de dysenterie à Balantidium chez le Macacus cynomolgus. Compt. Rend. Soc. Biol. Paris *64*, 878-880 (1908)
25. Peters, W.; P.C.C. Garnham, and R. Killick-Kendrick: Some intestinal protozoa of the orang-utan. Trans. roy. Soc. trop. Med. Hyg. *67*, 3 (1973)
26. Peters, W.; P.C.C. Garnham; R. Killick-Kendrick; N. Rajapaksa; W.H. Cheong, and F. Cadigan: Malaria of the orang-utan *(Pongo pygmaeus)* in Borneo. Phil. Trans. B *275*, 439-482 (1976)
27. Phillips, L.G. jr.: The parasitology of primates and carnivores. Amer. Ass. Zoo Vet. Ann. Proc. 1980, 116-119 (1980)
28. Poindexter, H.A.: A study of the intestinal parasites of the monkeys of the Santiago Island primate colony. Puerto Rico J. Publ. Hlth Trop. Med. *18*, 175-191 (1942)
29. Reardon, C.V. and B.F. Reininger: A survey of parasites in laboratory primates. Lab. Anim. Care *18*, 511-580 (1968)
30. Sano, M.; H. Kino; T.S. de Guzman; A.I. Ishii; J. Kino; T. Tanaka, and M. Tsurita: Studies on the examination of imported laboratory monkey, Macaca fascicularis for E. histolytica and other intestinal parasites. Int. J. Zoonoses *7*, 34-39 (1980)
31. Schmitt, J.: Die Balantidiase, unter besonderer Berücksichtigung eines enzootischen Auftretens bei Menschenaffen des Zoologischen Gartens Frankfurt am Main. Dtsch. tierärztl. Wschr. *71*, 68-70 (1964)
32. Smith, J.M. and E. Meerovitch: Primates as source of *Entamoeba histolytica*, their zymodeme status and zoonotic potential. J. Parasit. *7*, 751-756 (1985)
33. Teare, J.A. and M.R. Loomis: Epizootic of balantidiasis in lowland gorillas. J. Amer. vet. med. Ass. *181*, 1345-1347 (1982)
34. Toft, J.D. II: The pathoparasitology of the alimentary tract and parasites of nonhuman primates: A review. Vet. Path. *19* (Suppl. 7), 44-92 (1982)
35. van Riper, D.C.; P.W. Day; J. Fineg, and J.R. Prine: Intestinal parasites of recently imported chimpanzees. Lab. Anim. Care *16*, 360-363 (1966)
36. Walzer, P.D. and G.R. Healy: Balantidiasis; pp. 15-24 in "CRC Handbook Series in Zoonoses" (Steele, J.H., Series ed.), Section C: Parasitic Zoonoses, Vol. I (Jacobs, L. and P. Arambulo, Section eds.), CRC Press, Boca Raton (1982)
37. Young, R.J.; B.D. Fremming; R.E. Benson, and M.D. Harris: Care and management of a Macaca mulatta monkey colony. Proc. Anim. Care Panel *7*, 67-82 (1957)

6.3 Zoomastigophora (Flagellata)

6.3.1 Diplomonadina

6.3.1.1 Giardia Lamblia (G. Intestinalis, G. Duodenalis, Lamblia Intestinalis)

Giardiasis is an infection of the small intestines with a world wide distribution but with a somewhat confusing nomenclature and of unknown zoonotic potentials. At least 40 *Giardia* spp. have been proposed primarily on the basis of assumed host specificity, although there is evidence indicating that some *Giardia* are capable of parasitizing more than one host species (Meyer and Jarroll, 1982). Most of the proposed *Giardia* spp. cannot be distinguished morphologically, although Filice (1952) differentiated *G. duodenalis*, *G. muris* and *G. agilis* based on the shapes of the median bodies.

G. lamblia has occasionally been found in *M. mulatta* (Rowland and Vandenbergh, 1965), *Hylobates* sp., *Pan troglodytes*, *Gorilla gorilla*, *Pongo pygmaeus* (Anon, 1979) and *Callithrix geoffroyi* (own observations).

Agent
G. lamblia occurs in a trophozoite and a cyst form. The tear-drop-shaped trophozoites are of 9–21 × 6–12 µm size and 2–4 µm thick. They are binucleated, flagellated and possess an adhesive (sucker) disk, which occupies ¾ of the ventral, concave aspect of the parasite. The flagella not only propel the detached trophozoite very efficiently but also help in its adhesion by expelling fluid from the cavity under the disk by the beating of the two ventral flagella (Oweb, 1980a).

Two heavily stained median bodies are situated behind the sucker disk, they resemble the claw of a claw hammer in *B. duodenalis*, whereas in *G. muris* they are rounded and in *G. agilis* long, teardrop-shaped formations.

The factors inducing encystment are unknown, although it has been proposed that detachment from the gut wall may trigger it (Owen et al., 1979). After encystment the organism divides once. In vitro the encysted parasite needs a maturation period before excystment is possible (Bingham et al., 1979). The cysts are ellipsoid, smooth-walled, and of 9–12 × 6–8 µm in size. Two to four round nuclei can be recognized in the anterior part of the cyst, whilst the median bodies are visible in the posterior part. Giardia cysts are destroyed in water by halogenation at 20 °C, but not at 3 °C (Jarroll et al., 1980). They are also destroyed by 2 to 5% phenole or cresole. Boiling is the most simple and most reliable method for killing Giardia cysts in drinking water (Anon, 1980). Chlorination is the principal method for disinfecting larger bodies of water, but normal levels (0.4 ml/l free chlorine) are not sufficient (Anon, 1980).

G. lamblia can be cultivated on TPS (trypticase, Panmede (ox liver digest), serum) medium (Meyer, 1984).

Epidemiology
G. lamblia is more frequently found in warmer climates and in crowded, unsanitary environments (Yardley, 1976). Its incidence therefore is higher in developing

countries and in the tropics. Large outbreaks have been reported in orphanages and institutions for the mentally retarded (Owen, 1984). It contaminates surface waters causing Hiker's diarrhea even in the sparsely populated areas of western North America (Meyer and Jarroll, 1982), and major outbreaks of waterborn diarrhea (Anon, 1980; Meyer and Jarroll, 1982).

Infection usually follows the oral uptake of infective cysts. Excystment occurs at low pH (Bingham and Meyer, 1979), therefore most probably in the gastric juice. After excystment the *Giardia*-trophozoites remain in the small intestines, usually in the lower jejunum and ileum (Ackers, 1980). Transmission is accomplished by the fecal - oral route, by contaminated food or drinking water.

Pathogenesis

After excystation the trophozoites firmly attach themselves to the intestinal wall, probably aided by their sucker-like structure, the sucker-disk, on their ventral surface. Attachment occurs preferentially near the bases of the villi and the edges of Peyer's patch follicles, but it rarely occurs on the villous tips or to the M-cells (Owen et al., 1979). The intimate contact of the disk with the enterocyte can damage the microvilli and fuzzy coat, as well as reducing the activity of the brush-border enzymes. It may contribute to intestinal villous atrophy (WHO Scientific Working Group, 1980). The attached trophozoites feed on the intestinal contents, probably through micropinocytosis (Nemanic et al., 1979). The mechanisms rendering the parasites to pathogenic properties are unknown, although immunodeficiency (Kean et al., 1979; Owen, 1980b), secondary bacterial or viral infections (Tomkins et al., 1978) have been suggested. In the one *C. geoffroyi* the animal was also infected by *Campylobacter jejuni* (personal observations).

Clinical Symptoms

Nonhuman primates: Occasional mucoid diarrhea, weight loss and interference with the intestinal fat reabsorption have been observed (Burrows, 1972; Anon, 1979; own observations).

Man: Giardiasis is normally an acute diarrheal illness lasting approximately six weeks, followed by spontaneous recovery. In heavy infections steatorrhea, diarrhea with malabsorption, abdominal pain, flatulence, loss of appetite, vomitus, achlorhydria, anemia and nervous symptoms have been reported (Meyer and Jarroll, 1982). The malabsorption can be a very serious complication in children, immunodeficient patients or in individuals suffering from protein malnutrition (Smith, 1985). In chronic giardiasis, which lasted for several years, an increased frequency of constipation and upper gastrointestinal complaints was noted (Chester et al., 1985).

Pathology

In heavy infections of nonhuman primates and of man nonspecific inflammations and small ulcers may be present. In children the focal enteritis may resemble celiac sprue with villous atrophy, flattening of epithelial cells and dense mononuclear cell infiltrations (Yardley et al., 1964; Smith 1985).

Treatment

Metronidazole (750 mg in man, or 250 to 1000 mg in hominoids daily for 2 weeks) proved to be effective (Anon, 1979). In nonhuman primates Flagyl (Searle & Co.) also was successful (Phillips, 1980). The WHO Scientific Working Group (1980) estimates 80% cure rates with nitroimidazoles (metronidazole, tinidazole, nimorazole), nitrofuran derivatives (furazolidone) and acridine compounds (mepacrine, acranil).

6.3.1.2 Trypanosoma Cruzi

T. cruzi is the causative agent of Chagas' disease (American trypanosomiasis) of man in South- and Central America and in parts of the USA.

T. cruzi infects also a great variety of marsupials, chiroptera, edentates, lagomorphs, rodents, carnivores, artiodactyla and nonhuman primates, with opposums being the principal sylvatic reservoirs (Miles et al., 1979; Yaeger, 1982) (Fig. 29).

The following monkey-species have been reported as being naturally infected: *Ateles belzebuth hybridus, Ateles geoffroyi grisescens* (Marinkelle, 1966), *Ateles fusciceps* (Sousa et al., 1974), *Ateles paniscus, Cacajao calvus* (Dunn et al., 1963), *Cebus albifrons* (Dunn et al., 1963; Marinkelle, 1966; Bullock et al., 1967), *Cebus apella* (Dunn et al., 1963; Marinkelle, 1966; Travi et al., 1982), *Cebus apella versatus* (Barreto et al., 1966), *Alouatta caraya* (Funayama and Barreto, 1970; Travi et al., 1982), *Alouatta seniculus* (Yaeger, 1982), *Saimiri sciureus* (Garnham and Gonzales-Mogaburo, 1962; Dunn et al., 1963; Marinkelle, 1966; Sousa et al., 1974; Travi et al., 1982), *Saimiri sciureus boliviensis* (Ayala, 1961; Dunn et al., 1963), *Saguinus oedipus* (Werner et al., 1977), *Saguinus leucopus* (Marinkelle, 1966), *Saguinus nigricollis* (Dunn et al., 1963); *Callithrix jacchus* (Deane, 1962), *Callithrix geoffroyi* (DeMelorocha and Barreto, 1977), and *Callicebus personatus nigrifrons* (Albuquerque and Barreto, 1969).

Flagellates undistinguishable from *T. cruzi* have been reported from several Asian species (*Macaca fascicularis, M. mulatta, M. cyclopis, Nycticebus coucang, Hylobates pileatus*) (Malamos, 1935; Fulton and Harrison, 1946; Seneca and Wolf, 1955; Hoare, 1963; Dunn, 1964; Weinman and Wiratmadja, 1969; Kuntz et al., 1970; Zaman, 1972; Weinman, 1974; Cross et al., 1983). The question is still open to discussion whether those were really *T. cruzi* or were they possibly *T. conorrhini*, a trypanosome of rats in Asia (Hoare, 1963; Cross et al., 1983).

T. cruzi infections of nonhuman primates outside South- and Central America have been reported on several occasions in the United States (Davis, 1943 - cit. from Kasa et al., 1928; Cicmanec et al., 1974; Kasa et al., 1978).

Agent

T. cruzi is a pleomorphic trypanosome with a biphasic life cycle. In mammalian hosts *T. cruzi* occurs extracellularly as the trypomastigote (trypanosome) stage in various body fluids during the acute stage of the disease. In its trypomastigote stage, *T. cruzi* does not undergo division, but penetrates a tissue cell (e.g. myocardial cell) where it transforms into the amastigote (Leishmania) stage (Fig. 28). In the amastigote stage it multiplies by binary fission. Subsequently the amastigotes

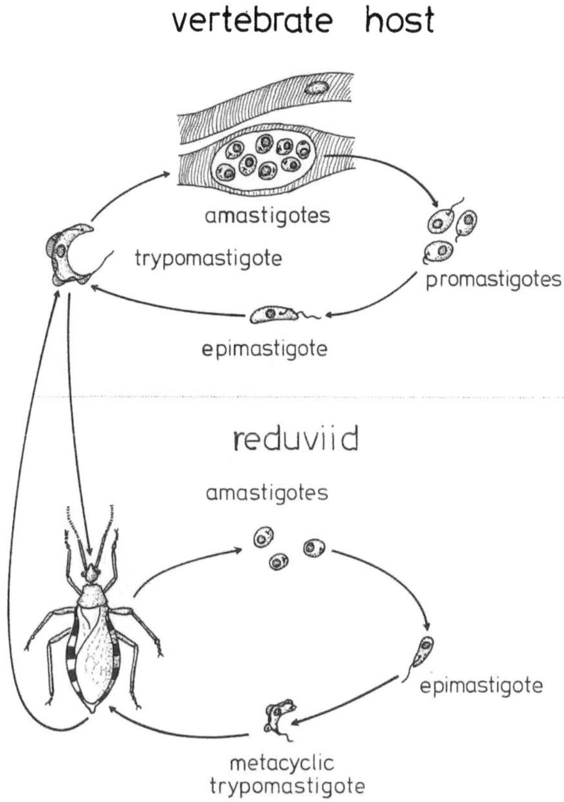

Fig. 28. Developmental cycle of *Trypanosoma cruzi*

elongate to resemble "promastigote" (leptomonas) stage. It then further transforms into a epimastigote (crithidia) stage and finally the parasite completes its mammalian cycle when it has reassumed the trypomastigote form. On rupture of the host cell the highly motile trypanosomes are released. They spread via the blood and other body fluids and may infect other cells of the same host or they may be taken up by insect vectors (reduviids) to undergo the second phase of the life cycle (Fig. 29). Other possible intermediate hosts may be some bedbugs and ticks (Edgcomb and Johnson, 1976), but these are probably of only marginal importance.

The nocturnal triatome bugs which live in palm trees and root thatchings made of palm tree leaves, or in bat-hollows, bird nests and armadillo burrows, or are adapted to human environment (D'Alessandro et al., 1984) become infected by ingestion of blood containing trypomastigotes after biting an infected mammalian host during the night (Ristic and Smith, 1974). Another mode of spread among the triatomids is feeding on the gut contents of another infected bug. In the midgut of the bug the ingested trypomastigotes change to amastigotes and epimastigotes with reproduction by binary fission in both stages. As the gut contents move towards the rectal portion metacyclic trypomastigotes, the infective stage, develop. Approximately 10 days after the infected blood meal, infective trypomastigotes are excreted in the bugs feces. The bugs frequently defecate while feeding, the feces

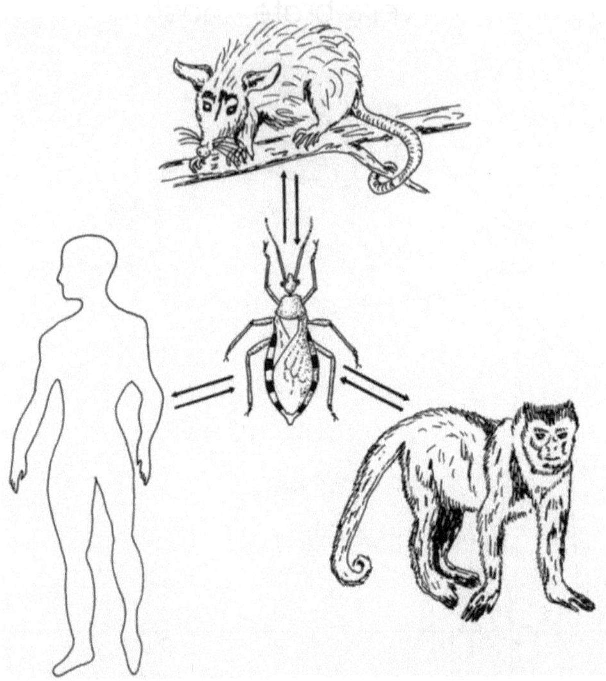

Fig. 29. Nonhuman primates in the life cycle of *Trypanosoma cruzi*

Table 36. Reduviids Naturally Infected with *Trypanosoma cruzi*

Cavernicola pilosa	Triatoma maculata
Dipetalogaster maximus	*Triatoma megista*
	Triatoma neotomae
Eratyrus cuspidatus	Triatoma nigromaculata
Eratyrus mucronatus	Triatoma nitida
Panstrongylus geniculatus	Triatoma longipes
Panstrongylus guntheri	Triatoma oswaldoi
Panstrongylus lignarius	Triatoma patagonica
Panstrongylus megistus	Triatoma phyllosoma
Panstrongylus rufotuberculatus	Triatoma phyllosoma intermedia
Psammolestes arthuri	Triatoma phyllosoma longipennis
Rhodnius pallescens	Triatoma phyllosoma mazzottii
Rhodnius pictipes	Triatoma phyllosoma pallidipennis
Rhodnius prolixus	Triatoma phyllosoma picturata
Triatoma barberi	Triatoma phyllosoma usingeri
Triatoma brasiliensis	Triatoma platensis
Triatoma carrioni	Triatoma protracta protracta
Triatoma delpontei	Triatoma protracta woodi
Triatoma dimidiata	Triatoma rubida
Triatoma dimidiata capitata	Triatoma rubida sonoriana
Triatoma dimidiata maculipennis	Triatoma rubida uhleri
Triatoma eratyrusiforma	Triatoma rubrofasciata
Triatoma gerstaeckeri	Triatoma rubrovaria
Triatoma guasayana	Triatoma sanguisuga
Triatoma hegneri	Triatoma sanguisuga ambigua
Triatoma infestans	*Triatoma sordida*

are rubbed into the puncture wounds set by the triatomines or possibly into the conjunctiva. Animals may also become infected by eating infected bugs, which may be an even more effective way of disease transmission in New World monkeys, all of which are partially insectivorous.

At least 53 species of triatomid (reduviid) bugs are known to be naturally infected with *T. cruzi* (Table 36).

Larvae, nymphs and adult bugs become infected when feeding on blood but once infected, the bug may remain infected for life (several years) (Edgcomb and Johnson, 1976).

In man congenital infection of the fetus may be a real hazard, affecting predominantly newborns with a gestational age of 26 to 37 weeks (Azogue et al., 1985).

The *T. cruzi* trypomastigotes can be distinguished from those of other trypanosomes by a distinct C or U shape when fixed. They are small, 16 to 20 μm in length, occurring in either a broad or a slender form. In the broad form the centrally located nucleus is subspherical, and stains reddish purple in Giemsa-stained blood-smears. The rather large, deeply purple staining kinetoplast is at the posterior terminal end. The axoneme of the flagellum originates close to the kinetoplast, it then extends outside the body of the trypomastigote to the anterior end of the flagellate. It is covered with a flagellar sheath, and attached by an undulating membrane. There is no significant length of free flagella extending beyond the anterior end of the trypomastigote. In the slender form the nucleus is rather elongated, with the kinetoplast being subterminally located.

The tissue form, the amastigote, is 3 to 5 μm in size and characterized by the kinetoplast, a prominent bar, which stains deeply with hematoxylin. The kinetoplast distinguishes *T. cruzi* amastigotes from Toxoplasma, Histoplasma etc.

Clinical Symptoms

Nonhuman primates: Despite the comparatively frequent natural *T. cruzi* infections in New World monkeys information on the clinical disease is extremely rare. Kasa et al. (1978) reported leukocytosis in a naturally infected rhesus monkey, presumably infected by *T. cruzi*. Experimental infections of rhesus monkeys induced transient signs of cardiac failure (edema, bilateral enlargement of cardiac shadow) or even death due to myocarditis (Szarfman et al., 1978).

Man: *T. cruzi* may induce acute or chronic infections. The acute infection, although occurring at any age, is seen particularly in children and may be fatal in infants, although even in children ⅔ of acute infections are asymptomatic (Teixeira, 1979). The incubation period varies from 1 week to 1 month in acute Chagas' disease. This is followed by an abrupt onset of fever, which may be intermittent, remitting or continuous. In about one third to one half of the cases the Romana sign, a unilateral peripalpebral edema, develops at the portal of entry in conjunctival infections. A nodular, painless, edematous lesion, known as chagoma, may be seen in 25% of cases. Other symptoms are general malaise, muscular pain, lymphadenitis, anorexia, diarrhea, vomitus, rashes and hepato-splenomegaly. Encephalitis or meningoencephalitis and occasionally acute myocarditis are fatal complications of Chagas' disease in infants.

Most infected individuals survive the acute stage, hence the majority of Chagas' disease patients have the chronic form of the disease, which may persist for life unless successfully treated. The symptoms of the chronic form are uncharacteristic, with weight loss, pseudomyxedemas, lymphadenopathy, anemia, monocytosis and general exhaustion. Myocarditis occurs in the majority of chronic cases with enlarged heart shadow, tachycardia, prolonged P-R or Q-T intervals in the ECG, primary T-wave changes, partial and complete A-V block, premature ventricular contraction, arrhythmia, dyspnea and signs of congestion. The myocarditis often results in premature death.

Megacolon and megaesophagus are frequently associated with the chronic form in Brazil, with megacolon being also frequently seen in chronic Chagas' disease in Paraguay, Chile, Bolivia, and Argentina. Both manifestations are rarely seen in other endemic areas (Yaeger, 1982).

Congenital Chagas' disease due to parazitation of the placenta usually causes abortion or stillbirth. In live births acute necrotizing encephalitis is seen.

Chronic Chagas' disease is characterized by chronic Chagasic cardiopathy as a consequence of severe myocarditis. The heart is heavily dilated, the right ventricular outflow tract is prominent. Microscopically, focal areas of chronic lymphocytic-plasma cellular myocarditis alternate with patches of interstitial fibrosis amongst hypertrophied fibers, but the parasites themselves are sometimes hard to demonstrate.

The megaviscera are associated with a profound decrease in the ganglion cells of the myenteric plexuses; thinning of the visceral wall with variable amounts of scarring and chronic myositis.

The CNS lesions consist of microglial nodules and occasional nonpurulent meningitis, but they are not a prominent feature of chronic Chagas' disease (Edgcomb and Johnson, 1976).

Treatment

Prevention of Chagas' disease has to focus on the interruption of the cycle by socioeconomic measures.

The drugs of choice for treatment of acute Chagas' disease are probably the imidazoles. Fexinidazole has a high activity against *T. cruzi* trypomastigotes in mice. Side effects after higher dose levels were anemia and weight loss (Raether and Seidenath, 1983). Ketoconazole affects amastigotes during the phase of generalized organ infestation better than most of the standard drugs do. However, because of its liver toxicity and its inhibitory effects on adrenal steroid synthesis the customary long-term treatment of Chagas' disease would require very careful medical observation (Raether and Seidenath, 1984).

6.3.1.3 Leishmania Brasiliensis

Members of the *L. brasiliensis*-complex *(L. b. panamensis?)* have been reported in Panamanian *Saguinus geoffroyi* and *Aotus trivirgatus*, which are considered to be secondary hosts (Marinkelle, 1980; Lainson, 1982). *L. b. panamensis* is of low pathogenicity to man, the cases of the disease seen in Panama are thought to be caused by another parasite. Because of the taxonomical problems of the *L. brasiliensis* and *L. mexicana* complexes with somewhat confusing references in the past, the geographic overlap of the subspecies and the undetermined, probably only marginal role of nonhuman primates in the life cycles of the American *Leishmania*-species these parasites will not be further discussed. For extensive information see Lainson (1982).

References

1. Ackers, J. P.: Giardiasis: Basic parasitology. Trans. roy. Soc. trop. Med. Hyg. *74*, 427-428 (1980)
2. Albuquerque, R. D. R. y M. P. Barreto: Estudos sobre reservatorios e vectores silvestres do *Tryoanosoma cruzi.* XXXII. Infeccao natural do simio *Callicebus nigrifrons* (Spix, 1823) pelo *T. cruzi.* Rev. Inst. Med. trop. S. Paulo *11*, 115-122 (1969)
3. Anon: Giardiasis in apes and zoo attendants, Kansas City, Missouri. CDC Vet. Publ. Hlth Notes, January, 7-8 (1979)
4. Anon: Waterborne giardiasis - California, Colorado, Oregon, Pennsylvania. CDC Morbid. Mortal. Wkly. Rep. *29*, 121-123 (1980)
5. Ayala, F. M.: Hallazgo de *Trypanosoma cruzi* Chagas', 1909 en el mono *Saimiri boliviensis* de la Amazonia Peruana. Rev. Gras. Malar. Doenc. trop. *13*, 99-105 (1961)
6. Azogue, E.; C. la Fuente, and C. Darras: Congenital Chagas' disease in Bolivia: epidemiological aspects and pathological findings. Trans. roy. Soc. trop. Med. Hyg. *79*, 176-180 (1985)
7. Barreto, M. P.; A. F. Siqueira; F. Ferrioli y J. R. Carvalheiro: Estudos sobre reservatories e vectores silvestres de *Trypanosoma cruzi.* Rev. Inst. Med. trop. S. Paulo *8*, 103-112 (1966)
8. Bingham, A. K.; E. L. Jarroll, jr.; E. A. Meyer, and S. Radalescu: Induction of *Giardia* excystation and the effect of temperature on cyst viability as compared by eosin exclusion and *in vitro* excystation; pp. 216-229 in "Proceedings of the Symposium on Waterborne Transmission of Giardiasis." (Jakubowski, W. and J. C. Hoff, eds.). U. S. Environmental Protection Agency, Cincinnati (1979)
9. Bingham, A. K. and E. A. Meyer: *Giardia* excystation can be induced *in vitro* in acidic solution. Nature *277*, 301-302 (1979)
10. Bullock, B. C.; R. H. Wolf, and T. B. Clarkson: Myocarditis associated with trypanosomiasis in a cebus monkey *(Cebus albifrons).* J. Amer. vet. med. Ass. *151*, 920-922 (1967)
11. Burrows, R. B.: Protozoa of the intestinal tract, pp. 2-28 in "Pathology of Simian Primates". Part II (Fiennes, R. N. T-W, ed.), Karger, Basel (1972)
12. Chester, A. C.; F. G. MacMurray; M. D. Restifo, and O. Mann: Giardiasis as a chronic disease. Dig. Dis. Sci. *30*, 215-218 (1985)
13. Cicmanec, J. L.; F. A. Neva; H. M. McClure, and W. F. Loeb: Accidental infection of laboratory-reared *Macaca mulatta* with *Trypanosoma cruzi.* Lab. Anim. Sci. *24*, 783-787 (1974)
14. Cross, J. H.; M. Y. K. Hsu, and C. K. Hung: Studies on trypanosomes in the Taiwan monkey. Southeast Asian J. Trop. Med. Publ. Hlth *14*, 536-542 (1983)
15. D'Alessandro, A.; P. Barreto; N. Saravia, and M. Barreto: Epidemiology of *Trypanosoma cruzi* in the oriental plains of Colombia. Amer. J. trop. Med. Hyg. *33*, 1084-1095 (1984)
16. Deane, L. M.: Infeccao natural do sagüi *Callithrix jacchus* por tripanosoma do tipo cruzi. Rev. Inst. Med. trop. S. Paulo *4*, 225-229 (1962)

17. Demelo Rocha, G. y M. P. Barreto: Estudos sobre reservatories e vectores silvestres do *Trypanosoma cruzi.* LXV: Infeccao natural do simio *Callithrix geoffroyi* (Humboldt, 1812) pelo *T. cruzi.* Rev. bras. Biol. *37,* 419-424 (1977)
18. Dunn, F. L.: Blood parasites of South East Asian primitive primates. J. Parasit. *50,* 214-216 (1964)
19. Dunn, F. L.; F. L. Lambrecht, and R. du Plessis: Trypanosomes of South American monkeys and marmosets. Amer. J. trop. Med. Hyg. *12,* 524-534 (1963)
20. Eberhard, M. and A. D'Alessandro: Congenital Trypanosoma cruzi infection in a laboratory-born squirrel monkey, *Saimiri sciureus.* Amer. J. trop. Med. Hyg. *31,* 931-933 (1982)
21. Edgcomb, J. H. and C. M. Johnson: American trypanosomiasis (Chagas' disease); pp. 244-251 in "Pathology of Tropical and Extraordinary Diseases. An Atlas." Vol. 1 (Binford, C. H. and D. H. Connor, eds.). Armed Forces Institute of Pathology, Washington, D. C. (1976)
22. Filice, F. P.: Studies on the cytology and life history of a *Giardia* from the laboratory rat. Univ. Calif. Publ. Zool. *57,* 53-143 (1952)
23. Fulton, J. D. and C. V. Harrison: An outbreak of *Trypanosoma cruzi* infection in Indian monkeys. Trans. roy. Soc. trop. Med. Hyg. *39,* 513-520 (1946)
24. Funayama, G. K. y M. P. Barreto: Estudos sobre reservatories e vectores silvestres do pelo *T. cruzi.* XLII. Infeccao natural do simio, *Alouatta caraya* (Humboldt, 1812) pelo *T. cruzi.* Rev. Inst. Med. trop. S. Paulo *12,* 257-265 (1970)
25. Garnham, P. C. C. and L. Gonzales-Mugaburu: A new trypanosome in *Saimiri* monkeys from Colombia. Rev. Inst. Med. trop. S. Paulo *4,* 79-84 (1962)
26. Hoare, C. A.: Does Chagas' disease exist in Asia? J. Trop. Med. Hyg. *66,* 297-299 (1963)
27. Hoare, C. A.: Does Chagas' disease exist in Asia? (Further contribution to this problem). J. Trop. Med. Hyg. *72,* 282-284 (1969)
28. Jarroll, E. L., jr.; A. K. Bingham, and E. A. Meyer: *Giardia* cyst destruction: effectiveness of six small-quantity water desinfection methods. Amer. J. trop. Med. Hyg. *29,* 8-11 (1980)
29. Kasa, T. J.; G. D. Lathrop; H. J. Dupuy; C. H. Bonney, and J. D. Toft II: An endemic focus of *Trypanosoma cruzi* infection in a subhuman primate research colony. J. Amer. vet. med. Ass. *171,* 850-854 (1978)
30. Kean, B. H.; D. C. Williams, and S. K. Luninais: Epidemic of amoebiasis and giardiasis in a biased population. Brit. J. Vener. Dis. *55,* 375-378 (1979)
31. Kuntz, R. E.; B. J. Myers, and T. S. McMurray: *Trypanosoma cruzi*-like parasites in the slow loris *(Nycticebus coucang)* from Malaysia. Trans. Amer. micr. Soc. *89,* 304-307 (1970)
32. Lainson, R.: Leishmaniasis; pp. 41-103 in "CRC Handbook Series in Zoonoses" (Steele, J. H., Series ed.), Section C: Parasitic Zoonoses, Vol. I (Jacobs, L. and P. Arambulo, Section eds.). CRC Press, Boca Raton (1982)
33. Malamos, B.: Über das Vorkommen von *Schizotrypanum cruzi* bei Affen in Niederländisch-Indien. Arch. Schiffs- u. Tropenhyg. *39,* 156-171 (1935)
34. Marinkelle, C. J.: Observations on human, monkey and bat trypanosomes and their vectors in Colombia. Trans. roy. Soc. trop. Med. Hyg. *60,* 109-116 (1966)
35. Marinkelle, C. J.: The control of leishmaniases. Bull. Wld Hlth Org. *58,* 807-818 (1980)
36. Meyer, E. A.: Culture of Giardia lamblia. Lancet (2), 527 (1984)
37. Meyer, E. A. and E. L. Jarroll: Giardiasis; pp. 25-40 in "CRC Handbook Series in Zoonoses" (Steele, J. H. Series ed.); Section C: Parasitic Zoonoses, Vol. I (Jacobs, L. and P. Arambulo, Vol. eds.), CRC Press, Boca Raton (1982)
38. Miles, M. A.; T. V. Barrett; R. H. Hoff, and K. E. Mott: Epidemiological studies of Chagas' disease in Brazil. Trans. roy. Soc. trop. Med. Hyg. *73,* 135-136 (1979)
39. Nemanic, P. C.; R. L. Owen; D. P. Stephens, and J. C. Mueller: Endosymbiosis and organelle distribution in *Giardia muris* and comparison with *Giardia lamblia.* J. Infect. Dis. *140,* 222-228 (1979)
40. Owen, R. L.: The ultrastructural basis of *Giardia* function. Trans. roy. Soc. trop. Med. Hyg. *74,* 429-433 (1980 a)
41. Owen, R. L.: The immune response in clinical and experimental giardiasis. Trans. roy. Soc. trop. Med. Hyg. *74,* 443-445 (1980 b)
42. Owen, R. L.; P. C. Nemanic, and D. P. Stevens: Ultrastructural observations in giardiasis in a murine model. I. Intestinal distribution, attachment and relationship to the immune system of *Giardia muris.* Gastroenterology *76,* 757-769 (1979)

43. Phillipps, L. G.: The parasitology of primates and carnivores. Amer. Ass. Zoo Vet. Ann. Proc. 1980, 116-119 (1980)

44. Raether, W. and H. Seidenath: The activity of fexinidazole (HOE 239) against experimental infections with *Trypanosoma cruzi*, trichomonads and *Entamoeba histolytica*. Ann. trop. Med. Parasit. *77*, 13-26 (1983)

45. Raether, W. and H. Seidenath: Ketoconazole and other potent antimycotic azoles exhibit pronounced activity against *Trypanosoma cruzi*, *Plasmodium berghei* and *Entamoeba histolytica* in vivo. Z. Parasitenkd. *70*, 135-138 (1984)

46. Ristic, M. and R. D. Smith: Zoonoses caused by hemoprotozoa; pp. 41-63 in "Parasitic Zoonoses. Clinical and Experimental Studies." (Soulsby, E. J. L., ed.). Academic Press, New York (1974)

47. Rowland, E. and J. G. Vandenbergh: A survey of intestinal parasites in a new colony of rhesus monkeys. J. Parasit. *51*, 294-295 (1965)

48. Seneca, H. and A. Wolf: *Trypanosoma cruzi* infection in the Indian monkey. Amer. J. trop. Hyg. *4*, 1009-1014 (1955)

49. Smith, P. D.: Pathophysiology and immunology of giardiasis. Ann. Rev. Med. *36*, 295-307 (1985)

50. Sousa, O. E.; R. N. Rossan, and D. C. Baerg: The prevalence of trypanosomes and microfilariae in Panamanian monkeys. Amer. J. trop. Med. Hyg. *23*, 862-868 (1974)

51. Szarfman, A.; F. S. Laranja; W. de Souza; L. Galvao Quintao; D. Gerecht, and G. A. Schmunis: Tissue reacting antibodies in a rhesus monkey with long-term *Trypanosoma cruzi* infection. Amer. J. trop. Med. Hyg. *27*, 832-834 (1978)

52. Teixeira, A. R. L.: Chagas' disease: trends in immunological research and prospects for immunoprophylaxis. Bull. Wld Hlth Org. 697-710 (1979)

53. Tomkins, A. M.; S. G. Wright; B. Drasar, and W. P. T. James: Bacterial colonization of jejunal mucosa in giardiasis. Trans. roy. Soc. trop. Med. Hyg. *72*, 33-36 (1978)

54. Travi, B.; A. M. Rutz; O. J. Colillas, e E. L. Segura: Trypanosomiasis en monos neotropicales. Medicina (Buenos Aires) *42*, 55-60 (1982)

55. Weinman, D.: Trypanosomes of primates in Indonesia. Trans. roy. Soc. trop. Med. Hyg. *64*, 649 (1970)

56. Weinman, D.: Trypanosomiasis in primates, human and subhuman, in India. Trans. roy. Soc. trop. Med. Hyg. *68*, 345-346 (1974)

57. Weinman, D. and N. S. Wiratmadja: The first isolates of trypanosomes in Indonesia and in history from primates other than man. Trans. roy. Soc. trop. Med. Hyg. *63*, 497-506 (1969)

58. Werner, H.; D. Springer und H. Köhler: Spontaninfektion von Marmoset-Affen (Saguinus oedipus) mit einem Trypanosoma cruzi-like Stamm, Springer, 1975. Isolierung und Identifizierung. Zbl. Bakt., I. Abt. Orig. A, *239*, 557-564 (1977)

59. WHO Scientific Working Group: Parasite-related diarrheas. Bull. Wld. Hlth Org. *58*, 819-830 (1980)

60. Yaeger, R. G.: American trypanosomiasis "Chagas disease"; pp. 105-119 in "CRC Handbook Series in Zoonoses" (Steele, J. H., Series ed.). Section C: Parasitic Zoonoses Vol. 1 (Jacobs, L. and P. Arambulo, Vol. eds.), CRC Press, Boca Raton (1982)

61. Yardley, J. H.; J. Tanako, and T. R. Hendrix: Epithelial and other mucosal lesions of the jejunum in giardiasis. Jejunal biopsy studies. Bull. John Hopkins Hosp. *115*, 389-406 (1964)

62. Zaman, V.: A trypanosome of the slow loris *(Nycticebus coucang)*. Southeast Asian J. Trop. Med. Publ. Hlth *3*, 22-24 (1972)

6.4 Sporozoa

6.4.1 Haemosporidia: Plasmodiidae

The malaria parasites and the closely related *Hepatocystis* spp. are a group of well organized, highly adapted blood parasites which probably evolved over several million years together with their hosts. The quartan malarias appeared during the tertiary period. 30 million years later, during the oligocene period, the now generating Cercopithecoidea acquired the ovale-group and the benign tertian malarias, possibly also *Plasmodium knowlesi.* During the pleistocene era, 2 million years ago, the *Laverania*-species evolved together with the hominids and pongids. However, since certain Laverania-type malarias *(P. coatneyi, P. fragile)* occur in macaques this would suggest an earlier evolutionary development of the Laveranias occurring before the split of cercopithecids and pongids (Garnham, 1963; Bruce-Chwatt, 1965).

Life cycle

The cycle of the malarias in the primate host is initiated by the inoculation of sporozoites by the female mosquito when she punctures the skin for blood sucking. The inoculated sporozoites reach the circulating blood and circulate for an hour at the most, though probably for much less (Bray and Garnham, 1982). Immediately on their first liver passage they invade the hepatocytes and initiate an exoerythrocytic life cycle of asexual multiplication (schizogony). Certain malarias develop into latent stages by arrest of the sporozoites for varying periods of time (Krotoski et al., 1980). The parasitized hepatocytes enlarge and when the schizont reaches maturity (after 5 to 15 days, depending on the species) finally rupture, liberating large numbers of merozoites. The hepatocellular merozoites are probably short lived and have to enter either another hepatocyte or an erythrocyte (in some species reticulocyte) within minutes (Bray and Garnham, 1982).

The result is either a secondary (tertiary etc.) exoerythrocytic cycle or, after invasion of erythrocytes, the merozoites divide by schizogony within the erythrocytes (Fig. 30). *Laverania* spp. absolve only one exoerythrocytic schizogony, other plasmodia pass two or more hepatocytic cycles, and finally in the *Hepatocystis* spp. the hepatocellular schizogony is a permanent event.

In the erythrocytic cycle the liberated merozoites first attach to the red cells by specialized organelles at their anterior end, which subsequently induce an invagination of the erythrocytic membrane by forced endocytosis (Bray and Garnham, 1982; Wunderlich et al., 1982). The erythrocytes themselves possess specific receptor sites (Miller et al., 1977). The resulting vacuole is at first connected to the outside by a channel, the inflowing Na+-ions and water induce a rotation of the merozoite and in most malaria species except the *Laverania* a swelling of the erythrocyte. In *P. knowlesi,* the quotidian malaria of South East Asian macaques it takes about 30 seconds from attachment to complete invasion of the erythrocyte (Dvorak et al., 1975; Miller, 1975). Once the merozoite has truly invaded the erythrocyte it multiplies again by one or more schizogonies with subsequent destruction

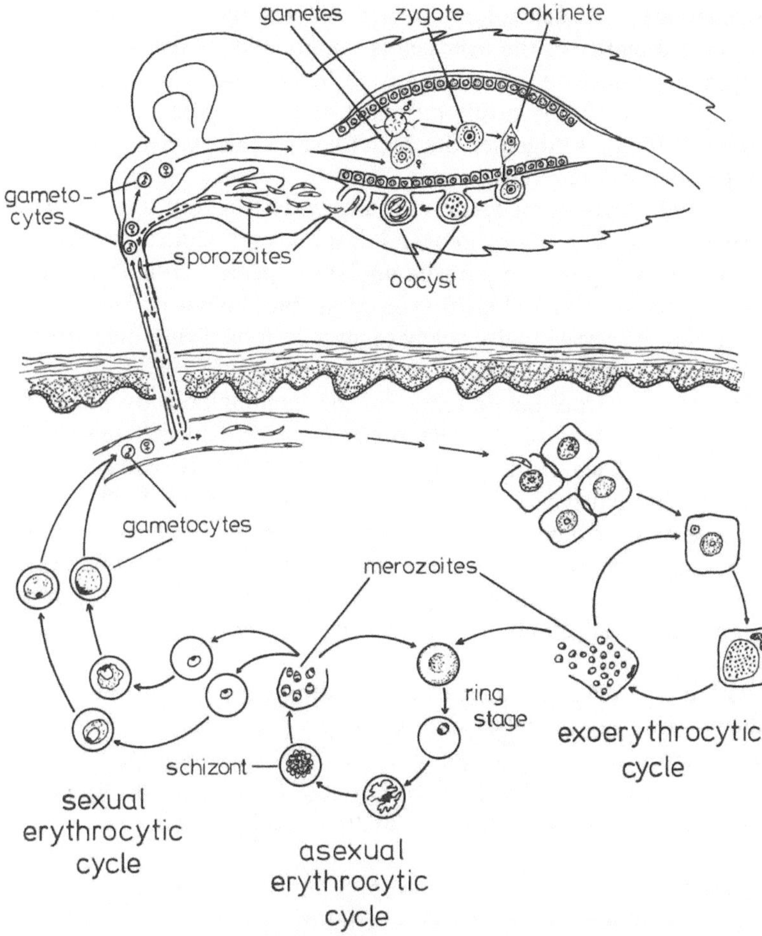

Fig. 30. Life cycle of *Plasmodium spp.*

of the erythrocyte, which takes 24 hs in quotidian, 48 hs in tertian, and 72 hs in quartan malarias. After one or more erythrocytic schizogonies some merozoites develop to sexual stages: macro- and microgametocytes and the vertebrate host becomes infectious to the next mosquito. With blood-sucking of a new *Anopheles* the macro- and microgametocytes reach the midgut of the mosquito. The microgametocytes then exflagellate within 2 to 30 min. and become microgametes. The macrogametocytes become macrogametes after digestion of the erythrocyte and after fertilization by the microgametes transform to zygotes. The zygotes penetrate the midgut epithelia as ameboid ookinetes and reach the haemocele of the mosquito under the protection of a thin cyst wall. The nucleus of the oocyst first divides meiotic, subsequently in rapid mitotic sequences. The so-formed sporozoites mature in 4 to 15 days. The zygote then ruptures, releasing the merozoites into the haemocele and induces a generalized infection of the mosquito, thereby entering the salivary glands. The mosquito is now infectious to a new vertebrate host.

Pathogenesis

The exoerythrocytic schizogony in hepatocytes elicits no response by the host as long as the integrity of the liver cell is maintained. With the rupture of the greatly enlarged hepatocyte a minor neutrophilic-lymphocytic-plasma cellular reaction results, followed by the proliferation of a few fibroblasts (Garnham et al., 1957; Maegraith, 1968). Altogether the hepatocytic schizogony is without recognizable harm to the vertebrate host.

The erythrocytic cycle usually results in destruction of an appreciable number of erythrocytes with morphological lesions and clinical symptoms of varying severity depending on the parasite and host species. Human malarias usually are associated with distinct clinical symptoms, but simian malarias may be clinically misread despite occasionally heavy organic lesions. Symptoms are most frequently observed in *P. knowlesi*-malaria.

The growing trophozoite feeds on the haemoglobin by phagotrophy, but the haemoglobin is incompletely metabolized, the remaining haematin forms the malaria pigment (haemazoin). In haemazoin the iron is oxidized and the globulin denatured and partly proteolysed. It is inert and has no effect on the tissues either when free or after phagocytosis (Maegraith, 1968). The haemazoin is deposited in several organs, most heavily in the spleen, in the Kupffer cells of the liver and in the renal glomerula. Glomerular excreted pigment is partially reabsorbed by tubular epithelia as a golden-brown dust in the cytoplasm of the tubular epithelial cells.

Red cell ATPase is increased; 2,3 diphosphorylglycerate is decreased; the Embden-Meyerhof glycolysis is altered by a pyruvate Kinase isoenzyme produced by the merozoites (Oehlschlegel et al., 1975); whilst erythrocytic and serum lipids are increased (Angus et al., 1971). Using *P. knowlesi* as a model, several hepatic lesions can be observed due to circulatory breakdown.

Treatment

The preferable treatment for acute malaria, except for *P. falciparum* in areas with prevalent drug resistance, is chloroquine, possibly combined with other drugs such as primaquine phosphate, quinine sulfate, sulfa-trimethoprim or tetra-

Table 37. Malaria-Parasites of Nonhuman Primates

Vivax-type (tertian)	Ovale-type (tertian)	Falciparum-type (tertian) (Laverania)	Malariae-type (quartan)	Others
P. cynomolgi	P. fieldi	P. falciparum	*P. brasilianum*	***P. knowlesi***
P. eylesi	P. simiovale	P. reichenowi	***P. inui (shortti)***	
P. gonderi		P. coatneyi	***P. rodhaini***	
P. hylobati		P. fragile	V. girardi	
P. jefferi				
P. pitheci				
P. schwetzi				
P. youngi				
P. simium				
P. silvaticum				

cycline (Conrad, 1981). Chloroquine is also effective in nonhuman primates (Ayala, 1978).

In nonhuman primates 21 malaria species have been reported so far (Table 37); those underlined are spontaneously or experimentally transmissible to man.

6.4.1.1 Vivax-Type Malarias

6.4.1.1.1 Plasmodium Cynomolgi (Incl. P. C. Bastianelli)

P. cynomolgi was first found and named by Mayer (1907) in some cynomolgus monkeys imported into Germany and *M. fascicularis* is considered to be the natural host of the type strain (Eyles et al., 1960; Eyles, 1963). *P. cynomolgi* is also found in *M. nemestrina* (Eyles et al., 1962a), *M. radiata* (Prakash and Chakrabarti, 1962), *M. cyclopis* (Inoki et al., 1951), *M. sinica* (Dissanaike, 1963), *M. mulatta* (Coatney et al., 1971), *Presbytis cristatus* (Eyles et al., 1962b) and *P. entellus* (Dissanaike et al., 1965).

P. cynomolgi has caused also spontaneous infections in man following accidental laboratory exposure. The first cases of that category affected Dr. Eyles himself and two days later his technician, both became probably infected via the mosquitoes that occasionally escaped into the room (Coatney et al., 1971). The same happened to two staff members of another laborotory (Schmidt et al., 1961) and to a technician working with infected *Anopheles* spp. (Cross et al., 1973); a student working with *P. cynomolgi* (Most, 1973) and two laboratory workers at the Institute of Tropical Diseases in Paris (Druilhe et al., 1980). Blacks are refractory to the infection (Coatney et al., 1971) as they are to vivax-malaria due to their Duffy-blood group-negative phenotype (Miller, 1977; Miller et al., 1977).

Agent

The asexual stages of *P. cynomolgi* develop preferably in reticulocytes and erythrocytes. They are rather small (below 2 μm), with oval or rounded nuclei and sometimes with only a whisp of cytoplasm. The ring forms are often located at the periphery of the host cell. Multiple infections with up to 4 parasites in one erythrocyte are common. The ring fills half to two thirds of the host cell. The parasitized erythrocytes become enlarged and dotted (Schüffner's dots), and a dark pigment appears. The older trophozoites look like vivax-parasites of man. Schizogony with nuclear divisions as in *P. vivax* results in 14 to 20 merozoites, the pigment becomes concentrated either peripherally or centrally in one or several masses.

The macrogametocytes are round with compact light blue cytoplasm. The pigment is scattered and heavy, whilst the nucleus is located excentrically and often has a centrally located deeply staining portion (Coatney et al., 1971).

The cycle length is 48 hs. The prepatent period in *M. mulatta* 7 to 16 days and in experimental infections of *Aotus trivirgatus* 16 to 19 days (Collins et al., 1975).

Vectors

The vectors of *P. cynomolgi* belong to the *Anopheles leurosphyrus*-group: *A. balabacensis introlatus*, *A. hackeri*, *A. hyrcanus sinensis*, all of which feed in both the canopy- and ground levels (Wharton et al., 1962; Warren and Wharton, 1963; Bennett and Warren, 1965; Warren et al., 1965). Experimental infections can be achieved in a wide variety of anophelines (Coatney et al., 1971). In experimental infections of *A. sacharovi*, *A. stephensi*, *A. l. atroparvus* and *A. gambiae* normal development of sporozoites occurred (Omar, 1968).

Geographical Distribution
The infection occurs naturally in Java, South India, East Pakistan, Cambodia, Philippines, Formosa, Malaya (Eyles, 1963), and the Nicobars (Kalra, 1980).

Clinical Symptoms
P. cynomolgi infections in nonhuman primates are very mild (Fiennes, 1967), with the host erythrocytes becoming enlarged with prominent stippling. In experimental infections the resulting parasitemia was less in *M. fascicularis* than in *M. mulatta* and *Cercopithecus aethiops*. In the latter two species the haematocrit values dropped transiently to approximately one-half of the pre-inoculation levels, then they returned to original levels between day 34 and 40 (Schmidt et al., 1977). Serial passages in quick successions may increase the virulence of the parasites to be fatal even in adult macaques (Voller, 1972).

Man: *P. cynomolgi bastianelli* infections are benign in caucasians, causing high fever, night sweats, dizziness, headaches, sore throat and a nonproductive cough (Cross et al., 1973). These symptoms are similar to but usually milder than those in vivax-malaria. *P. cynomolgi*, like *P. vivax* is unable to infect blacks except the rare Duffy-positive ones (Bray and Garnham, 1982). The incubation period in experimental human infections is 15 to 23 days with a prepatent period of 15 to 30 days (Beye et al., 1961).

6.4.1.1.2 Plasmodium schwetzi

P. schwetzi is a natural vivax-type plasmodium of gorillas and chimpanzees (Bray, 1963; Coatney et al., 1971). It can be transmitted to man as shown in human volunteers (Contacos et al., 1970).

Agent
A description of the agent will not be given since it has been transmitted to man only experimentally. Excellent descriptions are given by Garnham (1966) and Coatney et al. (1971).

Vectors
The natural vectors are unknown, experimental infections were achieved in *A. maculipennis atroparvus*, *A. gambiae*, *A. aztecus*, *A. b. balabacensis*, *A. freeborni*, and *A. maculatus* (Rodhain and Lassman, 1940; Bray, 1958; Garnham, 1966; Collins et al., 1969).

Geographical Distribution
The infection has been described as occurring in Cameroon, Guinea, Sierra Leone and the northern Congo (Bray, 1963; Coatney et al., 1971).

Clinical Symptoms
Nonhuman primates: *P. schwetzi* seems not to be able to evoke clinical symptoms even in young chimpanzees or in older splenectomized chimpanzees (Coatney et al., 1971).

6.4.1.1.3 Plasmodium Simium

P. simium is the vivax-type malaria parasite of New World monkeys. It occurs naturally in howler monkeys *(Alouatta fusca)* (Eyles, 1963) and woolly spider monkeys *(Brachyteles arachnoides)* (Deane et al., 1968, 1969).

The transmissibility to man is doubted, since the one reported case of a spontaneous infection (Deane et al., 1966a) could have been a *P. vivax* infection (Coatney et al., 1971) although in the Horto Florestal area human malaria has never been detected (Dean et al., 1965). Experimental human infections have so far failed (Coatney et al., 1971).

Agent
The agent will not be described because of the limited space and its uncertain zoonotic potentials, detailed descriptions are found in Garnham (1966) and Coatney et al. (1971).

Vectors
As natural vector *Anopheles (Kerteszia) cruzi* is suspected (Deane et al., 1984), experimentally *A. freeborni*, *A. stephensi*, *A. maculatus*, and *A. atroparvus* could be infected (Coatney et al., 1971). Deane et al. (1966b, 1969) suggested *A. cruzi* as the probable vector, as they were able to transmit *P. simium* by naturally infected wild caught *A. cruzi* to *Alouatta belzebul belzebul* (Deane et al., 1970, 1971).

Geographical Distribution
P. simium is restricted to a small area of Brazil (Rio Grande do Sul, Santa Catarina, and Sao Paulo states) (Coatney et al., 1971).

6.4.1.2 Malariae-Type Malarias

6.4.1.2.1 Plasmodium Brasilianum

P. brasilianum is the malariae-type (quartan) plasmodium of New World monkeys and rarely marmosets. Its host species are given in table 38.

Man can experimentally be infected via *A. freeborni* (Coatney et al., 1971) with equal success in both caucasian and black volunteers. Natural infections have not been reported so far.

Agent
A detailed description will not be given for reasons already stated. It should be mentioned, however, that band forms, characteristic in malariae-type malarias

Table 38. Host Species of *P. brasilianum*

Host Species	Reference
Alouatta fusca	Deane et al., 1969
Alouatta palliata	Clark, 1931
Alouatta seniculus	Serrano, 1967; Deane et al., 1969
Alouatta villosa	Dunn and Lambrecht, 1963; Porter et al., 1966
Ateles fusciceps	Clark, 1931; Porter et al., 1966
Ateles geoffroyi	Clark, 1931; Dunn and Lambrecht, 1963; Porter et al., 1966; Marinkelle and Grose, 1968
Ateles paniscus	Dunn and Lambrecht, 1963; Deane et al., 1969
Brachyteles arachnoides	Deane et al., 1969
Cacajao calvus	Gonder and von Berenberg-Gossler, 1908
Callicebus moloch	Renjifo and Peidrahita, 1949 - cit. from Marinkelle and Grose, 1968
Callicebus torquatus	Deane et al., 1969
Cebus albifrons	Dunn and Lambrecht, 1963; Marinkelle and Grose, 1968; Ayala, 1978
Cebus apella	Dunn and Lambrecht, 1963; Marinkelle and Grose, 1968; Deane et al., 1969
Cebus capucinus	Clark, 1931; Porter et al., 1966; Marinkelle and Grose, 1968
Chiropotes chiropotes	Deane et al., 1969
Lagothrix cana	Deane et al., 1969
Lagothrix infumata	Dunn and Lambrecht, 1963
Lagothrix lagotricha	Garnham et al., 1963; Marinkelle and Grose, 1968; Deane et al., 1969
Saguinus geoffroyi	Baerg, 1971
Saimiri boliviensis	Dunn and Lambrecht, 1963
Saimiri sciureus	Garnham et al., 1963; Dunn and Lambrecht, 1963; Marinkelle and Grose, 1968; Deane et al., 1969

appear particularly in the *Alouatta* spp. during the late trophozoite stage. Exoerythrocytic stages have been found on days 14, 18 and 21 after introduction of sporozoites (Sodeman et al., 1969).

Vectors

Deane et al. (1971) consider *A. cruzi* to act as the natural vector, but Coatney et al. (1971) found *A. freeborni* to be the most suitable anophelines in experimental studies. Other susceptible species were *A. stephensi* and *A. maculatus.*

Geographical Distribution

The geographical range of *P. brasilianum* extends from the Yucatan peninsula to northwestern Brazil and the lowland forests of Columbia and Peru (Eyles, 1963).

Clinical Symptoms

Nonhuman primates: *P. brasilianum*, although usually inducing only low grade parasitemia in natural infections, can produce severe symptoms and even death in adult monkeys (Voller, 1972). In experimental infections sharp peaks in temperature coincide with the 72 hr cycle of sporulation particularly in spider-, howler-, and night monkeys and in *S. geoffroyi* (Coatney et al., 1971; Ayala, 1978). Other symptoms include severe chills and anemia (Ayala, 1978).

Man: The symptoms in volunteers consisted mainly of headaches, loss of appetite and fever. The prepatent period was 63 days (Coatney et al., 1971).

6.4.1.2.2 Plasmodium Inui (P. Shortti)

P. inui is prevalent in *Macaca nigra* (Eyles and Warren, 1962), *M. cyclopis* (Hsieh, 1960), *M. fascicularis* (Donovan et al., 1983; Stokes et al., 1983; Sinha and Gajanana, 1984), *M. mulatta* (Schofield et al., 1985), *M. nemestrina* (Halberstaedter und von Prowazek, 1907), *M. radiata* (Mulligan and Swaminath, 1940), *Presbytis cristatus* (Collins et al., 1970) and *P. obscurus* (Eyles et al., 1962). *P. inui* is also infectious to man as shown in several experimental infections (das Gupta, 1938; Coatney, 1971).

Agent

For descriptions see Garnham (1966) and Coatney et al. (1971).

Vectors

P. inui has been isolated from *A. leucosphyrus* (Wharton et al., 1962) and *A. b. introlatus* (Warren and Wharton, 1963). In experimental infections a wide variety of anophelines proved to be susceptible (Coatney et al., 1971).

Geographical Distribution

P. inui is indigenous to Malaysia, Thailand, Pakistan and Indonesia (Eyles and Warren, 1962; Eyles et al., 1962; Eyles, 1963). Recently it has also been found in the Nicobar Islands (Sinha and Gajanana, 1984).

Clinical Symptoms:

Nonhuman primates: Spontaneous disease was observed in an adult cynomolgus monkey two weeks after stress and immunosuppression, as a severe regenerative anemia, hepatosplenomegaly, weakness, lethargy, weight loss and anorexia (Stokes, 1983). Clinical illness was also observed in several diabetic rhesus monkeys as an acute anemia (haematocrit in one animal 6%!), dyspnea and elevated pulse rate (Schofield et al., 1985). In 5 of 31 experimentally infected rhesus monkeys pallor, malaise, indifferent appetite and subnormal weight gain were observed, 2 monkeys died (Schmidt et al., 1980).

Man: Experimental infections of volunteers resulted in headache, malaise, muscular and joint pain, and loss of appetite. Chills were mild and of short duration, whilst the fever, if occurring at all, had a typical quartan pattern. Caucasian and black volunteers were equally susceptible (Coatney, 1971).

Pathology

Nonhuman primates: Necropsies revealed marked hepato-splenomegaly, clear, orange colored hydrops ascites and hydrothorax, and an intense slate-grey pigmentation of the viscera and the skin (Schmidt et al., 1980).

6.4.1.2.3 *Plasmodium Rodhaini*

P. rodhaini is a quartan malaria of chimpanzees and gorillas in Cameroon, Sierra Leone, Liberia, Angola and northern Congo (Reichenow, 1920; Rodhain and Dellaert, 1943; Bray, 1963) which is considered to be identical to *P. malariae* of man (Coatney et al., 1971).

Agent
Because of its uncertain taxonomical position no description of *P. rodhaini* will be given. Its morphology and cycle are essentially identical to those of *P. malariae* (Coatney et al., 1971).

Vector
A. gambiae can be experimentally infected with *P. rodhaini* (Bray, 1960).

Clinical Symptoms
Nonhuman primates: No clinical signs have been reported in any of the *P. rodhaini*-infected apes.

Man: Transmission of *P. rodhaini* to man, if met with success, resulted in moderate to low parasitemias and fever up to 40 °C.

6.4.1.3 Quotidian Malarias

6.4.1.3.1 *Plasmodium Knowlesi*

P. knowlesi is a quotidian malaria parasite primarily of *Macaca fascicularis*, but also of *M. nemestrina* and *Presbytis melalophos* (Sinton and Mulligan, 1932, 1933-cit. from Coatney et al., 1971; Lambrecht and Dunn, 1961; Eyles et al., 1962 a, b). Man can be infected both experimentally and spontaneously (Knowles and das Gupta, 1932; Chin et al., 1965; 1968).

Agent
The young ring forms may appear in large numbers in the circulating blood of monkeys and man. They possess prominent spherical nuclei which are often located inside the ring, and also accessory chromatin dots. Fully grown rings may occupy more than half of the host erythrocyte. They are often succeeded by band forms similar to those of *P. malariae.* The parasite subsequently loses its vacuole, shrinks, and pigment appears as dark grains. The nucleus enlarges and takes a deep red stain, while the cytoplasm stains a deep blue. The host erythrocyte shows a faint stippling ("Sinton and Mulligans' stippling"). The process of schizogony results in up to 16 merozoites, and with the pigment collecting into one or more yellowish-black masses.

With further development to sexual forms a striking color difference between macro- and microgametocytes appears. The mature macrogametocyte cytoplasm stains a distinctive blue, the excentrically placed nucleus takes a pink stain enclosing a heavier stained irregular area. The black pigment granules are irregularly scattered in the cytoplasm.

The microgametocyte stains pink throughout with the nucleus, which occupies about one half of the parasite, a darker shade. Jet black pigment granules are scattered throughout the cytoplasm. The asexual cycle in the blood lasts 24 hrs. Tissue forms can be seen as early as 92 hrs after infection.

Vectors

A. balabacensis introlatus and *A. hackeri* have been described as natural vectors (Warren and Wharton, 1963). Other anophelines such as *A. vagus*, *A. sinensis*, *A. maculatus*, *A. kochi*, *A. b. balabacensis* and *A. quadrimaculatus*, most of which are present in Malaysia, were also found to be susceptible (Coatney et al., 1971).

Geographical Distribution

The agent is common in Malaysia, the Philippines and Formosa (Yokagawa et al., 1941 - cit. from Coatney et al., 1971; Lambrecht and Dunn, 1961; Eyles, 1963; Collins, 1982).

Clinical Symptoms

Nonhuman primates: *P. knowlesi* produces only mild to moderate parasitemias and clinical signs in cynomolgus monkeys of Philippine origin, but distinctively more pronounced courses in cynomolgus monkeys of Malayan origin (Schmidt et al., 1977b). It usually induces a fulminating infection and often death in experimentally infected rhesus monkeys (Abildgaard et al., 1975; Schmidt et al., 1977b; Spangler et al., 1978). Due to the short (quotidian) periodity, its sequestration in inner organs and the rapid erythrocytic schizogony which builds up so quickly, no effective immunity can develop in the foreign hosts and an overwhelming parasitemia results in rhesus monkeys (Bray and Garnham, 1982). Three clinical types of infection are recognized (Maegraith, 1968):

a) the "lytic" with predominating haemolysis and with haemoglobinuria in the last 48 hrs of infection;
b) the acute shock with rapid onset of shock, cyanosis, cold extremities, low blood pressure and death, and
c) prolonged shock which is usually survived by the animals.

Sick animals become listless, lethargic, irritable and anorectic, the body temperature is noticeably increased (Collins, 1982). The haematocrit values drop to less than one half of the original levels (Abildgaard et al., 1975; Spangler et al., 1978). With progressing disease a marked anemia, rapid weight loss and dehydration become evident (Collins, 1982).

The phospholipid content of both the parasitized and nonparasitized erythrocytes is increased, indicating an alteration of the membrane structure (Angus et al., 1971).

Man: In volunteers the prepatent periods ranged from 9 to 12 days. The clinical symptoms consisted of malaise, headache and both anorexia, occasionally nausea and vomiting, and a fever of both remittent and intermittent types. Black volunteers showed less distinct symptoms than caucasians (Chin et al., 1968; Coatney et al., 1971) which has been related to the Duffy-negative phenotype (Miller, 1972; Miller et al., 1977).

Pathology

Nonhuman primates: In lytic infections the portal veins are dilated and filled with parasitized and nonparasitized erythrocytes, macrophages and polymorphs (Maegraith, 1968). Haemozoin and haemosiderin are deposited in swollen Kupffer cells. The hepatocytes may show signs of fatty degeneration, their glycogen disappears, and in advanced lesions centrolobular necrosis with karyorrhexis and pyknosis occur. The haemozoin-deposits cause discoloration of the organs. The cardiac ventricles in most cases contain smooth, grey-green mural thrombi. The renal lesions are dominated by tubular reabsorption of the glomerularly excreted pigment and haemoglobin, by haemoglobin casts and by tubular degeneration and necrosis due to intrarenal vasoconstriction (Maegraith, 1968; Abildgaard et al., 1975; Spangler et al., 1978).

In the acute shock the cellular lesions are less distinctive, the most prominent being congestions with dilatation of central veins and of the lobular centers (Maegraith, 1968). Intravascular coagulation, although suggested by the prolongation of prothrombin time and partial thromboplastin time, factor V and factor VIII, could not be demonstrated (Abildgaard et al., 1975).

References

1. Abildgaard, C.; J. Harrison; S. de Nardo; W. Spangler, and D. Gribble: Simian *Plasmodium knowlesi* malaria: Studies of coagulation and pathology. Amer. J. trop. Med. Hyg. *24*, 764-768 (1975)
2. Angus, M.G.N.; K.A. Fletcher, and B.G. Maegraith: Studies on the lipids of *Plasmodium knowlesi*-infected rhesus monkey *(Macaca mulatta)*. IV. Changes in erythrocyte lipids. Ann. trop. Med. Parasit. *65*, 429-439 (1971)
3. Angus, M.G.N.; K.A. Fletcher, and B.G. Maegraith: Studies on the lipids of *Plasmodium knowlesi*-infected rhesus monkeys *(Macaca mulatta)*. II. Changes in serum non-esterified fatty acids. Ann. trop. Med. Parasit. *65*, 155-167 (1971c)
4. Ayala, S.C.: Clinical malaria in a pet capuchin monkey. Vet. Med. small Anim. Clin. *73*, 217-218 (1978)
5. Baerg, D.C.: A naturally acquired infection of *Plasmodium brasilianum* in the marmoset, *Saguinus geoffroyi*. J. Parasit. *57*, 8 (1971)
6. Bennett, G.F. and McW. Warren: Transmission of a new strain of *Plasmodium cynomolgi* to man. J. Parasit. *51*, 79-80 (1965)
7. Beye, H.K.; M.E. Getz; G.R. Coatney; H.A. Elder, and D.E. Eyles: Simian malaria in man. Amer. J. trop. Med. Hyg. *10*, 311-316 (1961)
8. Bray, R.S.: Studies on malaria in chimpanzees. V. The sporogonous cycle and mosquito transmission of *Plasmodium vivax schwetzi*. J. Parasit. *44*, 46-51 (1958)
9. Bray, R.S.: Studies on malaria in chimpanzees. VIII. The experimental transmission and pre-erythrocytic phase of *Plasmodium malariae*, with a note on the host-range of the parasite. Amer. J. trop. Med. Hyg. *9*, 455-465 (1960)
10. Bray, R.S.: The malaria parasites of anthropoid apes. J. Parasit. *49*, 888-891 (1963)
11. Bray, R.S. and P.C.C. Garnham: The life cycle of primate malaria parasites. Brit. med. Bull. *38*, 117-122 (1982)
12. Bruce-Chwatt, L.J.: Pathogenesis and paleo-epidemiology of primate malaria. Bull. Wld Hlth Org. *32*, 363-387 (1965)
13. Chin, W.; P.G. Contacos; G.R. Coatney, and H.R. Kimball: A naturally acquired quotidian-type malaria in man transferable to monkeys. Science *149*, 865 (1965)
14. Chin, W.; P.G. Contacos; W.E. Collins; M.H. Jeter, and E. Alpert: Experimental mosquito-

transmission of *Plasmodium knowlesi* to man and monkey. Amer. J. trop. Med. Hyg. *17*, 355-358 (1968)
15. Clark, H.C.: Progress in the survey for blood parasites of the wild monkeys of Panama. Amer. J. trop. Med. Hyg. *11*, 11-20 (1931)
16. Coatney, G.R.: The simian malarias: Zoonoses, anthroponoses or both? Amer. J. trop. Med. Hyg. *20*, 795-803 (1971)
17. Coatney, G.R.; W.E. Collins; McW. Warren, and P.G. Contacos: The Primate Malarias. U.S. Dept. Health, Educ., Welf., NIH, Bethesda (1971)
18. Collins, W.E.: Simian malaria; pp.141-150 in "CRC Handbook Series in Zoonoses" (Steele, J.H., Series ed.), Section C: Parasitic Zoonoses, Vol. I (Jacobs, L. and P. Arambulo, Vol. eds.), CRC Press, Boca Raton (1982)
19. Collins, W.E.; T.C. Orihel; P.G. Contacos; M.H. Jeter, and L.S. Gell: Some observations on the sporogonic cycle of *Plasmodium schwetzi*, *P. vivax* and *P. ovale* in five species of *Anopheles*. J. Protozool. *16*, 589-596 (1969)
20. Collins, W.E.; J.C. Skinner; B.B. Richardson, and P.S. Stanfill: Transmission of *Plasmodium cynomolgi* to *Aotus trivirgatus* monkeys. J. Parasit. *61*, 146-148 (1975)
21. Collins, W.E.; McW. Warren; J.C. Skinner, and D.W. Alling: *Plasmodium inui*: Serological relationships of Asian isolates. Exp. Parasit. *27*, 507-515 (1970)
22. Contacos, P.G.; G.R. Coatney; T.C. Orihel; W.E. Collins; W.Chin, and M.H. Jeter: Transmission of *Plasmodium schwetzi* from the chimpanzee to man by mosquito bite. Amer. J. trop. Med. Hyg. *19*, 190-195 (1970)
23. Conrad, M.E.: Malaria; pp.373-380 in "Infectious Diseases" (Sanford, J.P. and J.P. Luby, eds.), Grune and Stratton (1981)
24. Cross, J.H.; M.-Y. Hsu-Kuo, and J.C. Lien: Accidental human infection with *Plasmodium cynomolgi bastianelli*. Southeast Asian J. trop. Med. publ. Hlth *4*, 481-483 (1973)
25. Das Gupta, B.M.: Transmission of *P. inui* to man. Proc. nat. Inst. Sci. India *4*, 241-244 (1938)
26. Deane, L.M.; M.P. Deane, and J.A. Ferreira Neto: Studies on transmission of simian malaria and on a natural infection of man with *Plasmodium simium* in Brazil. Bull. Wld. Hlth Org. *35*, 805-808 (1966a)
27. Deane, L.M.; M.P. Deane; J.A. Ferreiro Neto, and F.B. de Almeida: On the transmission of simian malaria in Brazil. Rev. Inst. Med. trop. Sao Paulo *13*, 311-319 (1971)
28. Deane, L.M.; J.A. Ferreira Neto; M.P. Deane, and I.P.S. Silveira: *Anopheles (Kertesia) cruzi*, a natural vector of the monkey malaria parasite, *Plasmodium simium* and *Plasmodium brasilianum*. Trans. roy. Soc. trop. Med. Hyg. *64*, 647 (1970)
29. Deane, L.M.; J.A. Ferreiro Neto; M. Okumura, and M.O. Ferreira: Malaria parasites of Brazilian monkeys. Rev. Inst. Med. trop. Sao Paulo *11*, 71-86 (1969)
30. Deane, L.M.; J.A. Ferreiro Neto, and J.G. Sitonio: Novo hospedeiro natural do *Plasmodium simium* e do *Plasmodium brasilianum*: o mono, *Brachyteles arachnoides*. Rev. Inst. Med. trop. Sao Paulo *10*, 287-288 (1968)
31. Deane, L.M.; M. Okumura; B. Hertha, and W.T. de Souza: Malaria de macacos des arredores de Sao Paulo. VI. Infeccao experimental de Macaco coata *Ateles paniscus* pelo *Plasmodium simium*. Rev. paul. Med. *68*, 181-182 (1966b)
32. Deane, L.M.; M. Paumgartten, and J. Ferreiro Neto: Studies on transmission of simian malaria and on a natural infection of man with *Plasmodium simium* in Brazil. Bull. Wld. Hlth Org. *35*, 805-808 (1965)
33. Dissanaike, A.S.: *Plasmodium* infection in Ceylon monkeys. Trans. roy. Soc. trop. Med. Hyg. *57*, 488-489 (1963)
34. Dissanaike, A.S.; P. Nelson, and P.C.C. Garnham: Two new malaria parasites, *Plasmodium cynomolgi ceylonensis* subsp. nov. and *Plasmodium fragile*, sp. nov. from monkeys in Ceylon. Ceylon J. med. Sci. (D), *14*, 1-9 (1965)
35. Donovan, J.C.; W.S. Stokes; R.D. Montrey, and H. Rozmiarek: Hematologic characterization of naturally occurring malaria *(Plasmodium inui)* in cynomolgus monkeys *(Macaca fascicularis)*. Lab. Anim. Sci. *33*, 86-89 (1983)
36. Druilhe, P.; J.F. Trape; J.P. Leroy; C. Godard, and M. Gentilini: Deux cas d'infection humaine accidentelle par *Plasmodium cynomolgi bastianellii*. Etude clinique et sérologique. Ann. Soc. belge Med. trop. *60*, 349-354 (1980)

37. Dunn, F.L. and F.L. Lambrecht: The hosts of *Plasmodium brasilianum* Gonder and von Berenberg-Gossler, 1908. J. Parasit. *49*, 316-319 (1963)
38. Dvorak, J.A.; L.H. Miller; W.C. Whitehouse, and T. Shiroishi: Invasion of erythrocytes by malaria merozoites. Science *187*, 748-750 (1975)
39. Eyles, D.E.: Symposium of simian malaria: The species of simian malaria: Taxonomy, morphology, life cycle and geographical distribution of monkey species. J. Parasit. *49*, 866-887 (1963)
40. Eyles, D.E.; G.R. Coatney, and M.E. Getz: Vivax-type malaria parasite of macaques transmissible to man. Science *131*, 1812-1813 (1960)
41. Eyles, D.E.; A.B.G. Laing, and C.G. Dobrovolny: The malaria parasites of the pig-tailed macaque, *Macaca nemestrina nemestrina* (Linnaeus), in Malaya. Ind. J. Malariol. *16*, 285-298 (1962b)
42. Eyles, D.E.; A.B.G. Laing; McW. Warren, and A.A. Sandosham: Malaria parasites of Malayan leaf monkeys of the genus *Presbytis*. Med. J. Malaysia *17*, 85-86 (1962a)
43. Eyles, D.E. and McW. Warren: *Plasmodium inui* in Sulawesi. J. Parasit. *48*, 739 (1962)
44. Fiennes, R.: Zoonoses of Primates. The Epidemiology and Ecology of Simian Diseases in Relation to Man. Weidenfeld and Nicolson, London (1967)
45. Garnham, P.C.C.: Distribution of simian malaria parasites in various hosts. J. Parasit. *49*, 905-911 (1963)
46. Garnham, P.C.C.: A new sub-species of *Plasmodium knowlesi* in the long-tailed macaque. J. trop. Med. Hyg. *66*, 156-158 (1963)
47. Garnham, P.C.C.: Malaria Parasites and Other Haemosporidia. Blackwell Sci. Publ. Oxford (1966)
48. Garnham, P.C.C.; J.R. Baker, and P.E. Nesbitt: Transmission of *Plasmodium brasilianum* by sporozoites and the discovery of an exoerythrocytic schizont in the monkey liver. Parasitologia *5*, 5-9 (1963)
49. Garnham, P.C.C.; R. Lainson, and W. Cooper: The tissue stages and sporogony of *Plasmodium knowlesi*. Trans. roy. Soc. trop. Med. Hyg. *51*, 384-396 (1957)
50. Gonder, R. und H.V. von Berenberg-Gossler: Untersuchungen über Malaria-Plasmodien der Affen. Malaria-Intern Arch. *1*, 47-56 (1908)
51. Halberstaedter, L. und S. von Prowazek: Untersuchungen über die Malariaparasiten der Affen. Arb. Kaiserl. Ges.amt. (Berlin), *26*, 37-43 (1907)
52. Hsieh, H.C.: Malaria parasites of the Taiwan monkey. Formosan Science *14*, 477-487 (1960)
53. Inoki, S.: Studies on exoerythrocytic schizogony of the malaria parasite. Exoerythrocytic forms of the Formosan monkey malaria parasite, *Plasmodium inui* var. *cyclopis*. Med. J. Osaka Univ. *2*, 45-53 (1951)
54. Jarcho, S.: Some observations on disease in prehistoric North America. Bull. Hist. Med. *38*, 1-18 (1964)
55. Kalra, N.L.: Emergence of malaria zoonosis of simian origin as natural phenomenon in Greater Nicobars, Audaman and Nicobar Islands - A preliminary note. J. Commun. Dis. *12*, 49-54 (1980)
56. Knowles, T. and B.M. das Gupta: A study of monkey-malaria, and its experimental transmission to man. Ind. Med. Gaz. *67*, 301-320 (1932)
57. Krotoski, W.A.; D.M. Krotoski; P.C.C. Garnham; R.S. Bray; R. Killick-Kendrick; C.C. Draper; G.A.T. Targett, and W.M. Guy: Relapses in primate malaria: discovery of two populations of exoerythrocytic stages. Preliminary note. Brit. med. J. *23*, 153-155 (1980)
58. Lambrecht, F.L. and F.L. Dunn: Isolation of *Plasmodium knowlesi* from Philippine macaques. Nature *191*, 117-118 (1961)
59. Maegraith, B.: Liver involvement in acute mammalian malaria with special reference to *Plasmodium knowlesi* malaria. Advanc. Parasitol. *6*, 189-231 (1968)
60. Marinkelle, C.J. and E.S. Grose: *Plasmodium brasilianum* in Colombian monkeys. Trop. geogr. Med. *20*, 276-280 (1968)
61. Mayer, M.: Über Malaria beim Affen. Med. Klin. *3*, 579-580 (1907)
62. Miller, L.H.: Transfusion malaria; pp. 241-266 in "Transmissible Disease and Blood Transfusion" (T.J. Greenwalt and G.A. Jamieson, eds.), Grune and Stratton, New York (1975)
63. Miller, L.H.: Hypothesis on the mechanism of erythrocyte invasion by malaria merozoites. Bull. Wld. Hlth Org. *55*, 157-162 (1977)

64. Miller, L.H.; F.M. McAuliffe, and S.J. Mason: Erythrocyte receptors for malaria merozoites. Amer. J. trop. Med. Hyg. *26*, 204-208 (1977)
65. Most, H.: *Plasmodium cynomolgi*, accidental human infection. Amer. J. trop. Med. Hyg. *22*, 157-158 (1973)
66. Mulligan, H.W. and C.S. Swaminath: Natural infection with *Plasmodium inui* in *Silenus sinicus* from South India. J. Malar. Inst. Ind. *3*, 603-604 (1940)
67. Oehlschlegel, F.J. jr.; B.J. Sander, and G.J. Brewer: Pyruvate kinase in malaria host-parasite interaction. Nature *255*, 345-347 (1975)
68. Omar, M.S.: Die Empfänglichkeit verschiedener Anopheles-Arten für den Erreger der Affenmalaria *Plasmodium cynomolgi bastianelli*. Z.Tropenmed. Parasitenk. *19*, 152-170 (1968)
69. Porter, J.A. jr.; C.M. Johnson, and L.de Sousa: Prevalence of malaria in Panamanian primates. J. Parasit. *52*, 669-670 (1966)
70. Prakash, S. and S.C. Chakrabarti: The isolation and description of *Plasmodium cynomolgi* (Mayer, 1907) and *Plasmodium inui* (Halberstaedter and Prowazek, 1907) from naturally occurring mixed infections in *Macaca radiata radiata* monkeys of the Nilgiris, Madras State, India. Ind. J. Malariol. *16*, 303-311 (1962)
71. Reichenow, E.: Über das Vorkommen der Malariaparasiten des Menschen bei den afrikanischen Menschenaffen. Zbl. Bakt., I. Abt. Orig. *85*, 207-216 (1920)
72. Rodhain, J.et P.Lassman: Le cycle schizogonique de *Plasmodium schwetzi* et l'évolution de ce parasite chez *Anopheles maculipennis* var. *atroparvus*. Ann. Soc. belge Med. trop. *20*, 179-180 (1940)
73. Rodhain, J. and R.Dellaert: L'infection à *Plasmodium malariae* du chimpanzé chez l'homme. Etude d'une première souche isolée de l'anthropoide *Pan satyrus verus*. Ann. Soc. belge Med. trop. *23*, 19-46 (1943)
74. Schmidt, L.H.; D.V. Cramer; R.N. Rossan, and J.Harrison: The characteristics of *Plasmodium cynomolgi* infections in various Old World primates. Amer. J. trop. Med. Hyg. *26*, 356-372 (1977a)
75. Schmidt, L.H.; R.Fradkin; J.Harrison, and R.N. Rossan: Differences in the virulence of *Plasmodium knowlesi* for *Macaca irus (fascicularis)* of Philippine and Malayan origin. Amer. J. trop. Med. Hyg. *26*, 612-622 (1977b)
76. Schmidt, L.H.; R.Fradkin; J.Harrison; R.N. Rossan, and W.Squires: The course of untreated *Plasmodium inui* infections in the rhesus monkey *(Macaca mulatta)*. Amer. J. trop. Med. Hyg. *29*, 158-169 (1980)
77. Schmidt, L.H.; R.Greenland, and C.S. Genther: The transmission of *Plasmodium cynomolgi* to man. Amer. J. trop. Med. Hyg. *10*, 679-688 (1961)
78. Schofield, L.D.; B.T. Bennett; W.E. Collins, and F.Z. Beluhan: An outbreak of *Plasmodium inui* in a colony of diabetic *rhesus monkeys*. Lab. Anim. Sci. *35*, 167-168 (1985)
79. Serrano, J.A.: Infeccion natural de un araguato *Alouatta seniculus straminea* por *Plasmodium brasilianum* en Venezuela. Acta cient. venez. *18*, 13-15 (1967)
80. Sinha, S. and A.Gajanana: First report of natural infection with quartan malaria parasite *Plasmodium shortti* in *Macaca fascicularis umbrosa (=irus)* of Nicobar Islands. Trans. roy. Soc. trop. Med. Hyg. *78*, 567 (1984)
81. Sodeman, T.M.; J.R. Held; P.G. Contacos; J.R. Jumper, and C.S. Smith: Studies of the exoerythrocytic stages of simian malaria. IV. *Plasmodium brasilianum*. J. Parasit. *55*, 963-970 (1969)
82. Spangler, W.L.; D.Gribble; C.Abildgaard, and J.Harrison: *Plasmodium knowlesi* malaria in the rhesus monkey. Vet. Path. *15*, 83-91 (1978)
83. Stokes, W.S.; J.C. Donovan; R.D. Montrey; W.L. Thompson; R.M. Wannemacher, jr. and H.Rozmiarek: Acute clinical malaria *(Plasmodium inui)* in a cynomolgus monkey *(Macaca fascicularis)*. Lab. Anim. Sci. *33*, 81-85 (1983)
84. Voller, A.: Plasmodium and Hepatocystis; pp.57-73 in "Pathology of Simian Primates", Part II (Fiennes, R.N.T.-W., ed.), Karger, Basel (1972)
85. Warren, McW.; W.H. Cheong; A.H.B. Omar, and A.A. Sandosham: Ecology of simian malaria in the monsoon forests of the northern Malayan states. 40 th. Ann. Meet. Amer. Soc. Parasit., Atlanta, Georgia, 12.-17. April, 1965; Abstr.: J. Parasit. *51* (Suppl. 2), 17 (1965)
86. Warren, McW. and R.H. Wharton: The vectors of simian malaria: Identity, biology, and geographical distribution. J. Parasit. *49*, 892-904 (1963)

87. Wharton, R.H.; D.E. Eyles; McW. Warren, and D.E. Moorhouse: *Anopheles leucosphyrus* identified as a vector of monkey malaria in Malaya. Science *137*, 758 (1962)
88. Wunderlich, F.; H. Stübig, and E. König: Intraerythrocytic development of Plasmodium knowlesi. Structure, temperature - and CA2+ - response of the host and parasite membranes. J. Protozool. *29*, 49-59 (1982)

6.4.2 Endodyococcidoria

6.4.2.1 Toxoplasma Gondii

T. gondii, discovered in a small North African rodent, the gundi *(Ctenodactylus gundii)* is found in a wide variety of birds and mammals including the primates. Its definitive hosts are felids, whilst other homoiothermic animals and man are the intermediate hosts. In the nonhuman primates the New World species are more frequently infected, but the infection affects also Old World monkeys, apes and prosimians (table 39). This may reflect the feeding behavior of the species affected. New World monkeys have generally much higher protein requirements than Old World species and are known to catch insects, birds and, at least in captivity, small rodents.

Baboons, macaques and chimpanzees prey occasionally on other animals or are apt to cannibalism. Purely herbivorous genera like the colobids, in contrast, so far have not been found to be infected by *T. gondii.*

T. gondii is a well known pathogen in man, causing severe disease especially in the newborn.

Agent and Life Cycle

T. gondii is basically an intestinal coccidian parasite of domestic, and probably also of wild cats, passing a sexual cycle only in the felids. After the cats ingest one of the infective forms of *T. gondii,* they invade the enterocytes following a prepatent period of 3 to 10 days and grow into large ellipsoid or round trophozoites located near to the brush border. The trophozoites multiply first by endodyogeny and later by schizogony resulting in the production of 5 to 32 merozoites. The enterocyte ruptures and the propagative cycle is repeated up to 5 times. Some of the merozoites finally enter a cycle of gametogony, leading to the development of micro- and macrogametocytes and of micro- and macrogametes. Fertilization of the macrogamete by the microgamete results in the formation of a zygote (oocyst) which protects itself by a thin but resistant wall, and which is unsporulated when shed. After a period of development in the external environment the oocyst transforms to a sporoblast which gives rise to two sporocysts (Fig. 31) and which is undistinguishable from the "small race" of *Isospora bigemina,* a common coccidian parasite of cats. Each sporocyst contains four sporozoites.

The fecally excreted oocysts, after the necessary developmental period, may be ingested by another cat or by an intermediate host, ideally a small rodent.

If ingested by another susceptible cat, the parasite may enter another sexual cycle, the ingestion by an intermediate host is invariably followed by an asexual cycle.

Table 39. Spontaneous *Toxoplasma gondii* Infections in Nonhuman Primates

Host Species	References
Prosimians:	
Nycticebus coucang	Ruch, 1959; Zaman and Goh, 1968; Zaman and Krishnamurti, 1969
Lemur catta	Sureau et al., 1962; Uilenberg and Ribot, 1969; Itakura and Nigi, 1968; Nigi and Itakura, 1968; Borst and van Knapen, 1984
Lemur variegatus	Uilenberg and Ribot, 1965
Old World Monkeys:	
Macaca mulatta	Nery-Guimares et al., 1971; Wong and Kozek, 1974
Macaca fascicularis	Pokorny et al., 1961
Macaca sp.	Levine, 1961
Papio ursinus	McConnell et al., 1973, 1974
Papio sp.	Levaditi and Schoen, 1933
Apes:	
Pan troglodytes	Schobert, 1970 - cit. from McClure and Guilloud, 1971
New World Monkeys:	
Callithrix jacchus	Hilgenfeld, 1965
Saguinus oedipus	Benirschke and Richart, 1960
Saguinus geoffroyi	Frenkel and Sousa, 1983
Saguinus illigeri	Ratcliffe, 1967 - cit. from Seibold and Wolf, 1971
Leontopithecus rosalia	Ratcliffe, 1967 - cit. from Seibold and Wolf, 1971
Callicebus moloch	Seibold and Wolf, 1971
Saimiri spp.	Ratcliffe and Worth, 1951; Ratcliffe, 1954; Döbereiner, 1955; McKissick et al., 1968; Anderson and McClure, 1982; Dickson et al., 1983; Borst and van Knapen, 1984
Cebus capucinus	Rodaniche, 1954
Cebus apella	Nery-Guimares et al., 1971
Cacajao rubicundus	Ratcliffe, 1954
Ateles geoffroyi	Ratcliffe and Worth, 1951; Ratcliffe, 1967 - cit. from Seibold and Wolf, 1971
Pithecia monachus	Ratcliffe, 1954; Ruch, 1959
Lagothrix sp.	Stolz, 1962; Ratcliffe, 1967; Benirschke and Low, 1970; Hessler et al., 1971
Aotus trivirgatus	Seibold and Wolf, 1971

The asexual multiplication in the heterologous (or homologous) host is by endodyogeny, a special form of schizogony, which is characterized by the formation of two daughter cells and their associated membranes within a mother cell. The nucleus of the mother organism buds out into each of the daughter cells. The daughter cells are originally surrounded by a single membrane, but as they break out of the depleted and dying mother organism, they acquire their outer membrane. The repetitive endodyogeny within a parasitized cell results in the production of a rosette-like pseudocyst of crescentic tachyzoites of 2 to 6 μm size. Finally the pseudocyst ruptures and the tachyzoites are released.

After the acute infection has subsided, a new asexual form, with formation of the cyst is initiated. The cyst has a dense wall produced by deposition of material of parasitic origin on the parasitophorous vacuole. The parasite continues to replicate by endodyogeny but on a slower rate (bradyzoite):

The crescentic bradyzoites are resistant to gastric juice, while the tachyzoites are susceptible. Thus the bradyzoite is adapted to transmission to carnivores.

After uptake of a bradyzoite by a susceptible cat the sexual cycle is completed in 3 to 10 days, the subsequent shedding of oocysts lasts usually one to two weeks. In contrast, after ingestion of tachyzoites it takes 19 to 48 days for completion of the cycle.

The asexual cycle is repeated in carnivores including cats or in omnivorous animals including the primates (Jacobs and Frenkel, 1982).

Cats can shed several millions oocysts, capable of infecting an equal number of heterologous hosts, and therefore cats are the true foci of toxoplasmosis.

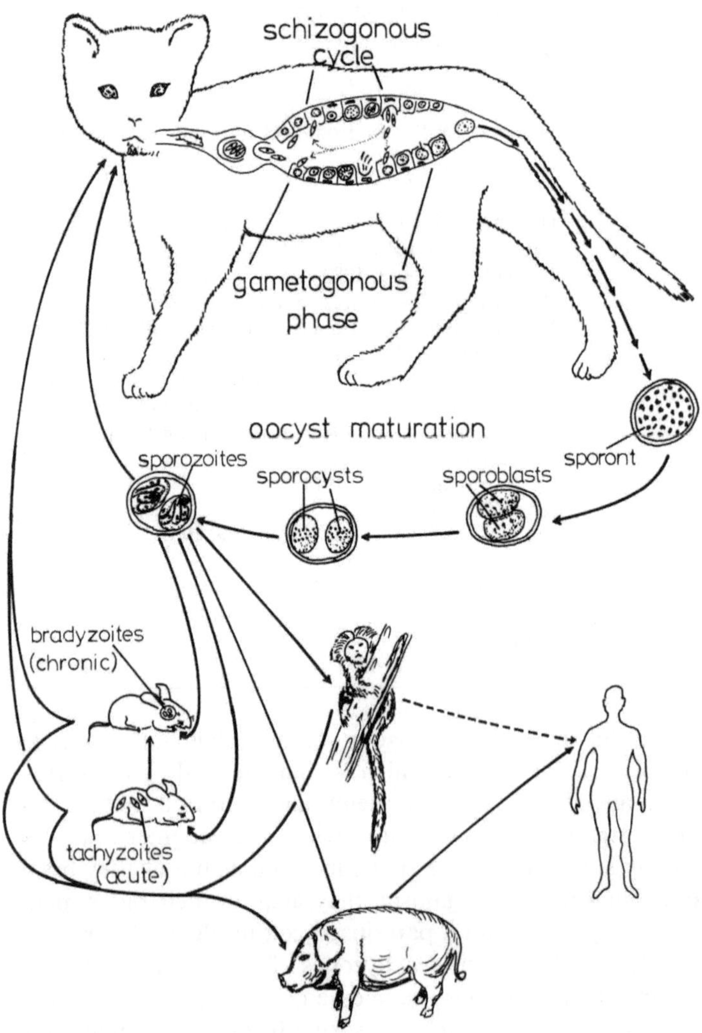

Fig. 31. Life cycle of *Toxaplasma gondii*

Heterologous hosts can be infected by ingestion of food contaminated by cat feces. That way even purely herbivorous animals such as kangaroos, chinchillas, ruminants etc. become infected, although carnivorous and omnivorous species are at a much greater risk. Carnivorous or omnivorous hosts can additionally become infected by ingestion of bradyzoites or tachyzoites from other heterologous hosts. Man, for example can be infected by direct contact with shedding cats, by uptake of oocysts from the environment (Sulzer et al., 1986) or by eating raw pork or mutton. With respect to nonhuman primates transmission of *T. gondii* to man could occur in populations depending on monkeys for food.

Transmission also occurs across the placenta in many mammalian species. Other, but less common ways of transmission are inhalation of tachyzoites or smear infections of the conjunctiva. Infected intermediate hosts falling prey to a cat complete the life cycle of *T. gondii*.

Geographical Distribution

Toxoplasmosis has been found all over the world, but more commonly in the temperate and moist tropical regions than in the polar zones or in hot deserts.

Epidemiology

The entire life cycle of *T. gondii* depends on felids. In areas without domestic or wild cats, such as certain Pacific Islands, toxoplasmosis is also lacking (Munday, 1972; Wallace et al., 1972; 1974 - all cit. from Jacobs and Frenkel, 1982).

Pathogenesis

Ingested tachyzoites or bradyzoites penetrate the intestinal epithelium in both homologous and heterologous hosts and induce a transient parasitemia during which every organ can be reached. Lymphoreticular tissues, the CNS and its extensions (retina), the muscle tissues, and in pregnant women the placenta and the fetus are particularly prone to the infection. The circulating tachyzoites attach to nucleated cells (in birds also to erythrocytes!) and penetrate the cellular wall, probably aided by enzymes produced by the rhoptries (club-like structures situated at their pointed anterior end; Beverley, 1974; Zaman, 1980). After entering the cell the tachyzoites are surrounded by cytoplasmic vesicles, the "parasitophorous vacuoles", whose membranes are both host and parasite derived. The tachyzoites induce lytic necrosis of the host cells and a surrounding lymphocytic-plasma cellular inflammation. The necroses may later be replaced by fibrosis which occurs particularly in muscular tissues, or by focal calcification as seen in the brain tissues.

Clinical Symptoms

Nonhuman primates: The symptoms in acute infections of nonhuman primates consist of fever and anorexia (Itakura and Nigi, 1968; McKissick et al., 1968; Nigi and Itakura, 1968; Hessler et al., 1971; Nery-Guimares et al., 1971) or depression (Dubey et al., 1985). Involvement of the respiratory tract is indicated by quick breathing or friction sounds (Benirschke and Richart, 1960; Itikura and Nigi, 1968; Nigi and Itakura, 1968; Dickson et al., 1983), and that of the digestive tract by vomitus and intermittent diarrhea (Stolz, 1962; McKissick et al., 1968; Wong and Kozek, 1974). Other monkeys were found dead without preceding illness (Seibold and Wolf, 1971; Anderson and McClure, 1982).

Man: Most *T. gondii* infections of adults and children are asymptomatic. Clinical illness ensues if the cellular damage by the parasites is not compensated by regeneration or if extensive inflammatory responses are initiated. The course may be acute, subacute or chronic. Acute toxoplasmosis appears with the proliferation of tachyzoites, leading to fever, malaise, rash, pneumonia, myocarditis, hepatitis, lymphadenopathy, and rarely encephalitis. The corresponding symptoms are usually terminated by the development of immunity. The most common kind of acute toxoplasmosis is manifested by lymphadenitis, fever, headache and myalgia. Other forms are typhus-like with myocarditis, meningoencephalitis and atypical pneumonia - these are often fatal. A third form involves primarily the CNS, frequently also with a fatal outcome, and a fourth type may be noncongenital retinochoroiditis (Faust et al., 1975).

The most serious consequences occur in newborns, because women who acquire the infection during the second or third trimester of pregnancy frequently transmit the parasites transplacentally to the child although they themselves remain asymptomatic. At birth or shortly thereafter these infants are found to have retinochoroiditis, cerebral calcification, hydrocephalus or microcephaly. Subacute toxoplasmosis is frequently seen in newborns from mothers who develop and transplacentally pass antibodies which delay the haematogenous spread of *T. gondii.* The illness in subacute toxoplasmosis of newborns tends to be prolonged, affecting especially the brain and eyes, both of which are less immunologically protected than other organs. Signs of encephalitis and retinochoroiditis have been noted in 69 to 80% of such cases (Jacobs and Frenkel, 1982).

Chronic toxoplasmosis with persisting cysts is usually asymptomatic, but rupture of cysts located in the retina may lead to a progressive inflammation and finally blindness.

Pathology

Nonhuman primates: The most common lesions are acute focal pneumonia (Benirschke and Richart, 1960; Itakura and Nigi, 1968; Nigi and Itakura, 1968; Hessler et al., 1971; Seibold and Wolf, 1971; Anderson and McClure, 1982; Borst and van Knapen, 1984); hydropericard or hydrothorax (Itakura and Nigi, 1968; McKissick et al., 1968; Nigi and Itakura, 1968; Dickson et al., 1983), or focal necroses in several organs, especially in the liver (Benirschke and Richart, 1960; McKissick et al., 1968; Nigi and Itakura, 1968; Hessler et al., 1971; Seibold and Wolf, 1971; Anderson and McClure, 1982). In lemurs extensive necrosis of the mesenteric lymphnodes predominated besides necroses of livers and spleens (Dubey et al., 1985). The rare CNS-lesions consist of microglial nodules (McKissick et al., 1968; Nigi and Itakura, 1968; Anderson and McClure, 1982); calcified cysts in the brain (Wong and Kozek, 1974); the presence of cysts surrounded by marked cellular infiltration (Zaman and Goh, 1968; Zaman and Krishnamurti; McConnell et al., 1973), or subacute encephalitis (Stolz, 1962). *Toxoplasma* cysts are also found in a variety of other organs.

Man: Neonatal toxoplasmosis: Most important are the lesions of the CNS (focal liquefaction and calcification, necrosis, microglial nodules, ependymitis ulcerosa and encephalomyelitis) and the eye (retinochoroiditis).

Postnatal toxoplasmosis: The most common form: *lymphadenitis toxoplasmotica* is characterized by an epitheloid cellular reticulosis with marked nuclear destruction and sinus histiocytosis.

In chronic toxoplasmosis persisting *Toxoplasma*-cysts may be accompanied by myocarditis and myositis (Frenkel, 1976).

Treatment

In man the combination of sulfadiazine and pyrimethamine (Daraprim®) is synergistic against the trophozoites (Frenkel, 1976; Filice and Remington, 1981), but the encysted forms are not affected (McCulloch and Remington, 1975).

Spiramycin is also effective and possibly the drug of choice in pregnant women (Hof, 1978). In experimentally infected squirrel monkeys it prevented losses, but the combination of sulfonamides with pyrimethamine or trimethoprim were more effective (Harper et al., 1985).

References

1. Anderson, D.C. and H.M. McClure: Acute disseminated fatal toxoplasmosis in a squirrel monkey. J.Amer. vet. med. Ass. *181*, 1363-1366 (1982)
2. Benirschke, K. and R.Richart: Spontaneous acute toxoplasmosis in a marmoset monkey. Amer. J.trop. Med. Hyg. *9*, 264-273 (1960)
3. Beverley, J.K.A.: Some aspects of toxoplasmosis, a world-wide zoonosis; pp.1-25 in "Parasitic Zoonoses. Clinical and Experimental Studies." (Soulsby, E.J.L., ed.). Academic Press, New York (1974)
4. Borst, G.H.A. and F.van Knapen: Acute acquired toxoplasmosis in primates in a zoo. J.Zoo Anim. Med. *15*, 60-62 (1984)
5. Dickson, J.; J.Fry; R.Fairfax, and T.Spence: Epidemic toxoplasmosis in captive squirrel monkeys *(Saimiri sciureus)*. Vet. Rec. *112*, 302 (1983)
6. Döbereiner, J.: Toxoplasmose espontontanea em macaco. Veterinariya (Sao Paulo) *9*, 44-54 (1955)
7. Dubey, J.P.; L.W. Kramer, and S.E. Weisbrode: Acute death associated with *Toxoplasma gondii* in ring-tailed lemurs. J.Amer. vet. med. Ass. *187*, 1272-1273 (1985)
8. Faust, E.C.; P.C. Beaver, and R.C. Jung: Animal Agents and Vectors of Human Disease. 4th. ed., Lea and Febiger, Philadelphia (1975)
9. Filice, G.A. and J.S. Remington: Toxoplasmosis; pp.380-383 in "Infectious Diseases" (Sanford, J.P. and J.P. Luby, eds.), Grune and Stratton, New York (1981)
10. Frenkel, J.K.: Toxoplasmosis; pp.284-300 in "Pathology of Tropical and Extraordinary Diseases". An Atlas. Vol.1 (Binford, C.H. and D.H. Connor, eds.). Armed Forces Inst. Path., Washington, D.C. (1976)
11. Frenkel, J.K. and O.E. Sousa: Antibodies to *Toxoplasma* in Panamanian mammals. J.Parasit. *69*, 244-245 (1983)
12. Harper. J.S. III; W.T. London, and J.L. Sever: Five drug regimens for treatment of acute toxoplasmosis in squirrel monkeys. Amer. J.trop. Med. Hyg. *34*, 50-57 (1985)
13. Hessler, J.R.; J.C. Woodard, and P.C. Tucek: Lethal toxoplasmosis in a woolly monkey. J.Amer. vet. med. Ass. *159*, 1588-1594 (1971)
14. Hilgenfeld, M.: Zur Toxoplasmose der weiblichen Geschlechtsorgane; pp.22-24 in VII. Internat. Symp. Erkr. Zootiere, Zürich-Basel (1965)
15. Hof, H.: Toxoplasma gondii-Infektion und Infektabwehr. Med. Klin. *73*, 888-899 (1978)
16. Itakura, C.and H.Nigi: Histopathological description on two spontaneous cases of toxoplasmosis in the monkey *(Lemur catta)*. Jap. J.vet. Sci. *30*, 341-346 (1968)
17. Jacobs, L. and J.K. Frenkel: Toxoplasmosis; pp.167-185 in "CRC Handbook Series in Zoo-

noses" (Steele, J. H., Series ed.), Section C: Parasitic Zoonoses. Vol. I (Jacobs, L. and P. Arambulo, Vol. eds.). CRC Press, Boca Raton (1982)
18. Levaditi, C. and R. Schoen: Présence d'un toxoplasme dans l'encepole du *Cynocephalus* babuin. Bull. Soc. Path. exot. *26*, 402-405 (1933)
19. Levine, N. D.: Protozoan Parasites of Domestic Animals and of Man. Burgess, Minneapolis (1961)
20. McClure, H. M. and N. B. Guilloud: Comparative pathology of the chimpanzee; pp. 103-272 in "The Chimpanzee", Vol. 4 (Bourne, G. H., ed.), Karger, Basel (1971)
21. McConnell, E. E.; P. A. Basson; V. de Vos; B. J. Myers, and R. E. Kuntz: A survey of diseases among 100 free-ranging baboons *(Papio ursinus)* from the Kruger National Park. Onderstepoort J. vet. Res. *41*, 97-168 (1974)
22. McConnell, E. E.; P. A. Basson; B. Wolstenholme; V. de Bos, and H. H. Malherbe: Toxoplasmosis in free-ranging chacma baboons *(Papio ursinus)* from the Kruger National Park. Trans. roy. Soc. trop. Med. Hyg. *67*, 851-855 (1973)
23. McCulloch, W. F. and J. S. Remington: Toxoplasmosis; pp. 805-821 in "Diseases Transmitted from Animals to Man", 6th ed. (Hubbert, W. T.; W. F. McCulloch, and P. R. Schnurrenberger, eds.), C. C. Thomas, Springfield (1975)
24. McKissick, G. E.; H. L. Ratcliffe, and A. Koestner: Enzootic toxoplasmosis in caged squirrel monkeys *Saimiri sciureus*. Path. Vet. *5*, 538-560 (1968)
25. Nery-Guimares, F.; A. J. Franken, y W. A. Chagas: Toxoplasmose em primatos nao humanos. I. Infeccoes naturais em *Macaca mulatta* e *Cebus apella*. Mem. Inst. Osw. Cruz *69*, 77-96 (1971)
26. Nigi, H. and C. Itakura: Spontaneous toxoplasmosis in *Lemur catta*. Primates *9*, 155-160 (1968)
27. Pokorny, J.; J. Hübner, and M. Zastera: Toxoplasma gondii z nekterych domacich i volue zigicich zvirat. Cs. Epidem. *10*, 323-329 (1961)
28. Ratcliffe, H. L.: Causes of death in the animal collection. Rept. Penrose Res. Lab., 6-16 (1954)
29. Ratcliffe, H. L. and C. B. Worth: Toxoplasmosis of captive wild birds and mammals. Amer. J. Path. *27*, 655-667 (1951)
30. Rodaniche, E. D.: Spontaneous toxoplasmosis in the whiteface monkey, *Cebus capucinus*, in Panama. Amer. J. trop. Hyg. *3*, 1023-1025 (1954)
31. Ruch, T. C.: Diseases of Laboratory Primates. W. B. Saunders, Philadelphia (1959)
32. Seibold, H. R. and R. H. Wolf: Toxoplasmosis in *Aotus trivirgatus* and *Callicebus moloch*. Lab. Anim. Sci. *21*, 118-120 (1971)
33. Stolz, G.: Spontaneous lethal toxoplasmosis in a woolly monkey. Schweiz. Arch. Tierheilk. *104*, 162-166 (1962)
34. Sulzer, A. J.; E. L. Franco; E. Takafuji; M. Benenson; K. W. Walls, and R. L. Greenup: An oocyst-transmitted outbreak of toxoplasmosis: Patterns of immunoglobulin G and M over one year. Amer. J. trop. Med. Hyg. *35*, 290-296 (1986)
35. Sureau, P.; J. P. Raynaud; C. Lopeire et E. R. Brygoo: Premier isolement de Toxoplasma gondii à Madagascar. Toxoplasmose spontanée et expérimentale de Lemur catta. Bull. Soc. Path. exot. *55*, 357-362 (1962)
36. Uilenberg, G. and I. J. Ribot: Note sur la toxoplasmose des lemuriens (Primates: *Lemuridae*). Rev. Elev. Med. Pays Trop. *18*, 247-248 (1965)
37. Wong, M. M. and W. J. Kozek: Spontaneous toxoplasmosis in macaques: A report of four cases. Lab. Anim. Sci. *24*, 273-278 (1974)
38. Zaman, V.: Atlas of Medical Parasitology. An Atlas of Important Protozoa, Helminthes and Arthropods, Mostly in Colour. MTP Press Ltd. Lancaster (1980)
39. Zaman, V. and T. K. Goh: Isolation of Toxoplasma gondii from the slow loris, Nycticebus coucang. Ann. trop. Med. Parasit. *62*, 52-53 (1968)
40. Zaman, V. and A. Krishnamurti: Distribution of *Toxoplasma* cysts in the central nervous system of slow loris, *Nycticebus coucang*. Acta trop. *26*, 167-172 (1969)

6.4.3 Eimeriorina

6.4.3.1 Cryptosporidium Spp.

Cryptosporidia are small coccidian parasites of little host specificity that inhabit the microvillous region of enterocytes primarily of the small intestines. They are known since 1907 (Tyzzer, 1910), but are difficult to demonstrate in clinical specimens unless specially stained (Bronsdon, 1984) or viewed by phase-contrast microscopy (Fayer, 1982). Until recently the cryptococci have been considered more or less as opportunists, but now they are realized as an important cause of gastroenteritis in man and several animal species (Current and Hayes, 1984; Perez-Schael et al., 1985). With respect to nonhuman primates they have been reported in squirrel monkeys (Bryant et al., 1983), and rhesus monkeys (Kovatch and White, 1972; Cockrell et al., 1974; Rodger and Bronson, 1983; Wilson et al., 1984).

Agent

Cryptosporidia differ from other coccidia because they develop just inside the microvillous region and not in the cell proper (Levine, 1980; Fayer, 1982). Like in other coccidia an asexual schizogony is followed by a sexual gametogony, and fertilized macrogametes develop into tiny oocysts of 2 to 6 μm diameter that are passed in the feces and that may be immediately infectious. The oocysts contain four sporozoites without sporocysts.

Although several species of *Cryptosporidium* have been named based on assumed host specificity (e.g. *C. rhesi*), recent transmission experiments suggest that at least mammalian *Cryptosporidium* is a single-species genus (Fayer, 1982). According to Levine (1984) the Cryptosporidia of mammals should be considered as *C. muris*, but Upton and Current (1985) proposed *C. parvum* to be a separate species.

Clinical Symptoms

Nonhuman primates: The most consistent sign is persistent diarrhea, which has been observed especially in juvenile or in immune deficient animals (Kovatch and White, 1972; Cockrell et al., 1974; Rodger and Bronson, 1983; Kuller et al., 1984; Wilson et al., 1984).

Man: In immunocompetent individuals, *Cryptosporidium* may rarely produce a short, transient, flu-like, gastrointestinal illness with low-grade fever and several days of diarrhea (Reese et al., 1982). In immunodeficient patients, however, prolonged, severe diarrhea may result (Meisel et al., 1976; Lasser et al., 1979).

Pathology

Nonhuman primates: In clinical cases a neutrophilic inflammation primarily of the distal small intestine with thickening, fusion and flattening of the villi, as well as necrosis and sloughing of enterocytes was apparent (Kovatch and White, 1972; Cockrell et al., 1974; Bryant et al., 1983; Wilson et al., 1984). Similar lesions may be found in the walls of the common bile duct and the pancreatic duct (Kovatch and White, 1972). The cryptosporidia which stain lightly basophilic with H. and E. are better demonstrated lining the microvillous area of the epithelium by Giemsa- or by toluidine blue staining techniques.

Treatment
No effective therapy is so far available (Current and Haynes, 1984).

References

1. Bronsdon, M.A.: Rapid dimethyl sulfoxide-modified acid-fast stain of *Cryptosporidium* oocysts in stool specimens. J. Clin. Microbiol. *19*, 952-953 (1984)
2. Bryant, J.L.; H.F. Stills, and C.C. Middleton: *Cryptosporidia* in squirrel monkeys *(Saimiri sciureus)*. Lab. Anim. Sci. *33*, 482 (1983)
3. Cockrell, B.Y.; M.G. Valerio, and F.M. Garner: Cryptosporidiosis in the intestines of rhesus monkeys *(Macaca mulatta)*. Lab. Anim. Sci. *24*, 881-887 (1974)
4. Current, W.L. and T.B. Haynes: Complete development of *Cryptosporidium* in cell culture. Science *224*, 603-605 (1984)
5. Fayer, R.: Other protozoa: *Eimeria, Isospora, Cystoisospora, Besnoitia, Hammondia, Frenkelia, Sarcocystis, Cryptosporidium, Encephalitozoon,* and *Nosema*; pp. 187-196 in "CRC Handbook Series in Zoonoses" (Steele, J.H., Series ed.), Section C: Parasitic Zoonoses, Vol. I (Jacobs, L. and P. Arambulo, Vol. eds.), CRC Press, Boca Raton (1982)
6. Kovatch, R.M. and J.D. White: Cryptosporidiosis in two juvenile rhesus monkeys. Vet. Path. *9*, 426-440 (1972)
7. Kuller, L.; W.R. Morton, and M.A. Bronsdon: Occurrence of Cryptosporidium in a nonhuman primate nursery. Amer. J. Primatol. *6*, 411 (1984)
8. Lasser, K.H.; K.J. Levin, and F.W. Ryning: Cryptosporidial enteritis in a patient with congenital hypogammaglobulinemia. Hum. Path. *10*, 234-240 (1979)
9. Levine, N.D.: Some corrections of coccidian (Apicomplexa: Protozoa) nomenclature. J. Parasit. *66*, 830-834 (1980)
10. Levine, N.D.: Taxonomy and review of the coccidian genus *Cryptosporidium* (Protozoa, Apicomplexa). J. Protozool. *31*, 94-98 (1984)
11. Meisel, J.L.; D.R. Perem; C. Melligro, and C.E. Rubin: Overwhelming watery diarrhea associated with a *Cryptosporidium* in an immunosuppressed patient. Gastroenterology *70*, 1156-1160 (1976)
12. Perez-Schael, I.; Y. Boher; L. Mata; M. Perez, and F.J. Tapia: Cryptosporidiosis in Venezuelan children with acute diarrhea. Am. J. trop. Med. Hyg. *34*, 721-722 (1985)
13. Reese, N.C.; W.L. Current; J.V. Ernst, and W.S. Bailey: Cryptosporidiosis of man and calf: A case report and results of experimental infections in mice and rats. Amer. J. trop. Med. Hyg. *31*, 226-229 (1982)
14. Rodger, R.F. and R.T. Bronson: A rhesus monkey with acquired immune deficiency syndrome and cryptosporidiosis. Lab. Anim. Sci. *33*, 482 (1983)
15. Tyzzer, E.E.: An extracellular coccidium, *Cryptosporidium muris* (Gen. et sp. nov.) of the gastric glands of the common mouse. J. med. Res. *23*, 487-509 (1910)
16. Upton, S.J. and W.L. Current: The species of *Cryptosporidium* (Apicomplexa: Cryptosporidiae) infecting mammals. J. Parasit. *71*, 625-629 (1985)
17. Wilson, D.W.; P.A. Day, and M.E.G. Brummer: Diarrhea associated with *Cryptosporidium* spp. in juvenile macaques. Vet. Path. *21*, 447-450 (1984)

7 Nematodes

7.1 Strongylina

7.1.1 Ancylostomatidae

Ancylostomatidae or "hookworms" are a group of nematodes which have in common the following characteristics: 1. Distinct lips are lacking; 2. The male posterior end bears a copulating bursa typically supported by 7 pairs of rays; 3. The female internal reproductive organs are paired with one set stretching anteriorly, the other one posteriorly from the vulva and vagina; 4. The anterior end of the adult worms of either sex is bent dorsally thus creating a hook; 5. They are small blood-sucking parasites of the small intestines.

Human hookworm infections are caused by either *Ancylostoma duodenale* or *Necator americanus*, both of which are found also in nonhuman primates.

7.1.1.1 Ancylostoma Duodenale

A. duodenale is spontaneously transmitted to man's closest relatives. It has been reported in chimpanzees, gorillas, gibbons, baboons, rhesus and mona monkeys (Benson et al., 1954; Young et al., 1957; Myers and Kuntz, 1965; Fiennes, 1967; Healy and Myers, 1973).

Agent
Adult *A. duodenale* are greyish-white to pink, cylindrical worms of approximately 1 cm length. They possess a well developed, cup-shaped buccal capsule which is heavily reinforced by four conspicious, slightly curved, subequal teeth on the ventral part of the mouth. In the depth of the buccal capsule there is an additional pair of small subventral teeth (Fig. 32).

The buccal capsule is dorsally lined by a plate with a median cleft.

The male worm, which measures 8 to 11 × 0.4 to 0.5 mm is provided with a bell-shaped copulatory bursa supported by fleshy rays and a pair of spicules (Fig. 32). The dorsal ray is not deeply cleft and the tip of each branch is tridigitate.

The slightly larger female measures 10 to 13 × 0.6 mm. Its posterior end tapers rather bluntly and is provided with an elongated process (mucron). The vulvar opening lies mid-ventrally at the beginning of the posterior third of the body.

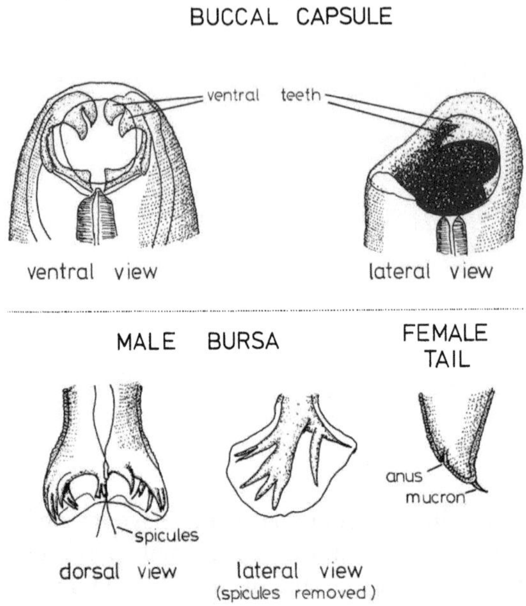

Fig. 32. *Ancylostoma duodenale*

The eggs are broadly ovoid, of 40 × 60 μm size. They have a thin, transparent shell and at fecal excretion are in the 2 to 8 cell stage.

Cycle

The hookworms have several larval stages incorporated into their life cycle. Embryonation of the liberated eggs to the first (rhabdidiform) larval stage takes place in 24 to 48 hours in moist, shaded environments at an optimal temperature of 25 °C. The rhabdidiform hookworm larvae may migrate beneath the surface of the soil and after two to three days molt to the second larval stage. The first- and second-stage larvae both feed on bacteria, organic matter etc. After a few more days the second-stage larvae undergo a second molt, or ecdysis, resulting in the infective, third, filariform larval stage. The third-stage larvae do not shed their last cuticle at molting but keep it as a sheath around them, which protects against environmental influences, but also keeps them from feeding. They have to rely entirely upon their own body reserves, but can stay alive for two to three months in soil. The filariform larvae are positively thermotrope and negatively geotrope. They move actively up to the highest point of the soil or vegetation, thereby using up their reserves. When they come into contact with the unprotected skin of a new host they penetrate the epidermis, enter the blood stream, and break through the pulmonary capillaries into the lung alveoles. Here they undergo a third molt, migrate up to the larynx where they are swallowed. When the larvae reach the small intestine they attach themselves to the intestinal wall, and, after a final molt on the 13th day, reach maturity. The females then excrete new eggs after 4 to 7 weeks, and the life cycle is completed (Fig. 33).

Fig. 33. Life cycle of *Ancylostoma duodenale*

Geographical Distribution

A. duodenale infections are found throughout the world with the exception of Australia, but predominate in the tropical and subtropical areas of South East Asia. In East Africa *A. duodenale* is limited to coastal regions below 200 feet altitude (Chunge et al., 1986).

Pathogenesis

The infective larvae of *A. duodenalis* usually gain entrance to the body through a hair follicle or an abraided area. The migration through the subcutaneous tissues may initiate an allergic reaction. When the larvae break through from the capillary into the alveolar system, minute haemorrhages and inflammatory responses occur.

The adult worms after reaching the upper small intestines not only attach to the intestinal mucosa, the injury which they cause to the host results from the lytic destruction of the villous tissues sucked into the buccal capsule, and from the blood loss due to active blood sucking by the worms and to oozing of blood from the sites of attachment caused by anticoagulants secreted by the worms. Each adult *A. duodenale* is estimated to cause the loss of 0.1 to 0.2 ml of blood per day. The total blood loss is compounded by the number of intestinal lacerations that each worm produces during its period of activity. The blood is not digested in the worm's intestine but is only somewhat depleted of oxygen (Katz et al., 1982). The anemia developing from the continous blood loss in hookworm disease begins as a haemorrhagic type (normocytic and normochromic) but progresses to the hypoferric type (microcytic and hypochromic) (Barriga, 1982).

Clinical Symptoms

Nonhuman primates: Hookworms are not common parasites of nonhuman primates (Orihel and Seibold, 1972) and reports on diseases in these animals are almost non-existent. Nonspecified hookworms caused a "pot-belly" appearance and dyspnea in a gorilla (Rewell, 1948). Ruch (1959) reported on cutaneous larva migrans-symptoms due to *A. duodenale.*

Man: The clinical picture of hookworm infections varies considerably depending on the general nutritional status of the patient and the severity of the infection.

Cutaneous creeping eruptions, known as "ground itch", are less common in *A. duodenale* than in *N. americanus.* They occur in tropical and subtropical regions particularly during the rainy season and affect especially the interdigital areas of the feet. These creeping eruptions typically consist of an initial intense itching and burning sensation, followed by edema, erythema and a papule, which finally transforms into a vesicle. The lesions disappear in about 10 days.

Passage of the lungs is occasionally indicated by an irritating cough and dyspnea, although these pulmonary symptoms usually are not observed before the worms reach the intestines. They have been observed especially in Japan as part of the "Wakana"-syndrome (Yoshida and Nakanishi, 1958) and in Australian troops in New Guinea during World War II (Sangster, 1946 - cit. from Belding, 1965).

Acute intestinal hookworm disease has a prodromal syndrome of nausea, headache and fatigue, followed by acute colic in the gastric region, flatulence, diarrhea, weight loss and dizziness. During this period there is an eosinophilic leukocytosis, whereas overt signs of anemia are usually lacking.

Chronic hookworm disease, which is the most common form, is dominated by the consequences of the permanent blood loss and thus resembles malnutrition. The most commonly observed symptoms are vague dyspepsia and weakness, while in heavy infections bloody-mucoid diarrhea alternating with obstipation may ocur. The anemia is of microcytic and hypochromic type and is accompanied by dyspnea, depression and a characteristic apathic "puffy face". The blood-eosinophilia reaches its peak. Affected children are frequently "pot-bellied" and often mentally and physically retarded (Faust et al., 1975; Barriga, 1982).

Pathology

Nonhuman primates: So far no reports on the morphology of the lesions caused by *A. duodenale* are available; they probably resemble those observed in man.

Man: During creeping eruption the larvae can be found in epithelial tunnels of the Malpighian layer, surrounded by degenerated epithelial cells and numerous eosinophilic leukocytes (Galliard, 1974).

The pulmonary migration induces petechial haemorrhages and leukocytic infiltration, followed by focal proliferation of fibroblasts in the damaged alveoli. Secondary infection by bacteria may lead to purulent lobar pneumonia.

The adults cause injury to the upper small intestinal villi by mechanical damage and lytic destruction at the point of attachment, so that the distal portion of affected villi is eroded and digested. Tissue edema may be present in heavy infections, sometimes even anasarca and hydrops ascites may be observed, due to the hypoproteinaemia initiated also by the blood loss.

The bone marrow is primarily normoblastic but may contain no haemosiderin.

Treatment

In nonhuman primates Tetramisol (12 to 16 mg/kg) or Mebendazol (2 × 15 mg/kg or 10 × 3 mg/kg) have been recommended (Göltenboth, 1976).

7.1.1.2 Necator Americanus

N. americanus is considered to be a more benign hookworm species in man than *A. duodenale* and has been reported in Old- and New World simian primates as well. Again chimpanzees seem to be the most commonly affected species (Gedoelst, 1916; Chandler, 1929; Riopelle and Daumy, 1962 - cit. from Orihel and Seibold, 1972; Kuntz and Myers, 1968/69; Orihel, 1971), but the infection has also been reported in gorillas (Looss, 1911 - cit. from Nelson, 1965); patas monkeys (Orihel, 1970; 1971), *Papio doguera* (Ruch, 1959; Yamashita, 1963; Kuntz and Myers, 1966); spider monkeys (Dollfus and Chabaud, 1955) and *Cebus capucinus* (Thatcher and Porter, 1968).

Agent
N. americanus differs from *A. duodenale* by its strongly reflexed anterior end and the structure of its small buccal capsule, which is provided with two ventral semilunar cutting plates, two poorly developed dorsal plates, one median dorsal tooth and a pair of short triangular lancets in its depth (Fig. 34).

The copulatory bursa of the male worm is symmetrical, long and wide, whilst the dorsal ray is deeply clefted and the tips of each branch are bidigitate. The spic-

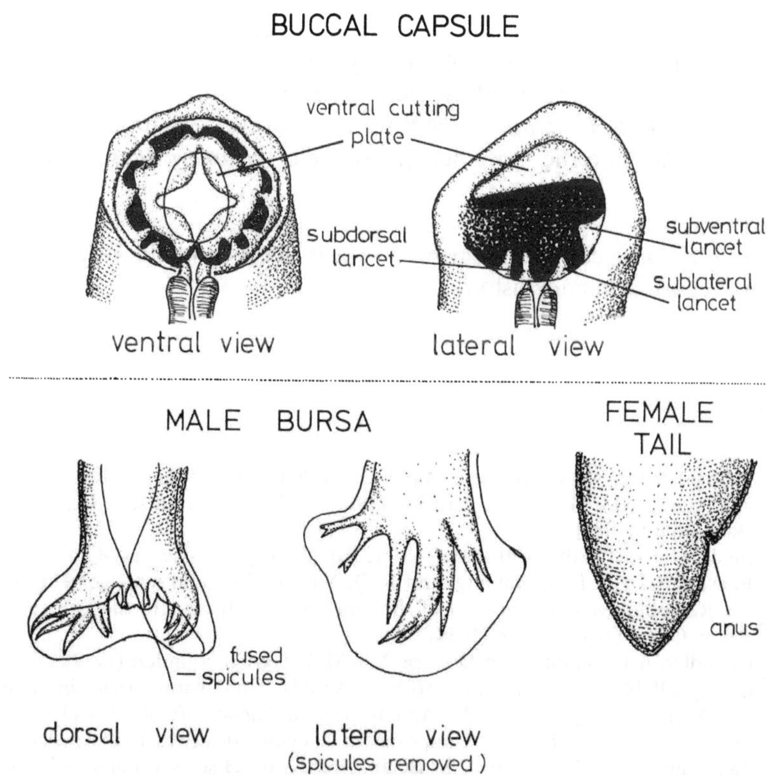

Fig. 34. *Necator americanus*

ules are long, delicate bristless, fused at their outer ends, and terminating in a barb or hook (Faust et al., 1975).

The posterior end of the female is rather sharply conical, the vulva is located at or somewhat anterior to the equatorial plane (Fig. 34). *Necator*-larvae can be differentiated from *Ancylostoma*-larvae by the existence of a gap at the esophageal-intestinal junction, by having a more pointed tail, clearer sheath striations, and a shorter body and tail length (Chunge et al., 1986).

Life Cycle and Epidemiology

The life cycle is identical to that of *A. duodenale* and so is its epidemiology.

Geographical Distribution

The geographical range of *N. americanus* is limited especially by the minimum environmental temperature, because *N. americanus* eggs quickly die at temperatures below 7 °C. It is therefore restricted to tropical rain forests and the subtropics. In East Africa it is ubiquitous (Chunge et al., 1986).

Clinical Symptoms

Nonhuman primates: A gorilla observed by Hamerton (1933 - cit. from McClure and Guilloud, 1971) developed intractable diarrhea for several months, the animal's condition deteriorated and it finally died.

Experimental infection of a chimpanzee with 1000 infective larvae resulted in slight anemia after 1 month of patency (Orihel, 1971).

Man: The symptoms of *N. americanus* infections in man are similar to those described for *A. duodenale*, except for the more common occurrence of cutaneous larva migrans and the more benign course of *N. americanus* infections.

Pathology

The morphological lesions in necatoriasis are essentially very much the same as those reported in ancylostomiasis.

References

1. Barriga, O.O.: Ancylostomiasis; pp. 3-24 in "CRC Handbook Series in Zoonoses" (Steele, J.H., Series ed.); Section C: Parasitic Zoonoses. Vol. II (Schultz, M.G., Vol. ed.), CRC Press, Boca Raton (1982)
2. Belding, D.L.: Textbook of Parasitology, 3rd ed., Appleton-Century-Crofts, New York (1965)
3. Benson, R.E.; B.D. Fremming, and R.D. Young: Care and management of chimpanzees at the Radiobiological Laboratory of the University of Texas and the United States Air Force. Anim. Care Panel *5*, 27-36 (1954)
4. Chandler, A.C.: Hookworm Disease. MacMillan & Co., London (1929)
5. Chunge, R.N.; P.N. Karumba, and E.O. Andala: Hookworm species in patients from Kenyatta National Hospital, Nairobi. Ann. trop. Med. Parasit. *80*, 147-148 (1986)
6. Dollfus, R.P. et A. Chabaud: Cinq espèces de nematodes chez un atele (Ateles ater (G. Cuvier 1823)) mort à la Menagerie du Muséum. Arch. Mus. Hist. Nat. Paris *3* (7), 27-40 (1955)
7. Faust, E.C.; P.C. Beaver, and R.C. Jung: Animal Agents and Vectors of Human Disease. 4th. ed. Lea and Febiger, Philadelphia (1975)

8. Fiennes, R.: Zoonoses of Primates. The Epidemiology and Ecology of Simian Diseases in Relation to Man. Weidenfeld and Nicolson, London (1967)
9. Galliard, H.: Larva migrans; pp.295-304 in "Parasitic Zoonoses. Clinical and Experimental Studies." (E.J.L. Soulsby, ed.), Academic Press, New York (1974)
10. Gedoelst, L.: Notes sur la faune parasitaire du Congo. Belge Rev. Zool. Africaine *5*, 1-90 (1916)
11. Göltenboth, R.: Menschenaffen, Affen, Halbaffen; pp.53-83 in "Zootierkrankheiten" (Klös, H.-G. and E.M. Lang, eds.), Pasey, Berlin (1976)
12. Healy, G.R. and B.J. Myers: Intestinal helminths; pp.265-296 in "The Chimpanzee", Vol.6 (Bourne, G.H., ed.), Karger, Basel (1973)
13. Katz, M.; D.D. Despommier, and R.Gwadz: Parasitic Diseases. Springer, New York (1982)
14. Kuntz, R.E. and B.J. Myers: Parasites of baboons (*Papio doguera*, Pucheran, 1856) captured in Kenya and Tanzania, East Africa. Primates *7*, 27-32 (1966)
15. Kuntz, R.E. and B.J. Myers: Parasitic protozoa, commensals and helminths of chimpanzees imported from the Republic of Congo; pp.184-190 in "Proc. 2nd Internat. Congr. Primatol. Atlanta, Georgia, 1968", Vol.3: Neurology, Physiology and Infectious Diseases (Hofer, H.O., ed.), Karger, Basel (1969)
16. McClure, H.M. and N.B. Guilloud: Comparative pathology of the chimpanzee; pp.103-272 in "The Chimpanzee", Vol.4 (Bourne, G.H., ed.), Karger, Basel (1971)
17. Myers, B.J. and R.E. Kuntz: A checklist of parasites reported for the baboon. Primates *6*, 137-194 (1965)
18. Nelson, G.S.: The parasitic helminths of baboons with particular reference to species transmissible to man; pp.441-469 in "The Baboon in Medical Research", Vol.1 (Vagtborg, H., ed.), Univ. Texas Press, Austin (1965)
19. Orihel, T.C.: The helminth parasites of nonhuman primates and man. Lab. Anim. Care *20*, 395-401 (1970)
20. Orihel, T.C.: *Necator americanus* infection in primates. J.Parasit. *57*, 117-121 (1971)
21. Orihel, T.C. and H.R. Seibold: Nematodes of the bowel and tissues; pp.76-103 in "Pathology of Simian Primates", Part II (Fiennes, R.N.T-W., ed.), Karger, Basel (1972)
22. Paciepnik, O.: Pasozyty jelitowe matp z ogrodu zoologicznege we Wroclawiu. Wiad. Parazyt. *22*, 289-296 (1976)
23. Rewell, R.E.: Diseases of tropical origin in captive wild animals. Trans. roy. Soc. trop. Med. Hyg. *42*, 17-25 (1948)
24. Riopelle, A.J. and O.J. Daumy: Care of chimpanzees for radiation studies; pp.205-227 in "Proceedings of the International Symposium on Bone Marrow Therapy and Chemical Protections in Irradiated Primates" (Balner, H., ed.). TNO, Rijswijk (1962)
25. Ruch, T.C.: Diseases of Laboratory Primates. Saunders, Philadelphia (1959)
26. Thatcher, V.E. and J.A. Porter, jr.: Some helminth parasites of Panamanian primates. Trans. Amer. micr. Soc. *87*, 186-196 (1968)
27. Yamashita, J.: Ecological relationships between parasites and primates. I. Helminth parasites and primates. Primates *4*, 1-96 (1963)
28. Yoshida, Y. and Y. Nakanishi: Experimental and clinical studies on the cortisone treatment of Wakana-disease. Jap. J. Parasit. *7*, 442-448 (1958)
29. Young, R.J.; B.D. Fremming; R.E. Benson, and M.D. Harris: Care and management of a *Macaca mulatta* monkey colony. Proc. Anim. Care Panel *7*, 67-82 (1957)

7.2 Rhabdisoidea

Strongyloididae are unique among nematode parasites in that they have alternate parasitic and free-living generations. Several species of *Strongyloides* occur in nonhuman primates, some of which are transmissible to man, e.g. the human species *S. stercoralis* and the monkey species *S. fülleborni*.

7.2.1 Strongyloides Stercoralis

S. stercoralis naturally infects man, pongids, hylobatids, and occasionally lower primates or dogs. Most infections in nonhuman primates have occurred in chimpanzees (Blacklock and Adler, 1922; Schmid, 1937; Desportes, 1945; Essbach, 1949; Liebegott, 1962; Lang, 1966; van Riper et al., 1966; Orihel and Seibold, 1972; Healy and Myers, 1973; Penner, 1981) and orangutans (Lang, 1966; McClure and Guilloud, 1971; McClure et al., 1973; Uemura et al., 1979; Schaller et al., 1984). Two cases have been reported in gorillas (Penner, 1981), with the infestation being occasionally found in gibbons (Urbain and Nouvel, 1944; Desportes, 1945; de Paoli and Johnson, 1978), captive *Papio ursinus* (Goldsmid, Rogers, 1978), *Erythrocebus patas* (Harper et al., 1982) and dogs (Katz et al., 1982).

Agent

The adult parasitic female is approximately 2 × 0,4 mm in size, with a slowly tapering anterior end and a short, conical posterior extremity (Fig. 35). It is colorless and is semitransparent with a finally striated cuticle. The short buccal capsule has 4 lips and there are four perioral papillae. The esophagus is long, nearly cylindrical and occupies the anterior quarter of the worm (Georgi, 1982). The intestine is composed of 40 uninucleate cells arranged in a dorsal and a ventral row. The vulva in the female is a transverse slit at the junction of the middle and posterior thirds of the body. The vagina is very short, whilst the ovaries are long and straight, and the two uteri are filled by a row of 8 to 12 thinshelled eggs.

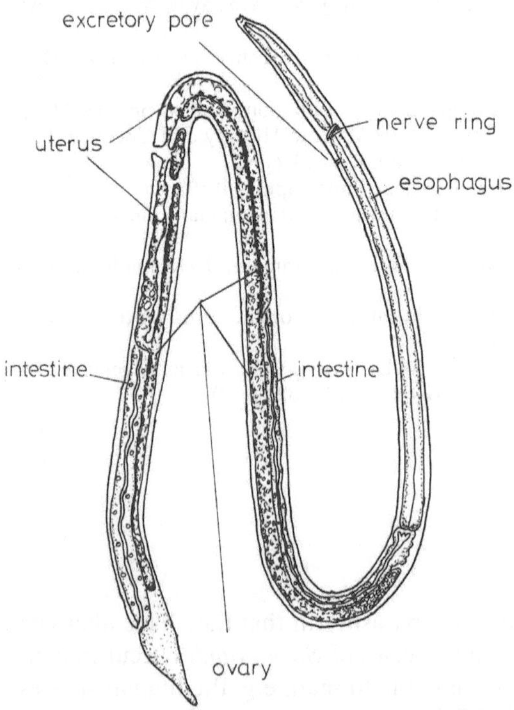

Fig. 35. *Strongyloides stercoralis* (♀, parasitic)

There is no parasitic male strongyloid.

Free living females are somewhat shorter and stouter than the parasitic ones (1.5 mm × 0.085 mm). They possess 2 well-developed lips and 4 papillae. The muscular rhabdidoid esophagus is divided by an isthmus into a corpus and a bulb which is equipped with a well-developed valvular apparatus. The intestine is composed of a straight double-row of 22 uninucleate cells. The reproductive system is didelphic-like in the parasitic females, with the vulva being situated near to the middle of the body and the vagina again is very short.

Adult free-living males are 1.2 × 0.055 mm in size. Their tails are longer and broader than those of the free-living females and are ventrally curved in fixed specimens. The reproductive organs consist of a testis, seminal vesicle, and vas deferens. The spicules are equal, short and ventrally curved at their ends. A gubernaculum is present, but they do not have a bursa.

The eggs of parasitic and free-living females are essentially similar in appearance but those of the parasitic females lack a vitelline membrane (the innermost of three layers of the egg shell), which is an indication of fertilization in eggs from free-living individuals (Georgi, 1982). The *in utero* embryonated eggs hatch immediately after liberation so that only free rhabdidiform larvae are excreted in the host's feces.

The rhabdidiform larvae are 150 to 390 μm long in fresh feces but they grow quickly to 800 μm length (Flynn, 1973). The long esophagus is rhabdidoid, similar to that of the free-living adults. Their intestine consists of 22 uninucleate cells in 2 rows.

The infective filariform larvae are 400 to 800 μm long, slender, and vigorously motile. The esophagus is cylindrical and very long (about 40% of the entire body length). The tail is composed of four small projections of the lateral alae, giving the impression of a bifurcated or trifurcated tail. The intestine is composed of 18 binucleate and 4 uninucleate cells.

Life cycle

The parasitic females produce only genotypically female eggs by mitotic (apomictic) parthenogenesis (Georgi, 1982). Sexual differentiation of the developing rhabdidoid larvae is produced by extrinsic intrahost factors operating during oogenesis and embryogenesis (duration and density of infection, species, age and immunologic status of the host - Georgi, 1982). In *S. stercoralis* the larvae hatch already in the host's large intestines and may be either fecally excreted or may complete the entire cycle including the necessary two molts to the infective larvae in the host's own digestive tract. The filariform larvae in the latter case either penetrate the intestinal mucosa or the perianal skin resulting in autoinfection and, by accumulation, in hyperinfection.

Fecally liberated rhabdidoid larvae may, after a short feeding period, develop directly within 20 to 30 hrs to infectious filariform larvae ready to infect new mammalian hosts, or they may enter an indirect cycle (Beach, 1936).

In the indirect free-living cycle the rhabdidiform larvae develop within 24 to 36 hrs to free-living male and female sexual stages. The free-living adult females produce 60 or more eggs which hatch in the soil within a few hours again to rhabdidiform larvae. This indirect cycle may be repeated in certain *Strongyloides* spe-

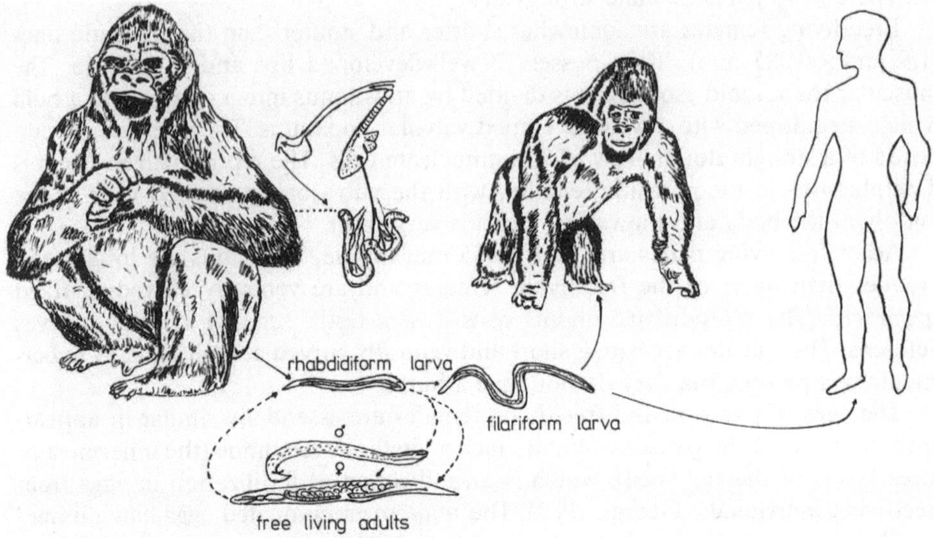

Fig. 36. Life cycle of *Strongyloides stercoralis*

cies e.g. *S. cebus*, or the so generated rhabdidiform larvae develop to infective filariform larvae (Fig. 36).

Both the directly and the indirectly generated filariform larvae can penetrate the skin of a new host within minutes. The majority of the larvae migrate via the venous circulation and the right side of the heart to the lung capillaries. Once there they molt again to postfilariform larvae which penetrate both the capillary and the alveolar walls. Subsequently they move actively up into the larynx and pharynx and are swallowed. On reaching the small intestine the larvae burrow into the mucosa and finally become parasitic female worms within 28 days after infection (Belding, 1965). The prepatent period is 11 to 18 days (Faust et al., 1934).

The migration of the infective larvae after burrowing through the skin may, however, be incomplete. Some larvae obviously stay in a dormant stage for an unknown time in unknown places in the host's body ready to reestablish the infection after successful treatment of the adults living in the intestines. Other larvae may develop to mature worms in the lung tissues leading to chronic bronchial or pulmonary strongyloidosis (Seabury et al., 1971).

Epidemiology

S. stercoralis is cross-transmissible between man and apes. Due to the active penetration of the skin by the larvae percutaneous infection via unprotected skin is more likely to occur than per-oral infection. The extremely rapid larval development within 1 or 2 days to the infective filariform stage permits infections not only from soil, but also from soiled, moist clothing (Neva, 1986).

Careless handling of infected animals or fecal material can lead to infection. Outbreaks of strongyloidosis among animal handlers have occurred (Georgi and

Sprinkle, 1974). Once the infection is established it may persist for decades. Significant numbers of ex-POWs from Southeast Asia were still infected nearly 40 years later (Gill and Bell, 1979; Grove, 1980).

Geographical Distribution

S. stercoralis is found throughout the tropical and subtropical parts of the world and in certain temperate zones. The prevalence in tropical regions may be as high as 85% (WHO Scientific Working Group, 1980).

Pathogenesis

The migrating filariform larvae damage the basement membranes by eliciting collagen-depolymerization (Lewert and Lee, 1954) but do not initiate an as extensive host response as hookworm-larvae do. Larvae burrowing through the pulmonary tissues cause intestinal and alveolar haemorrhages followed by usually minor cellular proliferations, the extensiveness of which depends on the number of larvae. Maturation of the larvae in the lung (chronic pulmonary strongyloidosis), however, is followed by a much more pronounced tissue response, involving a heavy participation by eosinophilic granulocytes (Kyle et al., 1948).

The parasitic females invade the mucosa, but they rarely go below the *muscularis mucosae.* They cause little or no damage to the host as long as they are present in small numbers, but even then they provoke a local inflammatory response which is probably immunologically mediated. The damage to the mucosa is, however, cumulative due to the continuous movements of the females within the intestinal tract. The affected tissues become increasingly nonfunctional. Epithelial sloughing occurs which may extend to produce mucosal damage down to the *muscularis mucosae* and there may be an eosinophilic infiltrate (Faust et al., 1975). In experimentally infected patas monkeys the intestinal mast cells and tissue histamine levels were increased. Steroid treatment resulted in reduction of the mast cell responses and in fatal hyperinfection (Harper et al., 1984; Barrett et al., 1985).

Clinical Symptoms

Nonhuman primates: The penetrating larvae may cause urticaria and toxemia in chimpanzees (Fineg et al., 1961). The symptoms of pulmonary passage are dyspnea (McClure et al., 1973; de Paoli and Johnson, 1978) or pneumonia (Penner, 1981).

The most common and most serious consequence of strongyloidosis in nonhuman primates is acute or persistent-chronic catarrhal to haemorrhagic-mucoid diarrhea (Blacklock and Adler, 1922; Essbach, 1949; de Paoli and Johnson, 1978). Anorexia and vomitus, depression, dehydration, loss of weight or paralytic ileus might be associated with the diarrhea, but in orangutans these symptoms are known to occur without diarrhea being apparent (McClure et al., 1973; Schaller et al., 1984). Haematological signs include leukocytosis, but eosinophilia is only rarely seen (de Paoli and Johnson, 1978).

Man: Penetration of the skin by the infective larvae is associated with a sharp, burning or picking sensation, and within 30 sec. the first inflammatory signs are evident. These lesions may be located on the feet, or, in autoinfections, in the perianal skin. In external autoinfections with invasion of the perianal skin a characteristic form of *larva migrans* develops. This starts in the perianal area and then moves out rapidly over the thighs and abdomen. The advancing point of the usually serpiginous lesion, which contains the larva, may move as fast as 10 cm/hr.

During the migratory phase the larvae may cause minor pulmonary symptoms that are associated with fever, dry cough, asthma-like symptoms, dyspnea or urticaria. This phase may be associated with a circulating eosinophilia, which, at the beginning of the intestinal phase comprises 50 to 75% of the entire leukocyte counts (Faust et al., 1975; Katz et al., 1982).

The intestinal symptoms may be apparent as a watery, mucous diarrhea (Katz et al., 1982) or they may closely resemble those of peptic ulcer (Faust et al., 1975). Prodromal symptoms are hunger pains, lassitude and abdominal pain, which may be diffuse or rather sharply localized to the right upper quadrant. Other symptoms are nausea, vomitus, dyspepsia, flatulence, fatigue and weight loss. In severe infections, a malabsorption syndrome with loss of large amounts of body fat and protein may be present (Mathies, 1981).

Autoinvasion is often fatal, as it may produce paralytic ileus, hepatitis, cholecystitis, myocarditis, pneumonia or tracheitis, with symptoms of abdominal pain, incurable watery diarrhea alternating with obstipation, dyspnea, cough, haemoptisis and low grade fever.

Pathology

Nonhuman primates: The most common gross lesions in gibbons and orangutans are multifocal or diffuse pulmonary haemorrhages and acute catarrhal to haemorrhagic necrotizing enterocolitis (McClure et al., 1973; de Paoli and Johnsen, 1978; Uemura et al., 1979; Harper et al., 1982). The trachea and bronchioles frequently contain blood-tinged, frothy mucous material. In chimpanzees the lung lesions consist of edema, and only occasionally of haemorrhages or even pneumonia (Schmid, 1937).

The intestinal lesions in the gibbons could generally be divided into those associated with adults or rhabdidiform larvae restricted to the mucosa, and those associated with the invading filariform larvae in primary or hyperinfections.

The lesions of the first group are characterized by multifocal erosive and ulcerative enteritis, with the parasites lying in interepithelial parasitic tunnels or in the lumina of intestinal glands. The villi are shortened and blunted, whilst eosinophilic infiltration is generally rare (de Paoli and Johnsen, 1978).

Lesions of the second group with the laryae being distributed throughout the entire intestinal wall and other organs (occasionally even the CNS), elicited host responses which varied from mild inflammatory cellular accumulations to multifocal necrotizing-ulcerative or granulomatous enteritis, possibly associated with granulomatous endolymphangitis in the small and large intestines (Schmid, 1937; Essbach, 1949; de Paoli and Johnsen, 1978).

The lesions in auto- or hyperinfections may be even more pronounced in the large intestine (Uemura et al., 1979).

The presence of larvae in the lymphatic system or in other internal organs, especially the liver, indicates an autoinfection (McClure et al., 1973).

Man: In man the cutaneous lesions consist of local erythema, whilst those of autoinvasions are typified by perianal dermatitis and urticaria. Fatal cases had marked pulmonary haemorrhages, and a dense parasitism of duodenal and jejunal mucosal crypts with only minor inflammatory responses around disintegrating larvae (Andrade and Gomez, 1964). In hyperinfections the larvae may be found throughout the lymph nodes, liver, lung and CNS; the latter responding with an inflammatory reaction (Matthies, 1981). Chronic pulmonary strongyloidosis results in lobular, chronic, eosinophilic bronchopneumonia (Kyle et al., 1948).

Treatment

Adult *S. stercoralis* can be effectively treated using thiabendazole. In the nonhuman primates dosages of 50 to 100 mg/kg/day for several days are recommended (Göltenboth and Klöppel, 1964; Cullum and Hamilton, 1965; Loew, 1968; Göltenboth und Klös, 1970; Vickers, 1971; Remfry, 1978; Karr et al., 1980; Philipps, 1980). Thiabendazole given orally does not affect the migrating or resting larvae, therefore the treatment has to be repeated twice at intervals of 2 to 4 weeks.

Other benzimidazoles are less effective (WHO Scientific Working Group, 1980).

In man nearly 100% cure rates were achieved using thiabendazole (Katz et al., 1982). The drug is given in dosages of 50 mg/kg over 2 or 3 successive days with repetition after 1 or 2 weeks (Faust et al., 1975). Possible side effects are nausea or vomiting (Mathies, 1981).

7.2.2 Strongyloides Fülleborni

S. fülleborni is a natural parasite of Asian and African monkeys. It has been found in *Macaca* sp., *M. mulatta, M. fascicularis, M. fuscata, M. arctoides, Papio* sp., *Cercopithecus* sp. and *Presbytis cristatus* (Habermann and Williams, 1957; Young et al., 1957; Ruch, 1959; Sasa et al., 1962; Guilloud et al., 1965; Rowland and Vandenberg, 1965; Hashimoto and Honjo, 1966; Reardon and Rininger, 1968; Ford and Speltie, 1972; Owen and Casillo, 1973; Arambulo et al., 1974; Kagei and Hasegawa, 1974; Brede et al., 1977; Goldsmid and Rogers, 1978; Wong and Conrad, 1978; Prosl and Tamer, 1979; Karr et al., 1980); chimpanzees (van Riper et al., 1966; File et al., 1976) and man (Buckley, 1946; Wallace et al., 1948; Brannon and Faust, 1949; Elsdon-Dew and Freedman, 1952; Goldsmid, 1968; Pampiglione and Ricciardi, 1971; Hira and Patel, 1977, 1980). In fact, almost all 10 cases of human strongyloidosis reported in Zambia were caused by *S. fülleborni* (Hira and Patel, 1977).

Agent

S. fülleborni, although grossly resembling *S. stercoralis* is distinguished from the latter by the absence of perioral lappets and especially by the shape of the genitalia in free-living females. In these the ovaries are spirally wound around the intestinal tract, whereas in *S. stercoralis* they are straight. Their anterior vulval lip overhangs the aperture and their body is markedly constricted just below the vulva (Hira and Patel, 1977).

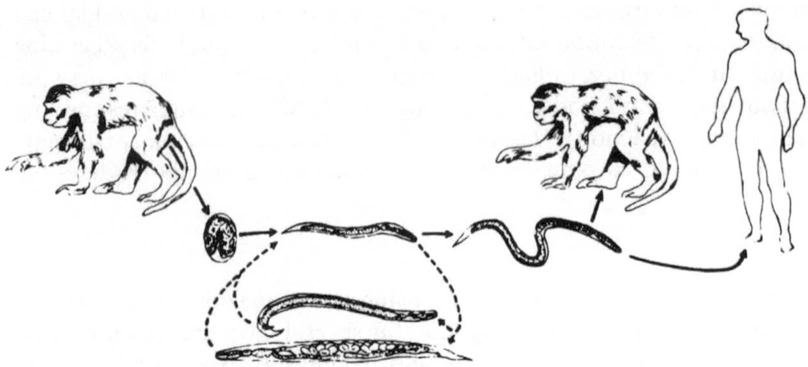

Fig. 37. Life cycle of *Strongyloides fülleborni*

The embryonated eggs do not hatch inside the host's intestinal tract. In freshly examined feces the thinshelled oval eggs contain the hooklike larvae, whilst in older feces the larvae may be hatched.

Life Cycle and Epidemiology

The life cycle of *S. fülleborni* resembles that of *S. stercoralis*, but since the rhabdidiform larvae do not hatch within the host, there is no internal cycle leading to auto- and hyperinfections.

S. fülleborni differs from *S. stercoralis* also by its tendency to pass repeated free-living cycles.

S. fülleborni can be transmitted to man from monkeys via the free-living larval stages (Fig. 37), and possibly also according to Hira and Patel (1980) from man to man.

Geographical Distribution

S. fülleborni is restricted to tropical and subtropical areas of Asia and Africa and does not naturally occur in the New World. Human infections so far have been limited to the rain forests in Zambia and Rhodesia (Hira and Patel, 1980).

Clinical Symptoms

Nonhuman primates: Heavy infections are often associated with diarrhea similar to that caused by *S. stercoralis*, but usually it runs a milder course and rarely causes the animal's death.

Man: The pathogenicity of spontaneous *S. fülleborni* in man is somewhat clouded by diagnostic problems and by the fact that it occurs in tropical countries where a multiplicity of parasitic infections are present. In experimental infections abdominal pain, diarrhea and anemia have been recorded, but again concurrent infections with other helminths or protozoans might have contributed to the clinical signs (Hira and Patel, 1977).

Pathology

The morphological lesions are essentially similar to those reported in *S. stercoralis* but are of a lesser severity and extension as autoinfection does not occur.

Treatment

The treatment of *S. fülleborni* is the same as that of *S. stercoralis.*

References

1. Andrade, Z.A. and M.C. Gomez: Pathology of fatal strongyloidiasis. Rev. Inst. Med. trop. S. Paulo *6*, 28-34 (1964)
2. Arambulo, P.V. III; J.B. Abass, and J.S. Walker: Silvered leaf monkeys *(Presbytis cristatus).* II. Gastrointestinal parasites and their treatment. Lab. Anim. Sci. *24*, 299-306 (1974)
3. Barrett, K.E.; F.A. Neva; A.A. Gam; J. Cicmanec; W.T. London, and D.D. Metcalfe: Modulation and function of mucosal mast cells in parasite infections of monkeys. Clin. Res. *33*, 160 A (1985)
4. Beach, T.D.: Experimental studies on human and primate species of Strongyloides. V. The free-living phase of the life cycle. Amer. J. Hyg. *23*, 243-277 (1936)
5. Belding, D.L.: Textbook of Parasitology. 3rd ed., Appleton-Century-Crofts, New York (1965)
6. Blacklock, B. and S. Adler: The pathologic effects produced by Strongyloides in a chimpanzee. Ann. trop. Med. Parasit. *16*, 99-106 (1922)
7. Brannon, M.J.C. and E.C. Faust: Preparation and testing of a specific antigen for diagnosis of human strongyloidiasis. Amer. J. trop. Med. Hyg. *29*, 229-239 (1949)
8. Brede, H.D.; L. Heber, and A. Steinbach: Comparative bacteriological survey of imported experimental monkeys. Arb. Paul-Ehrlich-Inst. *71*, 113-117 (1977)
9. Buckley, J.J.C.: A helminthological survey of Northern Rhodesia. J. Helminth. *21*, 111-174 (1946)
10. Cullum, J.E. and B.R. Hamilton: Thiabendazole as an anthelminthic in research monkeys. Amer. J. vet. Res. *26*, 779-780 (1965)
11. DePaoli, A. and D.O. Johnsen: Fatal strongyloidiasis in gibbons *(Hylobates lar).* Vet. Path. *15*, 31-39 (1978)
12. Desportes, C.: Sur *Strongyloides stercoralis* (Bavay, 1876) et sur les *Strongyloides* des primates. Ann. Parasit. hum. comp. *20*, 160-190 (1945)
13. Elsdon-Dew, R. and L. Freedman: Intestinal parasites in the Natal Bantu. S. Afr. J. Clin. Sci. *3*, 59-65 (1952)
14. Essbach, H.: Strongyloidose beim Schimpansen. Beitr. Path. Anat. *110*, 319-345 (1949)
15. Faust, E.C.; J.W. Wells; C. Adams, and T.D. Beach: Experimental studies in human and primate species of *Strongyloides.* III. The fecundity of Strongyloides females of the parasitic generation. Arch. Path. *18*, 605-625 (1934)
16. File, S.K.; W.C. McGrew, and C.E.G. Tutin: The intestinal parasites of a community of feral chimpanzees, *Pan troglodytes Schweinfurthi.* J. Parasit. *62*, 259-261 (1976)
17. Fineg, J.; W.E. Britz, jr.; J.E. Cook, and R.H. Edwards: Clinical observations and methods used in the treatment of young chimpanzees. Air Force Missile Development Center, Holloman Air Force Base/New Mexico, Report AFMDC-TR-61-R (1961)
18. Flynn, R.J.: Parasites of Laboratory Animals. Iowa State Univ. Press, Ames/Iowa (1973)
19. Ford, A.C. and T.M. Speltie: Some aspects of the parasitology of non-human primates. Ann. Rep. TNO 1972, 160-162 (1972)
20. Georgi, J.R. and C.L. Sprinkle: A case of human strongyloidosis apparently contracted from asymptomatic colony dogs. Amer. J. trop. Med. Hyg. *23*, 899-901 (1974)
21. Georgi, J.R.: Strongyloidiasis; pp. 257-267 in "CRC Handbook Series in Zoonoses" (Steele, J.H., Series ed.), Section C: Parasitic Zoonoses, Vol. II (Schultz, M.G., Vol. ed.), CRC Press, Boca Raton (1982)
22. Gill, G.V. and D.R. Bell: *Strongyloides stercoralis* infection in former Far East prisoners of war. Brit. Med. J. *2*, 572-574 (1979)
23. Goldsmid, J.M.: Studies on intestinal helminths in African patients at Harari Central Hospital, Rhodesia. Trans. roy. Soc. trop. Med. Hyg. *62*, 619-629 (1968)
24. Goldsmid, J.M. and S. Rogers: Parasitological study on chacma baboon *(Papio ursinus)* from northern Transvaal. J. S. Afr. vet. Ass. *49*, 109-111 (1978)

25. Göltenboth, R. and G. Klöppel: Die Bekämpfung der Endoparasiten bei Zootieren mit Thiabendazole (2(4-Thiazolyl)Benzimidazol). Kleintierpraxis *9*, 173-176 (1964)
26. Göltenboth, R. and H.G. Klös: Klinische Erfahrungen bei der Überwachung und Bekämpfung der Endoparasiten der Zootiere im Zoologischen Garten Berlin. Zool. Garten *39*, 91-100 (1970)
27. Grove, D.I.: Strongyloidiasis in allied ex-prisoners of war in Southeast Asia. Brit. Med. J. *1*, 598-601 (1980)
28. Guilloud, N.B.; A.A. King, and A. Lock: A study of the efficacy of thiabendazole and dithiazanine iodidepiperazine citrate suspensions against intestinal parasites in the *Macaca mulatta.* Lab. Anim. Care *15*, 354-358 (1965)
29. Habermann, R.T. and F.P. Williams, jr.: Diseases seen at necropsy of 708 *Macaca mulatta* (rhesus monkey) and *Macaca philippinensis* (cynomolgus monkey). Amer. J. vet. Res. *18*, 419-426 (1957)
30. Harper, J.S. III; J.M. Rice; W.T. London; D.L. Sly, and C. Middleton: Disseminated strongyloidiasis in *Erythrocebus patas.* Amer. J. Primatol. *3*, 89-98 (1982)
31. Hashimoto, I. and S. Honjo: Survey of helminth parasites in cynomolgus monkeys *(Macaca irus).* Jap. J. med. Sci. Biol. *19*, 218 (1966)
32. Healy, G.R. and B.J. Myers: Intestinal helminths; pp. 265-296 in "The Chimpanzee", Vol. VI (Bourne, G.H., ed.), Karger, Basel (1973)
33. Hira, P.R. and B.G. Patel: *Strongyloides fülleborni* infections in man in Zambia. Amer. J. trop. Med. Hyg. *26*, 640-643 (1977)
34. Hira, P.R. and B.G. Patel: Human strongyloidiasis due to the primate species *Strongyloides fülleborni.* Trop. geogr. Med. *32*, 23-29 (1980)
35. Kagei, N. and H. Hasegawa: On the fecal examination against helminthic ova of the Japanese monkeys, *Macaca fuscata* (Blyth, 1875), from Takasaki-yama, Oita prefecture - problems as parasitic zoonoses. Bull. Inst. Publ. Hlth *23*, 234-238 (1974)
36. Karr, S.L., jr.; R.V. Henrickson, and J.G. Else: A survey for intestinal helminths in recently wild-caught *Macaca mulatta* and results of treatment with mebendazole and thiabendazole. J. med. Primatol. *9*, 200-204 (1980)
37. Katz, M.; D.D. Despommier, and R. Gwadz: Parasitic Diseases. Springer, New York (1982)
38. Kyle, L.H.; D.G. McKay, and H.J. Sparling, jr.: Strongyloidiasis. Ann. intern. Med. *29*, 1014-1042 (1948)
39. Lang, E.M.: The care and breeding of anthropoids. Symp. Zool. Soc. London *17*, 113-125 (1966)
40. Lewert, R.M. and C.L. Lee: Studies on the passage of helminth larvae through host tissues. I. Histochemical studies on the extracellular changes caused by penetrating larvae. II. Enzymatic activity of larvae *in vitro* and *in vivo.* J. Infect. Dis. *95*, 13-51 (1954)
41. Liebegott, G.: Pericarditis verminosa (strongyloides) beim Schimpansen. Virchows Arch., A (Path. Anat.) *335*, 211-225 (1962)
42. Loew, F.F.: A review of some helminths of laboratory animals. Cornell Vet. *58*, 408-421 (1968)
43. Mathies, A.W. jr.: Intestinal roundworms; pp. 387-391 in "Infectious Diseases" (Sanford, J.P. and J.P. Luby, eds.), Grune and Stratton, New York (1981)
44. McClure, H.M. and N.B. Guilloud: Comparative pathology of the chimpanzee; pp. 103-272 in "The Chimpanzee", Vol. 4 (Bourne, G.H., ed.), Karger, Basel (1971)
45. McClure, H.M.; L.M. Strozier; M.E. Keeling, and G.R. Healy: Strongyloidiasis in two infant orangutans. J. Amer. vet. med. Ass. *163*, 629-632 (1973)
46. Neva, F.A.: Biology and immunology of human strogyloidiasis. J. Infect. Dis. *153*, 397-406 (1986)
47. Orihel, T.C. and H.R. Seibold: Nematodes of the bowel and tissues; pp. 76-103 in "Pathology of Simian Primates", Vol. II (Fiennes, R.N.T-W., ed.), Karger, Basel (1973)
48. Owen, D. and S. Casillo: A preliminary survey of the nematode parasites of some imported Old-World monkeys. Lab. Anim. *7*, 265-269 (1973)
49. Pampiglione, S. and M.L. Ricciardi: The presence of *Strongyloides fülleborni* von Linstow, 1905 in man in Central and East Africa. Parasitologia *13*, 257-268 (1971)
50. Penner, L.R.: Concerning threadworm *(Strongyloides stercoralis)* in great apes - lowland gorillas *(Gorilla gorilla)* and chimpanzees *(Pan troglodytes).* J. Zoo Anim. Med. *12*, 128-131 (1981)

51. Phillips, L.G. jr.: The parasitology of primates and carnivores. Amer. Ass. Zoo Vet. Ann. Proc. 1980, 116-119 (1980)
52. Prosl, H. and A. Tamer: Zur Parasitenfauna des Rhesusaffen (Macaca mulatta) und des Javaneraffen (Macaca irus). Zbl. Vet.-Med. B *26*, 696-709 (1979)
53. Reardon, L.V. and B.F. Rininger: A survey of parasites in laboratory primates. Lab. Anim. Care *18*, 577-580 (1968)
54. Remfry, J.: The incidence, pathogenesis and treatment of helminth infections in rhesus monkeys *(Macaca mulatta)*. Lab. Anim. *12*, 213-218 (1978)
55. Rowland, E. and J.G. Vandenberg: A survey of intestinal parasites in a new colony of rhesus monkeys. J. Parasit. *51*, 294-295 (1965)
56. Ruch, T.C.: Diseases of Laboratory Primates. W.B. Saunders, Philadelphia (1959)
57. Sasa, M.; H. Tanaka; M. Fukui, and A. Takata: Internal parasites of laboratory animals; pp. 195-214 in "The Problems of Laboratory Animal Disease" (Harris, R.J.C., ed.), Academic Press, New York (1962)
58. Schaller, K.; R. Göltenboth und B. Uphoff: Erfolgreiche Behandlung von Strongyloidose mit schwerem Verlauf bei einem weiblichen erwachsenen Orang-Utan *(Pongo pygmaeus abeli)*. Zeitschr. Kölner Zoo *27*, 151-156 (1984)
59. Schmid, F.: Parasiten als Todesursache bei Zootieren. Dtsch. tierärztl. Wschr. *45*, 797-798 (1937)
60. Seabury, J.H.; S. Abadie, and F. Savoy, jr.: Pulmonary strongyloidiasis with lung abscess. Ineffectiveness of thiabendazole therapy. Amer. J. trop. Med. Hyg. *20*, 209-211 (1971)
61. Uemura, E.; W.D. Houser, and C.J. Cupp: Strongyloidiasis in an infant orangutan *(Pongo pygmaeus)*. J. med. Primatol. *8*, 282-288 (1979)
62. Urbain, A. et J. Nouvel: Petite enzootie de strongyloidose observée sur des singes supérieurs: Gibbons à favoris blancs (*Hylobates concolor Leucogenis* Ogilby) et chimpanzés *(Pan troglodytes)*. Bull. Acad. Vet. France *17*, 337-341 (1944)
63. van Riper, D.C.; P.W. Day; J. Fineg, and J.R. Prine: Intestinal parasites of recently imported chimpanzees. Lab. Anim. Care *16*, 360-363 (1966)
64. Vickers, J.H.: Gastrointestinal diseases of primates. Curr. vet. Ther. *4*, 408-411 (1971)
65. Wallace, F.G.; R.D. Mooney, and A. Sanders: *Strongyloides fülleborni* infection in man. Amer. J. trop. Med. Hyg. *28*, 299-302 (1948)
66. WHO Scientific Working Group: Parasite-related diarrheas. Bull. Wld. Hlth Org. *58*, 819-830 (1980)
67. Wong, M.M. and H.D. Conrad: Prevalence of metazoan parasite infections in five species of Asian macaques. Lab. Anim. Sci. *28*, 412-416 (1978)
68. Young, R.J.; B.D. Fremming; R.E. Benson, and M.D. Harris: Care and management of a *Macaca mulatta* monkey colony. Proc. Anim. Care Panel *7*, 67-82 (1957)

7.3 Oesophagostomidae

The oesophagostomidae or nodular worms are among the most commonly found parasites in wild caught Old World nonhuman primates. The rare infections seen in New World monkeys, were most probably contracted in captivity after contact with infected Old World monkeys.

Oesophagostomidae are of only limited host specificity - even those of pigs, sheep, cattle or antilopes may be transmissible to man (Goldsmid, 1974).

The taxonomy of the oesophagostomids is plagued not only by the use of several synonyms, but also by the impossibility to classify larvae found in histological slides beyond the genus level.

7.3.1 Oesophagostomum Apiostomum

O. apiostomum, sometimes considered synonymous with *O. aculeatum* or *O. bifurcum* (Neveu-Lamaire, 1936; Sasa et al., 1962; Levine, 1968; Faust et al., 1975) is primarily a parasite of several Asian macaques (Habermann and Williams, 1957; Ruch, 1959; Tanaka et al., 1962; Levine, 1968; Reardon and Rininger, 1968; Vickers, 1969; Janagi, 1981). A few cases have been reported in *Presbytis cristatus,* which is also an Asian primate species (Arambulo et al., 1974); in chimpanzees (Hamerton, 1937) and man (Chabaud and Lariviere, 1958; Goldsmid, 1974).

Agent
The adult oesophagostomids look very much like hookworms except for their basically cylindrical or conical buccal capsule (in hookworms this is globose), which is armed by a crown of bristles or setae *(corona radiata)* instead of the teeth or cutting plates found in the hookworm.

In *O. apiostomum,* the buccal capsule is cylindrical and the *corona radiata* consists of 10 bristles or setae (Faust et al., 1975). In the males the bursa is symmetrical, whilst the spicules are long and equal. The females have two uterine branches extending forward from the vulva and ovijector (Faust et al., 1975) (Fig. 38).

The eggs of most oesophagostomids can also be easily confused with hookworm eggs. Those of *O. apiostomum* are of 60 to 63 × 27 to 40 μm size, and in fresh feces are in the early stages of cleavage.

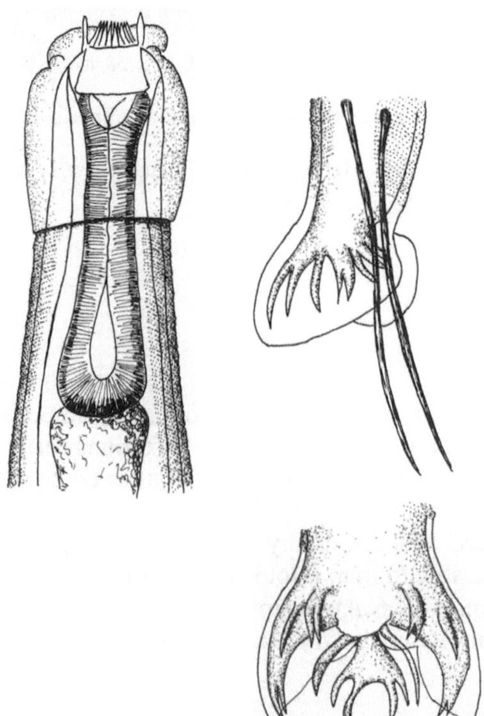

Fig. 38. *Oesophagostomum apiostomum*

Life Cycle

The eggs excreted in the host's feces develop quickly to the first larval stage and may hatch within 24 to 48 hrs. After two molts the sheathed, infective larvae develop, which after being swallowed by a host burrow into the submucosa of, primarily, the large intestine. The first two larval stages are very sensitive to drying and die rather quickly, in contrast the third, sheathed larva, which is unable to feed, may live for several months under favorable conditions. The third larvae after burrowing into the intestinal submucosa molt again to become unsheathed fourth larvae. These after approximately 5 to 8 days return to the intestinal lumen and, after one more molt become adult male and female worms (Healy and Myers, 1973).

Geographical Distribution

O. apiostomum is probably indigenous to India, China and the Philippines.

Pathogenesis

The first infection by oesophagostomids produces normally only small, fast healing lesions, whereas in preimmunized hosts the return of the fourth larval stage into the intestinal lumen is quite often blocked, the larvae being trapped in the tissues and later on die. Secondary bacterial infection of the migratory tracts and hyperergic responses lead to rapid necrosis and inflammation of the surrounding tissues (Elek and Durie, 1967). In primary and repeated infections the ecdysis of the fourth larvae is accompanied by transient intestinal haemorrhages (Bremner and Keith, 1970). At least in oesophagostomiasis of domestic animals these haemorrhages may result in marked anemia, with the concurrent leakage of protein from the mucosa into the intestinal lumen producing hypoproteinemia.

Clinical Symptoms

Nonhuman primates: Minor infections are normally not associated with noticeable clinical abnormalities. Heavy infections or concurrent overwhelming disease induce diarrhea, weight loss, weakness and high mortality.

Man: The most frequent symptoms are tumorlike, associated with abdominal pains and tenderness. The latter is often localized in the right lower quadrant and an abdominal mass may be palpable. The intestinal nodules may cause intussusception (Marcus, 1982).

Pathology

Nonhuman primates: The trapped fourth stage larvae produce semi-globose to lentil-shaped nodules of 2 to 4 mm diameter in the submucosal layer particularly of the large intestines. They have mostly greenish-grey to black, although sometimes pearl-like, color. The nodules usually project over the serosal surface, but the serosa itself usually remains smooth unless the nodule has perforated and caused peritonitis. On the cut surface the nodules appear as cysts filled by a brown, creamy material, which sometimes contains yellowish caseous material or foci of calcification. The nodules are connected to the intestinal lumen by a small ulcer (Habermann and Williams, 1958). The nodules are located in the *propria mucosa, muscularis mucosae* and submucosa in histological slides. Their centres comprise coagulated blood around the larva with many Charcot-Leyden crystals (McConnell et al., 1974). The amorphous centers are surrounded by a zone composed of macrophages, foreign body type giant cells, eosinophilic granulocytes, lymphocytes and monocytes and finally by a fibrous capsule. Secondary bacterial infections may alter the structure of the lesion to a more purulent type and may also cause the perforation of the nodule and subsequent purulent peritonitis (Orihel, 1970).

Man: The lesions in human oesophagostomiasis are described in *O. bifurcum*,, since most human infections occurred in Africa.

Treatment

The drug of choice is thiabendazole (2 × 50–100 mg/kg in *M. mulatta*) - (Arambulo et al., 1974; Remfry, 1978). Less effective are mebendazole and levamisole (Remfry, 1978).

7.3.2 Oesophagostomum Aculeatum

O. aculeatum is another Asian species (Faust et al., 1975) which is considered to be synonymous with *O. apiostomum* by Faust et al. (1975). Its normal hosts are Asian macaques (Ruch, 1959; Tanaka et al., 1962; Hashimoto and Honjo, 1966; Reardon and Rininger, 1968; Kagei and Hasegawa, 1974; Horii et al., 1982). A single human infection has been reported in Indonesia (Lie, 1949; Tan and Lie, 1953 - both cit. from Levine, 1968).

Agent

O. aculeatum differs from *O. bifurcum* by the length of its spicules and by the chitinous structure of its buccal capsule, which has the appearance of a conical stump in *O. aculeatum* (Prosl and Tamer, 1979).

The external leaf crown consists of 10 sharply pointed bristles alternating with 10 pairs of minute spines. The males are 8 to 10 × 0.3 to 0.35 mm. Their spicules are longer than 1,15 mm (Levine, 1968).

7.3.3 Oesophagostomum Bifurcum

O. bifurcum is a nodular worm primarily of African monkeys and apes, and it is also the oesophagostomid most frequently identified in man. It has been reported in several species of baboons (Myers and Kuntz, 1965; Kuntz and Moore, 1973; McConnell et al., 1974); *Theropithecus gelada* (Kuntz et al., 1967); *Erythrocebus patas, Cercopithecus* spp., *Cercocebus* spp. (Levine, 1968); chimpanzees (Healy and Myers, 1973), and Asian macaques (Bezubik and Furmaga, 1959; Bingham and Rabstein, 1964; Guilloud et al., 1965; Kuntz et al., 1968).

O. bifurcum has been recovered from several human patients in Africa (Haaf and van Soest, 1964).

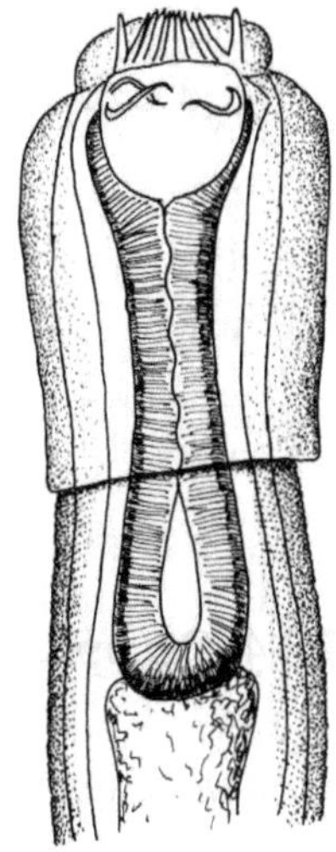

Fig. 39. *Oesophagostomum bifurcum*

Agent

O. bifurcum resembles *O. aculeatum*, but its oral capsule has the shape of a three-fold festoon with long, hooklike teeth (Fig. 39) rather than that of a stump conus, and in the males the spicules are less than 1.15 mm long (Levine, 1968). The males are 11.7 to 15.5 × 0.32 to 0.46 mm in size and have a gubernaculum measuring 0.13 to 0.14 × 0.025 to 0.03 mm. The dimensions of the females are 12.9 to 18.7 × 0.46 to 0.67 mm. The anus is located 0.17 to 0.24 mm before the posterior end (Bezubik and Furmaga, 1959).

The eggs are of 48 to 72 × 27 to 55 μm size and usually contain larvae (Kuntz et al., 1968; Healy and Myers, 1973).

Geographical Distribution

There is some diversity of opinion about the normal occurrence of *O. bifurcum*. Faust et al. (1975) claim that it is an African species, whilst Prosl and Tamer (1979) found it only in rhesus monkeys. It has been identified in man in Africa (Chabaud and Lariviere, 1958; Haaf and van Soest, 1964).

Clinical Symptoms and Pathology

Nonhuman primates: The symptoms are identical to those described for *A. apiostomum*.

Man: The symptoms resembled those of amebomas, appendicitis, colonic carcinoma or ileocaecal tuberculosis (Haaf and van Soest, 1964; Anthony and McAdam, 1972). The parasites are often found in ectopic sites (Chabaud and Lariviere, 1958).

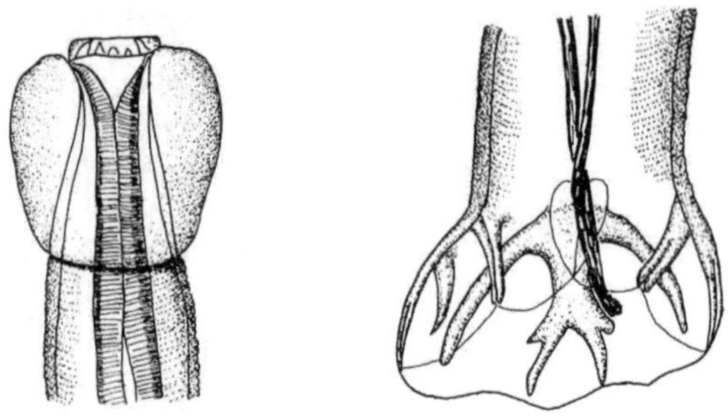

Fig. 40. *Oesophagostomum stephanostomum*

7.3.4 Oesophagostomum Stephanostomum

O. stephanostomum occurs in baboons, chimpanzees, wild living gorillas and man (Rousselot et Pelissier, 1952; Vuylestke, 1956; Sabater-Pi, 1963 - cit. from Fiennes, 1967; van den Berghe et al., 1964; Myers and Kuntz, 1965; Levine, 1968; Crestian et Crespeau, 1975). Human infections have been diagnosed in Manaos/Brazil (Levine, 1968) and in Africa (Chabaud and Lariviere, 1958).

Agent

O. stephanostomum has a prominent cephalic vesicle of 0.375 mm length and 0.42 mm diameter (Fig. 40) and a leaf-crown composed of 38 leaflike spines. The male bursa is formed by two lateral lobes united by a small median lobe (Fig. 40). The spicules measure 1.45 mm in length (Crestian and Crespeau, 1975).

Clinical Symptoms and Pathology

Nonhuman primates: *O. stephanostomum* seems to be the most hazardous nodular worm species in the great apes. It has been considered to be responsible for death in captive gorillas (Rousselot et Pelissier, 1952; Crestian et Crespeau, 1975). The animals suffered from chronic diarrhea and anemia or died quickly after a short period of diarrhea, anorexia and lethargy (Rousselot et Pelissier, 1975).

At necropsy anemia, cachexia, abdominal lesions, hydrothorax, hydropericardium and diphtheroid enteritis were noted (Crestian and Crespeau, 1975).

7.3.5 Ternidens Deminutus

T. deminutus is a close relative of both the oesophagostomids and the hookworms. It occurs in the large intestine of macaques (Bezubik and Furmaga, 1959; Kuntz et al., 1968; Owen and Casillo, 1973); African green monkeys (Levine, 1968); *Papio* spp. (Owen and Casillo, 1973; Goldsmid, 1974; McConnell et al., 1974; Goldsmid and Rogers, 1978); chimpanzees (Schmid, 1937) and gorillas (Leiper, 1908). It has also been found in man (Amberson and Schwarz, 1952; Goldsmid, 1967, 1974; Levine, 1968; Kilala, 1971). According to Goldsmid (1974) it is the most important zoonotic helminthic species in Zimbabwe.

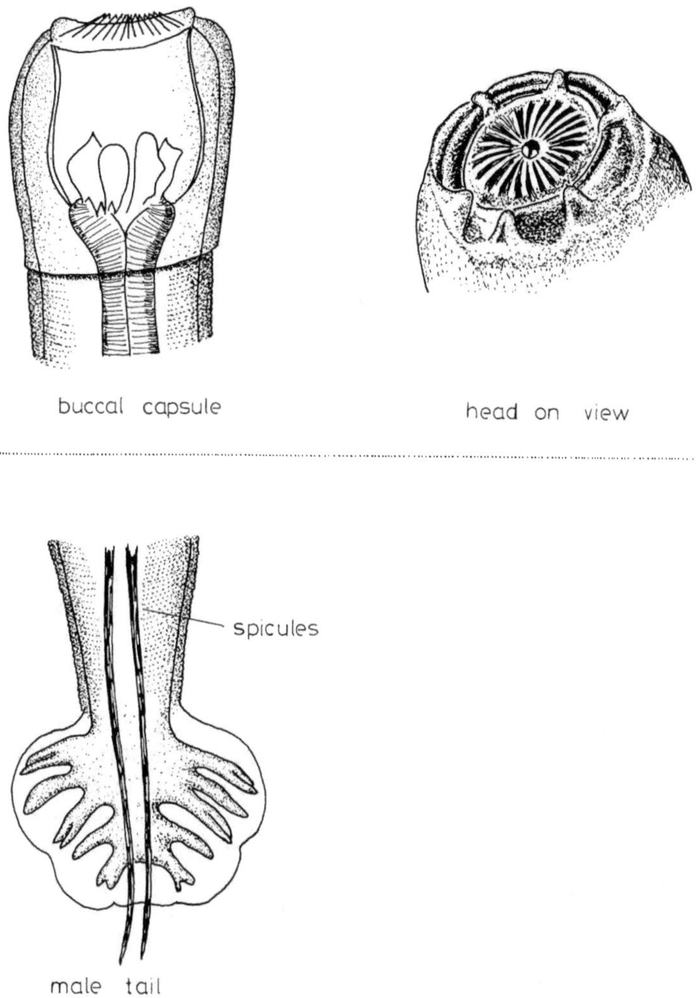

Fig. 41. *Ternidens deminutus*

Agent

T. deminutus are morphologically not only similar to hookworms in size and shape, but also to *Oesophagostomum* spp. particularly by the presence of a terminal subglobular buccal capsule armed by two leaf crowns *(corona radiata)* and with forked teeth in its depth (Goldsmid, 1967; Belding, 1965; Orihel and Seibold, 1972; Faust et al., 1975) - (Fig. 41).

Adult males have a size of 8 to 10 × 0.5 mm, their slender spicules are uniform with a length of 1.16 to 1.22 mm. The gubernaculum measures 0.11 to 0.13 × 0.036 to 0.04 mm.

Female worms are 12 to 15 × 0.6 mm in size (Healy and Myers, 1973). The distance between vulva and posterior end is 0.79 to 0.98 mm (Bezubik and Furmaga, 1959).

The eggs are also like those of hookworms but are somewhat larger, averaging 84 × 51 μm.

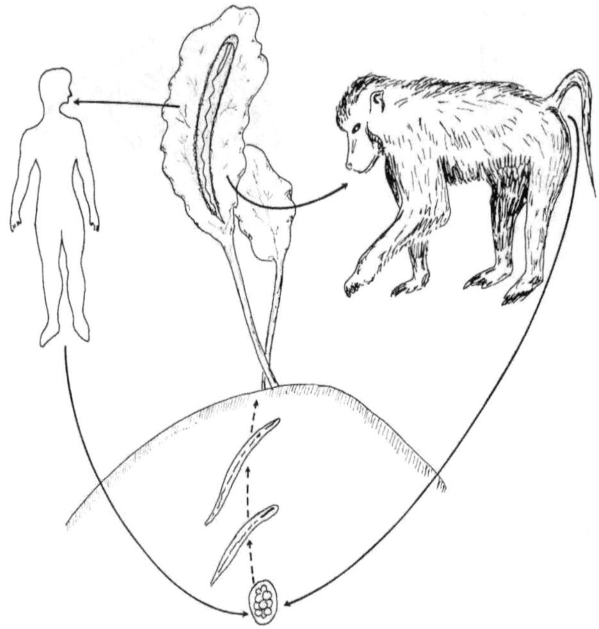

Fig. 42. Life cycle of *Ternidens deminutus*

Life Cycle

Within 2 to 3 days after excretion by the host in the faeces rhabdidiform larvae hatch from the eggs. These, after a first ecdysis after 24 hrs and a second ecdysis after 8 to 10 days become infective larvae, which are semirhabdidoid rather than filariform (Amberson and Schwarz, 1952; Faust et al., 1975). The infective third stage larvae, which are unable to penetrate the skin, move vertically up and down on available surfaces until they are finally ingested by a new host. From the intestinal lumen they burrow into the submucosa to undergo another molt. The fourth stage larvae break through to the intestinal lumen for a final molt to the adult stage.

Cross infections from nonhuman primates to man occur probably via contaminated vegetables (Fig. 42) but transmission by insects has also been considered as a possibility (Goldsmid, 1974). Infection of man by direct contact to fresh faeces is unlikely because of the time requirements of the cycle.

Geographical Distribution

Ternidens-infections in nonhuman primates occur in some parts of tropical Asia and Africa including the Comoro-Islands and the island of Mauritius, whilst those of man are limited to certain areas of Central and Southern Africa. Since almost all human infections occurred in black people special dietary customs have been considered to be responsible (Amberson and Schwarz, 1952; Anthony and McAdam, 1972; Flynn, 1973; Goldsmid, 1974).

Pathogenesis

T. deminutus causes not only ulceration and nodule formation in the gut wall similar to the oesophagostomids, but it is also bloodsucking like the hookworms (Faust et al., 1975).

Clinical Symptoms

Nonhuman primates: The symptoms of *T. deminutus*-parasitism in nonhuman primates are unknown, they probably resemble those of benign oesophagostomiasis.

Man: Most infections are symptomless (Spencer, 1973), only in heavy infections do anemia, weakness, somnolence, loss of appetite and obstipation occur (Amberson and Schwarz, 1952).

Pathology

Nonhuman primates: The lesions in nonhuman primates consist of cystic nodules or small ulcers in the colonic wall, frequently without the larvae being present (Healy and Myers, 1973).

Man: *T. deminutus,* although an inhabitant of the large intestines in nonhuman primates occasionally infects the entire intestinal tract in man (Belding, 1965). Heavy infections cause anemia, intestinal haemorrhages, occasionally mucoid-bloody enteritis or small, dark nodules and ulcers. These may lead to chronic adhesive peritonitis (Amberson and Schwarz, 1952).

References

1. Amberson, J.M. and E. Schwarz: *Ternidens deminutus* Railliet and Henry, a nematode parasite of man and primates. Ann. trop. Med. Parasit. *46*, 227-237 (1952)
2. Anthony, P.P. and I.W.J. McAdam: Helminthic pseudotumors of the bowel: Thirty-four cases of helminthoma. Gut *13*, 8-16 (1972)
3. Arambulo, P.V. III; J.B. Abass, and J.S. Walker: Silvered leaf-monkeys *(Presbytis cristatus).* II. Gastrointestinal parasites and their treatment. Lab. Anim. Sci. *24*, 299-306 (1974)
4. Belding, D.L.: Textbook of Parasitology. 3rd ed., Appleton-Century-Crofts, New York (1965)
5. Bezubik, B. and F. Furmaga: The helminth parasites in *Macacus rhesus* Audeb. from China. Acta Parasitol. Polon. *7*, 591-598 (1959)
6. Bingham, G.A. and M.M. Rabstein: A study of the effectiveness of thiabendazole in the rhesus monkey. Lab. Anim. Care *14*, 357-365 (1964)
7. Bremner, K.C. and R.K. Keith: *Oesophagostomum radiatum:* Adult nematodes and intestinal hemorrhage. Exp. Parasit. *28*, 416-419 (1970)
8. Chabaud, A.G. and M. Lariviere: Sur les oesophagostomes parasites de l'homme. Bull. Soc. Path. exot. *51*, 384-393 (1958)
9. Crestian, J. et F. Crespeau: Observation d'un cas d'oesophagostomose du chimpanzé. Rec. Mech. Vét. *151*, 13-18 (1975)
10. Elek, P. and P.H. Durie: The histopathology of the reaction of calves to experimental infection with the nodular worm, *Oesophagostomum radiatum.* II. Reaction of the susceptible host to infection with a single dose of larvae. Austral. J. Agricult. Res. *18*, 549-559 (1967)
11. Faust, E.C.; P.C. Beaver, and R.C. Jung: Animal Agents and Vectors of Human Disease. 4th ed. Lea and Febiger, Philadelphia (1975)
12. Fiennes, R.: Zoonoses of Primates. The Epidemiology and Ecology of Simian Disease in Relation to Man. Weidenfeld and Nicolson, London (1967)
13. Flynn, R.J.: Parasites of Laboratory Animals. Iowa State Univ. Press, Ames/Iowa (1973)

14. Goldsmid, J. M.: Ternidens deminutus Raillet and Henry (Nematoda). A diagnostic problem in Rhodesia. Centr. Afr. J. Med. *13*, 54–58 (1967)
15. Goldsmid, J. M.: *Ternidens deminutus* (Raillet and Henry, 1909), and hookworm in Rhodesia and a review of the treatment of human infections with *T. deminutus*. Centr. Afr. J. Med. *18*, Suppl. 11, 1–14 (1972)
16. Goldsmid, J. M.: The intestinal helminthozoonoses of primates in Rhodesia. Ann. Soc. belge Med. Trop. *54*, 87–101 (1974)
17. Goldsmid, J. M. and S. Rogers: Parasitological study on chacma baboon *(Papio ursinus)* from northern Transvaal. J. S. Afr. vet. Ass. *49*, 109–111 (1978)
18. Guilloud, N. B.; A. A. King, and A. Lock: A study of the efficacy of thiabendazole and dithiazine iodide-piperazine citrate suspensions against intestinal parasites in the *Macaca mulatta*. Lab. Anim. Care *15*, 354–358 (1965)
19. Haaf, E. and A. H. van Soest: Oesophagostomiasis in man in North Ghana. Trop. geogr. Med. *16*, 49–53 (1964)
20. Habermann, R. T. and F. P. Williams: Diseases seen at necropsy of 708 *Macaca mulatta* (rhesus monkey) and *Macaca philippinensis* (cynomolgus monkey). Amer. J. Vet. Res. *18*, 419–426 (1957)
21. Habermann, R. T. and F. P. Williams, jr.: The identification and control of helminths in laboratory animals. J. nat. Cancer Inst. *20*, 979–1009 (1958)
22. Hamerton, A. E.: Report on the deaths' occurring in the society's gardens during 1936. Proc. Zool. Soc. London *107*, 443–474 (1973)
23. Hashimoto, I. and S. Honjo: Survey of helminth parasites in cynomolgus *(sic)* monkeys *(Macaca irus)*. Jap. J. med. Sci. Biol. *19*, 218 (1966)
24. Healy, G. R. and B. J. Myers: Intestinal helminths; pp. 265–296 in "The Chimpanzee", Vol. 6 (Bourne, G. H., ed.), Karger, Basel (1973)
25. Horii, Y.; I. Imada; T. Yanagida; M. Usui, and A. Mori: Parasite changes and their influence on the body weight of Japanese monkeys *(Macaca fuscata fuscata)* of the Koshima troop. Primates *23*, 416–431 (1982)
26. Janagi, T. S.: A report on gastro-intestinal helminth parasites found in *Macaca fascicularis* in peninsular Malaysia. Malays. Appl. Biol. *10*, 99–100 (1981)
27. Kagei, N. and H. Hasegawa: On the fecal examination against helminthic-ova of the Japanese monkeys, *Macaca fuscata* (Blyth, 1875), from Takasaki-yama, Oita prefecture - Problems as parasitic zoonoses. Bull. Inst. Publ. Hlth *23*, 234–238 (1974)
28. Kilala, C. P.: *Ternidens deminutus* infecting man in Tanzania. E. Afr. med. J. *48*, 636–645 (1971)
29. Kuntz, R. E. and J. A. Moore: Commensals and parasites of African baboons (*Papio cynocephalus* L. 1766) captured in Rift Valley Province of Central Kenya. J. med. Primatol. *2*, 236–241 (1973)
30. Kuntz, R. E.; B. J. Myers; J. F. Bergner, jr., and D. E. Armstrong: Parasites and commensals of the Taiwan macaque (*Macaca cyclopis* Swinhoe, 1862). Formosan Science *22*, 120–136 (1968)
31. Kuntz, R. E.; B. J. Myers, and T. E. Vice: Intestinal protozoans and parasites of the gelada baboon (*Theropithecus gelada* Rüppel, 1835). Proc. Helminth. Soc. Washington *34*, 65–66 (1967)
32. Leiper, R. T.: The occurrence of a rare sclerostome of man in Nyassaland. J. trop. Med. Hyg. *11*, 181–184 (1908)
33. Levine, N. D.: Nematode Parasites of Domestic Animals and of Man. Burgess, Minneapolis (1968)
34. Marcus, L. C.: Oesophagostomiasis; pp. 221–223 in "CRC Handbook Series in Zoonoses" (Steele, J. H. Series ed.), Section C: Parasitic Zoonoses. Vol. II (Schultz, M. G., Vol. ed.), CRC Press, Boca Raton (1982)
35. McConnell, E. E.; P. A. Basson; V. de Vos; B. J. Myers, and R. E. Kuntz: A survey of diseases among 100 free-ranging baboons *(Papio ursinus)* from the Kruger National Park. Onderstepoort. J. vet. Res. *41*, 97–168 (1974)
36. Myers, B. J. and R. E. Kuntz: A checklist of parasites reported for the baboon. Primates *6*, 137–194 (1965)
37. Neveu-Lamaire, M.: Traité d'helminthologie médicale et vétérinaire. Vigot, Paris (1936)
38. Orihel, T. C.: The helminth parasites of nonhuman primates and man. Lab. Anim. Care *20*, 395–401 (1970)

39. Orihel, T.C. and H.R. Seibold: Nematodes of the bowel and tissues; pp. 76-103 in "Pathology of Simian Primates", Part II (Fiennes, R.N.-T-W., ed.), Karger, Basel (1972)
40. Owen, D. and S. Casillo: A preliminary survey of the nematode parasites of some imported Old-World monkeys. Lab. Anim. *7*, 265-269 (1973)
41. Prosl, H. und A. Tamer: Zur Parasitenfauna des Rhesusaffen (Macaca mulatta) und des Javaneraffen (Macaca irus). Zbl. Vet.-Med. B *26*, 696-709 (1979)
42. Reardon, L.V. and B.F. Rininger: A survey of parasites in laboratory primates. Lab. Anim. Care *18*, 577-580 (1968)
43. Remfry, J.: Incidence, pathogenesis and treatment of helminth infections in rhesus monkeys *(Macaca mulatta)*. Lab. Anim. *12*, 213-218 (1978)
44. Rousselot, R. and A. Pelissier: Pathologie du gorille. III. Oesophagostomose nodulaire à *Oesophagostomum stephanostomum* du gorille et du chimpanzé. Bull. Soc. Path. exot. *45*, 568-574 (1952)
45. Ruch, T.C.: Diseases of Laboratory Primates. Saunders, Philadelphia (1959)
46. Sandground, J.H.: Studies on the life-history of *Ternidens deminutus*, nematode parasite of man, with observations on its incidence in certain regions of South Africa. Ann. trop. Med. Parasit. *25*, 147-184 (1931)
47. Sasa, M.; H. Tanaka; M. Fukui, and A. Takata: Internal parasites of laboratory animals; pp. 195-214 in "The Problem of Laboratory Animal Disease" (Harris, R.J.C., ed.), Academic Press, New York (1962)
48. Schmid, F.: Parasiten als Todesursache bei Zootieren. Dtsch. tierärztl. Wschr. *45*, 797-798 (1937)
49. Spencer, H.: Nematode diseases I; pp. 457-509 in "Spezielle pathologische Anatomie", (Doerr, W.; G. Seifert and E. Uehlinger, eds.), Bd. 8: "Tropical Pathology" (Spencer, H., ed.), Springer, New York (1973)
50. Tanaka, H.M.; M. Fujui; H. Yamamoto; S. Hayama, and S. Kodera: Studies on the identification of common intestinal parasites of primates. Bull. Exp. Anim. *11*, 111-116 (1962)
51. van den Berghe, L.; M. Chardome, and E. Peel: The filarial parasites of the eastern gorilla in the Congo. J. Helminth. *38*, 349-368 (1964)
52. Vickers, J.H.: Diseases of primates affecting the choice of species for toxicologic studies. Ann. N.Y. Acad. Sci. *162*, 659-672 (1969)
53. Vuylestke, A.: Note sur quelques nematodes parasites avec description de neuf espèces nouvelles. Rev. Zool. et Botan. Africaines *53*, 441-477 (1956)

7.4 Trichostrongylids

7.4.1 Trichostrongylus Colubriformis

T. colubriformis is primarily a parasite of ruminants which occasionally infects nonhuman primates and man. It occurs in baboons (Belding, 1965; Flynn, 1973); geladas (Kuntz et al., 1967); macaques (Lapin and Yakovleva, 1960; Reardon and Rininger, 1968; Valerio et al., 1969; Flynn, 1973; Prosl and Tamer, 1979), and chimpanzees (Flynn, 1973; Healy and Myers, 1973). References to human infections are contained in Faust et al. (1975) and Tongson and Eduardo (1982).

Agent
Adult trichostrongylids are threatlike slender worms with a small mouth, which is lacking in a buccal capsule as well as cervical papillae and teeth (Levine, 1968; Faust et al., 1975).

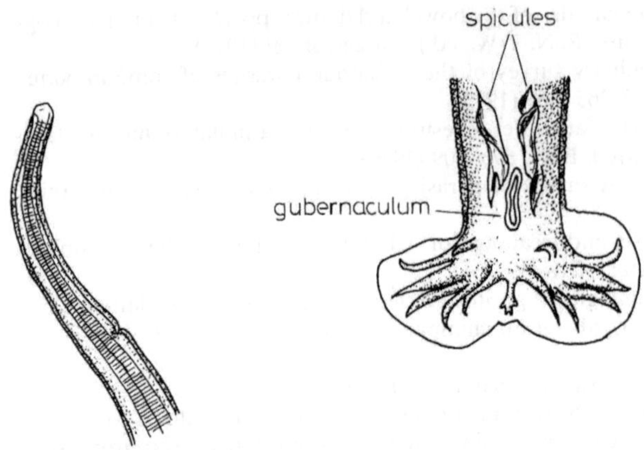

Fig. 43. *Trichostrongylus colubriformis*

The males are 4.3-7.7 mm long, with a large bursa which has large lateral lobes and a much reduced symmetrical dorsal lobe. The spicules are unequal, the left one being 136-171 μm long and the right one 123-154 μm long. They are dark brown, stout, ventrally bent and ridged (Fig. 43).

The females are 5.0-8.6 mm long. The vulva is situated somewhat behind the middle of the body and usually has prominent lips. The uteri are opposed (Amphidelph).

The thin-shelled eggs are segmented, 79-101 × 34-47 μm in size and tapered at both ends (Lapage, 1968; Levine, 1968; Flynn, 1973; Healy and Myers, 1973; Gibbons and Khalil, 1982).

Life Cycle

At the time of being passed with the host's feces, the eggs are in the morula stage. Under favorable conditions (18 to 21 °C) they hatch within 19 hrs to become first stage larvae, which then undergo the first ecdysis 25 to 28 hrs later at the optimal temperature of 27 °C. 40 hrs later the second molt occurs to produce the sheathed third stage larvae. The first and second stage larvae are very susceptible to drying, while the eggs and the third stage larvae are highly resistant especially at lower temperatures.

Exsheathing is in the anterior part of the small intestine in ruminants and most probably also in aberrant hosts. The larvae penetrate the mucosa and undergo the third molt about 4 days after ingestion. In 6 to 10 days after infection the fourth stage larvae returns to the intestinal lumen, molt and migrate to more distant parts of the intestines (Fig. 44) (Olsen, 1974).

Transmission to nonhuman primates and man is probably via contaminated food.

Clinical Symptoms

Nonhuman primates: Nothing exactly is known on the clinical effects of *T. colubriformis* in nonhuman primates. Diarrhea and eosinophilia in heavy infections have been suggested by Flynn (1973).

Man: Weak trichostrongyloidosis is symptomless, but heavy infections cause diarrhea and eosinophilia (Faust et al., 1975).

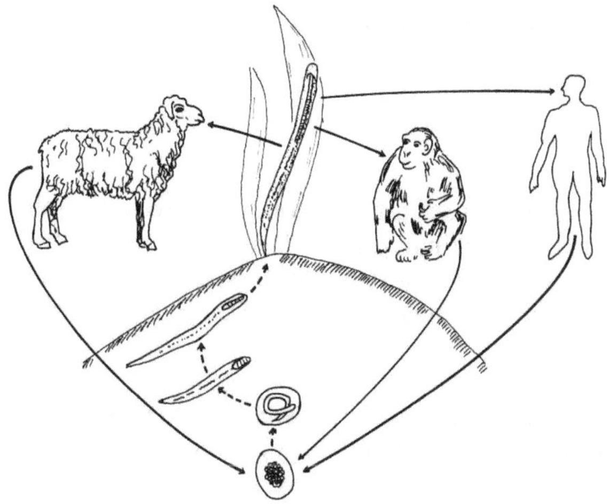

Fig. 44. Life cycle of *Trichostrongylus colubriformis*

Treatment

In nonhuman primates thiabendazol (100 mg/kg) was given twice with an interval of 2 weeks between them (Valerio et al., 1969).

References

1. Belding, D. L.: Textbook of Parasitology. 3rd ed., Appleton-Century-Crofts, New York (1965)
2. Faust, E. C.; P. C. Beaver, and R. C. Jung: Animal agents and vectors of human disease. 4th ed. Lea and Febiger, Philadelphia (1975)
3. Flynn, R. J.: Parasites of Laboratory Animals. Iowa State Univ. Press, Ames/Iowa (1973)
4. Gibbons, L. M. and L. F. Khalil: A key for the identification of genera of the nematode family Trichostrongylidae Leiper, 1912. J. Helminth. *56*, 185-233 (1982)
5. Healy, G. R. and B. J. Myers: Intestinal helminths; pp. 265-296 in "The Chimpanzee", Vol. 6 (Bourne, G. H., ed.), Karger, Basel (1973)
6. Kuntz, R. E.; B. J. Myers, and T. E. Vice: Intestinal protozoans and parasites of the gelada baboon (Theropithecus gelada Rüppel, 1835). Proc. Helminthol. Soc. Wash. *34*, 65-66 (1967)
7. Lapin, B. A. and L. A. Yakovleva: Comparative Pathology in Monkeys. C. C. Thomas, Springfield (1960)
8. Lapage, G.: Veterinary Parasitology. Olivier and Boyd, Edinburgh (1968)
9. Levine, N. D.: Nematode Parasites of Domestic Animals and of Man. Burgess, Minneapolis (1968)
10. Olsen, O. W.: Animal Parasites. Their Life Cycles and Ecology. 3rd ed., Univ. Park Press, Baltimore (1974)
11. Prosl, H. und A. Tamer: Zur Parasitenfauna des Rhesusaffen (Macaca mulatta) und des Javaneraffen (Macaca irus). Zbl. Vet.-Med. B *26*, 296-709 (1979)
12. Reardon, L. V. and B. F. Rininger: A survey of parasites in laboratory primates. Lab. Anim. Care *18*, 577-580 (1968)
13. Tongson, M. S. and S. L. Eduardo: Trichostrongylidosis; pp. 331-337 in "CRC Handbook Series in Zoonoses" (Steele, J. H., Series ed.), Section C: Parasitic Zoonoses, Vol. II (Schultz, M. G., Vol. eds.), CRC Press, Boca Raton (1982)
14. Valerio, D. A.; R. L. Miller; J. R. M. Innes; K. D. Courtney; A. J. Palotta, and R. M. Guttmacher: *Macaca mulatta:* Management of a laboratory breeding colony. Academic Press, New York (1969)

7.5 Metastrongylidae

7.5.1 Angiostrongylus Costaricensis (Morerastrongylus Costaricensis)

A. costaricensis is a metastrongyle of rodents (rats, cotton rats), which causes occasionally intestinal granulomas in man in Central America (Cespedes et al., 1967; Morera, 1967; Morera and Cespedes, 1971). Two *A. costaricensis*-infections were reported in moustached tamarins *(Saguinus mystax)* (Sly et al., 1982).

Agent
A. costaricensis closely resembles *A. cantonensis,* a parasite of rats in Southeast Asia, the Philippines, Hawaii, Tahiti, Guam and Okinawa (Spencer, 1973; Katz et al., 1982). Adult female angiostrongylids are 400 × 28 μm, whilst the males measure 350 × 23 μm in size.

Life Cycle
The adult *A. costaricensis* reside in the rodent (and monkey) mesenteric arteries supplying the small intestine and cecum. The female worms deposit eggs which hatch in the blood vessels. The larvae are passed in the faeces and are ingested by slugs *(Vaginulus plebius)* or fresh water snails. After being taken up by the mollusc,

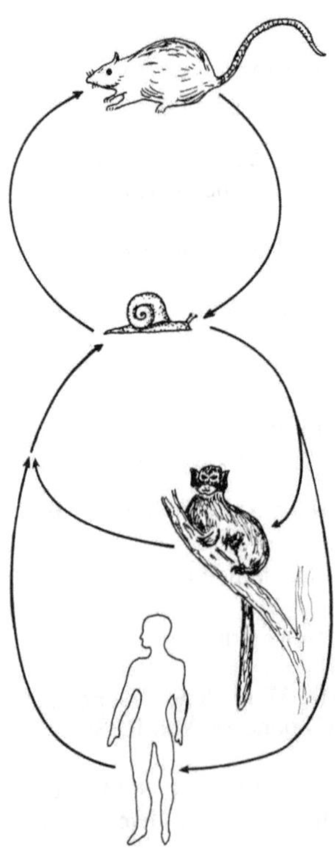

Fig. 45. Life cycle of *Angiostrongylus costaricensis*

in the slug the larvae become infective in about 3 weeks. Wild rodents and heterologous hosts are infected by eating the snails or vegetables contaminated with the mucus secretions of infected molluscs. After being taken up by a rat the worms reach maturity during the fourth week of infection (Faust et al., 1975). Infected monkeys could serve as a focus of infection of the environment.

Geographical Distribution

A. costaricensis is restricted to the Americas, especially to Costa Rica, Mexico, Venezuela, El Salvador, Colombia, Brazil and Peru (Katz et al., 1982; Sly et al., 1982). Although it has been found in cotton rats as far north as Texas (Ubelaker and Hall, 1979).

Pathogenesis

Larvae produced by adult parasites cause arteritis, thrombosis, and infarcts in the intestinal wall, possibly leading to necrotic ulceration, peritonitis, large granulomas or fistulas.

Clinical Symptoms

Nonhuman primates: In one of the animals an abdominal mass of 0.75 to 1.0 cm diameter could be palpated (Sly et al., 1982).

Man: Patients infected with *A. costaricensis* suffer from vomiting, right iliac fossa pain and peripheral eosinophilia, as well as exhibiting fever, abdominal masses may also be palpable (Spencer, 1973; Katz et al., 1982).

Pathology

Nonhuman primates: In both tamarins firm, white to greyish masses up to 2.5 cm diameter were firmly attached to the mesentery of the small intestine or surrounded the terminal ileum. They represented parasitic granulomas, consisting of eosinophils, macrophages, lymphocytes, multinucleated giant cells and loose fibrous tissue, that surrounded accumulations of nematode eggs.

Occasionally smaller microgranulomas surrounding eggs or larvae were found in other organs (renal glomeruli, mesenteric lymph nodes, liver, lungs, heart) (Sly et al., 1982).

Treatment

There is no effective treatment of the parasites or the granulomas in man. Thiabendazole is suggested to shorten the duration of symptoms, probably due to its anti-inflammatory properties. Corticosteroids are recommended in intense, painful inflammatory reactions (Katz et al., 1982).

References

1. Cespedes, R.; J. Salas; S. Mekbel; L. Trooper; F. Müllner y P. Morera: Granulomas entericos y linfaticos con intensa eosinofilia tisular producidos por un estrongilideo (Strongylata). I. Pathologia. Acta med. Costarricense *10*, 235-255 (1967)

2. Faust, E.C.; P.C. Beaver, and R.C. Jung: Animal Agents and Vectors of Human Disease. 4th ed. Lea and Febiger, Philadelphia (1975)
3. Katz, M.; D.D. Despommier, and R.Gwadz: Parasitic Disease. Springer, Heidelberg (1982)
4. Morera, P.: Granulomas entericos y linfaticos con intensa eosinofilia tisular producidos por un estrongilideo (Strongylata; Railliet y Henry, 1913). II. Aspecto parasitologico (nota previa). Acta med. Costarricense *10*, 257-265 (1967)
5. Morera, P. and R. Cespedes: *Angiostrongylus costaricensis* n.sp. (Nematoda: Metastrongyloidea), a new lungworm occurring in man in Costa Rica. Rev. Biol. Trop. *18*, 173-185 (1971)
6. Spencer, H.: Nematode Diseases I; pp. 457-509 in "Spezielle pathologische Anatomie" (Doerr, W.; G. Seifert, and E. Uehlinger, eds.). Vol. 8: "Tropical Pathology" (Spencer, H., ed.), Springer, Heidelberg (1973)
7. Sly, D.L.; J.D. Toft II; C.H. Gardner, and W.T. London: Spontaneous occurrence of *Angiostrongylus costaricensis* in marmosets *(Saguinus mystax)*. Lab. Anim. Sci. *32*, 286-288 (1982)
8. Ubelaker, J.E. and N.M. Hall: First report of *Angiostrongylus costaricensis* Morera and Cespedes 1971 in the United States. J. Parasit. *65*, 307 (1979)

7.6 Oxyurids

7.6.1 Enterobius Vermicularis

The human pinworm *E. vermicularis* has been reported by some authors as also occurring in nonhuman primates (Rewell, 1948; Sandosham, 1950 - cit. from Marcus, 1982; Ruch, 1959; Inglis, 1961; Honjo et al., 1963; Belding, 1965; Dunn, 1966; Hashimoto and Honjo, 1966; Bisseru, 1967; Fiennes, 1967; Healy and Myers, 1973; Keeling and McClure, 1974; Holmes, 1980). The close morphological similarities among the *Enterobius*-species and the remarkable host specificity of the oxyurids raise some doubt into those findings unless experimentally verified (Marcus, 1982). Skrjabin et al. (1974) suggest the species *E. vermicularis* to be restricted to the true human parasites.

Because of these uncertainties it seems unjustified to discuss the oxyurids at length.

References

1. Belding, D.L.: Textbook of Parasitology. 3rd ed. Appleton-Century-Crofts, New York (1965)
2. Bisseru, B.: Diseases of Man Acquired from his Pets. Heinemann, London (1967)
3. Dunn, F.L.: Patterns of parasitism in primates: Phylogenetic and ecological interpretation with particular reference to the hominoidea. Folia primatol. *4*, 329-345 (1966)
4. Fiennes, R.: Zoonoses of Primates. The Epidemiology and Ecology of Simian Diseases in Relation to Man. Weidenfeld and Nicolson, London (1967)
5. Hashimoto, I. and S. Honjo: Survey of helminth parasites in cynomolgus monkeys *(Macaca irus)*. Jap. J. med. Sci. Biol. *19*, 218 (1966)
6. Healy, G.R. and B.J. Myers: Intestinal helminths; pp. 265-296 in "The Chimpanzee", Vol. 6 (Bourne, G.H., ed.), Karger, Basel (1973)
7. Holmes, D.D.; S.D. Kosanke; G.L. White, and W.B. White: Fatal enterobiasis in a chimpanzee. J. Amer. vet. med. Ass. *177*, 911-913 (1980)

8. Honjo, S.; K. Muto; T. Fujiwara; Y. Suzuki, and K. Imaizumi: Statistical survey of internal parasites in cynomolgus monkeys *(Macaca irus)*. Jap. J. med. Sci. Biol. *16*, 217-224 (1963)
9. Inglis, W.G.: The oxyurids parasites (Nematoda) of primates. Proc. Zool. Soc. London *136*, 103-122 (1961)
10. Keeling, M.E. and H.M. McClure: Pneumococcal meningitis and fatal enterobiasis in a chimpanzee. Lab. Anim. Sci. *24*, 92-95 (1974)
11. Marcus, L.C.: Pinworm infections; pp. 251-253 in "CRC Handbook Series in Zoonoses" (Steele, J.H., Series ed.); Section C: Parasitic Zoonoses, Vol. II (Schulz, M.G., Vol. ed.), CRC Press, Boca Raton (1982)
12. Rewell, R.E.: Disease of tropical origin in captive wild animals. Trans. roy. Soc. trop. Med. Hyg. *42*, 17-25 (1948)
13. Ruch, T.C.: Diseases of Laboratory Primates. Saunders, Philadelphia (1959)
14. Skrjabin, K.I.; N.P. Shikhobalova, and E.A. Lagodovskaja: Oxyuridae of Animals and Man. Vol. 8, Part I: Essentials of Nematodology (Skrjabin, K.I., ed.), Israel Program for Scientific Translation, Jerusalem (1974)

7.7 Ascaridoidea

7.7.1 Ascaris Lumbricoides

Roundworm indistinguishable from *A. lumbricoides,* the human roundworm species, have been reported in chimpanzees, gorillas, orangutans and gibbons (Fiennes, 1959; Dunn and Greer, 1962; Healy and Myers, 1973) as well as several species of macaques (Habermann and Williams, 1957; Yamashita, 1963).

Agent

Ascarids are large whitish to reddish, cylindrical roundworms without lateral alae, that parasitize the intestinal tract. In *A. lumbricoides* the body tapers at both extremities to relatively blunt conical ends. The anterior end is provided with a large dorsal and 2 smaller subventral fleshy lips.

The lips have a dentigerous ridge on their inner surface formed by a row of triangular dentides.

The buccal cavity is small, leading to a cylindrical oesophagus and subsequently to a straight, simple intestine. The cuticle is finely striated and on each side there is a faint longitudinal line.

The male measures 12-31 cm × 0.3-0.5 cm, with its posterior end being somewhat curved ventrally. It is monorchic and lacks a gubernaculum. 2 spicules of equal length are on the dorsal side of the terminal part of the ejaculatory tract.

The size of the female worm is 20-35 cm × 0.3-0.6 cm. Its small vulva is ventrally located at about one third of the body length.

The thick-shelled eggs measure 65-75 × 35-50 µm, and the fertilized eggs contain a coarsely granular, spherical egg cell which does not completely fill the shell. The shell consists of a delicate, highly impenetrable, lipid containing vitelline membrane as the innermost layer; a relatively thick, colorless, glycogen containing, smooth middle layer and an outermost, coarsely mammilated, heavy albuminous covering, which is usually bile-stained when excreted in the hosts faeces.

Life Cycle

The adults live in the lumen of the upper small intestine. Each female produces approximately 200,000 eggs per day, which are fertilized but nonembryonated when passed in the host's faeces. Embryonation takes place in the soil within 2 to 4 weeks, depending on the ambient temperature. The eggs are able to survive for several months before even beginning embryonation (Bryan, 1977). They can survive putrefaction and withstand considerable desiccation, but are only moderately resistant to cold. At 22 to 33 °C the first stage larva develops in 9 to 13 days. One week later it molts to the motile, infective, second stage larva, still contained in the egg shell. The eggs remain viable in soil for months or for even longer than a year (Faust et al., 1975).

After the infective eggs are swallowed the larvae hatch in the duodenum and penetrate into the intestinal wall. After entering mesenteric venules or lymphatics they reach the liver and then move via the inferior vena cava and the right heart on to the lung capillaries where they perforate into the alveoli. There they molt again on the 9th day and subsequently move up the bronchi and trachea to the pharynx, where they are swallowed and reach the intestines again. They undergo two further molts and become sexually mature worms 8 to 12 weeks after exposure (Faust et al., 1975).

Transmission from nonhuman primates to man, although suspected has as yet not been substantiated. *A. lumbricoides* of human origin was, however, successfully transmitted to chimpanzees (Orihel, 1968 - cit. from Orihel and Seibold, 1972). It should be stressed that infection is normally via the contaminated environment rather than by direct contact to infected carriers due to the time required for larval development.

Geographical Distribution

A. lumbricoides infections in man occur in the entire tropical, subtropical and temperate belt of the globe, but not in dry desert areas or cold climates.

Pathogenesis

The first larval passage through the liver and lungs mechanically induces circumscribed haemorrhages along the migratory tracts, but in homologous hosts there are usually no other remarkable pathologic changes or clinical symptoms. Infections of paratenic (heterologous) hosts or repeated infection of homologous hosts even by only a few larvae may cause intense inflammatory responses in the liver and lungs. Eosinophilic infiltration and granulomas occur in the liver along the paths of the migrating larvae, which is accompanied by general portal inflammation. Subsequently focal fibrosis replaces the damage.

In the lungs the perforating larvae, after initially producing focal haemorrhages, induce a distinctive Ascaris pneumonitis with a high content of eosinophils in the inflammatory exudate.

In paratenic (heterologous) hosts the larvae may be transported to a variety of organs, e.g. the CNS.

The adult worms in the homologous host may cause no apparent damage in light enteric infections. A single worm may, however, produce serious disease if migrating to an ectopic site such as the common bile duct, the liver or the pancreas.

Intestinal infections can become serious events if in moderate or heavy infestations the intestinal lumen is completely or incompletely blocked by the tangled parasites. This occurs particularly in human and animal infants.

The nutritional demands of the worms can lead to significant nutritional impairment especially in infants with a low protein intake (Tripathy et al., 1971 - cit. from Faust et al., 1975).

Traumatic damage to the intestinal mucosa is caused by the frequent pinching off of small bits of intestinal mucosa. Other effects on the intestinal function may result from the degranulating effect induced in intestinal mast cells by a 2000-3000 dalton polypeptide produced by the helminths, which also is able to cause the destruction of mast cells (Uvnas and Wolf, 1967 - cit. from Murray, 1972).

Clinical Symptoms

Nonhuman primates: Specific clinical diseases due to *A. lumbricoides* have not been reported in nonhuman primates apart from the few fatal Ascarid infections in macaques (Habermann and Williams, 1957) and chimpanzees (Stam, 1960).

Man: The cardinal symptom of *Ascaris*-pneumonia during larval passage consists of dyspnea, often of asthmatic appearance, a dry or reproductive cough, rales, moderate fever and a high, but transient eosinophilia. The sputum contains many eosinophils, Charcot-Leyden crystals and occasionally larvae. X-rays show a scattered, shifting mottling of the lungs. All symptoms clear after a period lasting between a few days up to 2 weeks (Loeffler's syndrome).

Intestinal infections in infants may cause intermittent colic, loss of appetite, fretfulness, intussusception and occasional nervous symptoms.

The symptoms of extraintestinal ascariasis depend largely on the organs involved, they may appear as posthepatic jaundice, acute pancreatitis, etc.

Pathology

Nonhuman primates: Death in the few reported cases was associated with the presence of numerous worms causing blockage of the bowel, and migration of the worms to ectopic sites (Pillers, 1924; Habermann and Williams, 1957; Stam, 1960).

Man: Larval passage through the lung may induce only haemorrhages in primary infection or multifocal eosinophilic pneumonitis in repeated infections.

Adult worms filling parts of the intestines can cause pressure atrophy of the mucosa and occasionally the worms may perforate the intestinal wall causing peritonitis.

Invasion of ectopic sites induces inflammation, necrosis and/or functional disturbances of the organs involved. Secondary bacterial infections induce abscess-formation (Spencer, 1973).

Treatment

Piperacine citrate is still the drug of choice, although other, more modern anthelminthics are equally effective.

References

1. Barriga, O.O.: Ascariasis; pp. 55-61 in "CRC Handbook Series in Zoonoses" (Steele, J.H., Series ed.), Section C: Parasitic Zoonoses, Vol. II (Schultz, M.G., Vol. ed.), CRC Press, Boca Raton (1982)
2. Bryan, F.D.: Diseases transmitted by foods contaminated by waste water. J. Food Prot. *40*, 45-56 (1977)
3. Dunn, F. and W. Greer: Nematode resembling *Ascaris lumbricoides* L., 1758, from a Malaysian gibbon, *Hylobates agilis*, F. Cuvier, 1821. J. Parasit. *48*, 150 (1962)
4. Faust, E.C.; P.C. Beaver, and R.C. Jung: Animal Agents and Vectors of Human Disease. 4th ed., Lea and Febiger, Philadelphia (1975)
5. Fiennes, R.N.T-W.: Report of the society's pathologist for the year 1957. Proc. Zool. Soc. London *132*, 129-146 (1959)
6. Habermann, R.T. and F.P. Williams: Diseases seen at necropsy of 708 *Macaca mulatta* (rhesus monkey) and *Macaca philippinensis* (cynomolgus monkey). Amer. J. vet. Res. *18*, 419-426 (1957)
7. Healy, G.R. and B.J. Myers: Intestinal helminths; pp. 265-296 in "The Chimpanzee", Vol. 6, (Bourne, G.H., ed.), Karger, Basel (1973)
8. Murray, M.: Immediate hypersensitivity effector mechanisms. II. *In vitro* reaction; pp. 155-190 in "Immunity to Animal Parasites" (Soulsby, E.J.L., ed.), Academic Press, New York (1972)
9. Orihel, T.C. and H.R. Seibold: Nematodes of the bowel and tissues; pp. 76-103 in "Pathology of Simian Primates", Part II (Fiennes, R.N.T-W., ed.), Karger, Basel (1972)
10. Pillers, A.W.N.: *Ascaris lumbricoides* causing fatal lesions in a chimpanzee. Ann. trop. Med. Parasit. *18*, 101-102 (1924)
11. Spencer, H.: Nematode diseases I.; pp. 457-509 in "Spezielle pathologische Anatomie" (Doerr, W., G. Seifert, and E. Uehlinger, eds.), Vol. 8: Tropical Pathology (Spencer, H., Vol. ed.), Springer, New York (1973)
12. Stam, A.B.: Fatal ascaridosis in a dwarf chimpanzee. Ann. Parasit. hum. comp. *35*, 675 (1960)
13. Yamashita, J.: Ecological relationships between parasites and primates. I. Helminth parasites and primates. Primates *4*, 1-96 (1963)

7.8 Spirurida

7.8.1 Spirurina

The spirurina are tissue-dwelling nematodes belonging to several genera. Two species - *Gongylonema pulchrum* and *Abbreviata caucasica* - infect both man and nonhuman primates.

7.8.1.1 Gongylonema Pulchrum

The classification of the various Gongylonema species is far from satisfactory (Krull, 1969 - cit. from Cappucci et al., 1982), as the helminths of different hosts may be distinct species or one species with considerable host variation (Cappucci et al., 1982).

G. pulchrum, which is primarily a parasite of ruminants and swine is occasionally found in man (Cappucci et al., 1982) or in nonhuman primates such as

Macaca mulatta, M. sinica, Cebus sp., *Ateles* sp. (Yamaguti, 1961; Flynn, 1973; Abbott and Majeed, 1984) or chimpanzees (Paciepnik, 1976). Since some of these findings are related to histological slides they could possibly either be *G. macrogubernaculum* or *G. microgubernaculum*, both of which naturally occur in rhesus monkeys, *Cercopithecus* sp., baboons, talapoins, and capuchins (Yamaguti, 1961; Flynn, 1973).

Agent (Fig. 46)
G. pulchrum is a long and slender worm, but its actual size depends on the host. According to Cappucci et al. (1982) the males are 30 to 62 mm × 150 to 300 µm, with the females being distinctly longer, measuring 80 to 145 mm × 300 to 500 µm. The mouth is equipped with small dorsal and ventral lips which are cuticulinized at their external ends and possess a tooth on their inner surface. Additionally there are narrow, small lateral lips, two lateral and four submedian cephalic papillae. At the anterior end is a number of round or oval cuticular plaques irregularly arranged in longitudinal rows and on each side is a small lateral cervical papilla. The cervical alae are symmetrical and relatively broad. Other papillae are situated along the body, one in the anterior third and another one in the posterior third of the body.

The buccal capsule is short, narrow, cylindrical, and thick walled. The very long esophagus consists of a short and slender anterior part and a long and thick posterior part.

The male has a slightly twisted tail, asymmetrical caudal alae, a variable number of pre- and postanal papillae and a gubernaculum. The spicules are very unequal and dissimilar, the left one being slender and short, the right one much thicker and long.

The female has a conical, obtuse tail, a thick-walled vulva a little anterior to the anus, a long vagina and a very long ovejector.

The eggs produced by the ovioparous females are broadly ovoid, 50 to 70 µm × 25 to 37 µm and have a thick, transparent shell (Levine, 1968; Faust et al., 1975; Cappucci et al., 1982).

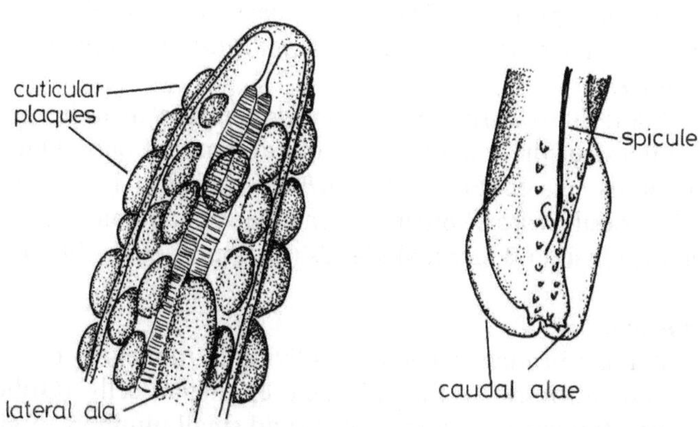

Fig. 46. *Gongylonema pulchrum* (redrawn from Levine 1968)

Life Cycle

Adult *G. pulchrum* dwell deeply within epithelia of the upper digestive tract including the mouth. The eggs are liberated together with sloughed off epithelial cells. When excreted by the host the eggs are embryonated, they have an indirect life cycle and so need to be ingested by insect intermediate hosts, such as coprophagous beetles (dung beetles) and cockroaches. The first stage larvae hatch in the insect's intestine, where they burrow into the haemal cavity, become encapsulated, develop to second stage larvae and finally mature into infective third stage larvae after about 4 weeks.

Transmission to a new host is by ingestion of infected insects, the normal way in animals, or by spontaneous emergence of larvae from insects that have fallen into water. That way man may become infected by drinking of contaminated water.

Larvae encapsulated in their intermediate hosts are freed probably in the stomach or the duodenum. The way in which they subsequently reach the upper digestive tract is unknown, presumably they migrate within the mucosal layers. Once they reach their point of destination - the oral or esophageal mucous membranes - they develop into the adult worms.

Geographical Distribution

G. pulchrum is reported to be widespread, but is more commonly encountered in the subtropics and tropics. In the USA the southeastern States are primarily affected (Cappucci et al., 1982). Human cases were diagnosed chiefly in Europe, Morocco, USSR, China, New Zealand and the USA (Faust et al., 1975).

Pathogenesis

Gongylonema spp. normally burrow and tunnel through the epithelia without reaching the underlying stromal layers (propria mucosa, submucosa) and thus usually do not come into contact with the host's vascular or immune system, so that they induce no remarkable responses by the host. Only occasionally they have been reported to induce chronic inflammation.

Clinical Symptoms

Nonhuman primates: There have been no clinical signs reported in nonhuman primates, all reported cases of *G. pulchrum* in monkeys or apes were incidental findings.

Man: Signs attributed to gongylonemiasis include spitting blood, tarry stools, numbness of the tongue, chest and abdominal pain. One case of pharyngitis and stomatitis related to the *G. pulchrum* infection has been reported (Faust et al., 1975). Adult worms sometimes emerge spontaneously from the lips, gums, tongue, palate, tonsil and lacrimal glands (Cappucci et al., 1982).

Pathology

Nonhuman primates: The adults living in tunnels in the Malpighian layers cause pressure atrophy to the adjacent epithelial cells (Orihel and Seibold, 1972). Vacated tunnels may contain eggs and small numbers of granulocytes. The organs affected are the oral cavity and the esophagus.

Man: *Gongylonema*-infections produce chiefly lesions of the mucosa and rarely of the submucosa, caused by the migration of the worms. The mostly affected mucosal membranes are those of the oral cavity. The worms have never been extracted from the human esophagus (Faust et al., 1975).

Treatment
Treatment is direct, involving the mechanical extraction of the worms or by antiseptics applied to the oral mucous membranes. Chemotherapy is not indicated.

7.8.1.2 Abbreviata Caucasica (Physaloptera Caucasica), Abbreviata Mordens

A. caucasica is a parasite of the digestive tract in orangutans, chimpanzees, baboons, mangabeys and macaques (Ruch, 1959; Yamaguti, 1961; Nelson, 1965; Levine, 1968; Skrjabin, 1969; Poinar and Quentin, 1972; Kuntz and Moore, 1973; McConnell et al., 1974; File et al., 1976) as well as in man (Ruch, 1959; Petter, 1960; Nelson, 1965; Levine, 1968).

Agent
A. caucasica is a relatively fleshy, pink worm (McConnell et al., 1974) whose mouth has, according to Levine (1968) 2 large, single lobed lateral pseudolabia (lips) with dentate margins. Each lip bears 1 externolateral tooth, 2 double submedian (subdorsal and subventral) teeth, 2 external papillae and 1 amphid (Fig. 47).

The male is 14–50 mm × 750 μm. Its tail has broad lateral alae which join ventrally in front of the anus. The ventral surface of the tail is studded with longitudinal or broken rows of tubercles, the spicules are uneven.

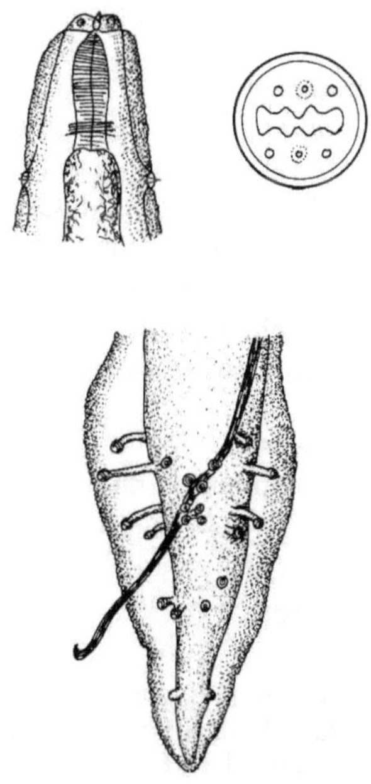

Fig. 47. *Abbreviata caucasica* (redrawn from Skrjabin 1969)

The female is 24–100 mm × 1 mm. Its vulva is usually in the anterior third of the body and leads to a common trunk which is connected to 2 or more uteri.

The eggs are oval, smooth, thick-shelled and embryonated. Each egg is surrounded by a sticky gelatinous deposit of 5 to 8 mm thickness (Poinar and Quentin, 1972).

Life Cycle

The embryonated eggs have to be ingested by an insect intermediate host, such as beetles, grasshoppers *(Schistocerca gregaria)* or cockroaches *(Blatella germanica)* (Petter, 1960; Poinar and Quentin, 1972). The eggs hatch in the intermediate host within 1 1/2 hrs after ingestion. The first stage larvae enter the wall of the midgut and migrate to the hindgut, where 16 days after exposure the first stage larvae are encapsulated. 10 days later they molt to the second stage larvae and 40 to 55 days after exposure the third stage larvae have developed, still contained in capsules attached to the hindgut of the insect (Poinar and Quentin, 1972).

Primates including man are infected by ingestion of the infected intermediate hosts. Consequently infected monkeys or apes may be the source of infection to the insects. The mode of development from the infective third stage larva to the adult within the vertebrate hosts is unknown, but finally the larvae reach the esophagus, stomach and small intestines where they become sexually mature worms.

Geographical Distribution

A. caucasica occurs in the Caucasus, in the southern USSR, Africa and the Arabian countries (Flynn, 1973), and also in South America (Levine, 1968).

Clinical Symptoms

No description of clinical symptoms of *A. caucasica* infections in nonhuman primates or man could be found.

Pathology

Nonhuman primates: The adult worms are usually well attached to the mucosa of the esophagus, stomach or small intestines. In baboons they are mostly found in the fundic region (McConnell et al., 1974). The lesions produced by the feeding worms are very small nodular ulcers, occasionally with peripheral haemorrhages. The inflammatory response is of a mixed cellular type with a large proportion of eosinophils. In the more pronounced lesions there is a proliferation of capillaries and connective tissue in addition to the inflammatory infiltration.

Treatment

No treatment is known.

References

1. Abbott, D.P. and S.K. Majeed: A survey of parasitic lesions in wild-caught laboratory-maintained primates: (rhesus, cynomolgus, and baboon). Vet. Path. *21*, 198–207 (1984)
2. Cappucci, D.T., jr.; J.K. Augsburg, and P.C. Klinck: Gongylonemiasis; pp. 181–192 in "CRC

Handbook Series in Zoonoses" (Steele, J. H., Series ed.). Section C: Parasitic Zoonoses, Vol. II (Schultz, M. G., Section ed.), CRC Press, Boca Raton (1982)
3. Coulanges, P.; H. Zeller; Y. Clerc; F. Rodhain et R. Albignac: Bactéries, virus, parasites, pathologie et pathologie expérimentale des lemuriens malgaches. Intérêt pour l'homme. Arch. Inst. Pasteur Madagascar *47*, 201-219 (1979)
4. Faust, E. C.; P. C. Beaver, and R. C. Jung: Animal Agents and Vectors of Human Disease. 4th ed., Lea and Febiger, Philadelphia (1975)
5. File, S. K.; W. C. McGrew, and C. E. G. Tutin: The intestinal parasites of a community of feral chimpanzees, *Pan troglodytes schweinfurthii*. J. Parasit. *62*, 259-261 (1976)
6. Flynn, R. J.: Parasites of Laboratory Animals. Iowa State Univ. Press, Ames/Iowa (1973)
7. Janagi, T.: A report on gastro-intestinal helminth parasites found in *Macaca fascicularis* in peninsular Malaysia. Malays. appl. Biol. *10*, 99-100 (1981)
8. King, N. W., jr.: Synopsis of the pathology of New World monkeys. First inter-American Conference on Conservation and Utilization of American Nonhuman Primates in Biomedical Research. PAHO, Washington D. C., 169-198 (1976)
9. Kumar, V.; W. de Meurichy; A.-M. Delahaye, and J. Mortelmans: Chemotherapy of helminthiasis among wild mammals. V. Gastric involvement of spider monkeys with Physaloptera sp. and chemotherapy of the infection. Acta zool. path. antverp. *76*, 191-199 (1981)
10. Kuntz, R. E. and J. A. Moore: Commensals and parasites of African baboons (*Papio cynocephalus* L. 1766) captured in Rift Valley Province of Central Kenya. J. med. Primatol. *2*, 236-241 (1973)
11. Levine, N. D.: Nematode Parasites of Domestic Animals and of Man. Burgess, Minneapolis (1968)
12. McConnell, E. E.; P. A. Basson; V. de Vos; B. J. Myers, and R. E. Kuntz: A survey of diseases among 100 free-ranging baboons *(Papio ursinus)* from the Kruger National Park. Onderstepoort J. vet. Res. *41*, 97-168 (1974)
13. Nelson G. S.: The parasitic helminths of baboons with particular reference to species transmissible to man; pp. 441-470 in "The Baboon in Medical Research" (Vagtborg, H., ed.), Univ. Texas Press, Austin (1965)
14. Orihel, T. C. and H. R. Seibold: Nematodes of the bowel and tissues; pp. 76-103 in "Pathology of Simian Primates", Part II (Fiennes, R. N. T-W, ed.), Karger, Basel (1972)
15. Paciepnik, O.: Pasozyty jelitowe malp z ogrodu zoologicznego we Wroctawiu. Wiad. Parazyt. *22*, 289-296 (19769
16. Petter, A. J.: La blatte germanique (*Blatella germanica* L.) hôte intermédiaire probable d'*Abbreviata caucasica* (Linstow, 1902) nematode parasite des primates et de l'homme. Compt. Rend. Soc. Biol. *154*, 87-90 (1960)
17. Poinar, G. O., jr. and J.-C. Quentin: The development of *Abbreviata caucasica* (von Linstow) (Spirurida: Physalopteridae) in an intermediate host. J. Parasit. *58*, 23-28 (1972)
18. Ruch, T. C.: Diseases of Laboratory Primates. Saunders, Philadelphia (1959)
19. Skrjabin, K. I.: Key to Parasitic Nematodes. Vol. I. Spirurata and Filariata. Israel Program for Scientific Translations, Jerusalem (1969)
20. Yamaguti, S.: Systema Helminthicum, Vol. III: The Nematodes of Vertebrates. Part I. Intersci. Publ. New York (1961)

7.9 Filariina

The filariae are tissue dwelling, long filariform worms with a complicated life cycle requiring insect vectors in which they undergo metamorphosis to become infective. Filariae are of varying host specificity, some restricted to one host species, others infecting several related species or genera and finally some filariae have almost no host specificity at all.

7.9.1 Dirofilaria Immitis

D. immitis is one of the species infecting a wide variety of hosts including occasionally man and nonhuman primates. In the latter it has been reported in orang-utan (Webber, 1955), chimpanzees (Sandosham, 1951 - cit. from Orihel and Seibold, 1972) and gibbons (Johnson et al., 1970). *D. immitis* is the causative agent of canine "heartworm disease", but in man and nonhuman primates it causes pulmonary dirofilariasis. Its reservoir hosts are primarily canids, whilst it is transmitted by several genera and species of mosquitoes (Tolbert and Johnson, 1982). Its geographical distribution is that of a cosmopolite with the USA being a known enzootic area (Barriga, 1982).

Because the infections of both man and nonhuman primates are nonpatent, usually resulting in granulomas containing dead worms *D. immitis* deserves no further attention within the scope of this chapter.

7.9.2 Loa Loa

Loa loa, causing ocular infections in man in West and Central Africa has also been thought to occur in nonhuman primates. However, the parasites recovered from *P. cynocephalus*, *Mandrillus leucophaeus*, *Cercopithecus mona* and *C. nictitans*, although morphologically similar to *Loa loa* of man have different biological characteristics and therefore other names *(L. L. gorillae, L. papionis)* have been suggested (Treadgold, 1920; van den Berghe et al., 1964). Because of these distinguishing features they will not be further discussed.

7.9.3 Onchocerca Volvulus

O. volvulus, the convoluted filariasis of man resulting in river blindness has been described once in a gorilla from the Congo (van den Berghe et al., 1964). However, according to Barriga (1982) *O. volvulus* seems not to be a uniform species but rather represents a number of distinct strains or biological variants with different biological properties. *O. volvulus* from the gorilla might not be a zoonotic filaria, although Scholtens (1975) considers the transmissibility between man and other primates as a real possibility in some localities. Experimentally infection has been induced by injection of third stage *O. volvulus* larvae (Duke, 1980) or by transplantation of adult worms (Prince et al., 1985).

Agent

The adults are transparent, white or cream colored and have a distinctly striated cuticle. Both extremities are blunt. The mouth is surrounded by 2 circles of 4 papillae and a pair of lateral papillae between these 2 circles (Hunter et al., 1964).

The males are 19–4 × 0.13–0.21 mm and tightly coiled. The much larger females are 330–500 × 0.27–0.45 mm.

The diurnal microfilaria are unsheathed. The cells of both the extremities and the region of the excretory pore lack nuclei.

Life Cycle

The microfilariae liberated from the adult female remain in the superficial cutaneous lymphatic vessels and connective tissues and only rarely enter the blood circulation. They are transmitted by black flies or gnats. In Africa the main vectors are *Simulium sanctipauli*, *S. yahense*, *S. soubrense* and *S. damnosum* (Kuhlow, 1978), whereas in the New World other species are encountered. In the Amazonas region of Venezuela the vectors probably are *Simulium pintoi* (Takaoka et al., 1984), in Guatemala *S. ochraceum* and *S. metallicum* (Kuhlow, 1978; Schiller et al., 1984).

In the gnat the microfilariae move directly to the thoracic muscles, there they molt twice into rhabdidiform and then filariform larvae within a minimum of 6 days. Thereafter they migrate down to the tip of the gnat's proboscic and are infectious (Faust et al., 1975).

Geographical Distribution

O. volvulus is primarily found in equatorial Africa.

The limited foci in Central and South America, which occur in altitudes between 600 and 1500 m probably represent introductions from Africa. Lately *O. volvulus* has also been found in Colombia and Brazil and in the southeastern part of the Arabian peninsula (Kuhlow, 1978).

Pathogenesis

Adult *O. volvulus* usually cause cellular reactions which finally result in the fibrotic encapsulation of the entangled worms. Almost invariably *O. volvulus* provokes an acute sensitization, leading to eczematoid licheniform or erysipelatoid lesions, dermatitis and eosinophilia. The migrating larvae may produce parasitic dermatitis (Faust et al., 1975).

Clinical Symptoms

Nonhuman primates: Clinical signs were not described in the one case in the gorilla (van den Berghe et al., 1964).

Man: Ocular microfilaria cause the disease known as river blindness, which consists of conjunctivitis, lacrimation and photophobia. In advanced cases circumcorneal congestion, iritis, and punctate keratitis occur.

Cutaneous onchocerciasis caused by adult worms takes the form of tumorlike solid nodules of 2 to 3 cm diameter in the subcutaneous tissues. These occur in African patients primarily on the trunk, while in Central American patients they are most numerous on the head and thorax. The nodules normally cause little inconvenience unless they are located in vicinity of joints or a troublesome pruritus develops. Microfilariae are usually abundant within and around the subcutaneous nodules.

Pathology

Nonhuman primates: In the one case in the gorilla, one adult female and three adult male worms were extracted from a small, flat nodule on the latero-medial side near the fourth rib (van den Berghe et al., 1964). In experimental infections of chimpanzees via *S. damnosum* the adult worms were found immediately adjacent to the posterior surfaces of the capsules of the hip joints (Duke, 1980).

Man: The adult worms first induce a local inflammatory reaction, which is followed by fibrous encapsulation, containing a relatively low contribution of inflammatory cells among which the comparatively high content of eosinophilic granulocytes is conspicious.

Dying parasites or secondary bacterial infections may result in abscesses. Ocular microfilaria produce congestion, haemorrhage, chronic iritis or iridocyclitis, punctate keratitis, synechias, pigment deposits and corneal opacities.

7.9.4 Brugia Malayi (Wucheri M., Filaria M.)

B. malayi occurs in two different strains, the periodic (nocturnal) strains limited to man, and subperiodic strains which infect both man and a variety of animals (Dissanaike, 1979). *B. malayi* is one of the filaria with only limited host specificity, naturally infecting not only nonhuman primates and man, but also wild and domestic cats, civets and the pangolins (Grove, 1982).

The subperiodic strains are true zoonotic helminths whose hosts are (among others) Asian monkeys of several species. It has been reported in *Macaca fascicularis, M. nemestrina, M. mulatta, Presbytis obscurus, P. melalophus, P. cristatus* (Laing et al., 1960; Sandosham et al., 1962; Mullin et al., 1972; Dissanaike, 1979; Masbar et al., 1981) and *Nycticebus coucang* (Belding, 1965; Flynn, 1973).

Agent

B. malayi is a fine, white, threadlike worm. The mouth, which lacks lips is surrounded by two rows of small papillae (Fig. 48). The males are 22-23 mm in length × 88 μm in diameter. Their posterior end is tightly coiled. Their spicules are unequal in length with the longer one having a stout basal, and a fine distal portion. They have 10 caudal papillae and a small naviculate gubernaculum (Hunter et al., 1964). The females are 55-60 mm long and 160 μm in diameter. They are didelphic and the vulva is situated 0.92 mm distal from the anterior end.

The microfilaria are 177-260 × 5-6 μm in size and have a subperiodic or nocturnal periodicity. They are sheathed, leaving an anterior cephalic space about twice as long as wide, and they are very much characterized by the presence of two nuclei at the caudal extremity (Belding, 1965).

Geographical Distribution

B. malayi occurs in Southeast Asia. Its geographical limits and the number of human patients are still extending (Kuhlow, 1978).
Subperiodic *B. malayi* occurs in swamp forests of Indonesia (Java, Kalimantan, Sumatra), peninsular Malaya, the Philippines and Thailand.

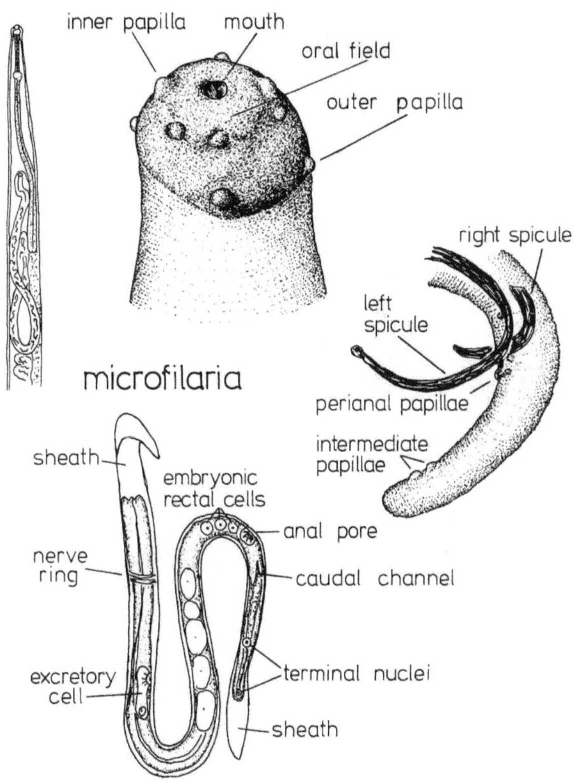

Fig. 48. *Brugia malayi*

Life Cycle and Vectors (Fig. 49)

The adult *Brugia* live within the lymphatic system and release the microfilariae into the circulating blood. In nocturnally periodic malayan filariasis, which is restricted to man, microfilariae are found in the peripheral blood at night and almost disappear during the day, as the microfilariae retreat to the lung during that time. In the zoonotic subperiodic strains the microfilariae are found both by day and night with a peak in the evening hours.

The circulating microfilariae are ingested by mosquitoes at blood feeding - in the case of zoonotic subperiodic strains mainly by the jungle living *Mansonia bonneae* and *M. dives* (Cabrera and Rozeboom, 1964; Grove, 1982). In suburban areas of Southeast Asia one of the major transmitters is *Culex pipieus fatigans* (Kuhlow, 1978). The microfilaria reach the midgut of the mosquitoes where they exsheath. They penetrate the intestinal wall, pass into the haemocele and migrate to the thoracic muscles where they molt twice to become infective larvae.

8 days after the infected blood meal, the infective larvae migrate to the base of the mosquitoes proboscis and are ready to emerge at the next feed. They are released from the tip of the labella during feeding and enter the new host through the puncture wound. Once they reach the tissues they migrate over the next few days through the lymphatics to the perinodal sinuses of the lymphnode draining the area. They migrate to the afferent lymphatic against the flow of the lymph, molt

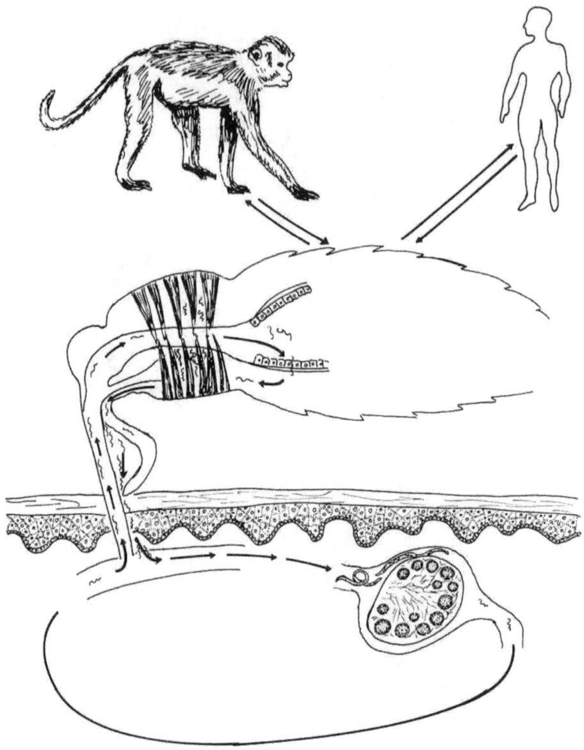

Fig. 49. Life cycle of *Brugia malayi*

twice and develop to adult worms. The adults may live in the lymphatics for many years. The females after mating produce large numbers of microfilariae.

Pathogenesis

In Malayan filariasis the circulating microfilariae are not pathogenic, whereas microfilariae trapped in the lungs cause tropical pulmonary eosinophilia. The most serious consequence, however, is blockage of lymphatics by the adult worms or by the resulting inflammatory reaction.

Clinical Symptoms

Nonhuman primates: No clinical symptoms have been reported.

Man: Many patients are entirely asymptomatic despite circulating microfilariae. In those patients exhibiting signs of disease the first symptoms are painful, tender, enlarged superficial inguinal lymph nodes on one side, periodic fever (lymphadenitis) and lymphangitis, often accompanied by general malaise, backache and headache. The acute "filarial fever" usually declines after a few days or weeks, but may recur.

In chronic cases the destruction or obliteration of the lymphatics causes lymphedema, most markedly in the lower limbs. Persistant lymphedema results in elephantiasis (which in contrast to Bancroftian filariasis rarely affects the genitals!) (Grove, 1982).

Tropical pulmonary eosinophilia is usually of insidious onset, being seen as dry cough, wheezing and dyspnea. The cough may become paroxysmal and productive and is sometimes accompanied by haemoptisis.

Pathology

Nonhuman primates: Apparently all reports of *B.malayi* in nonhuman primates are incidental findings not related to any morphological lesions.

Man: The adult *B.malayi* develop most commonly in the lymphatic system of the legs and lower trunk, causing edema, congestion, eosinophilic and mononuclear cellular infiltration of the lymphatics. Live worms are often coated by a fibrinoid deposit, while disintegrating adults are usually surrounded by a granulomatous reaction.

In chronic elephantiasis the calcified worms are surrounded by fibrous tissues and the epidermis is thickened by hyperkeratosis and acanthosis (Grove, 1982).

Tropical pulmonary eosinophilia discloses eosinophilic granulomas, sometimes around microfilaria. Other similar granulomas may be found in the liver and lymph nodes.

7.9.5 Dipetalonema Perstans (Acanthocheilonema P.)

D.perstans causes abdominal filariasis primarily in man (Belding, 1965; Grove, 1982), but occasionally it also infects chimpanzees, gorillas (Yamaguti, 1961), orangutans (Chase and Degaris, 1938) and *Lagothrix* sp. (Chiba, 1961).

Agent

Adult *D.perstans* are long, cylindrical, creamy-white worms with a smooth cuticle. The unarmed and bluntly rounded anterior end possesses a flat quadrangular cuticular shield, two large lateral papillae and two pairs of submedian papillae (Fig. 50). The caudal extremity is somewhat ventrally curved in both sexes.

Male worms average 45 mm in length. They possess 4 pairs of preanal papillae and one postanal pair. The spicules are rodlike and unequal.

Female worms are 70 to 80 mm long. Their posterior end is bifurcated to form a pair of triangular nonmuscular flaps.

The microfilaria are unsheathed and measure approximately 200 by 4.5 μm. The nuclei extend into the tip of the blunt tail. *D.perstans* microfilariae are nonperiodic and are found in the blood stream.

Vectors and Life Cycle

The intermediate hosts and transmitters in Africa are primarily several species of night-biting gnats *(Culicoides milneiousteni, C. grahami).* The microfilariae develop in the arthropod's thoracic muscles and are transmitted during blood meals (Hawking and Worms, 1961).

Geographical Distribution

The principal endemic area is the Congo, but it is also found in a belt across Africa extending from Senegal and Angola in the West to Kenya and Sudan in the

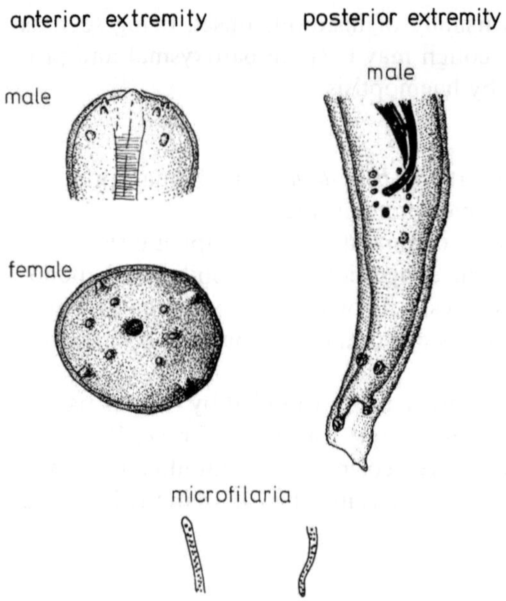

Fig. 50. *Dipetalonema perstans* (redrawn from Chabaud 1952)

East, from Algeria and Tunis in the North to Zimbabwe in the South (Belding, 1965; Faust et al., 1975; Grove, 1982). In the Americas it is found in Trinidad, Central America and the east coast of South America as far south as northern Argentina.

D. perstans is most commonly found in rain forest areas (Grove, 1982).

Clinical Symptoms

Nonhuman primates: The affected woolly monkey appeared to suffer from a wasting disease and anemia (Chiba, 1961).

Man: *D. perstans* infections of man are usually nonpathogenic. In some cases subcutaneous swellings, pruritus and urticaria, joint pain and abdominal pain, edema of the lower limbs and scrotum, eosinophilia and vertigo have been attributed to the infection (Hunter et al., 1964; Grove, 1982), but the evidence is inconclusive (Grove, 1982).

Pathology

Nonhuman primates: In the one Lagothrix chronic hepatic congestion, bile duct proliferation, chronic splenitis and chronic arteritis was associated with the *D. perstans* infection (Chiba, 1961).

Man: Adult *D. perstans* are found at the root of the mesentery, in the perirenal and retroperitoneal tissues, in the pleural cavity and pericardium of man (Levine, 1968).

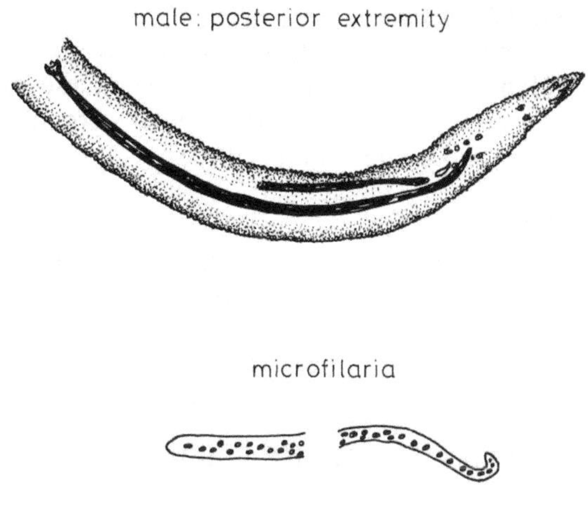

Fig. 51. *Dipetalonema streptocerca*

7.9.6 Dipetalonema Streptocerca (Acanthocheilonema S.)

D. streptocerca is very common in man, chimpanzees and gorillas in some parts of Africa (Yamaguti, 1961; van den Berghe et al., 1964; Levine, 1968; Faust et al., 1975).

Agent

Adult *D. streptocerca* are smaller than those of *D. perstans.* Males are 17 mm × 46 µm in dimension. They possess one pair of subterminal caudal papillae, a bifid caudal end and a small gubernaculum. The left spicule is bluntly pointed (Webber, 1955) (Fig. 51).

The females are 27.5 mm long and 81 µm wide. They possess one pair of minute caudal appendages.

The microfilariae are unsheathed, of 239.4–243.8 × 5.4–5.6 µm size. The first 4 nuclei are oval, followed by a single line of 8 irregularly shaped nuclei. The tail is curved, giving the worm a walking stick appearance (Fig. 51) (van den Berghe et al., 1964).

The microfilariae of *D. streptocerca* live in the dermis (Meyers et al., 1972) being found primarily in the skin of the neck, chest, torso and arms (Belding, 1965).

Vector

Culicoides grahami serves as both the intermediate host and vector with the microfilariae developing in the thoracic muscles (Hawking and Worms, 1961).

Geographical Distribution

D. streptocerca is found in Ghana, Zaire and the Gold Coast (Levine, 1968; Faust et al., 1975).

Clinical Symptoms
Nonhuman primates: No symptoms have been described.

Man: In rare cases cutaneous edema and elephantiasis of the skin occur or as in one case, a reddish-brown papular dermatitis of the arms and scapular region has been reported (Belding, 1965).

Pathology
The adults found in the subcutaneous tissues and peritoneal cavity (Flynn, 1973) apparently cause little or no pathology except those reported in the few clinical cases.

7.9.7 Dipetalonema Rodhaini

D. rodhaini is primarily a filarian parasite of chimpanzees and dwarf chimpanzees (Peel et Chardome, 1946) but microfilariae indistinguishable from *Microfilaria rodhaini* have been found in man during a systemic investigation in Gabon (Richard-Lenoble, 1982).

Agent
The adult males are 19 mm × 53 μm in size. They lack subterminal caudal appendages and a gubernaculum. The end of the tail is pointed. The left spicule has a double pointed end and measures 367 μm in length, the right spicule is 143 μm long (Webber, 1955).

The females are 25 mm × 86 μm in size. The microfilariae are 300 - 332 × 2-2.5 μm and unsheathed. The end of the tail is not curved. The nuclei are quincuncially arranged and not so clearly lined in pairs as, for example, in *D. streptocerca* (Webber, 1955; van den Berghe et al., 1964). They are found in dermal tissues rather than in the blood.

Vectors
The exact vector is unknown, although a *Culicoides*-species is suspected (Richard-Lenoble et al., 1982).

Geographical Distribution
D. rodhaini has been described in chimpanzees in Zaire, the microfilariae in man in Gabon.

Clinical Symptoms
No clinical signs have been described in either man or animals.

Pathology
The adult worms are found in the subcutis and body cavities according to Flynn (1973), probably without inducing any damage.

References

1. Barriga, O.O.: Dirofilariasis; pp.93-109 in "CRC Handbook Series in Zoonoses" (Steele, J.H., Series ed.), Section C: Parasitic Zoonoses, Vol. II (Schultz, M.G., Vol. ed.), CRC Press, Boca Raton (1982)
2. Belding, D.L.: Textbook of Parasitology III. ed. Appleton-Century-Crofts, N.Y. (1965)
3. Cabrera, B.D. and L.E. Rozeboom: Filariasis in Palawan, Philippine Islands. Nature *202*, 725-726 (1964)
4. Chase, R.E. and C.F. Degaris: Anomalies of venae superiores in an orang. Amer. J. Phys. Anthrop. *24*, 61-65 (1938)
5. Chiba, T.: Two autopsy cases of abdominal filaria infection of woolly monkey. Acta Path. Japan. *2*, 279-280 (1961)
6. Dissanaike, A.S.: Zoonotic aspects of filarial infections in man. Bull. Wld. Hlth Org. *57*, 349-357 (1979)
7. Duke, B.O.L.: Observations on *Onchocerca volvulus* in experimentally infected chimpanzees. Tropenmed. Parasitenkd. *31*, 41-54 (1980)
8. Faust, E.C.; P.C. Beaver, and R.C. Jung: Animal Agents and Vectors of Human Disease. 4th ed., Lea and Febiger, Philadelphia (1975)
9. Flynn, R.: Parasitology of Laboratory Animals. Iowa State Univ. Ames/Iowa (1973)
10. Grove, D.I.: Filariasis; pp.123-146 in "CRC Handbook Series in Zoonoses" (Steele, J.H., Series ed.), Section C: Parasitic Zoonoses, Vol. II (Schultz, M.G., Vol. ed.), CRC Press, Boca Raton (1982)
11. Hawking, F. and M.Worms: Transmission of filaroid nematodes. Ann. Rev. Entomed. *6*, 413-432 (1961)
12. Hunter, G.W.; W.W. Frye, and J.C. Swartzwelder: A Manual of Tropical Medicine. 3rd ed., W.B. Saunders, Philadelphia (1964)
13. Johnsen, D.O.; D.D. Gould; P.Tanticharoenyos; C.L. Diggs, and W.L. Wooding: Experimental infection of gibbons with *Dirofilaria immitis*. Trans. roy. Soc. trop. Med. Hyg. *64*, 937-938 (1970)
14. Kuhlow, F.: Zur Ökologie der Filariosen. Z. Parasitenkd. *57*, 285-290 (1978)
15. Laing, A.B.G.; J.F.B. Edeson, and R.H. Wharton: Studies on filariasis in Malaya: The vertebrate hosts of *Brugia malayi* and *B.pahangi*. Ann. Trop. Med. Parasitol. *54*, 92-99 (1960)
16. Levine, N.D.: Nematode Parasites of Domestic Animnals and of Man. Burgess Publ. Co., Minneapolis (1968)
17. Masbar, S.; J.R. Palmieri; H.A. Marwoto; Purnomo, and F.Darwis: Blood parasites of wild and domestic animals from South Kalimantan (Borneo), Indonesia. S.E. Asian J.Trop. Med. Publ. Hlth *12*, 42 46 (1981)
18. Meyers, W.M.; D.H. Connor; L.E. Harmann; K.Fleshman; R.Moris, and R.C. Neafie: Human streptocerciasis. A clinico-pathologic study of 40 Africans (Zairians) including identification of the adult filaria. Amer. J.trop. Med. Hyg. *21*, 528-545 (1972)
19. Mullin, S.W.; T.J. Dondero, jr.; S.Sivanandam, and R.Dewey: Filarial parasites of Malaysian monkeys. S.E. Asian J.Trop. Med. Publ. Hlth *3*, 548-551 (1972)
20. Orihel, T.C. and H.R. Seibold: Nematodes of the bowel and tissues; pp.76-103 in "Pathology of Simian Primates", Part II (Fiennes, R.N.T-W, ed.), Karger, Basel (1972)
21. Peel, E.et M.Chardome: Sur des filarides de chimpanzés "*Pan paniscus*" et "*Pan satyrus*" au Congo belge. Ann. Soc. Belge Med. Trop. *26*, 117-156 (1946)
22. Prince, A.M.; E.J. Albiez, and M.C. van den Ende: Onchocerca volvulus: transplantation of adult worms into chimpanzees. Trop. Med. Parasit. *36*, 105-108 (1985)
23. Richard-Lenoble; M.Kombila et O.Bain: Foyer de filariose humaine au Gabon à microfilaire dermique indifférenciable de *Microfilaria rodhaini*. Ann. Parasitol. (Paris) *57*, 506 (1982)
24. Sandosham, A.A.; R.H. Wharton; McW. Warren, and D.E. Eyles: Microfilariae in the rhesus monkey *(Macaca mulatta)* from East Pakistan. J.Parasitol. *48*, 489 (1962)
25. Schiller, E.L.; J.L. Petersen; D.Shirazian, and H.F. Marroquin: Morphogenesis of larval *Onchocerca volvulus* in the Panamanian black fly, *Simulium quadrivittatum*. Amer. J.trop. Med. Hyg. *33*, 410-413 (1984)
26. Scholtens, R.G.: Filarial infections - With a note on dracunculiasis; pp.572-583 in "Diseases

Transmitted from Animals to Man." 6th ed. (Hubbert, W.T.; W.F. McCulloch, and P.R. Schnurrenberger, eds.), C.C. Thomas, Springfield (1975)
27. Takaoka, H.; H. Suzuki; S. Noda; I. Taja; M.G. Basanez, and L. Yarzabal: Development of *Onchocerca volvulus* larvae in *Simulium pintoi* in the Amazonas region of Venezuela. Amer. J.trop. Med. Hyg. *33*, 414–419 (1984)
28. Tolbert, R.H. and W.E. Johnson, jr.: Potential vectors of *Dirofilaria immitis* in Macon County, Alabama. Amer. J.vet. Res. *43*, 2054–2056 (1982)
29. Treadgold, C.H.: On a filaria, *Loa papionis* n.sp., parasitic in *Papio cynocephalis*. Parasitology *12*, 113–135 (1920)
30. van den Berghe, L.; M. Chardome, and E. Peel: The filarial parasites of the eastern gorilla in the Congo. J. Helminthol. *38*, 349–368 (1964)
31. Webber, W.A.F.: Filarial parasites of primates: a review. I. Dirofilaria and Dipetalonema. Ann. trop. Med. Hyg. Parasitol. *49*, 123–141 (1955)
32. Yamaguti, S.: Systema Helminthum, Vol. III. The Nematodes of Vertebrates. Part I. Intersci. Publ. Inc. N.Y. (1961)

7.10 Dracunculoidea

Dracunculoidea are among the largest known phasmid nematodes, with the adult females reaching a length of more than a meter.

7.10.1 Dracunculus Medinensis

D. medinensis, the human "guinea"- or "dragon"-worm has been known since antiquity. *D. medinensis* has also been described in a variety of animals including *Cercopithecus aethiops*, *Papio hamadryas* and *Macaca mulatta* (Kale, 1982).

Agent
D. medinensis is an elongated cylindrical worm with a smooth whitish cuticle, a bluntly rounded anterior end and a curved posterior end. Adult females are 500 to 1200 mm long and 0.9 to 2.0 mm wide. The mouth is triangular and opens in the middle of a cuticularized shield surrounded by two rings of papillae (Fig. 52). The uteri, oviducts and ovarian tubules are paired. The tightly coiled embryo-distended uteri fill the greater part of the body and obliterate an atrophied nonfunctional gut.

The males are much smaller, measuring 12 to 25 × 0.4 mm and as a rule are rarely found. The posterior end is coiled, 4 pairs of preanal and 6 pairs postanal papillae are present (Fig. 52). The esophagus occupies about one third of its length and consists of a short anterior muscular portion and a wide glandular posterior part. The spicules are subequal, 0.49 to 0.73 mm long, the gubernaculum is 0.2 mm long. A pair of large, lateral phasmids is located posterior to the anus.

Geographical Distribution
D. medinensis occurs primarily in southern Asia, Africa, New Guinea and the USSR between the equator and the tropic of cancer. Recently a human case was

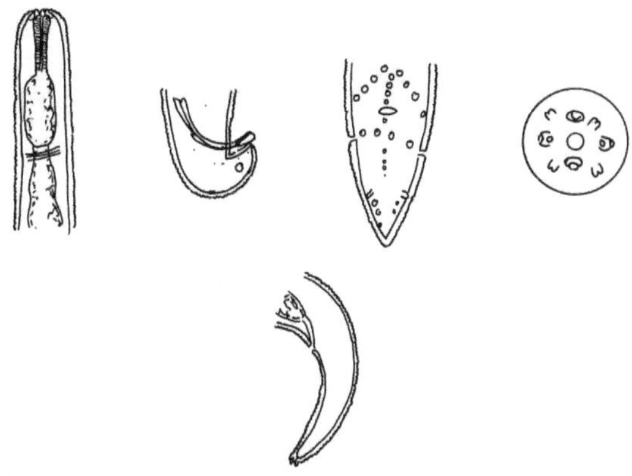

Fig. 52. *Dracunculus medinensis* (redrawn from Olsen 1974)

reported in Japan (Kobayashi et al., 1986). It is highly endemic in tropial Africa and the western half of India (Hunter et al., 1964). Foci of infection in northern regions of South America and in some of the Caribbean Islands have ceased to exist (Spencer, 1973).

Vector and Life Cycle

Dracunculoidea are seasonally transmitted by drinking water containing infected copepods (*Cyclops* spp., most frequently *Cyclops leuckarti* and *C. hyalinus*) which are found in shallow surface waters, stagnant ponds, water holes, step wells, slow-flowing brooks and streams (Kale, 1982). In the one case in Japan raw Loaches, probably imported from endemic areas were suspected (Kobayashi et al., 1986).

As the female *Dracunculus* matures in the vertebrate host she migrates to the subcutaneous tissues usually of the lower extremities, finally inducing a blister which bursts within a few days. On immersion of the female *Dracunculus* in fresh water a loop of the uterus is prolapsed and ruptures, liberating large numbers of delicate, wiry rhabdidoid larvae. The female withdraws into the subcutaneous tissues to repeat the process upon subsequent water contacts until all larvae have been released, following which she dies.

Larvae wriggling around in the water attract copepods which ingest them. The activity of the larvae, with the aid of its dorsal tooth, enables them to burrow through the copepod's gut wall and into the haemocele in 1 to 4 hrs. The infected copepods become relatively inactive and sink to the botton.

At temperatures above 19 °C the larvae molt twice to become infective third-stage larvae within about 2 weeks. Infection of vertebrate hosts occurs by ingestion of water containing the infected *Cyclops* (Fig. 53).

The larvae are freed from the copepod due to the latter being digested in the new host's duodenum. Within 13 hrs they are to be found in the duodenal wall and they can be seen in the abdominal arteries for up to 12 days. By day 15 they have migrated into the thoracic and abdominal muscles. Subsequently they move across the subcutaneous connective tissues to the axial, inguinal and retroperitoneal regions and during this period they obviously become sexually mature and in the females insemination has already occurred. In rhesus monkeys, the females

Fig. 53. Life cycle of *Dracunculus medinensis*

begin moving towards the extremities of the limbs between 8 and 10 months after infection (Olsen, 1974). In man the cycle is completed within about 12 months (Muller, 1972).

Clinical Symptoms

Nonhuman primates: No symptoms have been reported.

Man: The infection is symptomless during most of the time. Only a few hours before the appearance of the skin lesions are there pronounced symptoms such as erythema, urticarial rash with intense pruritus, nausea, vomiting, diarrhea, severe dyspnea, giddiness and sometimes syncope. A reddish papule of 7 to 15 mm diameter with a domelike vesicular center and indurated margins subsequently appears. As the anterior extremity of the worm reaches the skin, intense itching or burning sensations are frequently experienced and within 1 to 2 days the lesions burst. The lesions are most frequently located in the lower extremities.

Pathology

The blisters before bursting are usually sterile and contain macrophages, eosinophils and neutrophils in a yellowish fluid. Only after opening may secondary infection with the development of sepsis in the worm's tunnel occur (Faust et al., 1975).

Control and Treatment

The only effective method of control is the prevention of transfer of larvae from infected vertebrate hosts to water pools by the appropriate hydrotechnical measures.

The clinical signs can be treated by antihistamines. The mature female may be mechanically extracted under aseptic conditions in hospitals. The worms also respond to chemotherapy using diethylcarbamazide, thiabendazole, or with the present drug of choice, metronidazole (250 mg three times daily for 1 week - Faust et al., 1975).

References

1. Faust, E.C.; P.C. Beaver, and R.C. Jung: Animal Agents and Vectors of Human Disease. 4th ed. Lea and Febiger, Philadelphia (1975)
2. Hunter, G.W.; W.W. Frye, and J.C. Swartzwelder: A Manual of Tropical Medicine. 3rd ed., Saunders, Philadelphia (1964)
3. Kale, O.O.: Dracontiasis; pp. 111-122 in "CRC Handbook Series in Zoonoses" (Steele, J.H., Series ed.), Section C: Parasitic Zoonoses, Vol. II (Schultz, M.G., Vol. ed.), CRC Press, Boca Raton (1982)
4. Kobayashi, A.; K. Katakura; A. Hamada; T. Suzuki; Y. Hataba; N. Tashiro, and A. Yoshida: Human case of dracunliasis in Japan. Amer. J. trop. Med. Hyg. *35*, 159-161 (1986)
5. Levine, N.D.: Nematode Parasites of Domestic Animals and of Man. Burgess Publ. Co., Minneapolis (1968)
6. Muller, R.: Maintenance of *Dracunculus medinensis* (L.) in the laboratory and observations on experimental infections. Parasitology *64*, 107-116 (1972)
7. Olsen, O.W.: Animal Parasites. Their Life Cycles and Ecology. 3rd ed. Univ. Park Press, Baltimore (1974)
8. Spencer, H.: Nematode diseases II (Filarial diseases); pp. 511 - 559 in "Spezielle pathologische Anatomie" (Doerr, W.; G. Seifert, and E. Uehlinger, Series ed.), Bd. 8 "Tropical Pathology" (Spencer, H., Vol. ed.), Springer, New York (1973)

7.11 Aphasmidea

7.11.1 Trichinelloidea

7.11.1.1 Capillaria Hepatica

C. hepatica is a common parasite primarily of rodents and occasionally of other mammals. It is not uncommonly found in nonhuman primates, having been reported in *Gorilla gorilla* (Paciepnik, 1976); *Pan troglodytes* (Yamashita, 1963; Garner and Stookey, 1968; Orihel and Seibold, 1972); *Macaca mulatta, M. fascicularis, Ateles* sp. and *Cebus* sp. (Lubinsky, 1956; Habermann and Williams, 1957; Ruch, 1959; Abbott and Majeed, 1984); *Lagothrix lagotricha* and *Cacajao rubicundus* (own findings, unpublished). At least 23 human cases have been reported (Faust et al., 1975; Banzon, 1982).

Agent (Fig. 54)
The adult worms are thin, hairlike and rarely found.
The male worm ranges from 24 to 40 mm in length and from 0,07 to 0,1 mm in breadth, and is provided with a chitinized spicule 500 µm in length. The spicule

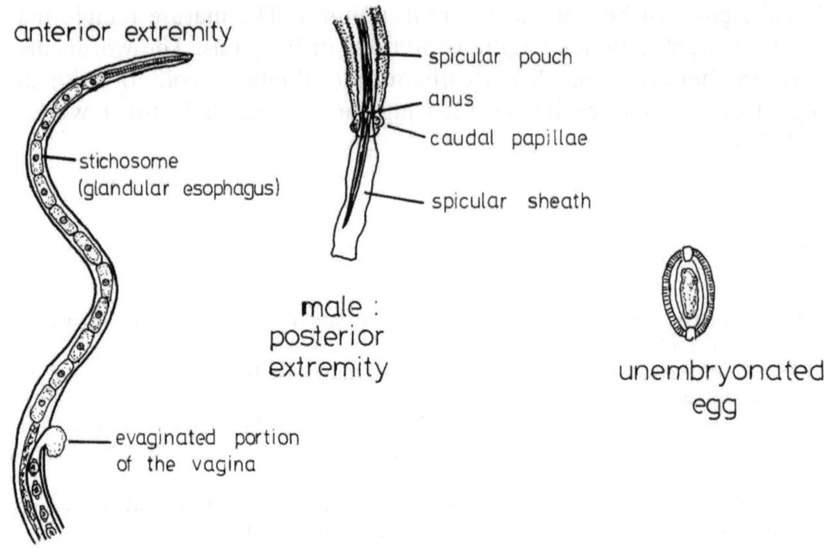

Fig. 54. *Capillaria hepatica*

seems to lie entirely within the posterior cloaca and therefore is difficult to see, as it is endorsed by a membranous protrusible spicular sheath that expands distally into a funnel-shaped dilatation. The posterior end is blunt and bears a pair of subventral lobes.

The females measure 53 to 78 mm in length and 0.11 to 0.2 mm in width. They possess a producible vulvar opening slightly below the end of the esophagus.

The eggs are barrel-shaped and 51–67 × 30–35 μm in size. They are thick walled with distinctive radial striations due to perforating minute pores and have a mucous plug at each end. They are quite resistant to adverse environmental conditions, being able to withstand freezing temperatures and desiccation for weeks. They are able to develop in rat pellets and remain alive for 7 months. They have been shown to survive in rat livers kept unfrozen in a refrigerator for 2 years (Olsen, 1974).

Life Cycle (Fig. 55)

The infection is maintained in nature by small rodents; in North America primarily by *Peromyscus maniculatus* and *Clethrionomys gapperi*; in Asia and other parts of the world by *Rattus rattus, R. norvegicus* or other *Rattus*-species (Banzon, 1982).

The adult female worms are embedded in the liver and are hardly ever seen because of their short life span. The eggs are clustered in the worm's channels around the worms and are usually trapped in fibrous tissues. Only early in heavy infections are some eggs passed via the bile ducts and appear in the faeces. The eggs are uncleaved at deposition in the liver and are arrested at the 2 to 8 cell-stage while remaining in it. Development starts again only after the eggs have been liberated from the liver by death and decomposition of the host, or when the host falls prey to another animal (or man). In the latter case the freed eggs do not develop

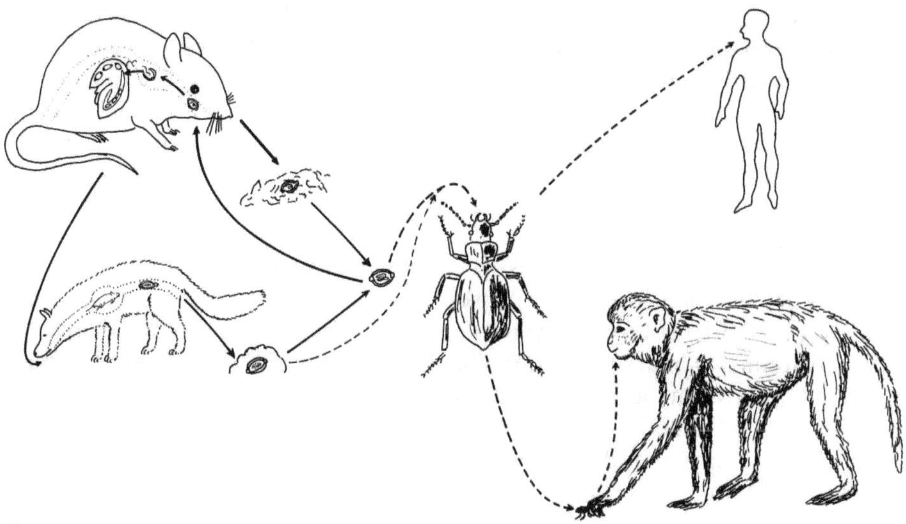

Fig. 55. Life cycle of *Capillaria hepatica*

until passed in the faeces where they become infective. Mobedi and Arfan (1971) suggested that ground beetles aid in the transmission of *C. hepatica.* After feeding infected liver to *Carabus* sp. and *Catalus* sp. the beetles started excreting eggs developed to the morula stage at days 4 and 5.

The development of the eggs in soil to the embryonic stage takes 4 to 5 weeks under favorable conditions of temperature (room temperature) and moisture. No molt or further development occurs in the eggs.

Infection of a new host is by swallowing fully developed eggs in contaminated soil eaten playfully by children or taken up together with vegetables. The eggs reach the intestinal tract and hatch in the cecum within 24 hrs. The first stage larvae burrow into the cecal wall within 6 hrs. They then enter the portal vein and arrive in the liver within 52 hrs after infection. Here they undergo the first molt during days 3 and 4 after infection to form the sheathed second stage larvae. The second molt follows at day 5. The third molt takes place approximately at days 9 to 12, with the resulting fourth stage larvae retaining their third-stage cuticle. The fourth stage larvae finally develop to the sexual stages and the ensheathing cuticle is lost by day 18 by the males and by day 20 by the female worms. The life span of males is around 40 days, that of females around 2 months. Gravid females appear 21 days after infection (Olsen, 1974).

Geographical Distribution

C. hepatica has a cosmopolitan distribution with varying infection rates in the small rodent populations.

Pathogenesis

The migrating larvae and adult worms first induce focal necrosis with degeneration of hepatocytes accompanied by pyknosis or karyorrhexis of their nuclei (Lämmler et al., 1974).

About 19 days after infection the host's responses begin in and around non-egg bearing tunnels as cellular infiltration and fibrosis, followed by calcification. Dead worms disintegrate or calcify. Egg-bearing lesions are slower to calcify or fail to calcify at all. The cellular response consists first of neutrophilic exudates, followed by eosinophils, lymphocytes, monocytes, plasma cells and occasionally multinucleated giant cells.

Clinical Symptoms
Nonhuman primates: Infected monkeys usually are found dead without preceding symptoms or they died from other causes.

Man: In mild to severe *C. hepatica*-infections the disease is characterized by fever, hepatomegaly, marked eosinophilia, leukocytosis and hypergammaglobulinemia. Undernourishment, anorexia, abdominal enlargement and edema are also seen. Secondary bacterial infection may complicate the clinical picture. The lesions of capillariasis tend to progress and may lead to death, although death may also in certain cases be attributed to the secondary infection (Banzon, 1982).

Pathology
Nonhuman primates: The clusters of eggs are irregularly distributed throughout the liver. Erratic eggs can be found in other organs such as the lungs or the spleen. The clusters are surrounded by a necrotic zone. In own experiences in woolly monkeys and uakaris widespread necrosis of the liver tissue predominated. In most cases a granulomatous-fibrotic response mostly originating from the portal vein system leads to enlargement, congestion or even fibrosis of the liver. The egg deposits are surrounded by dense cellular infiltrates and distinctive fibrosis with or without focal calcification. In several cases a fatal diffuse hepatitis results from the capillariasis (Ruch, 1959; Fiennes, 1967).

Treatment
There is no known therapy (Faust et al., 1975).

7.11.1.2 Anatrichosoma Cutaneum (Trichosoma C.)

A. cutaneum is a small *Capillaria*-like worm that causes a creeping eruption in man in Southeast Asia and Japan (Morishita and Tani, 1960; Le-Van-Hoa et al., 1963). The parasite is indigenous to several Asian monkey species, it has been found particularly in wild-living and captive *Macaca mulatta* (Swift et al., 1922; Ruch, 1959; Allen, 1960; Reardon and Rininger, 1968; Remfry, 1978; Karr et al., 1979). It has occasionally also been found in *M. fascicularis, M. nemestrina, M. maurus, M. niger, M. assamensis, Presbytis obscurus* (Harwell and Dalgard, 1979). *Anatrichosoma* found in African species such as *Erythrocebus patas, Cercopithecus aethiops, Miopithecus talapoin, Cercocebus* sp. (Orihel, 1970) may or may not be *A. cutaneum.*

Agent
Anatrichosomes are mostly found in histological slides with no species diagnosis being possible.

The genus *Anatrichosoma* consists of small, threadlike nematodes. Their esophageal region is shorter than the remainder of the body, occupying 1/6 to 1/3 of the body length.

The male *A. cutaneum* is unknown (Chitwood and Smith, 1958). The males of other *Anatrichosoma* spp. lack a cephalic cuticular inflation, a spicule and a copulatory sheath. They are as long or even longer than the corresponding females (Chitwood and Smith, 1958). Female *Anatrichosoma* spp. have a body which tapers at both ends. They have a cephalic cuticular inflation and the body has a constriction at the base of the esophageal region, posteriorly to which, the prominent vulva opens. Appendages are lacking (Chitwood and Smith, 1958).

The embryonated eggs are elliptoid to barrel-shaped, 56-70 × 37-49 μm in size and have bipolar opercula (Conrad and Wong, 1973).

Life Cycle

The life cycle of *Anatrichosoma* spp. is unknown, direct transmission via swallowed nasal mucus has been suggested (Conrad and Wong, 1973; Remfry, 1978; Harwell and Dalgard, 1979).

Clinical Symptoms

Nonhuman primates: In the nasal mucosal form in *M. mulatta*, mild to severe nasal discharge has been reported, whereas cutaneous infections resulted in creeping eruptions of the palms and soles with marked irritation and concurrent regional lymphadenopathy (Harwell and Dalgard, 1979).

Man: *A. cutaneum* causes a typical creeping eruption in the skin with superficial inflammatory galleries (Morishita and Tani, 1960; Le-Van-Hoa et al., 1963).

Pathology

The parasites are located deep in the stratum Malpighi near the basal layer and cause pressure atrophy of adjacent cells with moderate thickening of the epidermis and occasionally focal inflammation of the underlying *lamina propria* or dermis.

The female worms migrate through the stratified squamous epithelia and deposit eggs in the tunnels created by their movement. The eggs are finally sloughed together with the superficial epithelium, and therefore swabbing of the nasal passages with a dampened, cotton-tipped applicator stick is the diagnostic method of choice (Orihel and Seibold, 1972).

Treatment

Fenbendazole (10-25 mg/kg once daily for 3-10 days) has been recommended (Harwell and Dalgard, 1979).

References

1. Abbott, D. P. and S. K. Majeed: A survey of parasitic lesions in wildcaught, laboratory-maintained primates: (Rhesus, cynomolgus, and baboon). Vet. Path. *21*, 198-207 (1984)
2. Allen, A. M.: Occurrence of the nematode, *Anatrichosoma cutaneum*, in the nasal mucosae of *Macaca mulatta* monkeys. Amer. J. Vet. Res. *21*, 389-392 (1960)

3. Banzon, T.: *Capillaria hepatica* infection; pp. 67-71 in "CRC Handbook Series in Zoonoses" (Steele, J. H., Series ed.), Section C: Parasitic Zoonoses, Vol. II (Schultz, M. G., Vol. ed.), CRC Press, Boca Raton (1982)
4. Chitwood, M. B. and W. N. Smith: A redescription of *Anatrichosoma cynamolgi* Smith and Chitwood, 1954. Proc. Helminthol. Soc. Washington *25*, 112-117 (1958)
5. Conrad, H. D. and M. M. Wong: Studies of *Anatrichosoma* (Nematoda: Trichinellida) with description of *Anatrichosoma rhina* sp. n. and *Anatrichosoma nacepobi* sp. n. from the nasal mucosa of *Macaca mulatta*. J. Helminthol. *47*, 289-302 (1973)
6. Faust, E. C.; P. C. Beaver, and R. C. Jung: Animal Agents and Vectors of Human Disease, 4 th ed., Lea and Febiger, Philadelphia (1975)
7. Fiennes, R.: Zoonoses of Primates. The Epidemiology and Ecology of Simian Diseases in Relation to Man. Weidenfeldt and Nicolson, London (1967)
8. Garner, F. M. and J. L. Stookey: Diseases of Nonhuman Primates. Armed Forces Inst. Pathology, Washington D. C. (1968)
9. Habermann, R. T. and F. P. Williams, jr.: Diseases seen at necropsy of 708 *Macaca mulatta* (rhesus monkey) and *Macaca philippinensis* (cynomolgus monkey). Amer. J. Vet. Res. *18*, 419-426 (1957)
10. Harwell, G. and D. Dalgard: Clinical *Anatrichosoma cutaneum* dermatitis in non-human primates. Amer. Ass. Zoo Vet. Ann. Proc. 1979, 83-86 a (1979)
11. Karr, S. L.; R. V. Henrickson, and J. G. Else: A survey for *Anatrichosoma* (Nematoda: Trichinellida) in wild-caught *Macaca mulatta*. Lab. Anim. Sci. *29*, 789-790 (1979)
12. Lämmler, G.; H. Zahner; R. Vollerthun, and R. Rudolph: Egg production and host reaction in *Capillaria hepatica* from infection in *Mastomys natalensis*; pp. 327-341 in "Parasitic Zoonoes. Clinical and Experimental Studies." (Soulsby, E. J. L., ed.), Academic Press, New York (1974)
13. Le-Van-Hoa, Duong-Hong-Mo et Nguyen-Luuvien: Premier cas de capillariose cutanée humaine. Bull. Soc. Path. Exot. *56*, 121-126 (1963)
14. Lubinsky, G.: On the probable presence of parasitic liver cirrhosis in Canada. Canad. J. Comp. Med. *20*, 457-465 (1956)
15. Mobedi, I. and F. Arfan: Probable role of ground beetles in the transmission of *Capillaria hepatica*. Parasitology *57*, 1144-145 (1971)
16. Morishita, K. and T. Tani: A case of *Capillaria* infection causing cutaneous creeping eruption in man. J. Parasitol. *46*, 79-83 (1960)
17. Olsen, O. W.: Animal Parasites. Their Life Cycles and Ecology. Univ. Park Press, Baltimore (1974)
18. Orihel, T. C. and H. R. Seibold: Nematodes of the bowel and tissues; pp. 76-103 in "Pathology of Simian Primates, Part II" (Fiennes, R. N. T-W, ed.), Karger, Basel (1972)
19. Paciepnik, O.: Pasozyty jelitowe malp z ogrodu zoologicznego we Wrocławia. Wiad. Parazyt. *22*, 289-296 (1976)
20. Reardon, L. V. and B. F. Rininger: A survey of parasites in laboratory primates. Lab. Anim. Care *18*, 577-580 (1968)
21. Remfry, J.: The incidence, pathogenesis and treatment of helminth infections in rhesus monkeys *(Macaca mulatta)*. Lab. Anim. *12*, 213-218 (1978)
22. Ruch, T. C.: Diseases of Laboratory Primates. Saunders, Philadelphia (1959)
23. Swift, H. F.; R. H. Boots, and C. P. Miller: A cutaneous nematode infection in monkeys. J. Exp. Med. *35*, 599-620 (1922)
24. Yamashita, J.: Ecological relationships between parasites and primates. I. Helminth parasites and primates. Primates *4*, 1-96 (1963)

8 Trematodes

8.1 Dicrocoeliidae

8.1.1 Dicrocoelium Dendriticum (D. Lanceolatum, D. Macaci)

D. dendriticum is known as the small liver fluke or "lancet fluke" of sheep and other herbivorous mammals. Human infections (not to be confused with the passive passage of eggs after eating infected liver!) have been diagnosed several times (Hunter et al., 1964; Spencer, 1973; Faust et al., 1975). The parasite has also been found in *Macaca fuscata, M. speciosa, Pan troglodytes, Pongo pygmaeus, Semnopithecus* sp., *Cercocebus* sp., *Cercopithecus* sp. and *Papio* sp. (Kuntz, 1972; Healy and Myers, 1973).

Agent
The lancet-shaped, flat, thin, transparent flukes range in size from 5–15 × 1.5–2.5 mm and are covered by a smooth cuticle. They have two suckers in the anterior third of their body, with the oral sucker being slightly smaller than the ventral one (Fig. 56). The acetabulum is in the anterior half of the body. Two large testes are situated in tandem immediately behind the ventral sucker and the small ovary is positioned immediately behind the posterior testis. The median body area behind the ovary is filled by uterine loops. The ascending and descending uterine branches traverse one another. Vitellaria composed of minute follicles occupy the middle third of the lateral parts of the body.

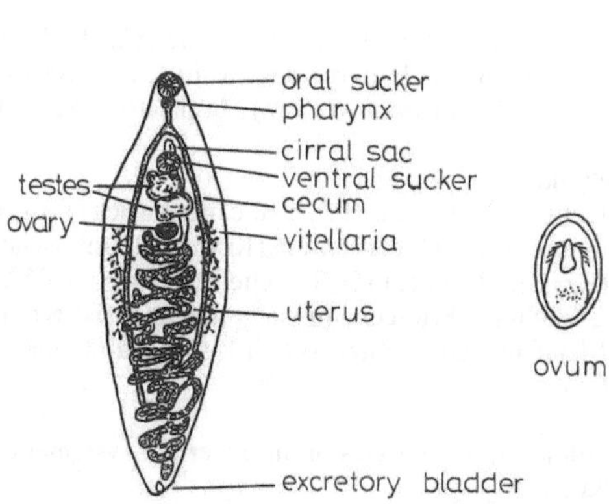

Fig. 56. *Dicrocoelium dentriticum*

The dark-brown eggs are asymmetrical-ovoidal, 36–45 × 22–30 μm in size, thickshelled and operculated. They contain a mature miracidium when passed in the faeces (Fig. 56) (Olsen, 1974; Rosicky and Groschaft, 1982).

Life Cycle (Fig. 57)
The eggs are laid in the bile ducts by the hermaphrodite fluke and are voided with the host's faeces after passing through the gallbladder and the intestines. The miracidia hatch after the infective eggs have been ingested by a suitable land snail (Table 40).

The liberated miracidia migrate through the gut wall to the hepatopancreas of the snail, where they transform into mother sporocysts within 1 or 2 months. Daughter sporocysts are produced by parthenogenesis from the germ mass inside the mother sporocysts. Finally cercariae develop in the germ cells, again by parthenogenesis. The entire parthenogenetic phase in the terrestrial snails requires 3

Table 40. List of Snail Intermediate Hosts of *D. Dendriticum*

Cochlicopidae
Cochlicopa lubrica (Muell.)

Pupillidae
Abida frumentum (Drap.)

Enidae
Chondrula tridens (Muell.); Jaminia potaniniana (=Subzebrinus potaninianus) (Ancey); Ena obscura (Muell.); Jaminia sieversi (Mousson); Zebrina hohenackeri (L. Pfr.); Zebrina detrita (Muell.); Subzebrinus sogdianus (Mart.); Pseudonapaeus miser (Mart.); Ch. microtraga (?); Pseudochondrula brevior (Mouss.); P. schelkovnikovi (Rosen)

Clausilidae
Armenica brunea (Rossm.); Laciniaria varnensis (?)

Zonitidae
Zonitoides nitidus (Muell.); Oxychilus derbentinus (Bttg.)

Ariophantidae
Macrochlamys cassida (?); M. monticola (Pfr.)

Eulotidae
Eulota lantzi (Lindh.); E. macci (Gerstf.); E. duplocincta (Mart.); E. phaeozona (Mart.); E. fruticum (Muell.); E. paricincta (Mart.); E. almaatini (Skv.); E. rubens (Mart.); E. semenovi (Mart.); E. skwortzowi (Tzv.); E. plectotropis (=Cathaica plectotropis) (Mart.)

Helicidae
Helicella candicans (Pfr.); H. candaharica (Pfr.); H. itala (L.); H. candidula (Stunder); H. crenimargo (Pfr.); H. derbentina (Kryn.); H. unifasciata (Poiret); H. virgata (Da Costa); H. krynickii (Kryn.); H. protea (?); H. schelkonickovi; Theba carthusiana (Muell.); Zenobiella rubiginosa (A. Schm.); Fruticocampylaea narzanensis (Kryn.); Trichia eichwaldi (Pfr.); E. selecta (Klika); Metafruticicola pratensis (Pfr.); Cepaea nemoralis (L.); Helix vulgaris (Rossm.); H. lucorum (L.)

In addition, two species of an uncertain systematic position: Trochoidea pyramidata and Choclicella acuta

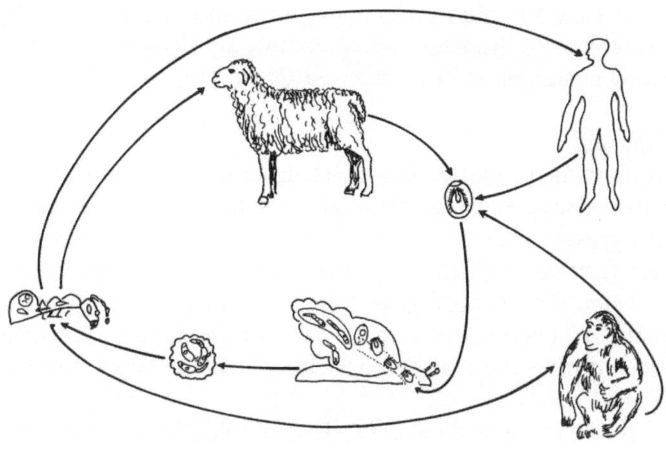

Fig. 57. Life cycle of *Dicrocoelium dendriticum*

to 5 months. The fully developed cercariae migrate to the molluscs respiratory chamber where they are enclosed in slime balls. When the ambient temperature drops the slime balls containing up to 500 cercariae are forcibly expelled through the respiratory pore.

If the slime balls are than eaten by ants (*Formica fusca* and other species) the cercariae migrate through the intestinal tract and encyst in the ant's gaster, thorax and head. The neurophysiological effects of the metacercarial metabolites cause the infected ants to climb vegetation where they become comatose and are thus easily swallowed by grazing animals (or accidentally by man).

When ants infected with metacercariae are eaten by the definitive host, the metacercariae are released in the duodenum and the young flukes migrate via the main bile duct or via the portal veins within 1 hour to the liver (Olsen, 1975; Rosicky and Groschaft, 1982).

Geographical Distribution

D. dendriticum is a common parasite in Europe, North Africa, Asia, Canada, USA, South America and Cuba (Rosicky and Groschaft, 1982).

Pathogenesis

Metacercariae migrating via the blood system may cause minute injuries in the intestinal epithelium, whereas flukes infecting the ductus choledochus may cause pressure atrophy of the ductal epithelia which may gradually develop into desquamative, catarrhal inflammation of the ductus choledochus.

Clinical Symptoms

Nonhuman primates: Kuntz (1972) assumes a symptomatology similar to that of *Clonorchis.*

Man: Dicrocoeliasis generally initiates either a prolonged period of constipation or diarrhea, with nausea and vomiting. These may occasionally be accompanied by abdominal discomfort and pain in the abdomen radiating to the collateral shoulder.

The pain in the liver is continuous and intensifies mainly at night. Other signs are subicterus, hepato- and splenomegaly or initially a leukocytosis, with an associated eosinophilia (Rosicky and Groschaft, 1982).

Pathology

Nonhuman primates: No exact description of morphological lesions in primate dicrocoeliasis is given, although a similarity to the lesions of *Clonorchis*-infections is suggested (Kuntz, 1972). These consist of a proliferation of biliary epithelium with formation of fibrous tissues around the sites of infection.

Man: The pathological lesions conform to that described for infected ruminants. In sheep infected by *D. dendriticum* minute subcapsular liver haemorrhages occur during the acute stage. Other haemorrhages can be found in the vascular intima.

In the chronic stage the lesions develop to a chronic catarrhal inflammation of the biliary ducts with marked fibrosis, which may lead to cirrhosis. The inflammatory exudate, containing lymphocytes, histiocytes and eosinophils, concentrates around the hepatic triads and leads to interstitial hepatitis.

Treatment

The drugs of choice in animals are Hetol, Hetolin and thiabendazol (Rosicky and Groschaft, 1982).

References

1. Faust, E.C.; P.C. Beaver, and R.C. Jung: Animal Agents and Vectors of Human Disease. 4th ed. Lea and Febiger, Philadelphia (1975)
2. Healy, G.R. and B.J. Myers: Intestinal helminths; pp. 265-296 in "The Chimpanzee" (Bourne, G.H., ed.), Vol.6, Karger, Basel (1973)
3. Hunter, G.W.; W.W. Frye, and J.C. Swartzwelder: A Manual of Tropical Medicine. 3rd ed., Saunders, Philadelphia (1964)
4. Kuntz, R.E.: Trematodes of the intestinal tract and biliary passages; pp.105-123 in "Pathology of Simian Primates", Part II (Fiennes, R.N.T-W, ed.), Karger, Basel (1972)
5. Olsen, O.W.: Animal Parasites. Their Life Cycles and Ecology. 3rd ed., Univ. Park Press, Baltimore (1974)
6. Rosicky, B. and J. Groschaft: Dicrocoeliosis; pp.33-52 in "CRC Handbook Series in Zoonoses" (Steele, J.H., Series ed.), Section C: Parasitic Zoonoses, Vol. III (Hillyer, G.V. and Hopla, Vol. eds.), CRC Press, Boca Raton (1982)
7. Spencer, H.: Fascioliasis, heterophyiasis and other fluke diseases; pp.597-610 in "Spezielle pathologische Anatomie" (Doerr, W.; G. Seifert, and E. Uehlinger, eds.), Bd.8: Tropical Pathology (Spencer, H., Vol. ed.), Springer, New York (1973)

8.2 Fasciolidae

8.2.1 Fasciola Hepatica

F. hepatica, the common liver fluke of ruminants can occasionally infect man (Bendezu et al., 1982; Boray, 1982; Knobloch et al., 1985) and uncommonly rhesus monkeys (Bezubik and Furmaga, 1959; Wong and Conrad, 1978) and *M. fascicularis* (Bezubik and Furmaga, 1960). Nonspecified *Fasciola* were reported in cynomolgus monkeys by Hashimoto and Honjo (1966).

Agent
F. hepatica is leaf-shaped with a narrow cephalic cone on the broad anterior end. It is 18–32 × 7–14 mm in size, flat, greyish-brown colored with a dark outline of the highly dendritic blood-filled ceca. The muscular oral sucker which surrounds the mouth of the parasite is smaller than the ventral one. Its tegument is covered with scales. The digestive tract consists of two main branches of the ceca extending to the posterior extremity of the body. Each branch has an elaborate system of secondary and tertiary branches (Fig. 58).

F. hepatica is hermaphroditic. It has a branched ovary, which lies a short distance posterior from the ventral sucker towards the right side. Its testes are extensively branched, filling most of the body behind the ovary. Numerous vitelline glands extend on both sides.

The eggs are large (130–148 × 60–90 μm), oval, of yellowish color and operculated on one end (Fig. 58).

Life Cycle
The adult flukes lay an average of 8000 to 25.000 unembryonated eggs/day which are passed out of the liver via the common bile duct and are voided with the host's faeces. For further development the presence of water and of suitable intermediate hosts is necessary.

The eggs hatch only when free in water. The miracidia when released are ciliated and have a proboscid structure at the anterior end. They are capable of sur-

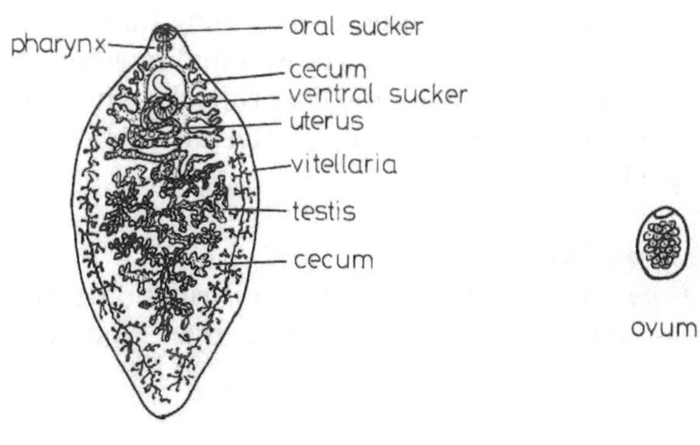

Fig. 58. *Fasciola hepatica*

Table 41. Intermediate Snail Hosts of *Fasciola Hepatica*

Geographical area	Principal or obligatory hosts	Alternate or facultative hosts
Europe (including most of U.S.S.R.)	L. truncatula	L. palustris s.l.
Africa	L. truncatula L. columella	
North America	L. humilis L. bulimoides L. cubensis	L. columella
South America	L. viator L. diaphana	L. columella
Asia	L. truncatula L. viridis	
Australia	L. tomentosa	L. columella[a] L. viridis[a]
New Zealand	L. tomentosa L. columella L. truncatula[a]	
Pacific region		
Hawaii	L. viridis	
Papua New Guinea	L. viridis (ollula)	
Philippines		
Japan		

[a] Very restricted distribution

viving up to 24 hrs free in water, where they search for a suitable intermediate host - a snail belonging to the family Lymnaeidae (Table 41).

The miracidia burrow into the snail, lose their cilia and transform into a sac-like mother sporocyst, each one of which produces numerous mother rediae. The mother rediae migrate to the snail's hepatopancreas, where they produce a great many daughter rediae, each one of which again producing a great many cercariae.

The cercariae emerge from the snail and quickly attach to objects in the water, particularly plants. They drop their tails and secrete a protecting double walled cyst around themselves. Metacercariae are infective 24 hrs after encystment and if eaten by a suitable new host - preferentially a ruminant - they excyst in the duodenum after activation of the young flukes in the rumen. Upon escaping from the cyst they penetrate the gut wall and migrate transabdominally to the liver and penetrate again through the liver capsule. After a period of migration and feeding in the parenchyma the flukes enter the bile ducts, where they grow and mature (Fig. 59).

Geographical Distribution

F. hepatica is cosmopolitan in its distribution except that it is not found in desert areas, nor in the eastern USA other than Florida.

Fig. 59. Life cycle of *Fasciola hepatica*

Clinical Symptoms

Nonhuman primates: In experimental infections of macaques with the "Japanese species" of *Fasciola* the animals died within 48 to 80 days after showing only general signs such as rough hair coat, anemia, inappetence, emaciation, prostration and ascites (Tomimura et al., 1975) and elevation of SGOT-, SGPT-, SLDH-, SLAP- and SAP-levels (Takemoto et al., 1977).

Man: Early symptoms consist of right upper quadrant abdominal pain, fever, hepatomegaly, coughing, biliary colic and vomiting. These may be accompanied by jaundice, generalized abdominal rigidity, diarrhea, fever, profuse sweating, eosinophilia and urticaria. Later profound systemic toxemia and macrocytic anemia, empyema of the gallbladder, cholecystitis or cholelithiasis may develop (Faust et al., 1975).

Pathology

Nonhuman primates: Early lesions during the migratory stage in experimental infection of macaques with the "Japanese species" of *Fasciola* consisted of emaciation, bloody ascites, hydrothorax and hydropericardium, and locally, of fibrinous and fibrous perihepatitis, haemorrhage and haematomas throughout the liver. During the maturation stage (80–160 days p.i.) fibrous perihepatitis, linear or stellate scars throughout the liver (especially in the left lobe), cholangitis and pericholangitis dominate the morphological lesions (Tomimura et al., 1975).

Man: Fascioliasis of man is usually a mild disease. In rare cases of heavy infections necrosis, hyperplasia and dilatation of the biliary passages are accompanied by leukocytic infiltration (Hunter et al., 1964).

Treatment

An efficient drug for treatment of human fascioliasis is dihydroemetin, given at doses of 1.5 mg/kg daily for 10 days (Boray, 1982).

Praziquantel is of varying efficacy in man, laboratory and domestic animals (Wegner, 1984). Only 21% of the patients stopped excreting *F. hepatica*-eggs despite repeated treatment with different doses of praziquantel (Knobloch et al., 1985).

8.2.2 Fasciolopsis Buski

F. buski, one of the largest flukes, and normally an intestinal parasite of pigs and man in the Far East (Rim, 1982), has also been reported twice in rhesus monkeys (Dieterich, 1955 - cit. from Hartmann, 1961; Hartmann, 1961).

Agent
The adult flukes are large, flat, fleshy worms of 2 to 7 cm length, 2 cm breadth and 0.5 to 3 mm thickness, and are covered by a spinose integument. *F. buski* lacks a cephalic cone. The oral sucker measures about 0.5 mm in diameter, whilst the nearby ventral sucker (acetabulum) has a diameter of 2 to 3 mm. The ceca are unbranched. The branched testes are situated in tandem position in the posterior half of the body. The branched ovary lies on the right side. The highly branched uterus occupies the anterior half of the body (Fig. 60). The large unembryonated eggs look like those of *F. hepatica* and measure 130–140 × 80–85 μm.

Life Cycle
F. buski occurs normally in the small intestine and sometimes the stomach of man and pigs, and also occasionally of dogs. Each worm is able to deposit 25.000 eggs daily, which are voided with the host's faeces. When contaminated faeces are deposited in stagnant freshwater the immature eggs develop to miracidia within 3 to 7 weeks at 27 to 32 °C. The miracidia then hatch, swim vigorously around in the water and on contact with a suitable small planorbid snail *(Segmentina hemisphaerula, S. trochoideus, Hippentis cantori, H. umbilicalis)* penetrate the soft tissues of the mollusc. Within a period of a few weeks sporocysts, first and second gener-

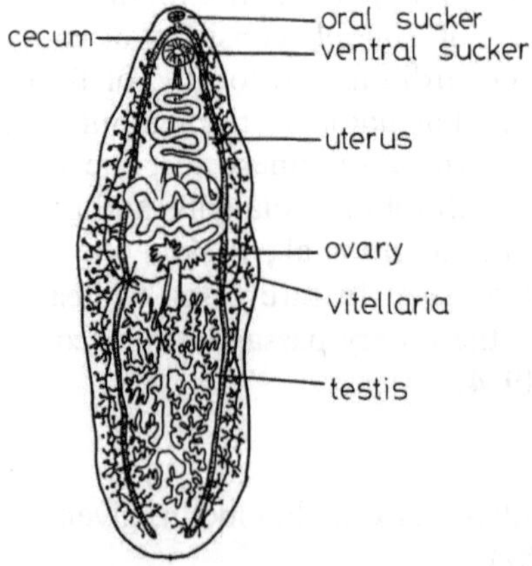

Fig. 60. *Fasciolopsis buski*

ation rediae and finally vigorous cercariae are produced. The cercariae emerge from the snail, swim around until they find a suitable water plant, usually the seedpods of red water caltrop *(Trapa natans)*, the bulb of water chestnuts *(Eliocharis tuberosa)* or water hyacinths *(Eichhornia crassipes)*. The cercariae then attach themselves to the plants, encyst and mature in 3 to 4 weeks. Water caltrops and water chestnuts are mayor food resources in the Far East, with water caltrops being cultivated in artificial ponds fertilized by nightsoil (Hunter et al., 1964). When the contaminated plants are eaten or the seeds are opened or peeled off with the teeth the cysts are ingested. The metacercariae excyst in the duodenum, where they become attached to the adjacent intestinal wall and develop into adult worms in about 3 months (Rim, 1982).

Geographical Distribution

Most human infections occur in the mainland China, but *F. buski* is also endemic with a much lower prevalence in Taiwan, Thailand, Vietnam, Laos, Bangladesh and India. In Indonesia and Burma it has been reported in pigs but not in man (Rim, 1982).

Clinical Symptoms

Nonhuman primates: No clinical disease was reported in the few *F. buski*-infected monkeys.

Man: The incubation period lasts 30 to 40 days, but the infection in man is not always associated with noticeable symptoms. If clinical signs are observed they may consist of gripping epigastric hunger pains especially in the morning hours and diarrhea, both of which mimic peptic ulcers. Ascites, asthenia, generalized abdominal pain, capricious appetite, anorexia, nausea and vomiting may be observed in more advanced case.

Pathology

Nonhuman primates: Small erythematous foci at the sites of attachment were found in the stomach (Hartmann, 1961).

Man: The parasites cause local inflammation, erosion and frequently ulceration at the sites of attachment. Occasionally gland abscesses develop in the small intestinal or gastric mucosa. Heavy infections may cause acute intestinal obstruction, whilst the toxic metabolites of the parasites can produce edema of the face, abdomen and lower extremities (Faust et al., 1975).

Treatment

In man thymol, carbon tetrachloride, tetrachorethylene (0,1-0,12 ml/kg) and hexylresorcinol crystoid (0,4-1,0 g) have been used (Rim, 1982). *F. buski* responds well to praziquantel (Wegner, 1984).

References

1. Bendezu, P.; A. Frame, and G. V. Hillyer: Human fascioliasis in Corozal, Puerto Rico. J. Parasitol. *68*, 297-299 (1982)
2. Bezubik, B. and S. Furmaga: The helminth parasites in *Macacus rhesus* Audeb. from China. Acta Parasitol. Polon. *7*, 591-598 (1959)
3. Bezubik, B. and S. Furmaga: The parasites in *Macacus cynomolgus* L. from Indonesia. Acta Parasitol. Polon. *8*, 335-444 (1960)
4. Boray, J. C.: Fascioliasis; pp. 71-88 in "CRC Handbook Series in Zoonoses" (Steele, J. C., Series ed.), Section C: Parasitic Zoonoses, Vol. III (Hillyer, G. C. and C. E. Hopla, Vol. eds.), CRC Press, Boca Raton (1982)
5. Faust, E. C.; P. C. Beaver, and R. C. Jung: Animal Agents and Vectors of Human Disease. 4th ed. Lea and Febiger, Philadelphia (1975)
6. Hartmann, H. A.: The intestinal fluke (Fasciolopsis buski) in a monkey. Amer. J. Vet. Res. *22*, 1123-1126 (1961)
7. Hashimoto, I. and S. Honjo: Survey of helminth parasites in cynomolgus monkeys *(Macaca irus)*. Jap. J. Med. Sci. Biol. *19*, 218 (1966)
8. Hunter, G. W.; W. W. Frye, and J. C. Swartzwelder: A Manual for Tropical Medicine. 3rd ed., Saunders, Philadelphia (1964)
9. Knobloch, J.; E. Delgado; A. Alvarez; G. U. Reymann, and R. Bialek: Human fascioliasis in Cajamarca/Peru. I. Diagnostic methods and treatment with praziquantel. Trop. Med. Parasit. *36*, 88-90 (1985)
10. Rim, H. J.: Fasciolopsiasis; pp. 89-97 in "CRC Handbook Series in Zoonoses" (Steele, J. H., Series ed.), Section C: Parasitic Zoonoses, Vol. III (Hillyer, G. V. and C. E. Hopla, Vol. eds.), CRC Press, Boca Raton (1982)
11. Takemoto, Y.; M. Yokata; H. Yoshida; S. Yamagami; T. Kotani, and T. Tomimura: Experimental fascioliasis in monkeys. II. Studies on liver function tests and electrophoretic patterns of serum protein in monkeys infected with the "Japanese species" of Fasciola. Bull. Univ. Osaka Prefect. Ser. B *29*, 32-41 (1977)
12. Tomimura, T.; T. Kotani; Y. Takemoto; M. Yokata; S. Yamagami, and H. Yoshida: Experimental fascioliasis in monkeys. I. Parasitological, clinical and pathological observations on monkeys infected with the "Japanese species" of Fasciola. Jap. J. Vet. Sci. *37*, 391-406 (1975)
13. Wegner, D. H. G.: The profile of the trematodicidal compound praziquantel. Drug Res. *34* (II), 9b, 1132-1136 (1984)
14. Wong, M. M. and H. D. Conrad: Prevalence of metazoan parasitic infections in five species of Asian macaques. Lab. Anim. Sci. *28*, 412-416 (1978)

8.3 Opisthorchiidae

8.3.1 Clonorchis Sinensis (Opisthorchis S.)

C. sinensis, the Chinese or oriental liver fluke in man and fish-eating mammals in the Far East (Rim, 1982) has occasionally been reported in *M. fascicularis* (Habermann and Williams, 1957; Kuntz, 1972) and *M. cyclopis* (Kuntz, 1972).

Agent

The adults are lanceolate, elongated, flat and flabby, transparent, golden brown to pinkish worms of varying size. Depending on the age and number of parasites, the host species, and their location in the bile ducts, the size of *C. sinensis* ranges from

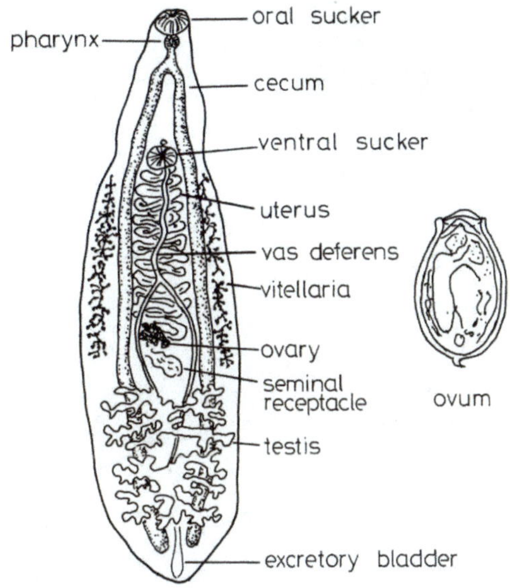

Fig. 61. *Clonorchis sinensis*

4 to 25 mm length and from 1,5 to 5 mm breath with a thickness of approximately 1 mm. The cuticle is smooth and shiny, through which all the important internal structures of the worm can be seen (Fig. 61). The globose oral sucker is directed anteriorly and is slightly larger than the acetabulum. The ceca are simple and unbranched. The deeply lobed testes lie in tandem position at the posterior end, with the small, slightly lobate ovary lying in front of the anterior testis. A loosely coiled uterus fills much of the middle and anterior body. The minutely follicular vitelline glands lie externally to the ceca.

The eggs are thick-shelled, narrowing a little behind the operculum, yellowish brown and ovoid (Fig. 61). The operculum is distinctly convex. At the thickened posterior end is a small comma-shaped median protuberance. The surface of the egg-shell has an arabesque polygonal pattern caused by membranous extuberances, the eggs are embryonated when passed in the faeces.

Life Cycle (Fig. 62)
Mature *C. sinensis* live typically in the bile passages, most frequently in the more distal tributaries under the liver capsule. The eggs deposited in the biliary passages are voided with the host's faeces. The miracidia contained in the eggs do not hatch unless the eggs are ingested by a suitable species of aquatic snails *(Parafossarulus manchouriensis, Bulimus fuchsianus, Alocinma longicornis, Hua ningpoensis).* The miracidia, after hatching in the snails and penetrating the mollusc's gut wall, become sporocysts in the perirectal region and within 14 days after being ingested have transformed into redia. The redia in turn produce 3 to 50 cercariae which complete their development in the interhepatic lymph space and finally leave the snail. The free cercariae survive for approximately 1 day in the water, after which they perish unless they encounter a freshwater fish (mostly belonging to the Cyprinidae, Gobiidae, Anabantidae or Salmonidae) within that time. The cer-

Fig. 62. Life cycle of *Clonorchis sinensis*

cariae penetrate beneath the scales and encyst either on the scales, in the muscles or in the subcutaneous tissues, thereby losing their tails. In 35 days the cercariae develop into encysted metacercariae and when an infected fish is eaten raw (smoked, pickled, dried) by the definitive host (man, animal) the parasite excysts in the host's duodenum. The liberated larvae migrate via the ampulla of Vater and the common bile duct to the distal biliary radicles to become mature worms within 2 weeks (Hunter et al., 1964; Rim, 1982).

Geographical Distribution

C. sinensis is largely confined to the Far East, being present in Japan, southern Korea, mainland China, Taiwan, and North Vietnam (Rim, 1982). It has also been reported in Chinese populations of the USA, Cuba, India and in native Hawaiians (Hunter et al., 1964).

Clinical Symptoms

Nonhuman primates: No symptoms have been reported.

Man: Early infections are characterized by fever up to 40 °C with or without chills, eosinophilia, leukocytosis and epigastric pains. The prodromal symptoms may be followed by diarrhea and progressive hepatomegaly, occasional episodes of jaundice and in the more advanced cases by liver cirrhosis, ascites, anasarca, cachexia and occasionally extreme jaundice.

Pathology

The adult *C. sinensis* in the distal bile passages causes proliferation and desquamation of the biliary epithelia with interstitial hepatitis and fibrosis, which are usually limited to the hepatic triads, and dense periductular fibrosis. In advanced cases pressure necrosis of the adjacent hepatic parenchyma occurs.

Treatment

Hexachloroparaxylol has been used successfully in man, hetol (Hoechst) was equally successful in animals (Faust et al., 1975). Today praziquantel seems to be the drug of choice (Wegner, 1984).

References

1. Faust, E.C.; P.C. Beaver, and R.C. Jung: Animal Agents and Vectors of Human Disease. 4th ed. Lea and Febiger, Philadelphia (1975)
2. Habermann, R.T. and F.P. Williams, jr.: Diseases seen at necropsy of 708 *Macaca mulatta* (rhesus monkey) and *Macaca philippinensis* (cynomolgus monkey). Amer. J. Vet. Res. *18*, 419-426 (1957)
3. Hunter, G.W.; W.W. Frye, and J.C. Swartzwelder: A Manual of Tropical Medicine. 3rd ed. Saunders, Philadelphia (1964)
4. Kuntz, R.E.: Trematodes of the intestinal tract and biliary passages; pp.104-123 in "Pathology of Simian Primates", Vol. II (Fiennes, R.N.T-W, ed.), Karger, Basel (1972)
5. Rim, H.J.: Clonorchiasis; pp.17-32 in "CRC Handbook Series in Zoonoses" (Steele, J.H., Series ed.), Section C: Parasitic Zoonoses, Vol. III (Hillyer, G.V. and C.E. Hopla, Vol. eds.), CRC Press, Boca Raton (1982)
6. Wegner, D.H.G.: The profile of the trematodicidal compound praziquantel. Drug. Res. *34* (II), 9b, 1132-1136 (1984)

8.4 Heterophyidae

8.4.1 Metagonimus Yokogawai

M. yokogowai is a small trematode living in the small intestines of dogs, cats, pigs, mice, fish-eating birds (pelikans) and man (Hunter et al., 1964) and which has also been apparently found in *Cercopithecus* sp. (Cosgrove, 1966).

Agent

M. yokogawai is a minute pyriform worm with a broadly rounded posterior end (Fig. 63) and of 1-2,5 × 0,4-0,75 mm size. The cuticle is covered by minute spines. The acetabulum is deflected to the right-hand side and is much larger than the oral sucker.

A single ovoid ovary is situated centrally, whilst the two testes lie diagonally in the posterior part of the body. The winding uterus fills part of the median and posterior body, whilst the vitelline glands are situated laterally in the posterior half of the body. The ceca are unbranched.

The eggs measure 26,5-28 × 15,5-17 μm. They have a conspicious conical operculum and contain a mature miracidium (Fig. 63).

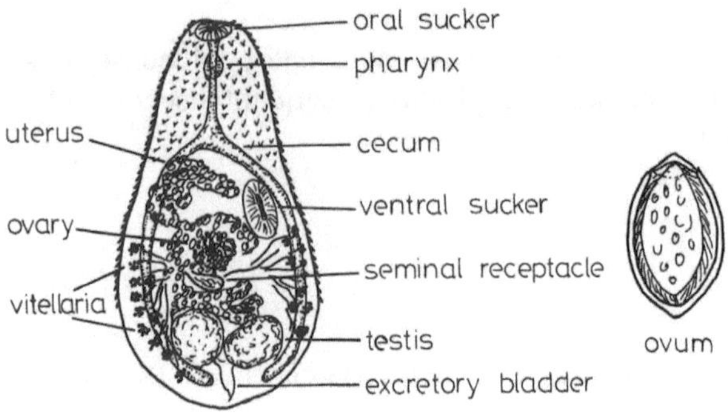

Fig. 63. *Metagonimus yokogawai*

Life Cycle

The eggs have to be ingested by an appropriate operculate snail *(hiara granifera, Semisulcospira libertina).* The miracidium transforms into a sporocyst followed by two successive generations of rediae and finally be the development of cercariae, within the soft extraintestinal tissues. The cercariae escape from the snail, attach to fresh-water fishes and encyst under the skin. Infection of the definitive host is by eating uncooked infected fish (Hunter et al., 1964; Faust et al., 1975; Yamaguchi, 1981).

Geographical Distribution

M. yokogawai is present in Japan, mainland China, Taiwan, Malaya, Korea, the maritime provinces of the USSR and northern Siberia, the Balkan peninsula and has also been reported from human cases in Spain (Faust et al., 1975; Yamaguchi, 1981).

The Cercopithecids, which are African monkey species, must have acquired the infection during captive life by feeding raw fish.

Clinical Symptoms

Nonhuman primates: No clinical symptoms were reported.

Man: Most infections go unnoticed, but in heavy infections persistent mild to moderate mucous diarrheas may occur (Faust et al., 1975).

Pathology

The small flukes are found attached to the intestinal mucosa of the mucosal folds, but they also may actually invade the intestinal mucosa. The most common sites of infection are the duodenum and the jejunum. The flukes cause superficial erosion and excess production and secretion of mucus in superficial infections and granulomatous infiltrations in deep invasions. Eggs transported to erratic sites may cause similar responses in other organs (Hunter et al., 1964; Faust et al., 1975).

Treatment

Praziquantel is as effective as niclosamide (Yamaguchi, 1981; Wegner, 1984).

References

1. Cosgrove, G.E.: The trematodes of laboratory primates. Lab. Anim. Care *16*, 23-39 (1966)
2. Faust, E.C.; P.C. Beaver, and R.C. Jung: Animal Agents and Vectors of Human Disease. 4th ed. Lea and Febiger, Philadelphia (1975)
3. Hunter, G.W.; W.W. Frye, and J.C. Swartzwelder: A Manual of Tropical Medicine. Saunders, Philadelphia (1964)
4. Wegner, D.H.G.: The profile of the trematocidal compound Praziquantel. Drug Res. *34* (II), 9b, 1132-1136 (1984)
5. Yamaguchi, T.: Color Atlas of Clinical Parasitology. Lea and Febiger, Philadelphia (1981)

8.5 Echinostomatidae

Echinostomatidae are characterized by a spinose collar mounted on a circumoral disc.

8.5.1 Echinostoma Ilocanum (Euparyphium I.)

E. ilocanum is primarily a parasite of rats in Indonesia which occasionally infects other mammalian species including man. *E. ilocanum* is relatively common in parts of the Philippine population and human infections have also been reported in Java, India, China and Celebes (Ulmer, 1975). The parasite is reported to occur occasionally in "monkeys" (Yamaguti, 1958; Ulmer, 1975).

Agent
The adult flukes measure up to 6.5 mm in length and 1 to 1.35 mm in breath. They are attenuate at both ends, reddish-grey in color and are covered by plaque-like scales. The small circumoral disc is surrounded by 49 to 51 spines (Faust et al., 1975; Rim, 1982). The rounded testes are deeply lobulated and lie in tandem in the last quarter of the body (Fig. 64).

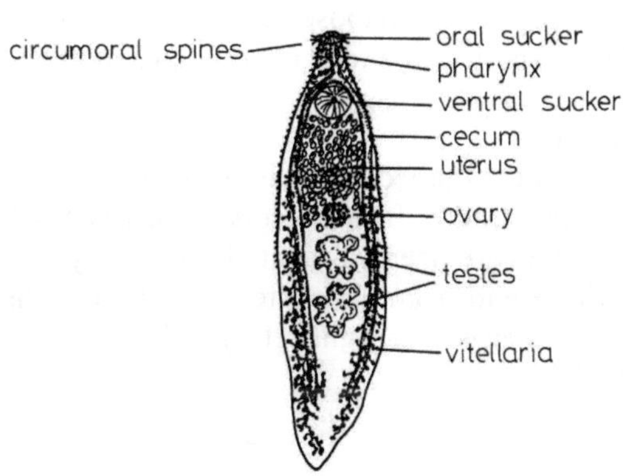

Fig. 64. *Echinostoma ilocanum*

The ovary is transversely ovoid and is positioned just in front of the anterior testis. The uterine coils are placed between the unbranched ceca, the anterior testis and the acetabulum. The vitellaria consist of medium-sized follicles in the lateral fields along the entire body except for its anterior quarter (Rim, 1982).

The eggs are large, straw colored and ovoid, measuring 83-116 × 58-69 μm. They have a small operculum and are unembryonated when discharged in the host's faeces (Faust et al., 1975).

Life Cycle

The eggs mature in 6-15 days in fresh water and subsequently the miracidia hatch. The first intermediate hosts are freshwater snails *(Gyraulus convexiusculus, G.prashadi, Hippeutis umbilicalis)*, the second intermediate hosts are again snails *(Vivipara javanicus, V.burranghina, V.rudipellis, Pila conica, P. ampullacea, G.prashadi, G.convexiusculus, Amphipepla quadrasi, Burinus hungerfordianus, Lymnaea peregra, Planorbis umbilicatus, Tiara asperata, Contradeus contradeus)*, which, when eaten raw, transmit the infection to the definitive host.

Geographical Distribution

E.ilocanum is relatively common in Luzon, other Philippine islands, Java, and Canton/China (Faust et al., 1975). Cases have also been reported from India and Celebes (Ulmer, 1975).

Clinical Symptoms

The usual symptoms in heavy infections of man are colic and diarrhea (Faust et al., 1975).

Pathology

The mature worms induce ulceration and inflammatory responses at the sites of attachment (Faust et al., 1975).

8.5.2 Echinostoma Malayana (Euparyphium M.)

E.malayana is a natural parasite of rats and pigs. It is fairly common in man, cat, dog, pig and mongoose in Southeast Asia (Ulmer, 1975; Rim, 1982) and has also been reported in a feral cynomolgus monkey caught in Malaysia (Janagi, 1981).

Agent

The lancet shaped, elongated worms measure 6,5 to 9,3 mm in length and 2 to 2,3 mm in breadth. Its head collar consists of 43 spines. The testes are situated in tandem and are deeply indented. The ovary is globular or slightly flattened. The vitellaria extend from the posterior margin of the acetabulum (Rim, 1982).

The eggs are slightly larger than those of *E.ilocanum*, measuring 106-131 × 69-80 μm.

Life Cycle

The first intermediate host is *Lymnaea leuteola*, the second one is either a snail *(L. leuteola, Indoplanorbis exustus, Gyraulus convexiusculus)* or fish *(Barbus stigma)* (Rim, 1982).

Geographical Distribution

Human infections have been reported in Singapore and Kuala Lumpur (Malaysia), Chiang-mai (Thailand), North Sumatra and in the population around Batang on the Chinese-Tibetan border (Rim, 1982). The one *M. fascicularis* found infected in the wild was obtained from a contractor who captured the monkeys in the Malaysian states of Kedah, Joho and Malacca (Janagi, 1981).

References

1. Faust, E.C.; P.C. Beaver, and R.C. Jung: Animal Agents and Vectors of Human Disease. 4th ed. Lea and Febiger, Philadelphia (1975)
2. Janagi, T.S.: A report on gastro-intestinal helminth parasites found in *Macaca fascicularis* in peninsular Malaysia. Malays. appl. Biol. *10*, 99-100 (1981)
3. Rim, H.J.: Echinostomiasis; pp. 53-69 in "CRC Handbook Series in Zoonoses" (Steele, J.H., Series ed.), Section C: Parasitic Zoonoses, Vol. III (Hillyer, G.V. and C.E. Hopla, Vol. eds.), CRC Press, Boca Raton (1982)
4. Ulmer, M.J.: Other trematode infections; pp. 646-677 in "Diseases Transmitted from Animals to Man." 6th ed. (Hubbert, W.T.; W.F. McCulloch, and P.R. Schnurrenberger, eds.), C.C. Thomas, Springfield (1975)
5. Yamaguti, S.: Systema Helminthum. Vol. 1: The Digenetic Trematodes of Vertebrates. Interscience, N.Y. (1958)

8.6 Paragonimidae

Paragonimus spp. are lung flukes with low host specificity. Final hosts are several mammalian species including man.

8.6.1 Paragonimus Westermani

P. westermani, the Oriental lung fluke, infects predominantly carnivores, especially wild felids, but it has also been found in domestic animals and monkeys (Nelson, 1975). It is the most important *Paragonimus*-species infecting man in Southeast Asia. Paragonimiasis has also been found in nonhuman primates of the same area with *P. westermani* being reported in cynomolgus monkeys (or crab-eating macaques - *M. fascicularis* -) (Sandosham, 1950, 1954; Hashimoto and Honjo, 1966; Murphy et al., 1979) and Taiwanese macaques - *M. cyclopis* - (Miyazaki, 1982).

Agent

P. westermani is a plump, ovoidal, reddish-brown trematode, 7,5 to 12 mm in length, 4 to 6 mm in breadth and 3,5 to 5 mm thick. Its cuticle is covered by scale-like spines. The two suckers are almost equal (Fig. 65) and 0,75 to 0,8 mm in diameter.

The lobated testes lie side by side in the posterior half of the body, with the club- or fingerlike lobated ovary (Fig. 65) lying slightly behind and on one side of the acetabulum. The uterus is located opposite to the ovary. The vitellaria are dark bodies scattered along the periphery, whilst the long undulating ceca reach the posterior end of the body.

The eggs of *P. westermani* are broadly ovoidal, thick-shelled, golden brown and range from 80 to 118 μm length × 48 to 60 μm width (Hunter et al., 1964; Faust et al., 1975). They are unembryonated when laid and possess a somewhat flattened operculum (Fig. 65). The nonoperculated end is somewhat thickened.

Life Cycle

Adult *P. westermani* reside in parasitic caverns in the lung tissues of the definitive hosts. They deposit their eggs in the fibrous capsules and these usually accumulate around the worm. The eggs on reaching the airways are coughed up, swallowed and passed in the faeces, then if they reach clear, cool, running water, they embryonate in 16 days. The miracidia hatch out slowly, swim in the water, enter into a snail host *(Semisulcospira libertina, Brotia asperata, B. costula episcopalis, Thiara granifera, Synerca lutea, Hua toucheana)* and within 3 to 5 months develop into mother sporocysts, rediae and finally the cercariae which escape from the snail and swim free in the water. The cercariae attack fresh water crabs, crayfish or shrimp (Hunter et al., 1964), invading their viscera, muscles or soft tissues and developing into encysted metacercariae which are infective for man after a developmental period of several weeks. When the metacercariae-containing crustacea are eaten by the definitive host excystation occurs in the duodenum. The young

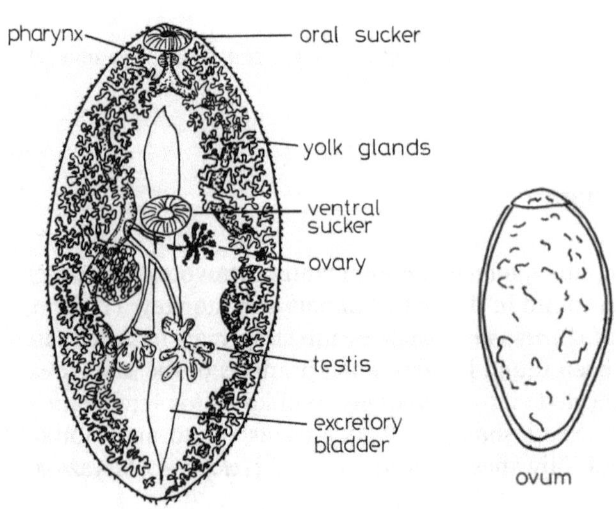

Fig. 65. *Paragonimus westermanni*

flukes migrate through the intestinal wall, the peritoneal cavity, the diaphragm and finally the pulmonary tissues. After about 3 weeks migration they finally settle down in the lung, usually in pairs. Maturity is reached in 5 to 6 weeks after ingestion of infected crabs. The complicated non-channelled migratory routes from the duodenum to the lung provide opportunities for the worms to wander to ectopic sites where they may encyst and remain infective, this happens particularly in paratenic hosts (Yokogawa, 1982). Wild boars, pigs, small mammals and also monkeys are such unfavorable hosts (Kim, 1984). If the flesh of such an animal is eaten by another definitive host, usually a larger carnivorous animal, the flukes may complete their development in the new host (Yokogawa, 1982; Kim, 1984).

Geographical Distribution

According to Faust et al. (1975) *P. westermani* is heavily endemic in China, Korea, Japan, the Philippines and Taiwan, occasional reports have been made from Manchuria, Nepal and Thailand. However, taxonomical difficulties may require a correction. *Paragonimus* other than *P. westermani* have been reported in the same areas, *P. pulmonalis* has been separated from *P. westermani.*

Pathogenesis

In experimental infections the young flukes travelling through the intestinal wall and the abdominal cavity caused bouts of acute fibrinoexudative peritonitis, ascites, and turbid or haemorrhagic pleural fluid at 20 days p.i. and during that time maximum liver damage and diaphragmatic injuries developed. The peritoneal damage regressed after day 45.

The flukes reaching the lungs are often found inside the bronchial lumina. They induce pulmonary haemorrhages and acute exudative pneumonitis. The gradually maturing flukes become encysted by zones of active inflammatory exudates and collagenous fibrous tissues (Choi, 1984).

Clinical Symptoms

Nonhuman primates: In the spontaneously infected *M. fascicularis* a nonproductive cough and mild dyspnea were observed, whilst auscultation revealed harsh lung sounds and in X-rays consolidation in the cranial and caudal lobes of both lungs, occasionally with cavitation, were visible (Murphy et al., 1979).

Man: In man two major manifestations of paragonimiasis can be distinguished:

a) Pulmonary paragonimiasis: Early in the infection almost all patients show pleurisy and occasionally pneumothorax. The symptoms of chronic infections are cough and haemoptysis, with occasionally difficulties in breathing.

b) Cerebral paragonimiasis: Cerebral paragonimiasis is the most common form of extrapulmonary paragonimiasis in man, it is probably caused by adult worms migrating into the brain from the lung (Yokogawa, 1982). The symptoms are similar to those of Jacksonian type epilepsy, with ophthalmological signs (hemianopsia, optic atrophy, papilledema) being common (Choi, 1984).

Pathology

Nonhuman primates: In the *M. fascicularis* thick-walled cysts containing reddish-brown flukes were found in both lungs. Histologically the flukes were surrounded by a thick wall of fibrous connective tissue containing diffuse infiltrations of plasma cells and eosinophils. The lining epithelia were partially hyperplastic or metaplastic, whilst in other areas were devoid of epithelia (Murphy et al., 1979).

Man: Parasites which fail to leave the peritoneal cavity may continue their development in the omentum, causing a chronic peritonitis with the flukes being possibly present in the liver and the pancreas.

Pulmonary paragonimiasis begins as burrow formation: small tortuous tunnels of haemorrhagic destruction without cellular reactions or surrounding wall formation. Such burrows are found particularly in the more superficial regions of the lung. With further development of the flukes the lesion may heal and their progress may be arrested with a zone of necrosis and granulation tissue around the flukes, or abscesses or cysts of 1 cm diameter may be formed. The cysts harbour 1 or 2 mature flukes, although older cysts often contain only eggs. The cysts are also filled by a viscous, pinkish-brown fluid containing debris of necrotic eosinophils and Charcot-Leyden crystals. The cyst walls may be very thin, but with age they become heavily fibrotic.

Cerebral paragonimiasis is most often localized in the temporal lobes. The migrating flukes cause necrosis and eventually liquefaction of white and grey matter forming creamy cysts varying in size from that of a soybean to the palm of a child's hand (Yokogawa, 1982) which are surrounded by a zone of gliosis several millimetres thick (Gibson, 1973).

Treatment

Todays drug of choice is probably Praziquantel with a cure rate of 100% at a dosage of 3 × 25,0 mg/kg (Wegner, 1984). Bithimol (2,2'thiobis(4,6-dichloro phenol)) is also effective against *Paragonimus* (Faust et al., 1975), but is of higher toxicity.

8.6.2 Paragonimus Africanus

P. africanus is one of two African lung fluke species occurring in man and has also been observed in natural infections of nonhuman primates. Its principal hosts are Viverridae (civet cats), whilst other common natural hosts are dogs, mongooses *(Crossarchus obscurus)*, and swamp mongooses *(Atilax paludinosus)*. In nonhuman primates it has been found in drills *(Mandrillus leucophaeus)* and once in a potto *(Perodicticus potto)* (Sachs and Voelker, 1975; Voelker and Sachs, 1977).

Agent

Adult worms are characterized by a large oral sucker, a delicately branched ovary and profusely divided testes, with either single or grouped cuticular spines (Miyazaki, 1982).

The eggs measure 91,6 × 48,3 μm.

Life Cycle
The life cycle closely resembles that of *P. westermani*. Snail hosts are probably *Potadoma freethii*, whilst the crab hosts are *Sudanonaulus africanus* and *S. pelli* (Miyazaki, 1982).

Geographical Distribution
P. africanus so far has only been recorded in the Bakossi region of West Cameroon (Sachs and Voelker, 1980).

Clinical Symptoms
Nonhuman primates: In X-rays of experimental infections of rhesus monkeys hilar dilatation, streaky, patchy or diffuse lung shadows, cavities, atelectases and pleural reactions have been reported particularly in the lower parts of the lung (Volkmer, 1977).

Man: All patients who had eggs of *P. africanus* suffered from chronic cough and haemoptysis (Yokogawa, 1982).

Pathology
Nonhuman primates: The lesions in the drills and potto consisted of pulmonary caverns and fibrous connections between the thoracic wall and the diseased lung (Voelker and Sachs, 1977). Histopathological examination of experimentally infected rhesus monkeys revealed parasitic caverns with intense plasma cellular infiltration of the cyst walls, egg-granulomas, peribronchitis, and pleuritis (Racz et al., 1977).

References

1. Choi, W.Y.: Paragonimus westermani: Pathogenesis and clinical features. Drug Res. 34 (II), 9b, 1184-1185 (1984)
2. Faust, E.C.; P.C. Beaver, and R.C. Jung: Animal Agents and Vectors of Human Disease. 4th ed., Lea and Febiger, Philadelphia (1975)
3. Gibson, J.B.: Clonorchiasis and paragonimiasis; pp. 611-632 in "Spezielle pathologische Anatomie" (Doerr, W.; G. Seifert und E. Uehlinger, eds.), Bd. 8: Tropical Pathology (Spencer, H., Vol. ed.), Springer, New York (1973)
4. Hashimoto, I. and S. Honjo: Survey of helminth parasites in cynomolgus monkeys (Macaca irus). Jap. J. Med. Sci. Biol. *19*, 218 (1966)
5. Hunter, G.W.; W.W. Frye, and J.C. Swartzwelder: A Manual of Tropical Medicine. 3rd ed. Saunders, Philadelphia (1964)
6. Kim, D.C.: Paragonimus westermani: Life cycle, intermediate hosts, transmission to man and geographical distribution in Korea. Drug Res. *34* (II), 9b, 1180-1183 (1984)
7. Miyazaki, I.: Paragonimiasis; pp. 143-164 in "CRC Handbook Series in Zoonoses" (Steele, J.H., Series ed.), Section C: Parasitic Zoonoses, Vol. III (Hillyer, G.V. and C.E. Hopla, Vol. eds.), CRC Press, Boca Raton (1982)
8. Murphy, J.C.; J.G. Fox, and M. Shaley: *Paragonimus westermani* infection in a cynomolgus monkey. J. Amer. Vet. Med. Ass. *175*, 981-984 (1979)
9. Nelson, G.S.: Paragonimiasis; pp. 641-645 in "Diseases Transmitted from Animals to Man". 6th ed. (Hubbert, W.T.; W.F. McCulloch, and P.R. Schnurrenberger, eds.), C.C. Thomas, Springfield (1975)
10. Racz, P.; K. Tenner; J. Voelker, and H.H. Schumacher: Histopathologische Veränderungen bei

Rhesus-Affen nach experimenteller Infektion mit afrikanischen Lungenegeln (Paragonimus africanus und P. uterobilateralis). Tropenmed. Parasit. *28*, 149-157 (1977)
11. Sachs, R. and J. Voelker: A primate, Mandrillus leucophaeus, as natural host of the African lung fluke Paragonimus africanus in West-Cameroon. Tropenmed. Parasit. *26*, 205-206 (1975)
12. Sachs, R. and J. Voelker: Paragonimus lung fluke infection in wild primates, with special reference to food habits of African drill (Mandrillus leucophaeus); pp. 147-152, XXII. Internat. Symp. Erkr. Zootiere, Arnhem (1980)
13. Sandosham, A. A.: *Paragonimus sp.* from the Kra monkey from Malaya. Trans. Roy. Soc. trop. Med. Hyg. *44*, 5 (1950)
14. Sandosham, A. A.: Malayian parasites. XV. Seven new worms from miscellaneous hosts. Studies from Inst. Med. Res. Fed. Malaya *26*, 213-226 (1954)
15. Voelker, J. und R. Sachs: Affen als natürliche und experimentelle Endwirte afrikanischer Lungenegel (Paragonimus africanus, P. uterobilateralis). Tropenmed. Parasit. *28*, 137-144 (1977)
16. Volkmer, K.-J.: Röntgenmorphologische Befunde beim Rhesus-Affen nach experimenteller Infektion mit afrikanischen Lungenegeln (Paragonimus africanus und P. uterobilateralis). Tropenmed. Parasit. *28*, 145-148 (1977)
17. Wegner, D. H. G.: The profile of the trematocidal compound praziquantel. Drug Res. 34 (II), 9b, 1132-1136 (1984)
18. Yokogawa, M.: Paragonimiasis; pp. 123-142 in "CRC Handbook Series in Zoonoses" (Steele, J. H., Series ed.). Section C: Parasitic Zoonoses, Vol. III (Hillyer, G. V. and C. E. Hopla, Vol. eds.), CRC Press, Boca Raon (1982)

8.7 Schistosomatidae

Schistosomes are digenic trematodes which remain *in copula* during much of their life within the vascular system of vertebrate hosts. In man the schistosomes cause diseases known as bilharziasis after Theodor Bilharz.

8.7.1 Schistosoma Mansoni

S. mansoni is fundamentally a parasite of man on the African continent, but the importation of infected African slaves to the neotropics enabled the introduction of the infection to Brazil, Surinam, Venezuela and the West Indies (Faust et al., 1975).

Natural *S. mansoni* infections, some of zoonotic background have also been found in nonhuman primates of endemic areas, being reported in *Papio* spp. in Kenya, Rhodesia, Uganda, South Africa, Ethiopia and Tanzania (Nelson, 1965; Strong et al., 1965; Kuntz and Myers, 1966; Fenwick, 1969; Fuller et al., 1979; Pitchford et al., 1973; Goldsmid, 1974); *Cercopithecus* spp. in Kenya, Ethiopia and St. Kitts (Cameron, 1928; Fuller et al., 1979; Graber, 1979; Else et al., 1982); *E. patas* in Nigeria (Kuntz et al., 1977) and *Pan troglodytes* from Sierra Leone (Hsü and Hsü, 1968; Renquist et al., 1975). It has also been found in a *Macaca radiata* from India (Wong and Conrad, 1978).

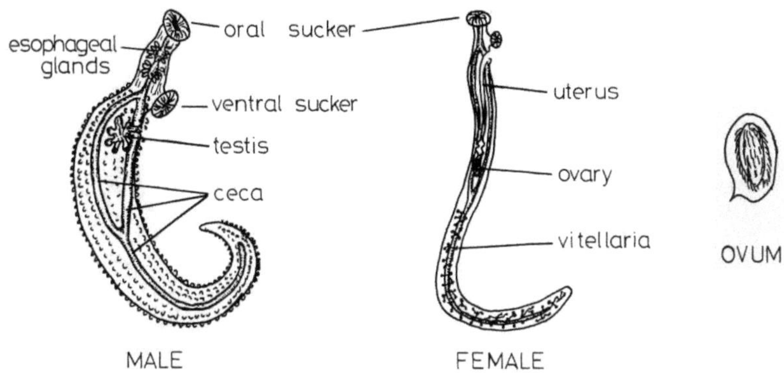

Fig. 66. *Schistosoma mansoni*

Agent
Schistosomes are more or less cylindrical trematodes. Adult male *S. mansoni*, are 6.4 to 9.9 mm long and usually less than 2 mm thick, with a cuticula that is provided with warty papillae (Fig. 66). They possess 6 to 9 minute, lobed testes which are positioned in a grape-like cluster closely behind the acetabulum (Faust et al., 1975). The ceca join to form a terminal limb (Cosgrove, 1966).

Female *S. mansoni* are 7.2 to 17 mm long and slender worms, with a very short uterus. They live permanently in the gynecophoric canal of the male (Fig. 66).

The eggs are elongated-ovoid, 114–175 × 45–68 μm in size, rounded at both ends and bear a lateral spine (Fig. 66). They are embryonated before penetrating the vascular wall.

Life Cycle (Fig. 67)
Adult *S. mansoni* usually reside continuously mating in the tributaries of the inferior mesenteric vein, mostly in the colonic vessels or the portal veins. The females produce approximately 300 eggs every day for a life span which lasts for several years (Katz et al., 1982). The eggs pass through the birth pore within the acetabulum and when the worm's sucker attaches to the endothelial surface, the eggs penetrate the vascular wall. This process is aided by proteolytic enzymes secreted by the miracidium contained in the egg. The eggs tend to accumulate in the perivascular tissues (submucosa, propria mucosae) before breaking out into the intestinal lumen. Worms moving to another site of attachment in the vein raise their posterior sucker from the endothelial surface and the eggs can be swept away to the liver.

Faecally excreted eggs must be deposited in fresh water in order to enable the miracidia to hatch within a few hours. The miracidia subsequently enter the soft, fleshy parts of a snail of the genus *Biomphalaria (B. pfeifferi, B. choanomphala, B. alexandria, B. sudanica, B. glabrata, B. stramina, B. tenagophila)* or in the neotropics *Australorbis* spp. and *Tropicorbis* spp. (Hunter et al., 1964). They invade first the snail's lymph spaces and subsequently the hepatopancreas. The miracidia transform to mother and daughter sporocysts and finally to male and female cercariae. The cercariae emerge from the snail, accumulate on the surface of the

Fig. 67. Life cycle of *Schistosoma mansoni*

water due to their positive phototropism and negative geotropism, and swim actively around. They must contact a definitive host within 8 hrs, otherwise the miracidia use up their glycogen reserves and die.

The cercariae penetrate the skin through a hair follicle which requires approximately half an hour. They penetrate the stratum corneum and burrow laterally by secreting a lytic enzyme from a pair of anterior cephalic glands. After shedding their tail the cercariae transform into schistosomulas while still in the dermis. Two days later the schistosomulas penetrate the remaining layers of the epidermis, migrate via the vascular system to the lung, the heart and the liver parenchyma and mature to adult worms (Spencer and Gibson, 1972; Katz et al., 1982).

Epidemiology

The primary host of *S. mansoni* is man, but nonhuman primates may act as a reservoir host. In northern Tanzania fifteen people were heavily infected with *S. mansoni* after bathing in a rock pool fed by the Ndala river. The ravine above the pool was a roosting site of baboons *(P. anubis)* with frequent defecation into the river waters. The baboons were subsequently discovered to be naturally infected by *S. mansoni* and the river hosted two species of snails including *Biomphalaria pfeifferi* (Fenwick, 1969). Another, similar incident occurred in Ethiopia where more than 50 tourists became infected after swimming in the Mui River/Ethiopia. Wild *P. anubis* and *C. aethiops* which were found infected were considered to be included in maintaining the infection, although the primary source of infection was most probably man (Fuller et al., 1979).

Geographical Distribution

According to Hillyer (1982) *S. mansoni* occurs in the following areas:

Africa:	Angola, Botswana, Burundi, Cameroon, Central African Republic, Chad, Congo (Brazzaville), Dahomey, Egypt, Ethiopia, Gambia, Ghana, Guinea, Ivory Coast, Kenya, Liberia, Libya, Malawi, Mali, Mozambique, Niger, Nigeria, Republic of South Africa, Rwanda, Senegal, Sierra Leone, Sudan, Swaziland, Tanzania, Togo, Uganda, Burkina Faso (Upper Volta), Zaire, Zambia, Zimbabwe.
Asia:	Israel, Saudi Arabia, Yemen.
South America:	Brazil, Surinam, Venezuela.
West Indies:	Antigua, Dominican Republic, Guadeloupe, Martinique, Puerto Rico, St. Lucia, St. Marten, Vieques.

Pathogenesis

The penetrating cercariae cause no major reactions in non-sensitized hosts, but the schistosomulas induce immunological reactions seen as a transient dermatitis, and during pulmonary passage as pulmonary symptoms.

The adult schistosomes are partially immunologically inert and protected against defense mechanisms by the incorporation of host proteins such as 2-macroglobulin antigenic determinants in their integument (Smithers and Terry, 1969; Damian et al., 1973). The adult schistosomes with intact integument therefore cause little damage except transient endothelial injuries by the suckers and some destruction of erythrocytes. The nutritional requirements of Schistosomes are served by the host's haemoglobin, whose globin portion the Schistosomes digest (Lawrence, 1973). Dead worms, however, can induce severe tissue reactions with necrosis of the vessel wall.

The actual lesions of schistosomiasis or bilharziasis are caused by the eggs. The eggs secrete a soluble antigen which induces delayed hypersensitivity reactions and they also secrete proteolytic enzymes which erode the interstitial tissues and cause inflammation and haemorrhage (Warren, 1973). Granulomatous responses are also caused by the cytotoxic and inflammatory effects of phospholipids contained in the miracidia within the egg shells (Smith et al., 1971). The extrusion of the eggs out of the vascular system into the intestinal lumen is therefore usually associated with haemorrhage, but some haemorrhages may also be caused by simple blockage of smaller veins. Eggs not expelled are subject to the immunological defense mechanisms in the host as first evidenced by Hoeppli-phenomenon seen around some eggs and resulting in marked chronic inflammation with the final result of fibrous scars.

Clinical Symptoms

Nonhuman primates: The clinical disease of the naturally infected chimpanzee was characterized by anemia, ascites, hepatomegaly, emaciation, anorexia and death 91 days after arrival (Renquist et al., 1975).

Experimentally infected *C. aethiops* became ill about 8 weeks after exposure with soft, bloody stools, anorexia and decreased activity (Cheever and Duvall, 1974. In experimentally infected *M. mulatta* a significant reduction of plasma cholinesterase was observed (Hass, 1976). Percutaneously infected *P. anubis* responded with a two stage eosinophilia with the first peak occurring towards the end of the prepatent period and the second rise following the onset of patency (Sturrock et al., 1977).

Man: The penetration and post-penetration stage frequently causes a "cercarial dermatitis" or "swimmers itch" within 24 hrs which may last for 2 or 3 days as an itching papular rash. In sensitized individuals systemic symptoms develop which include eosinophilia, fever, chills, sweating, diarrhea and cough.

The acute phase may resemble typhoid fever with generalized lymphadenopathy, hepatomegaly and splenomegaly and toxic symptoms which may even be fatal in heavy infections. The majority of cases, however, go unnoticed during the acute phase.

It is the chronic stage, directly related to the intensity of the infection which causes the fundamental clinical illness. Portal fibrosis, which in *S. mansoni*-infections develops more slowly than in *S. japonica*, causes hepatomegaly, occasionally haematemesis from ruptured esophageal varices, and hydrops ascites. Schistosome-eggs washed into the lungs, a relatively common occurrence in *S. mansoni*, cause obstructive lung disease and eventually *cor pulmonale*. Eggs washed into the CNS induce focal inflammatory reactions, which in *S. mansoni* infections are more likely to develop in the spinal cord than in the brain (Katz et al., 1982).

Pathology

Nonhuman primates: In the spontaneous, fatal infection of a chimpanzee the following lesions were noted: portal fibrosis and hepatomegaly due to egg-containing granulomas, haemorrhages and erosions of colonic mucosal epithelial cells, and pulmonary granulomas around Schistosome-eggs trapped in smaller arterioles (Renquist et al., 1975).

In spontaneous infections of apparently "healthy", wild living baboons the adult Schistosomes were found in the intrahepatic branches of the portal vein. Some eggs caused polymorphonuclear inflammation and granulomas of the portal fields, others did not elicit any responses (Strong et al., 1965).

In the experimentally infected baboons the colonic serosa was studded with firm nodules of 0.1-0,3 cm diameter which were most numerous in the transverse colon. The mesenteric lymph nodes and the liver became enlarged and 6-7 weeks after exposure numerous granulomas were present. Occasionally phlebitis of intestinal venules could be observed (Cheever and Powers, 1972).

Finally in the experimentally infected chimpanzees "pipe-stem fibrosis" of the liver and in one case esophageal varices were produced (von Lichtenberg and Sadun, 1968).

Man: The adult *S. mansoni* are mostly found in venules of the ileocolic and colic radicles of the superior and inferior mesenteric veins, particularly in the sigmoid branches (Spencer and Gibson, 1973). They elicit very little host response as long as they are alive and intact.

The eggs are distributed in all layers of the bowel, but are found chiefly in or close to the submucosa and so are the granulomas or focal scars. The inflammatory responses may be of a more diffuse type - the mucosa is at first congested, edematous and focally haemorrhagic, then in later stages it is chronically inflamed and thickened, occasionally with polyp-production.

A focal inflammatory reaction occurs also in the liver during the acute stage in relation to the presence of the schistosomules, but the most important hepatic lesions are caused by eggs flushed into the organ via the portal veins. The eggs trapped in smaller intrahepatic branches of the portal veins cause endophlebitis, whilst those extruded in the portal fields induce granulomas and fibrosis, which in chronic schistomiasis may appear as "clay-pipestem fibrosis". The portal tracts are markedly and irregularly thickened in "pipe stem-fibrosis", the fibrosis is not related to the bile ducts (Spencer and Gibson, 1973). Extensive portal scarring causes splenomegaly and ascites.

Treatment

Schistosomes are highly susceptible to praziquantel (Wegner, 1984). In experimentally infected baboons 9-acridanone-hydrazone compounds as single oral doses of 50 mg/kg body-weight resulted in the reduction or elimination of faecal egg production (Sturrock et al., 1985).

8.7.2 Schistosoma Haematobium

S. haematobium is predominantly an infection of the urinary tract in man.

Natural *S. haematobium* infections have occurred in a chimpanzee (de Paoli, 1965), in baboons (Purvis, 1965; Taylor et al., 1972), *Cercopithecus mitis* (Nelson, 1960; Else et al., 1982), *C. aethiops* and *Cercocebus* sp. (Kuntz, 1972).

Agent

S. haematobium resembles *S. mansoni* very closely. The males are 10 to 15 × 1 mm in size, their cuticle is covered with small tubercles and they possess 4 to 5 subglobose testes immediately behind the acetabulum.

The female's cuticle possesses minute papillae over the extremities, its ovary is located posterior to the middle of the body and the uterus is correspondingly long.

The eggs are elongated, ovoid, 112–117 × 40–70 μm in size and are characterized by a terminal spine at the posterior end (Faust et al. (1975).

Life Cycle

The life cycle of *S. haematobium* is essentially very similar to that of *S. mansoni*, except that the adult *S. haematobium* migrate from the mesenteric to the pelvic veins and the eggs are flushed out via the urinary system rather than in the faeces.

Intermediate hosts are *Bulinus* spp. *(B. africanus, B. truncatus, B. tropicus, B. forskali)*.

Geographical Distribution

The geographical range of *S. haematobium* partially corresponds to that of *S. mansoni* but includes also greater parts of North Africa and Asia, whereas it is not found in the New World. It has been reported (from Hillyer, 1982):

Africa: Algeria, Angola, Botswana, Cameroon, Central African Republic, Chad, Congo (Brazzaville), Dahomey, Egypt, Ethiopia, Gabon, Gambia, Ghana,

Guinea, Ivory Coast, Liberia, Libya, Madagascar, Malawi, Mali, Mauretania, Mauritius, Morocco, Mozambique, Niger, Nigeria, Portuguese Guinea, Republic of South Africa, Senegal, Sierra Leone, Somalia, Sudan, Swaziland, Tanzania, Togo, Tunisia, Uganda, Burkina Faso (Upper Volta), Zaire, Zambia, Zimbabwe.

Asia: India, Iran, Iraq, Israel, Lebanon, Saudi Arabia, Syria, Turkey, Yemen.

Clinical Symptoms

Nonhuman primates: In the experimentally infected chimpanzees small red papules appeared at the sites of penetration and persisted for a few days. Later in the infections haematuria was observed (Sadun et al., 1970). The experimentally infected baboons suffered from hydronephrosis due to chronic thickening of the urinary bladder (Webbe et al., 1974).

Man: Urinary tract involvement is first indicated by frequent urination with or without scalding sensation along the urethra during and after voiding. One of the most characteristic signs is haematuria.

The chronic inflammation of the urinary bladder can result in bladder cancer, whilst obstruction of the ureter leads to hydronephrosis.

Pathology

The most characteristic lesions of *S. haematobium* infections in nonhuman primates (de Paoli, 1965; Webbe et al., 1974) and man (Katz et al., 1982) are chronic granulomas and inflammation along the urinary tract, particularly in the urinary bladder, which in experimental infections of chimpanzees were present 6 months after exposure (Sadun et al., 1970). 11-24 months p.i. calcification and pseudotumors were observed in the bladder of *Hylobates lar* (Kuntz et al., 1975) and in *Cebus apella* (Cheever et al., 1976). In man the calcification of the eggs in the bladder mucosa gives a sandy appearance (von Lichtenberg et al., 1971). A late consequence of the infection, especially in Egypt, is bladder cancer.

Treatment

A 90% reduction of egg excretion of school children with urinary schistosomiasis was obtained with a single dose of 10 mg/kg of metrifonate (Tswana and Mason, 1985).

References

1. Cameron, T. W. M.: A new definitive host for *Schistosoma mansoni*. J. Helminthol. *6*, 219-222 (1928)
2. Cheever, A. W. and R. H. Duvall: Single and repeated infections of grivet monkeys with *Schistosoma mansoni*: Parasitological and pathological observations over a 31-month period. Amer. J. trop. Med. Hyg. *23*, 884-894 (1974)
3. Cheever, A. E.; R. E. Kuntz; J. A. Moore; G. T. Bryan, and R. R. Brown: Animal model: Proliferative urothelial lesions in nonhuman primates infected with *Schistosoma haematobium*. Amer. J. Pathol. *84*, 673-676 (1976)
4. Cheevers, A. W. and K. G. Powers: *Schistosoma mansoni* infection in rhesus monkeys: comparison of the course of heavy and light infections. Bull. Wld. Hlth Org. *46*, 301-310 (1972)

5. Cosgrove, G. E.: The trematodes of laboratory primates. Lab. Anim. Care *16*, 23-39 (1966)
6. Damian, R.T.; N.D. Greene, and W.J. Hubbard: Occurrence of mouse 2-macroglobulin antigenic determinants on *Schistosoma mansoni* adults, with evidence on their nature. J. Parasitol. *59*, 64-73 (1973)
7. de Paoli, A.: *Schistosoma haematobium* in the chimpanzee - a natural infection. Amer. J. trop. Med. Hyg. *14*, 561-565 (1965)
8. Else, J.G.; M. Satzger, and R.F. Sturrock: Natural infections of *Schistosoma mansoni* and *S. haematobium* in Cercopithecus monkeys in Kenya. Ann. trop. Med. Parasit. *76*, 111-112 (1982)
9. Faust, E.C.; P.C. Beaver, and R.C. Jung: Animal Agents and Vectors of Human Disease. 4th ed., Lea and Febiger, Philadelphia (1975)
10. Fenwick, A.: Baboons as reservoir hosts of *Schistosoma mansoni*. Trans. roy. Soc. trop. Med. Hyg. *63*, 557-567 (1969)
11. Fuller, G.K.; A. Lemma, and T. Haile: Schistosomiasis in Omo National Park of Southwest Ethiopia. Amer. J. trop. Med. Hyg. *28*, 526-530 (1979)
12. Goldsmid, J.M.: The intestinal helminth zoonoses of primates in Rhodesia. Ann. Soc. belge Med. trop. *54*, 87-101 (1974)
13. Graber, M.: Helminths and helminthiasis of different domestic and wild animals of Ethiopia. Bull. Anim. Hlth Prod. Afr. *23*, 57-86 (1975)
14. Hass, D.K.: Effect of *Schistosoma mansoni* on plasma cholinesterase activity in rhesus monkeys. Experentia *32*, 1390-1391 (1976)
15. Hillyer, G.V.: Schistosomiasis; pp. 177-210 in "CRC Handbook Series in Zoonoses" (Steele, J.H., Series ed.), Section C: Parasitic Zoonoses, Vol. III (Hillyer, G.V. and C.E. Hopla, Vol. eds.), CRC Press, Boca Raton (1982)
16. Hsü, S.Y. Li and H.F. Hsü: A chimpanzee naturally infected with *Schistosoma mansoni*: Its resistance against a challenge infection of *S. japonicum*. Trans. roy. Soc. trop. Med. Hyg. *62*, 901-902 (1968)
17. Hunter, G.W.; W.W. Frye, and C.J. Swartzwelder: A Manual of Tropical Medicine. 3rd ed., Saunders, Philadelphia (1964)
18. Katz, M.; D.D. Despommier, and R.W. Gwadz: Parasitic Diseases. Springer, New York (1982)
19. Kuntz, R.E.: Trematodes of the intestinal tract and biliary passages; pp. 104-123 in "Pathology of Simian Primates", Part II (Fiennes, R.N.T-W, ed.), Karger, Basel (1972)
20. Kuntz, R.E.; A.W. Cheever; B.J. Myers; S.W. Young, and J.A. Moore: Calcification of the bladder and papillary tumours of the bladder and ureters in gibbons *(Hylobates lar)* infected with *Schistosoma haematobium* (Iran). Trans. roy. Soc. trop. Med. Hyg. *69*, 494-502 (1975)
21. Kuntz, R.E.; T.C. Huang, and J.A. Moore: Patas monkey *(Erythrocebus patas)* naturally infected with *Schistosoma mansoni*. J. Parasitol. *63*, 166-167 (1977)
22. Kuntz, R.E. and B.J. Myers: Parasites of baboons (*Papio doguera* Pucheran, 1856) captured in Kenya and Tanzania, East Africa. Primates *7*, 27-32 (1966)
23. Lawrence, J.D.: The ingestion of red blood cells by *Schistosoma mansoni*. J. Parasitol. *59*, 60-63 (1973)
24. Nelson, G.S.: Schistosome infections as zoonoses in Africa. Trans. roy. Soc. trop. Med. Hyg. *54*, 301-316 (1960)
25. Nelson, G.S.: The parasitic helminths of baboons with particular reference to species transmissible to man; pp. 441-470 in "The Baboon in Medical Research", Vol. I (Vagtborg, H., ed.), Univ. Texas Press, Austin (19659
26. Nelson, G.S.: Schistosomiasis; pp. 620-640 in "Diseases Transmitted from Animals to Man". 6th ed. (Hubbert, W.T., W.F. McCulloch, and P.R. Schnurrenberger, eds.), C.C. Thomas, Springfield (1975)
27. Pitchford, R.J.; P.S. Visser; U. de V. Pieenaar, and E. Young: Observations on the ecology of *Schistosoma matthei* Veglia and Le Roux 1929 in portion of the Kruger National Park and the surrounding area. J.S. Afr. Vet. Ass. *44*, 405-420 (1973)
28. Purvis, A.J.; I.R. Ellison, and E.L. Husting: A short note on the findings of Schistosomes in baboon *(Papio rhodesiae)*. Centr. Afr. J. Med. *11*, 368 (1965)
29. Renquist, D.M.; A.J. Johnson; C.J. Lewis, and D.J. Johnson: A natural case of *Schistosoma mansoni* in the chimpanzee *(Pan troglodytes versus)*. Lab. Anim. Sci. *25*, 763-768 (1975)

30. Sadun, E.H.; F.von Lichtenberg; A.W. Cheever; D.G. Erickson, and R.L. Hickman: Experimental infection with *Schistosoma haematobium* in chimpanzees. Parasitologic, clinical, serologic, and pathological observations. Amer. J.trop. Med. Hyg. *19*, 427-458 (1970)
31. Smith, T.M.; H.L. Lucia; B.L. Doughty, and F.C. von Lichtenberg: The role of phospholipids in schistosome granulomas. J. Infect. Dis. *123*, 629-639 (1971)
32. Smithers, S.R. and J.R. Terry: The immunology of schistosomiasis. Adv. Parasitol. *7*, 41-93 (1969)
33. Spencer, H.and J.B. Gibson: Schistosomiasis; pp.561-595 in "Spezielle pathologische Anatomie" (Doerr, W., G.Seifert und E. Uehlinger, eds.), Bd.8: "Tropical Pathology". (Spencer, H., Vol. ed.), Springer, New York (1973)
34. Strong, J.P.; J.H. Miller, and H.C. McGill: Naturally occurring parasitic lesions in baboons; pp.503-512 in "The Baboon in Medical Research", Vol. I (Vagtborg, H., ed.), Univ. Texas Press, Austin (1965)
35. Sturrock, R.F.; A.E. Butterworth; S.D. Karamsadkar; R.Kimani; M.Imbuga, and V.Houba: Eosinophilia in Kenyan baboons *(Papio anubis)* induced by *Trichinella spiralis* and *Schistosoma mansoni.* Parasitology *75*, 89-100 (1977)
36. Sturrock, R.F.; M.Otieno; E.R. James, and G.Webbe: A note on the efficacy of a new class of compounds, 9-acridanone-hydrazones, against *Schistosoma mansoni* in a primate - the baboon. Trans. roy. Soc. trop. Med. Hyg. *79*, 129-131 (1985)
37. Taylor, M.G.; G.S. Nelson, and B.J. Andrews: A case of natural infection of *S.haematobium* in a Senegalese baboon (*Papio* sp.). Trans. roy. Soc. trop. Med. Hyg. *66*, 16-17 (1972)
38. Tswana, S.A. and P.R. Mason: Eighteen-month follow-up on the treatment of urinary schistosomiasis with a single dose of metrifonate. Amer. J.trop. Med. Hyg. *34*, 746-749 (1985)
39. von Lichtenberg, F.; G.M. Edington; I.Nwabuedo; J.R. Taylor, and J.H. Smith: Pathologic effects of schistosomiasis in Ibadan, Western State of Nigeria. II. Pathogenesis of lesions of the bladder and ureters. Amer. J.trop. Med. Hyg. *20*, 244-254 (1971)
40. von Lichtenberg, F.and E.H. Sadun: Experimental production of bilharzial pipestem fibrosis in the chimpanzee. Exp. Parasit. *22*, 264-278 (1968)
41. Warren, K.S.: The pathology of schistosome infections. Helminth. Abstr. Series A *42*, 591-633 (1973)
42. Webbe, G.; C.James, and G.S. Nelson: *Schistosoma haematobium* in baboon *(Papio anubis).* Ann. trop. Med. Parasitol. *68*, 187-203 (1974)
43. Wegner, D.H.G.: The profile of the trematocidal compound praziquantel. Drug Res. *34* (11), 9b, 1132-1136 (1984)
44. Wong, M.M. and H.D. Conrad: Prevalence of metazoan parasite infections in five species of Asian macaques. Lab. Anim. Sci. *28*, 412-416 (1978)

8.8 Paramphistomidae

Paramphistomidae are thick, heavily muscular flukes with a large acetabulum situated near the posterior extremity of the body.

8.8.1 Gastrodiscoides Hominis

G.hominis is a relatively common parasite of man in Southeast Asia. Pigs and deer *(Tragulus napu)* are considered to be reservoir hosts (Faust et al., 1975), but infections have also been reported in *Macaca mulatta* (Graham, 1960; Sasa et al., 1962; Reardon and Rininger, 1968; Fox and Hall, 1970; Zwicker and Carlton, 1972;

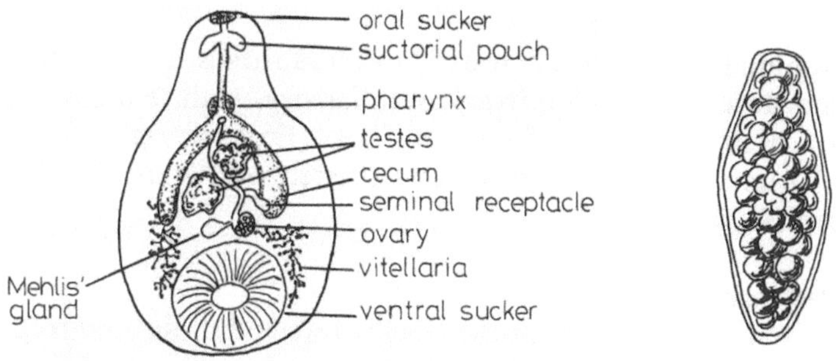

Fig. 68. *Gastrodiscoides hominis*

Prosl and Tamer, 1979; Hubbard and Butcher, 1983); *M. fascicularis* (Honjo et al., 1963; Hashimoto and Honjo, 1966; Herman, 1967; Reardon and Rininger, 1968) and orangutans (Ulmer, 1975). Nonspecified *Gastrodiscoides* have been found in 1.6% of wild living *Presbytis cristatus* (Palmieri et al., 1980).

Agent

The pyriform, reddish-orange parasites are 6-10 × 4-5 mm in size. They are characterized by a huge acetabulum with a posterior notch (Fig. 68), which occupies most of the ventral portion of the worms. The anterior end is cone-shaped and bears a small ventroterminal sucker. The lobed testes lie in tandem, with the ovary being situated between the posterior testis and the acetabulum. The vitellaria extend bilaterally from the level of the posterior testis to the level of the ventral sucker.

The operculated eggs are ovoid to spindle-shaped with bluntly rounded ends (Fig. 68) and 150-170 × 60-70 μm in size. They are unembryonated when passed in the host's faeces. They have a characteristic pale, greenish-grey color.

Life Cycle

The life cycle of *G. hominis* is largely unknown (Buckley, 1964; Spencer, 1973). The eggs hatch after 2 to 3 weeks in water (Spencer, 1973), but the intermediate hosts have as yet not been determined.

Indoplanorbis exustus has been suggested as vector (Le Roux, 1958 - cit. from Buckley, 1964), whilst experimentally *Helicorbis coenosus* was proved to be a possible vector (Dutt and Srivastava, 1966). Infection of the final host via metacercaria encysted on water plants or their fruits has been assumed (Buckley, 1964).

Geographical Distribution

G. hominis has been reported in man in Bengal, Bangladesh, Assam, "Indochina", and the Philippines (Buckley, 1964; Faust et al., 1975). It has also been found in a mouse deer in Malaya and in rats in Java (Herman, 1967). A *Gastrodiscoides* sp. was found in feral silvered leaf monkeys in Borneo/Indonesia (Palmieri et al., 1980).

Clinical Symptoms

Nonhuman primates: A heavily infected rhesus monkey became listless, anorectic and developed a bloody-mucoid diarrhea, with tenesmus and a rectal prolapse. The animal died after 5 days of illness (Fox and Hall, 1970).

Man: Heavy infections are associated with mucoid colitis and diarrhea (Spencer, 1973).

Pathology

Nonhuman primates: In the heavily infected rhesus monkey reported by Fox and Hall (1970) the mucosa was hyperemic, partially necrotic and infiltrated by neutrophils, lymphocytes, eosinophils and plasma cells. The ileocecal valve was intussuscepted into the colon. In less severe infections, the attachment of the worms to the mucosa results in short, white, papilliform mucosal projections of about 2 mm diameter and 8 mm length (Zwicker and Carlton, 1972). Histologically these plugs are infiltrated by macrophages, plasma cells and lymphocytes.

Man: The pathogenic effects in man are largely unknown. Mukerji (1945 - cit. from Buckley, 1964) considered *G. hominis* as a cause of ill health in many people, even causing death especially in children.

8.8.2 Watsonius Watsoni

W. watsoni is believed to be an indigenous parasite of monkeys with rather heavy infections having been noted in baboons (Deschiens, 1940; Pick and Deschiens, 1947) and a rhesus monkey (Anon, 1973). According to Kuntz (1972) the parasite has also been reported in *Macaca fascicularis* and *Cercopithecus aethiops.*

Occasionally human infections occur (Pick and Deschiens, 1947).

Agent

Adult *W. watsoni* are translucent, orange, pear-shaped worms of 2 to 10 mm length and 4 mm breadth. They possess a ventroterminally located ventral sucker. The genital pore opens anterior to the intestinal bifurcation (Fig. 69).

The vitellaria extend from the level of the intestinal bifurcation to the acetabulum (Ruch, 1959; Cosgrove, 1966; Kuntz, 1972).

The eggs are light yellow, ovoid, operculated and of 122–130 × 75–80 μm size.

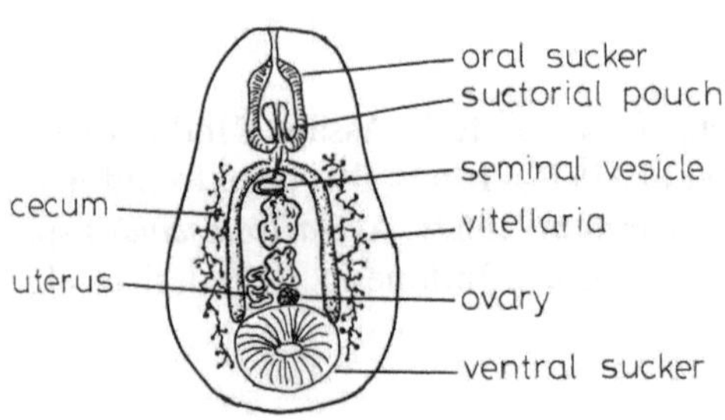

Fig. 69. *Watsonius watsoni*

Life Cycle

The life cycle of *W. watsoni* is unknown, but probably includes a snail intermediate host.

Geographical Distribution

W. watsoni is found in West- and Central Africa (Myers and Kuntz, 1965; Ulmer, 1982), but unless the Asian macaques acquired their infections in captivity from African species (which necessitates the presence of the intermediate hosts!) *W. watsoni* seems to be present in some Asian countries as well.

Clinical Symptoms

Nonhuman primates: Heavy infection caused diarrhea and severe enteritis in the baboons and a *M. mulatta* with passage of worms in the stools (Ruch, 1959; Anon, 1983).

Man: A heavy infection in a Nigerian patient was associated with severe diarrhea and death from starvation (Pick and Deschiens, 1947).

Pathology

The adult *W. watsoni* lodge in the large intestines as saprophytes, feeding on intestinal contents and fecal debris (Ulmer, 1975). The attachment of the ventral suckers to the intestinal wall, however, produces mechanical damage to the intestinal mucosa.

References

1. Anon: CDC Primate Zoonoses Surveillance Rep. No. 10, April-June 1972, p. 3, February (1973)
2. Buckley, J. J. C.: The problem of *Gastrodiscoides hominis* (Lewis and McConnell, 1876) Leiper, 1913. J. Helminthol. *36*, 1-6 (1964)
3. Cosgrove, G. E.: The trematodes of laboratory primates. Lab. Anim. Care *16*, 23-39 (1966)
4. Deschiens, R.: Relation de quatre cas d'infestation par *Watsonius watsoni* chez le papion. Bull. Soc. Pathol. Exot. *33*, 396-400 (1940)
5. Dutt, E. C. and H. D. Srivastava: The intermediate host of the cercaria of *Gastrodiscoides hominis* (Preliminary report). J. Helminthol. *40*, 45-53 (1966)
6. Fox, J. G. and W. C. Hall: Fluke *(Gastrodiscoides hominis)* infection in a rhesus monkey with related intussusception of the colon. J. Amer. Vet. med. Ass. *157*, 714-716 (1970)
7. Graham, G. L.: Parasitism in monkeys. Ann. N. Y. Acad. Sci. *85*, 842-860 (1960)
8. Hashimoto, I. and S. Honjo: Survey of helminth parasites in cynomolgus monkeys *(Macaca irus)*. Jap. J. Med. Sci. Biol. *19*, 218 (1966)
9. Herman, L. H.: *Gastrodiscoides hominis* infections in two monkeys. Vet. Med. Small. Anim. Clin. *62*, 355-356 (1967)
10. Hubbard, G. B. and W. I. Butcher: Passengers: Trematodiasis and typhlitis caused by *Gastrodiscoides hominis*. Lab. Anim. *12*, 12 and 14 (1983)
11. Kuntz, R. E.: Trematodes of the intestinal tract and biliary passages; pp. 104-123 in "Pathology of Simian Primates", Part II (Fiennes, R. N. T-W, ed.), Karger, Basel (1972)
12. Myers, B. J. and R. E. Kuntz: A checklist of parasites reported for the baboon. Primates *6*, 137-194 (1965)
13. Palmieri, J. H.; Purnomo, V. H. Lee; D. T. Dennis, and D. A. Marwoto: Parasites of the silvered leaf monkey, *Presbytis cristatus* Eschscholtz 1921, with a note on a Wucheria-like nematode. J. Parasitol. *66*, 170-171 (1980)

14. Pick, F. and R. Deschiens: La distomatose à *Watsonius watsoni* (Conyngham, 1904) Stiles et Goldberger, 1910 chez le papion. Bull. Soc. Path. Exot. *40*, 202-211 (1947)
15. Prosl, H. und A. Tamer: Zur Parasitenfauna des Rhesusaffen (Macaca mulatta) und des Javaneraffen (Macaca irus). Zbl. Vet. Med. B., *26*, 696-709 (1979)
16. Reardon, L.V. and B.F. Rininger: A survey of parasites in laboratory primates. Lab. Anim. Care *18*, 577-580 (1968)
17. Ruch, T.C.: Diseases of laboratory primates. Saunders, Philadelphia (1959)
18. Sasa, M.; H. Tanaka; M. Fukui, and A. Takata: Internal parasites of laboratory animals; pp. 195-214 in "The Problems of Laboratory Animal Disease" (Harris, R.J.C., ed.), Academic Press, New York (1962)
19. Spencer, H.: Fascioliasis, heterophyiasis and other fluke diseases; pp. 597-610 in "Spezielle pathologische Anatomie" (Doerr, W., G. Seifert, and E. Uehlinger, eds.), Bd. 8: "Tropical Pathology" (Spencer, H., ed.), Springer, New York (1973)
20. Ulmer, M.J.: Other trematode infections; pp. 646-677 in "Diseases Transmitted from Animals to Man", 6th ed. (Hubbert, W.T.; W.F. McCulloch, and P.R. Schnurrenberger, eds.), C.C. Thomas, Springfield (1975)
21. Zwicker, G.M. and W.W. Carlton: Fluke *(Gastrodiscoides hominus)* infection in a rhesus monkey. J. Amer. Vet. med. Ass. *161*, 702-703 (1972)

9 Cestodes

9.1 Intestinal Forms

9.1.1 Bertiella Studeri

B. studeri is an anoplocephalid tapeworm of primarily nonhuman primates, which infrequently infects man (Blanchard, 1913; Adams and Webb, 1933; Desowitz et al., 1961; Koivastik, 1963; D'Allessandro et al., 1963; Stunkard et al., 1964; Fogh and Seaton, 1967; Richard-Lenoble et al., 1986).

Nonhuman primate hosts include *M. mulatta* (Bezubik and Furmaga, 1959; Reardon and Rininger, 1968; Valerio et al., 1969; Prosl und Tamer, 1979; Eberhard, 1981); *M. fascicularis* (Bezubik and Furmaga, 1960; Tanaka et al., 1962); *M. fuscata* (Tanaka et al., 1962); *M. cyclopis* (Kuntz et al., 1968; Kuntz and Myers, 1969); *Papio* spp. (Round, 1964; Kuntz and Myers, 1966, 1967; Myers and Kuntz, 1967; Kuntz and Moore, 1973; McConnell et al., 1974; Goldsmid and Rogers, 1978); *Cercopithecus aethiops* (Graber, 1975); *C. mona mona* (Crockett, 1985); *Hylobates hoolock* (Stiles and Hassal, 1929 - cit. from Fiennes, 1967); *Pan troglodytes* (Myers and Kuntz, 1972; Richard-Lenoble et al., 1986) and *Pongo pygmaeus* (Stiles and Hassal, 1929 - cit. from Fiennes, 1967).

Agent
The adult worms measure 10–30 × 1 cm in size. They have a subglobose scolex, which bears a rudimentary unarmed rostellum and 4 prominent cup-shaped suckers (Round, 1964). The proglottids are much wider than they are long (8 ×). The genital pores are irregularly alternating. The segments contain only one set of male and female reproductive organs with the ovaries being multilobate and slightly poral. Gravid proglottids are filled with eggs, which are enclosed in a sac-like uterus (Turner, 1975).

The eggs are thin-shelled, of 38–45 μm diameter and contain an embryo which is surrounded by a pyriform inner shell with a bicornate protrusion.

Life Cycle
Bertiella spp. require oribatid mites as intermediate hosts. Experimentally *Scholoribates laevigatus* and *Galumna* spp. were able to transmit *B. studeri* (Stunkard, 1940; Jueco, 1982a). Gravid segments of the tapeworm and eggs excreted by the definitive host develop in the arthropods into cysticercoids within less than 2

months (Jueco, 1982). Infection of monkeys and man is by ingestion of the infective mites.

Geographical Distribution

Human infections have been reported from Mauritius, East Africa, Gabon, India, Indonesia, St. Kitts-Island, Brazil, Cuba, Argentina, Paraguay, USA, Britain (Jueco, 1982). In feral or quarantinized nonhuman primates the parasites were recovered from animals from Indonesia, China, Kenya, and Tanzania (Bezubik and Furmaga, 1959, 1960; Kuntz and Myers, 1966, 1967; Myers and Kuntz, 1967).

Clinical Symptoms

Nonhuman primates: In the animals the infection is tolerated without recognizable clinical signs.

Man: Most *B. studeri* infections occurred in children, but only rarely caused any symptoms such as anorexia, weight loss, dyspepsia, abdominal pain, and intermittent congestion and diarrhea (for references see Turner, 1975).

Pathology

B. studeri causes no apparent lesions in man or animal.

Treatment

Nemural (4–5 mg/kg) has been found effective in nonhuman primates.

9.1.2 Bertiella mucronata

B. mucronata is an anoplocephalid tapeworm of New World monkeys, being reported in *Alouatta caraya, Callicebus personatus, Cebus apella* (Dunn, 1963; Pope, 1968) but only rarely in the highly insectivorous marmosets such as *Saguinus leucopus* (Porter, 1972). The parasite has rarely been reported in man (D'Allessandro et al., 1963).

The morphology and the life cycle are very similar to that of *B. studeri*, clinical symptoms and morphological lesions are unknown.

9.1.3 Railletina Demerariensis

Members of the genus Railletina are primarily parasites of rodents, some occur also in nonhuman primates, and one species *(R. demerariensis)* is also reported in man (Jueco, 1982b). In monkeys *R. demerariensis* occurred in *Alouatta caraya* (Thatcher and Porter, 1968).

Agent

Railletina spp. are about 60 cm long and possess a very tiny, subglobular armed scolex with four suckers. The hooks are hammer-shaped and arranged in two alternating rows. The anterior proglottids are broader than long, whereas the gravid segments are longer than wide. Genital pores are unilateral and in the anterior part of the proglottids. The eggs are spindle-shaped, packed in capsules, each capsule containing up to 4 eggs (Jueco, 1982).

Life Cycle
The exact life cycle is unknown, transmission by flies, beetles etc. has been suggested (Dunn, 1963).

Geographical Distribution
R. demerariensis has been reported from Panama (Thatcher and Porter, 1968), South America and the West Indies (Flynn, 1973).

9.1.4 Hymenolepis Nana

H. nana is primarily a rodent parasite, there are also a number of human infections reported (Faust et al., 1975; Turner, 1975; Jueco, 1982c). Benson et al. (1954), Ruch (1959) and Middleton et al., 1964) referred to it in rhesus monkeys, chimpanzees and squirrel monkeys, Brack and Niemitz (1984) recovered *H. nana* from wild-caught *Tarsius bancanus.*

Agent
H. nana is a small, slender tapeworm of only 25 to 40 mm length and a maximum breadth of less than 1 mm. The globular scolex bears an armed rostellum with 20 to 30 small hooklets and 4 cup-shaped suckers (Fig. 70). The trapezoidal proglottids are generally broader than long (Fig. 70).

The eggs are greyish-hyalin, oval, 44–62 × 30–55 μm in size, they contain two membranous shells, the inner one provided with 2 poles. From each pole originate 4 to 8 long threadlike filaments which extend into the space between the two shells. The oncosphere contained in the egg possesses three pairs of small hooklets.

Life Cycle
H. nana (and the closely related *H. fraterna*) are the only directly transmitted tapeworms (Fig. 71).

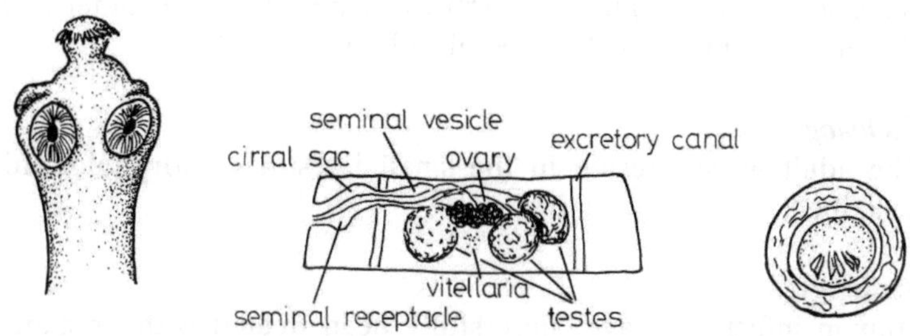

Fig. 70. *Hymenolepis nana*

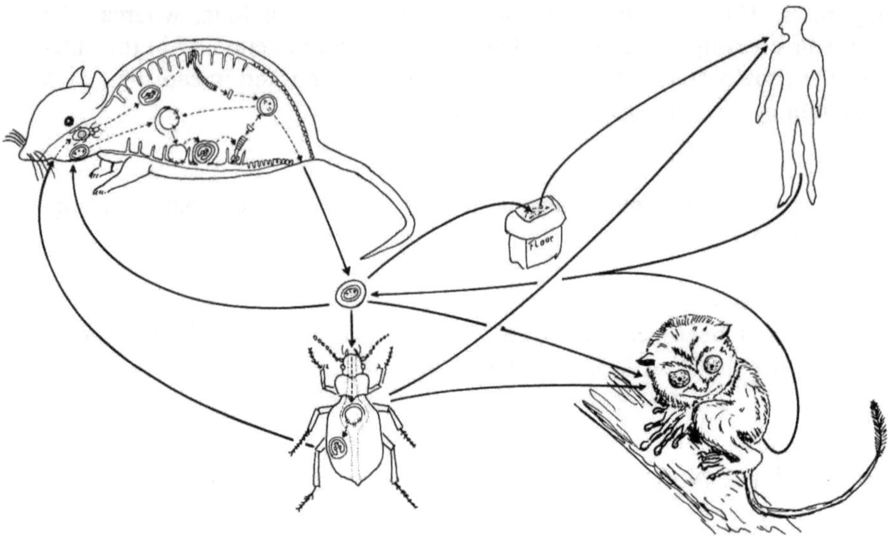

Fig. 71. Life cycle of *Hymenolepis nana*

Ingested eggs hatch in the duodenum of the definitive host, the embryos penetrate the intestinal villi, undergo larval development and become adult tapeworms in about 2 weeks time in the intestinal lumen (Loew, 1967).

Transmission via arthropods (fleas and beetles) can happen but is unusual. In such indirect transmission the eggs develop into cysticercoid larvae in the insect's body.

Geographical Distribution

H. nana is cosmopolitan in distribution. Poor hygiene and overcrowding favor the infection, in man children in warm climates are more frequently infected (Faust et al., 1975).

Clinical Symptoms

Nonhuman primates: Heavy infections may result in catarrhal enteritis (Habermann and Williams, 1958) but most infections go clinically unnoticed.

Man: Heavy infections are invariably pathogenic, causing vomiting and diarrhea, abdominal pain, anorexia and CNS-symptoms. General signs may be headache, insomnia, weight loss, pruritus especially in perianal areas, urticaria and choreiform symptoms (Faust et al., 1975).

Pathology

The adult worms reside in the small intestines, morphological lesions are not described.

Treatment

Human infections have successfully been treated with praziquantel in repeated oral doses (2 × 25–40 mg/kg in 10 days interval) (Farid et al., 1974; de Castro e de

Rezende, 1985) or in doses of 400 mg daily for 4 consecutive days (de Castro et al., 1985).

9.1.5 Hymenolepis Diminuta

H. diminuta like *H. nana* is primarily a rodent parasite which also caused numerous human infections (see Faust et al., 1975; Jueco, 1982d). The parasite has been reported in *Pan troglodytes, Papio hamadryas, Macaca fuscata, M. cyclopis* (Tscherner and Movcan, 1984), *Tupaia glis* (Schmidt and File, 1977) and *Tarsius bancanus* (Brack and Niemitz, 1984).

Agent
H. diminuta is somewhat larger than *H. nana*, measuring 20–30 × 3–4 mm. The knob-like scolex possesses four cup-like suckers and a deep apical suctorial pocket. The rostellum is rudimentary and unarmed (Fig. 72). The proglottids which number more than a thousand, are much wider than they are long and have unilateral genital pores (Fig. 72).

Usually three testes are arranged in a straight line across the segment, they are separated by the bilobed ovary. The terminal gravid proglottids disintegrate while still attached to the strobila.

Mature eggs are spherical or slightly oval, 62–88 × 52–81 μm in size with a thick straw colored outer membrane and an inner membrane, which often has two polar projections. The space between the shells is filled by a hyaline or granular albuminous substance. The oncosphere is provided with 3 pairs of hooklets (Fig. 72).

Life Cycle
H. diminuta in contrast to *H. nana* requires an insect intermediate host, such as meal worm (*Tenebrio molitor, T. obscurus*), meal moth (*Pyralis farinalis*), flea (*Xenopsylla cheopis*), earwig (*Anisolobis annulipes*) or occasionally other arthropods (Faust et al., 1975; Rau, 1979; Jueco, 1982d). Eggs ingested by the insect develop into a cysticercoid larva in their haemocoel in 18 to 20 days (Loew, 1967). After ingestion of an infective insect the worms develop in 16 to 19 days.

Geographical Distribution
H. diminuta is a cosmopolitan.

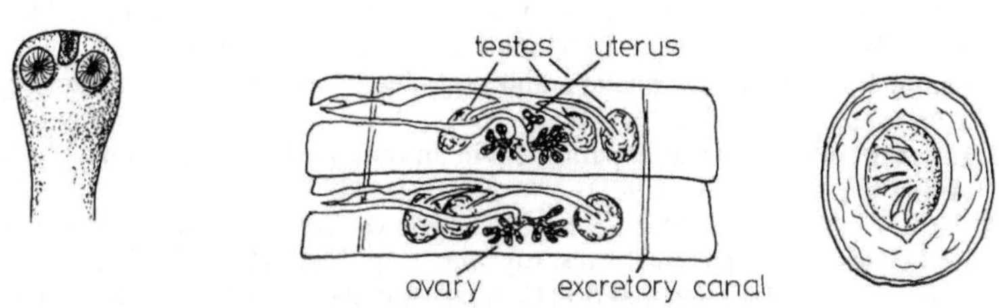

Fig. 72. *Hymenolepis diminuta*

Clinical Symptoms

Nonhuman primates: No clinical signs have been observed despite the interference of *H. diminuta* with the host's intestinal water-, glucose-, thiamine-, pyridoxine- and electrolyte transport (Pappas and Read, 1972a, b; Roberts and Mong, 1973; Mettrick and Jackson, 1979) and its inhibition of pancreatic lipase (Ruff and Read, 1973; Schroeder and Pappas, 1980).

Man: *H. diminuta* seemingly causes no clinical symptoms (Faust et al., 1975; Jueco, 1982d).

Pathology

No lesions have so far been reported in either man or nonhuman primates.

Treatment

Fenbendazole (50 mg/kg) has successfully been used in experimentally infected rats (McCracken and Taylor, 1983).

References

1. Adams, A.R.D. and L. Webb: Two further cases of human infestation with *Bertiella studeri* (Blanchard, 1891) (Stiles and Hassal, 1902) with some observations on the probable synonymy of the specimen previously recorded from man. Ann. Trop. Med. Parasitol. *27*, 471-475 (1933)
2. Benson, R.E.; B.D. Fremming, and R.J. Young: Care and management of chimpanzees at the radiobiological laboratory of the University of Texas and the United States Air Force. Proc. Anim. Care Panel *5*, 27-36 (1954)
3. Bezubik, B. and S. Furmaga: The helminth parasites in *Macacus rhesus* Audeb. from China. Acta Parasitol. Polon. *7*, 591-598 (1959)
4. Bezubik, B. and S. Furmaga: The parasites in *Macacus cynomolgus* L. from Indonesia. Acta Parasitol. Polon. *8*, 335-444 (1960)
5. Blanchard, R.: *Bertiella satyri* de l'Orang-outang est aussi parasite de l'homme. Bull. Acad. Med. *3*, 286-296 (1913)
6. Brack, M. and C. Niemitz: The parasites of wild-caught tarsiers (Tarsius bancanus); pp. 77-84 in "Biology of Tarsiers" (Niemitz, C., ed.), Gustav Fischer Verlag, Stuttgart (1984)
7. Crockett, E.C.: *Bertiella studeri* in a mona monkey *(Cercopithecus mona mona)* in Nigeria. Vet. Rec. *116*, 268 (1985)
8. D'Allessandro, B.A.; P.C. Beaver, and R.M. Pallares: *Bertiella* infection in man in Paraguay. Amer. J. trop. Med. Hyg. *12*, 193-198 (1963)
9. de Castro, M. de L.M.; J.R. Machado e Silva e I.L.M. Sampaio: Tentativa de erradicacao da himenolepiase em uma comunidade semifechada, empregando-se praziquantel. Rev. Inst. Med. trop. Sao Paulo *27*, 27-32 (1985)
10. de Castro, M. de L.M. e G.L. de Rezende: Estudo da eficacia terapeutica do praziquantel e das alteracoes morfologicas nos ovos de *Hymenolepis nana* apos sua administracao em dois esquemas posologicos. Rev. Inst. Med. trop. Sao Paulo *27*, 40-45 (1985)
11. Desowitz, R.S.; H.J. Wong, and M.A. Fernando: The first record of human infections with *Bertiella studeri* in Malaya. J. Helminthol. *35*, 207-208 (1961)
12. Dunn, F.L.: Acanthocephalans and cestodes of South American monkeys and marmosets. J. Parasitol. *49*, 717-722 (1963)
13. Eberhard, M.L.: Intestinal parasitism in an outdoor breeding colony of *Macaca mulatta.* Lab. Anim. Sci. *31*, 282-285 (1981)
14. Farid, Z.; N. Ayad El-Masry, and C.K. Wallace: Treatment of *Hymenolepis nana* with a single oral dose of praziquantel. Trans. roy. Soc. trop. Med. Hyg. *78*, 280 (1984)
15. Faust, E.C.; P.C. Beaver, and R.C. Jung: Animal Agents and Vectors of Human Disease. 4th ed. Lea and Febiger, Philadelphia (1975)

16. Fiennes, R.: Zoonoses of Primates. The Epidemiology and Ecology of Simian Diseases in Relation to Man. Weidenfeld and Nicolson, London (1967)
17. Flynn, R.J.: Parasites of Laboratory Animals. Iowa State Univ. Press, Ames (1973)
18. Fogh, S. and Seaton, D.R.: A case of infection with *Bertiella studeri* from the Yemen. Trans. roy. Soc. trop. Med. Hyg. *61*, 19 (1967)
19. Goldsmid, J.M. and S. Rogers: A parasitological study of the chacma baboon *(Papio ursinus)* from the Northern Transvaal. J. South Afr. Vet. Ass. *49*, 109-111 (1978)
20. Graber, M.: Helminths and helminthiasis of different domestic and wild animals of Ethiopia. Bull. Anim. Hlth Prod. Afr. *23*, 57-86 (1975)
21. Habermann, R.T. and F.P. Williams, jr.: The identification and control of helminths in laboratory animals. J. Nat. Cancer Inst. *29*, 979-1009 (1958)
22. Jueco, N.L.: *Bertiella* infections; pp. 215-216 in "CRC Handbook Series in Zoonoses", (Steele, J.H., Series ed.), Section C: Parasitic Zoonoses, Vol. I (Jacobs, L. and P. Arambulo, Vol. eds.), CRC Press, Boca Raton (1982a)
23. Jueco, N.L.: *Railletina* infection; pp. 289-291 in "CRC Handbook Series in Zoonoses", (Steele, J.H., Series ed.), Section C: Parasitic Zoonoses, Vol. I (Jacobs, L. and P. Arambulo, Vol. eds.), CRC Press, Boca Raton (1982b)
24. Jueco, N.L.: *Hymenolepis nana* infection; pp. 283-287 in "CRC Handbook Series in Zoonoses" (Steele, J.H., Series ed.), Section C: Parasitic Zoonoses, Vol. I (Jacobs, L. and P. Arambulo, Vol. eds.), CRC Press, Boca Raton (1982c)
25. Jueco, N.L.: *Hymenolepis diminuta* infection; pp. 279-282 in "CRC Handbook Series in Zoonoses" (Steele, J.H., Series ed.), Section C: Parasitic Zoonoses, Vol. I (Jacobs, L. and P. Arambulo, Vol. eds.), CRC Press, Boca Raton (1982d)
26. Koivastik, T.: Anoplocephalid tapeworm infection in a child. Lancet *83*, 63-65 (1963)
27. Kuntz, R.E. and J.A. Moore: Commensals and parasites of African baboons (*Papio cynocephalus* L. 1766) captured in Rift Valley Province of Central Kenya. J. med. Primatol. *2*, 236-241 (1973)
28. Kuntz, R.E. and B.J. Myers: Parasites of baboons (*Papio doguera* Pucheran, 1856) captured in Kenya and Tanzania, East Africa. Primates *7*, 27-32 (1966)
29. Kuntz, R.E. and B.J. Myers: Parasites of the Kenya baboon: Arthropods, blood protozoa and helminths (Kenya, 1966). Primates *8*, 75-82 (1967)
30. Kuntz, R.E. and B.J. Myers: A checklist of parasites and commensals reported for the Taiwan macaque (*Macaca cyclopis* Swinhoe, 1862). Primates *10*, 71-80 (1969)
31. Kuntz, R.E.; B.J. Myers; J.F. Bergner, jr., and D.E. Armstrong: Parasites and commensals of the Taiwan macaque (*Macaca cyclopis* Swinhoe, 1862). Formos. Sci. *22*, 120-136 (1968)
32. Loew, F.M.: A review of some helminths of laboratory animals. Cornell Vet. *LVIII*, 408-421 (1967)
33. McConnell, E.E.; P.A. Basson; V. de Vos; B.J. Myers, and R.E. Kuntz: A survey of diseases among 100 free-ranging baboons *(Papio ursinus)* from the Kruger National Park. Onderstepoort J. vet. Res. *41*, 97-168 (1974)
34. McCracken, R.O. and D.D. Taylor: Biochemical effects of fenbendazole on *Hymenolepis diminuta in vivo.* Internat. J. Parasitol. *13*, 267-272 (1983)
35. Mettrick, D.F. and D.J. Jackson: Vitamin absorption in the *in vivo* intestine of normal and infected (*Hymenolepis diminuta*: cestoda) rats. J. Helminthol. *53*, 213-272 (1979)
36. Middleton, C.C.; T.B. Clarkson, and F.M. Garner: Parasites of squirrel monkeys *(Saimiri sciureus).* Lab. Anim. Care *14*, 335 (1964)
37. Myers, B.J. and R.E. Kuntz: Parasites of baboons: Arthropods, blood protozoa and helminths (Kenya, 1966). Primates *8*, 75-82 (1967)
38. Myers, B.J. and R.E. Kuntz: A checklist of parasites and commensals reported for the chimpanzee *(Pan).* Primates *13*, 433-471 (1972)
39. Pappas, P.W. and C.W. Read: Thiamine uptake by *Hymenolepis diminuta.* J. Parasitol. *58*, 235-239 (1972a)
40. Pappas, P.W. and C.P. Read: The absorption of pyridoxine and riboflavin by *Hymenolepis diminuta.* J. Parasitol. *58*, 417-421 (1972b)
41. Pope, B.L.: Parasites; pp. 204-208 in "Biology of the Howler Monkey *(Alouatta caraya)*", (Malinow, M.R., ed.), Bibl. Primatol. *7*, Karger, Basel (1968)
42. Porter, J.A.: Parasites of marmosets. Lab. Anim. Sci. *22*, 503-506 (1972)

43. Prosl, H. and A. Tamer: Zur Parasitenfauna des Rhesusaffen (Macaca mulatta) und des Javaneraffen (Macaca irus). Zbl. Vet. Med. B *26*, 696–709 (1979)
44. Rau, M. E.: The frequency distribution of *Hymenolepis diminuta* cysticercoids in natural, sympatic population of *Tenebrio molitor* and *T. obscurus*. Internat. J. Parasitol. *9*, 85–87 (1979)
45. Reardon, L. V. and B. F. Rininger: A survey of parasites in laboratory primates. Lab. Anim. Care *18*, 577–580 (1968)
46. Richard-Lenoble, D.; M. Kombila; M. L. Maganga, and G. Affre: *Bertiella* infection in a Gabon-born girl. Amer. J. trop. Med. Hyg. *35*, 134 (1986)
47. Roberts, L. S. and F. N. Mong: Developmental physiology of cestodes. XIII. Vitamin B6 requirement of *Hymenolepis diminuta* during in vitro cultivation. J. Parasitol. *59*, 101–104 (1973)
48. Round, M. C.: A note on a species of *Bertiella* from the dog-faced baboon *(Papio doguera)* in Kenya. J. Helminthol. *38*, 77–82 (1964)
49. Ruch, T. C.: Diseases of Laboratory Primates. W. B. Saunders, Philadelphia (1959)
50. Ruff, M. D. and C. P. Read: Inhibition of pancreatic lipase by *Hymenolepis diminuta*. J. Parasitol. *59*, 105–111 (1973)
51. Schmidt, G. D. and S. File: *Tupaiataenia quentini* gen. et sp. n. (Anoplocephalidae: Linstowiinae) and other tapeworms from the common tree shrew, *Tupaia glis*. J. Parasitol. *63*, 473–475 (1977)
52. Schroeder, L. L. and P. W. Pappas: Trypsin absorption by *Hymenolepis diminuta*. J. Parasitol. *66*, 49–52 (1980)
53. Stunkard, H. W.: The morphology and life-history of the cestode, *Bertiella studeri*. Amer. J. trop. Med. *20*, 305–333 (1940)
54. Stunkard, H. W.; T. Koivastik, and G. R. Healy: Infection of a child in Minnesota by *Bertiella studeri* (Cestoda: Anoplocephalidae). Amer. J. trop. Med. *13*, 402–409 (1964)
55. Tanaka, H.; M. Fukui; H. Yamamoto; S. Hayama, and S. Kodera: Studies on the identification of common intestinal parasites of primates. Bull. Exp. Anim. *11*, 111–116 (1962)
56. Thatcher, V. E. and J. A. Porter, jr.: Some helminth parasites of Panamanian primates. Trans. Amer. Microscop. Soc. *87*, 186–196 (1968)
57. Tscherner, W. and A. T. Movcan: Zum Bandwurm-Befall der Affen; pp. 413–416 in 26. Internat. Symp. Erkr. Zootiere, Brno (1984)
58. Turner, J. A.: Other cestode infections; pp. 708–744 in "Diseases Transmitted from Animals to Man." 6th ed. (Hubbert, W. T.; W. F. McCulloch, and P. R. Schnurrenberger, eds.), C. C. Thomas, Springfield (1975)
59. Valerio, D. A.; R. L. Miller; J. B. M. Innes; D. A. Courtney; A. J. Pallotta, and R. M. Guttmacher: Macaca mulatta. Management of a Laboratory Breeding Colony. Academic Press, New York (1969)

9.2 Cestodes: Tissue Forms

Man and nonhuman primates can serve as intermediate hosts to several tapeworms of other animals or man. The type of larval stage depends on the type of tapeworm, according to Toft and Ekstrom (1980) they can be classified as

a) solid larvae = Sparganum (Pseudophyllidea)
b) bladder larvae (Cyclophyllidea) as Cysticercus (Taenia spp.), Coenurus (Multiceps spp.), Hydatid (Echinococcus spp.), Tetrathyridium (Mesocestoides spp.).

9.2.1 Sparganum

Spargana are the plerocercoid stages of Diphyllobotriidae (*Spirometra* spp., *Diphyllobotrium* spp.). Spargana have been reported in a wide variety of mammals, reptiles, amphibians, and birds (Daly, 1982). There are numerous reports on human infections (Daly, 1982) and on Spargana in nonhuman primates. The simian host species reported are baboons (Myers and Kuntz, 1967; Kuntz et al., 1970); *Cercopithecus aethiops* (Morton, 1969; Kuntz et al., 1970); *Macaca fascicularis, M. nemestrina, M. arctoides* (Wong and Conrad, 1978), *Callithrix argentatus melanurus* (Dunn, 1963), *Saguinus fuscicollis* (Cosgrove et al., 1968; Porter, 1972), *Saguinus geoffroyi* (Thatcher and Porter, 1968), *S. illegeri, S. mystax, S. oedipus, C. goeldii* (Porter, 1972), *Tupaia glis* (Schmidt and File, 1977).

Agent

The spargana of all pseudophyllideans look alike and can only be differentiated by feeding them to a definitive host. Spargana are flat, slender, white worms of up to several cm in length, 1 to 3 mm width and a ribbon-like structure. The scolex may be subglobose or bulbous with a cup or slot shaped groove on the crown (Fig. 73). Sometimes the bothria of the adult worm are indicated by invaginations. In "proliferating" spargana the larvae reproduce in the intermediate host by uncontrolled division with metastases in virtually every tissue. The "proliferating" larvae lack scolices.

Life Cycle (Fig. 74)

The adult diphyllobothrids live in the intestines of a canine or feline carnivore. They produce operculated eggs which are excreted with the host's faeces. For further development the eggs have to reach water, where the ciliated coracidia hatch. Free swimming coracidia must be eaten by a copepod crustacean (*Cyclops* sp.) where they invade the haemocel of the copepod and develop into vermiform propercoids. If the infective copepod is swallowed during drinking the procercoid penetrates the intestinal villi of the second intermediate host and is carried to connective, muscular, or adipose tissues to transform into the long slender sparganum.

If the tissue containing such spargana is eaten by a definitive host the cycle is completed, but if it is eaten by another paratenic host the plerocercoid penetrates the intestine and establishes itself in the tissues of the new intermediate host.

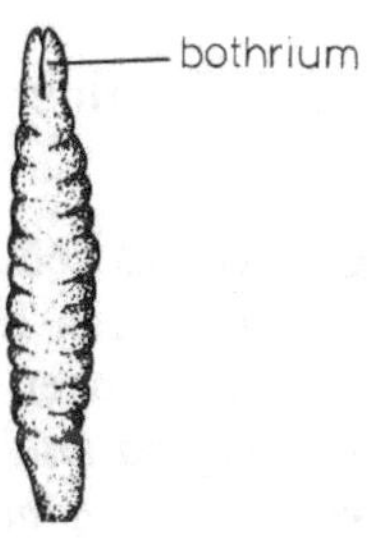

Fig. 73. Sparganum

Fig. 74. Life cycle of *Spirometra sp.*

Geographical Distribution
Spirometra spp. and spargana are world-wide distributed.

Clinical Symptoms
Nonhuman primates: Spargana rarely cause any discomfort unless they occur in large numbers or impair the animal's movement or sensual functions. Usually the spargana are incidental findings at necropsy.

Man: Human sparganosis usually occurs as a painful, inflamed, sometimes migrating nodule in the subcutaneous tissues (Daly, 1982).

Pathology
Nonhuman primates: In experimental infections of rhesus monkeys extensive hemorrhage and tissue damage in the intestinal mucosa and subperitoneal haemorrhages were seen 3 weeks p.i. (Opuni and Muller, 1975).

In spontaneous infections spargana were found in the abdominal cavity and muscle tissues (Myers and Kuntz, 1967; Morton, 1969) or in the subcutis (Thatcher and Porter, 1968).

Man: Most spargana are situated in the subcutaneous tissues but nevertheless occasionally have been found in the internal organs, the eyes or floating free in the abdominal cavity (Daly, 1982).

9.2.2 Cysticercus Cellulosae

Cysticercus cellulosae is the larval stage of *Taenia solium*, one of the human tapeworm species. The principal intermediate hosts are pigs, but occasionally the cysticerci have been found in *Macaca mulatta* (Vickers and Penner, 1968), *Hylobates lar* (Sagartz and Tingpalapong, 1974), baboons (Flynn, 1973), *Ateles geoffroyi, Cercopithecus cephus, C. aethiops* and *Macaca cyclopis* (Pawlowski, 1982).

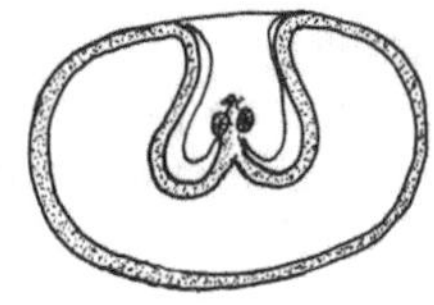

Fig. 75. Cysticercus cellulosae

Agent
The *T. solium* cysticerci appear as cysts, 5 to 20 mm in diameter. They are filled with a thin, clear fluid and contain an invaginated scolex (Fig. 75). The *T. solium* cysticercus develops in a cavity lined by epitheloid cells which are derived from a lymphatic vascular wall.

Adult worms are usually up to 3 mm long. The scolex bears a double crown of 22 to 32 rostellar hooks. 150 to 200 testes are scattered throughout the proglottid. The ovary is composed of two large lateral lobes and a median accessory lobe (Olsen, 1974).

Life Cycle
Adult *T. solium* produce 250.000 eggs daily in the gut of an infected person, which are expelled with the host's faeces. When swallowed by swine or another suitable mammalian intermediate host including man himself, the successive action of gastric and intestinal juices causes the breakdown of the embryophore and liberation of the oncosphere. The oncospheres penetrate the duodenal wall, enter the hepatic portal circulation and are carried throughout all of the body. Upon reaching a suitable site the circulating oncospheres leave the blood vessels and develop in 2 to 3 months into cysticerci. The most common sites of development are the muscular tissues, whereas another favorite site is the brain.

After eating uncooked infected pork (or other intermediate host's flesh) the cycle is completed with the development of the adult worms in a new human being.

Geographical Distribution
T. solium is worldwide distributed.

Clinical Symptoms
The cysticerci cause little effect in nonhuman primates (Myers, 1972), whereas in man they may cause serious disease depending on the tissues involved and the number of cysticerci present. Cysticerci in the subcutaneous and muscular tissues may cause pain and inconveniences if they encroach on nerve endings or become calcified. Cysticerci in the eye may result in uveitis, iritis, and atrophy of the choroid membranes. Cysticercosis of the meninges most frequently causes epilepsy, that of the 4th ventricle results in hydrocephalus, headaches, diplopia, dizziness etc.

Adult *T. solium* infections of man are usually symptomless, but sometimes there may be diarrhea, abdominal cramps and weight loss (Faust et al., 1975).

Pathology
Nonhuman primates: In the rhesus monkeys and the gibbon the cysticerci were found particularly in the brain (Vickers and Penner, 1968; Sagartz and Tingpalapong, 1974), but caused cellular responses (gliosis, eosinophils) in only one animal. The cysticerci dwelling in the heart, muscles or subcutis are usually well tolerated by the host.

9.2.3 Coenurus

Coenuri are the larval stages of several *Multiceps* species, those of nonhuman primates usually of *M. serialis.* The adult cestodes are parasites of canides and felides, whereas the coenuri develop principally in sheep, goats, lagomorphs or other herbivores. Coenuri have also been described in nonhuman primates, especially in *Theropithecus gelada* (Clark, 1969; Kaufmann, 1973; Leith and Scatterfield, 1974; Graber, 1975; Schürer, 1979). Single reports of coenuri in *Macaca radiata* (Wong and Conrad, 1978), *Cercopithecus mona mona* (Movcan et al., 1977 - cit. from Tscherner and Movcan, 1984) and *Hylobates lar* (Lau et al., 1973) have been reported. About 51 human cases of coenurosis have been reported (Arambulo, 1982).

Agent
Coenuri are glistening white polycephalid vesicular larval stages of several centimeters in diameter, which usually develop in the connective tissues or brains of ruminants or occasionally other mammals. The coenuri are characterized by secondary buds which project from the mother cyst wall to the inside or outside, each one becoming a scolex (Fig. 76). The adults living in dogs or other carnivorous animals are up to 70 cm long and 3 to 5 mm wide. Their scolex is armed with a rostellum which bears a double row of hooks. The mature proglottids are wider than long.

Life Cycle
Infection of the intermediate host is by ingestion of eggs passed in the faeces of an infected carnivorous animal, whereas infection of the definitive host is by ingestion of tissues containing coenuri.

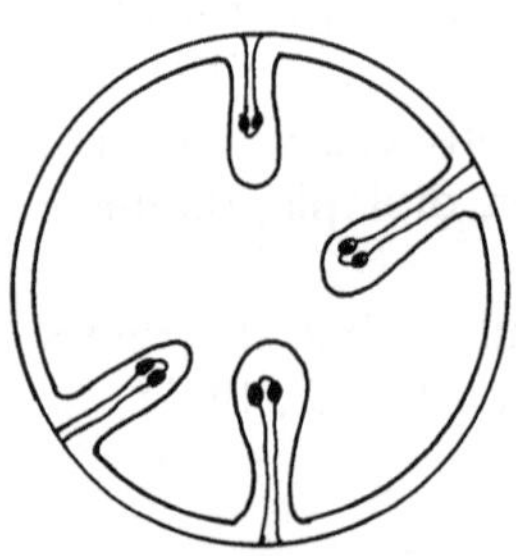

Fig. 76. Coenurus

Geographical Distribution
Multiceps spp. and coenuri are worldwide distributed.

Clinical Symptoms
Nonhuman primates: The symptoms vary according to the site of infection. In subcutaneous coenurosis slowly enlarging tumorlike, fluctuant masses are observed (Clark, 1969; Leith and Scatterfield, 1974), infections of the arm caused self mutilation (Kaufman, 1973). Retrobulbar coenurosis causes progressing proptosis of the infected eye, papilledema and dilated retinal veins (Lau et al., 1973).

Pathology
Nonhuman primates: The coenuri appear as cysts or multicystic masses, which destroy the surrounding tissues by pressure atrophy. The cyst wall may additionally be infiltrated by lymphocytes and eosinophils (Lau et al., 1973).

9.2.4 Hydatid

Hydatid cysts are the larval stages of *Echinoccus* spp., parasites of wild and domestic canides. The hydatid cysts occur in a wide variety of domestic and wild mammals as well as in man. As far as nonhuman primates are concerned, hydatids have been reported in *M. mulatta* (Allen, 1957; Healy and Hayes, 1963; Ilievski and Esber, 1969; Houser and Paik, 1971; Palotay and Uno, 1975; Wong and Conrad, 1978; Parker et al., 1979); *M. fascicularis, M. radiata* (Wong and Conrad, 1978); *M. sinica* (Dissanaike, 1958 - cit. from Healy and Hayes, 1963); *Erythrocebus patas* (Fowler et al., 1974); *Mandrillus sphinx* (Boever and Britt, 1975); *Mandrillus leucophaeus* (Powers et al., 1966); *Gorilla gorilla gorilla* (O'Grady et al., 1982); *Pongo pygmaeus* (Bernstein, 1972); *Lemur catta* (Palotay and Uno, 1975); *Lemur mongoz* (Ratcliffe, 1942 - cit. from Allen, 1957), and *Galago crassicaudatus* (Palotay and Uno, 1975).

Agent
The appearance of the hydatid cyst depends on the *Echinococcus*-species as well as on the site of infection. *E. granulosus*-cysts are unilocular (although commonly multiple), *E. granulosus* growing in bony tissues (osseous hydatid) grows rather diffusely, *E. multilocularis*, the causative agent of alveolar hydatidisis, multiplies by exogenous budding, thus producing large clusters of cysts.

E. granulosus cysts are of varying size and shape depending on the age of the cyst and its size. Each cyst consists of an external laminated cuticula; an interior germinative membrane which buds off the daughter cysts; a milky or clear hydatid fluid; new budding daughter cysts lined by the germinative membrane and the free daughter cysts (Fig. 77).

E. multilocularis in contrast proliferate by exogenous budding leading to large aggregates of small cysts (Hunter et al., 1964).

The adult *Echinococcus*-tapeworms, which are attached to the intestinal mucosa of carnivorous animals are very small worms, rarely exceeding 10 mm in

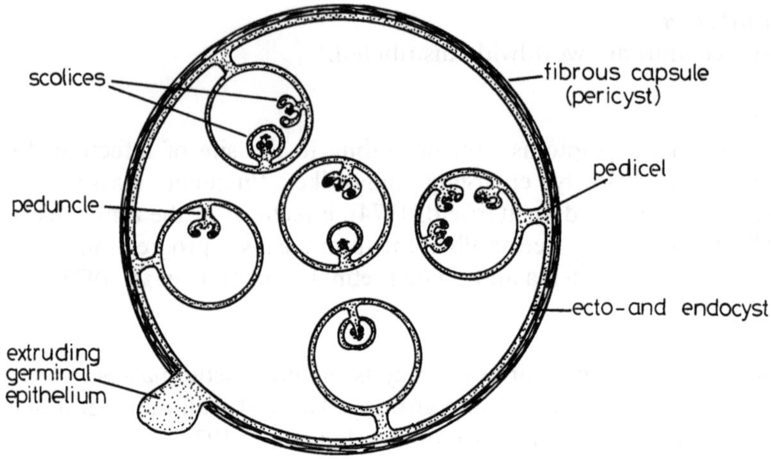

Fig. 77. Hydatid cyst

length. They consist of an armed scolex, which is provided with 4 suckers and a double row of rostellar hooklets. The scolex is followed by a short neck, one immature and one or two mature proglottids. Usually there are only a few adult worms present, but occasionally there may be hundreds or thousands attached to the small intestines of a single host (Brack und Lamina, 1968).

Life Cycle

The adult *Echinococcus*-tapeworms live in the small intestines of dogs, foxes, wolves etc. Gravid proglottids disintegrate in the small intestine so that only eggs appear in the faeces. When mature eggs are swallowed by a suitable intermediate host they hatch in the intestine. The hexacanth embryos penetrate the intestinal mucosa and migrate via the blood circulation and lymphatic vascular system to the liver and other viscera, particularly the lung. Here they develop rather slowly into unilocular *(E. granulosus)*, or by exogenous budding into alveolar *(E. multilocularis)* hydatids. The life cycle is completed if such an infected animal is eaten by a predator or if the organs discarded at meat-processing are fed raw to a dog.

Geographical Distribution

E. granulosus has a world wide distribution with the highest prevalence in a belt ranging from the Mediterranian through the Southern USSR into India and Western China, while *E. multilocularis* is restricted to the northern hemisphere, apparently due to its special host requirements.

Clinical Symptoms

Nonhuman primates: The clinical signs may be indicative of a tumor or tuberculosis, the actual symptoms depending on the organs involved. Ecchinococcosis of the liver results in abdominal distention, whilst high alkaline phosphatase and LDH-activity are indicative of the loss of liver function (Boever and Britt, 1975). Osseous (spinal) hydatidosis of a patas monkey was characterized by ataxia, handstand walking and pelvic limb paralysis (Fowler et al., 1974). Pulmonary hydatidosis may lead to sudden death following excretion, possibly from anaphylactic shock resulting from the rupture of a cyst (Allen, 1957).

Man: The disease in man depends on the organ involved, the type of echinococcosis and the size of the cyst. Cyst rupture, often a result of trauma, may cause mild to severe or even fatal anaphylactoid reactions. Symptoms are more pronounced in *E. multilocularis* because of the continuing growth of the parasites with invasion and destruction of the organ (Schantz, 1982).

Pathology

Nonhuman primates: The cysts are most frequently found in the abdominal and pelvic cavity, where they adhere to the mesentery or the omentum. The most commonly affected organs are the liver and the lung. The cysts ranged in size from 0.5 to 15 cm in diameter in the reported cases.

The growing cysts cause cavitation of adjacent parenchyma by compression atrophy. In hepatic hydatidosis connective tissue originating from the cyst wall proliferates into the liver lobule, causing cirrhosis in advanced stages.

Ruptured cysts are surrounded and infiltrated by histiocytes, lymphocytes and multinucleated giant cells.

Man: Most hydatids in man develop in the liver, as it is the first capillary filter. In heavily endemic areas in South America pulmonary echinococcosis is prevalent (Schenone et al., 1971 - cit. from Faust et al., 1975).

9.2.5 Tetrathyridium

Tetrathyridia are the second intermediate stages (= third larval stages) of *Mesocestoides* spp., whose adult forms occur in carnivores and occasionally in man (Chandler, 1942; Fain and Herin, 1959; Yamaguchi, 1981). Tetrathyridia have been found in a variety of amphibia, reptilia, mammals and birds. In nonhuman primates, it occurs in *Macaca fascicularis* (Reardon and Rininger, 1968); *M. sylvana* (Priemer, 1980); *Cercopithecus aethiops* (Priemer, 1980); *C. mitis* (Ezzat and Gaafar, 1951); *C. nictitans* (Priemer, 1980) and *Papio* sp. (Reid and Reardon, 1976).

Agent

Tetrathyridia resemble spargana, in that they have a long, slender body with an invaginated scolex. This in comparison, is provided with 4 suckers at the anterior end (in spargana bothrial slits!). The body of the tetrathyridia is unsegmented, and it varies in size from a few millimeters to several centimeters.

The adult tapeworms are 30 to 250 cm long. Their scolex bears 4 suckers and is unarmed. In the proglottids the sexual pore is near the midline.

Life Cycle

The life cycle requires 2 intermediate hosts. The first intermediate hosts are arthropods (oribatid mites in *M. corti* - Priemer, 1980), the second mammals, snakes etc.

Tetrathyridia may even develop in their final host species (Priemer, 1980). The tetrathyridia may asexually multiply in the abdominal cavity of mice (Specht and Voge, 1965). The natural pathway from the gut lumen to the liver is by direct penetration (Specht and Widmer, 1972).

Geographical Distribution
Mesocestoides sp. have been reported from USA, Europe and Asia (Flynn, 1973).

Clinical Symptoms
Nonhuman primates: No clinical signs have been reported in tetrathyriasis.

Man: Case reports indicate that abdominal pain and severe diarrhea may accompany the infection with adult *Mesocestoides* (Turner, 1975).

Pathology
The tetrathyridia are usually found in the abdominal organs. They induce compression atrophy of the adjacent parenchyma and the formation of a connective tissue capsule around with a minimum of inflammation.

References

1. Allen, A. M.: Pulmonary hydatid disease in a rhesus monkey. Arch. Pathol. *64*, 148-151 (1957)
2. Arambulo, P. III: The cestode zoonoses; pp. 209-214 in "CRC Handbook Series in Zoonoses" (Steele, J. H., Series ed.), Section C: Parasitic Zoonoses, Vol. 1 (Jacobs, L. and P. Arambulo, Vol. eds.), CRC Press, Boca Raton (1982)
3. Bernstein, J. J.: An epizootic of hydatid disease in captive apes. J. Zoo Anim. Med. *3*, 16-19 (1972)
4. Boever, W. J. and J. Britt: Hydatid disease in a mandrill baboon. J. Amer. vet. Med. Ass. *167*, 619-621 (1975)
5. Brack, M. und J. Lamina: Massenbefall mit *Echinococcus granulosus* bei einem Hunde (Kurzmitteilung). Dtsch. Tierärztl. Wschr. *75*, 95-96 (1968)
6. Chandler, A. C.: First record of a case of human infection with tapeworms of the genus Mesocestoides. Amer. J. trop. Med. *22*, 493-496 (1942)
7. Clark, J. D.: Coenurosis in a gelada baboon *(Theropithecus gelada)*. J. Amer. vet. Med. Ass. *155*, 1258-1263 (1969)
8. Cosgrove, G. E.; B. Nelson, and N. Gengozian: Helminth parasites of the tamarin, *Saguinus fuscicollis.* Lab. Anim. Care *18*, 654-656 (1968)
9. Crosby, W. R.; M. H. Ivey; W. L. Shaffer, and D. D. Holmes: Echinococcus cysts in the Savannah baboon. Lab. Anim. Care *18*, 395-397 (1968)
10. Daly, J. J.: Sparganosis; pp. 293-312 in "CRC Handbook Series in Zoonoses" (Steele, J. H., Series ed.), Section C: Parasitic Zoonoses, Vol. I (Jacobs, L. and P. Arambulo, Vol. eds.), CRC Press, Boca Raton (1982)
11. Dunn, F. L.: Acanthocephalans and cestodes of South American monkeys and marmosets. J. Parasitol. *49*, 717-722 (1963)
12. Ezzat, M. A. and S. M. Gaafar: *Tetrathyridium* sp. in a Syke's monkey *(Cercopithecus albigorus)* from Giza Zoological Gardens. Egypt. J. Parasitol. *37*, 392-394 (1951)
13. Fain, A. and V. Herin: Notes à propos d'un cas d'infestation humaine par un Mesocestoide à Astrida (Ruanda-Urundi). Présence de ce cestode chez le chat et découverte de larves du type *Tetrathyridium* chez divers vertebres. Ann. Soc. Belge Med. Trop. *34*, 893-900 (1954)
14. Flynn, R. J.: Parasites of Laboratory Animals. Iowa State Univ. Press, Ames (1973)
15. Fowler, M. E.; S. Silverman; T. A. Schultz, and N. Baker: Osseous hydatidosis in a patas monkey. J. Amer. vet. med. Ass. *165*, 840-844 (1974)

16. Graber, M.: Helminths and helminthiasis of different domestic and wild animals of Ethiopia. Bull. Anim. Hlth Prod. Afr. *23*, 57-86 (1975)
17. Healy, G.R. and N.R. Hayes: Hydatid diseases in rhesus monkeys. J.Parasitol. *49*, 837 (1963)
18. Houser, W.D. and S.K. Paik: Hydatid disease in a macaque. J.Amer. vet. med. Ass. *159*, 1574-1577 (1971)
19. Hunter, G.W.; W.W. Frye, and J.C. Swartzwelder: A Manual of Tropical Medicine. 3rd ed. Saunders, Philadelphia (1964)
20. Ilievski, V.and H.Esber: Hydatid disease in a rhesus monkey. Lab. Anim. Care *19*, 199-204 (1969)
21. Kaufmann, P.: Coenurosis in gelada baboons. Pp.231-233 in "American Association of Zoo Veterinarians: Annual Proceedings". Privately publ., Topeka, Kansas (1973)
22. Kuntz, R.E.; B.J. Myers, and A.A. Katzberg: Sparganosis and "proliferative" spargana in vervets *(Cercopithecus aethiops)* and baboons (*Papio* sp.) from East Africa. J.Parasitol. *56*, 196-197 (1970)
23. Lau, D.T.; W.J. Casey, and D.M. Jones: Coenurosis in a whitehanded gibbon. J.Amer. vet. med. Ass. *163*, 633-635 (1973)
24. Leith, J.D. and W.C. Scatterfield: Coenurus cysts of *Taenia multiceps* in a baboon, *Theropithecus gelada.* J.Zoo Anim. Med. *5*, 32-34 (1974)
25. Morton, H.L.: Sparganosis in African green monkeys *(Cercopithecus aethiops).* Lab. Anim. Care *19*, 253-255 (1969)
26. Myers, B.J.: Echinococcosis, coenurosis, cysticercosis, sparganosis, etc.; pp.124-143 in "Pathology of Simian Primates", Part II (Fiennes, R.N.T-W, ed.), Karger, Basel (1972)
27. Myers, B.J. and R.E. Kuntz: Parasites of baboons taken by the Cambridge Mwanza expedition (Tanzania, 1965). East Afr. Med. J. *44*, 322-324 (1967)
28. Myers, B.J.; R.E. Kuntz, and T.W. Vice: Hydatid disease in captive primates. J.Parasitol. *51*, 1019-1020 (1965)
29. O'Grady, J.P.; C.H. Yeager; G.N. Esra, and W.Thomas: Ultrasonic evaluation of echinococcosis in four lowland gorillas. J. Amer. vet. med. Ass. *181*, 1348-1350 (1982)
30. Olsen, O.W.: Animal Parasites. Their Life Cycles and Ecology. Univ. Park Press, Baltimore (1974)
31. Opuni, E.K. and R.L. Muller: Studies on *Spirometra theileri* (Baer, 1925) n.comb.2. Pathology of experimental plerocercoid infections. J.Helminthol. *49*, 121-127 (1975)
32. Palotay, J.L. and H.Uno: Hydatid disease in four nonhuman primates. J.Amer. vet. med. Ass. *167*, 615-618 (1975)
33. Parker, G.A.; C.J. Gilmore, and C.R. Roberts: Diagnostic exercise. Lab. Anim. Sci. *29*, 457-458 (1979)
34. Pawlowski, Z.S.: Taeniasis and cysticercosis; pp.313-348 in "CRC Handbook Series in Zoonoses" (Steele, J.H., Series ed.), Section C: Parasitic Zoonoses, Vol.1 (Jacobs, L.and P.Arambulo, Vol. eds.), CRC Press, Boca Raton (1982)
35. Porter, J.A.: Parasites of marmosets. Lab. Anim. Sci. *22*, 503-506 (1972)
36. Powers, R.D.; R.A. Price; R.P. Houk, and R.H. Mattlin: Echinococcosis in a drill baboon. J.Amer. vet. med. Ass. *149*, 902-925 (1966)
37. Priemer, J.: Funde von Bandwurmlarven der Gattung *Mesocestoides* (Tetrathyridien) im Tierpark Berlin. Milu *1/2*, 252-260 (1980)
38. Reardon, L.V. and B.F. Rininger: A survey of parasites in laboratory primates. Lab. Anim. Care *18*, 577-580 (1968)
39. Reid, W.A. and M.J. Reardon: *Mesocestoides* in the baboon and its development in laboratory animals. J.med. Primatol. *5*, 345-352 (1976)
40. Sagartz, J.W. and M.Tingpalapong: Cerebral cysticercosis in a white handed gibbon. J.Amer. vet. med. Ass. *165*, 844-845 (1974)
41. Schantz, P.M.: Echinococcosis; pp.231-277 in "CRC Handbook Series in Zoonoses" (Steele, J.H., Series ed.), Section C: Parasitic Zoonoses, Vol.1 (Jacobs, L.and P.Arambulo, Vol. eds.), CRC Press, Boca Raton (1982)
42. Schmidt, G.D. and S.File: *Tupaiataenia quentini* gen. et sp. n. (Anoplocephalidae: Linstowiinae) and other tapeworms from common tree shrews, *Tupaia glis.* J.Parasitol. *63*, 473-475 (1977)

43. Schürer, U.: Bandwurmbefall *(Multiceps multiceps)* bei einem Blutbrustpavian *(Theropithecus gelada).* Zool. Garten *49*, 80-81 (1979)
44. Schwartz, B.: Occurrence of larval tapeworms belonging to the genus *Tetrathyridium* in a baboon *(Papio procarius).* J. Parasitol. *15*, 67 (1928)
45. Specht, D. and M. Voge: Asexual multiplication of *Mesocestoides* tetrathyridia in laboratory animals. J. Parasitol. *51*, 268-272 (1965)
46. Specht, D. and E. A. Widmer: Response of mouse liver to infection with tetrathyridia of *Mesocestoides* (Cestoda). J. Parasitol. *58*, 431-437 (1972)
47. Thatcher, V. E. and J. A. Porter, jr.: Some helminth parasites of Panamanian primates. Trans. Amer. Microscop. Soc. *87*, 186-196 (1968)
48. Toft, J. D. II and M. E. Ekstrom: Identification of metazoan parasites in tissue sections; pp. 369-378 in "The Comparative Pathology of Zoo Animals" (Montali, R. J. and G. Migaki, eds.), Smithson. Inst. Press, Washington (1980)
49. Tscherner, W. und A. T. Movcan: Zum Bandwurm-Befall der Affen. 26. Internat. Symp. Erkr. Zootiere, Brno, 1984; pp. 413-418, Akademie Verlag, Berlin (1984)
50. Turner, J. A.: Other cestode infections; pp. 708-744 in "Diseases Transmitted from Animals to Man.". 6th ed. (Hubbert, W. T.; W. F. McCulloch, and P. R. Schnurrenberger, eds.), C. C. Thomas, Springfield (1975)
51. Vickers, J. H. and L. R. Penner: Cysticercosis in four rhesus brains. J. Amer. vet. med. Ass. *153*, 868-871 (1968)
52. Wong, M. M. and H. D. Conrad: Prevalence of metazoan parasite infections in five species of Asian macaques. Lab. Anim. Sci. *28*, 412-416 (1978)
53. Yamaguchi, T.: Color Atlas of Clinical Parasitology. Lea and Febiger, Philadelphia (1981)

9.3 Acanthocephala

Acanthocephalans form a unique phylum due to their structures and habits. They are unsegmented. They possess a resistant cuticle with an underlying syncytial hypodermis and a muscle layer like the nematodes, but they lack a mesothelium and a digestive tract like the cestodes. Instead they absorb nutriments through their body surface, with the absorptive surface being increased by extention of the cuticle over the trunk spines (Crompton and Lee, 1965). The cuticle is penetrated by pores.

All acanthocephalans possess a spinous retractile proboscis which serves for attachment to the host's intestinal wall. They all are bisexual with the males being considerably smaller than the females. Acanthocephalans are parasites particularly of fish and birds. A few are dangerous parasites of monkeys, one species *(Moniliformis moniliformis)* has also been reported occasionally in man.

9.3.1 Moniliformis Moniliformis (M. Dubius)

M. moniliformis is primarily a parasite of wild rodents, especially rats, but has also been found in dogs and cats (Noble and Noble, 1976). The parasite has also been occasionally reported in primates: *Pan troglodytes* (Baylis, 1929 - cit. from Ruch, 1959; Yamashita, 1963); *Aotus trivirgatus* (Thatcher and Porter, 1968); *Tarsius bancanus* (Brack and Niemitz, 1984) and man (Beck, 1959 - cit. from Faust et al., 1975; Moayedi et al., 1971).

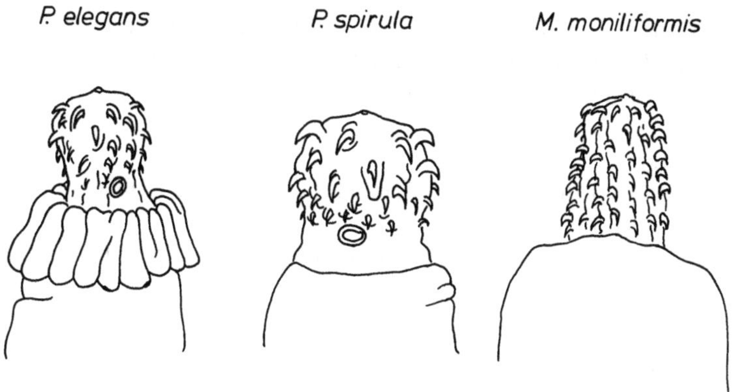

Fig. 78. Acanthocephalans of nonhuman primates: proboscises

Agent

The adult worms live in the small intestine or proximal colon of rats, etc. The males range in size from 40 to 145 μm, the females from 100 to 300 mm. The creamy-white worms resemble a beaded chain because of their annular thickenings. The proboscis is cylindrical or claviform and armed with 12-14 irregular longitudinal rows of 9-12 hooks (Fig. 78).

The thick-shelled eggs are 85-125 × 50-62 μm in size. Their outer shell is lightly sculptured with ridges and grooves (Schmidt, 1972).

Life Cycle and Vectors

All acanthocephalans require intermediate hosts. Those of aquatic animals being usually molluscs, whilst those of terrestrial animals are certain arthropods. The vectors of *M. moniliformis* are cockroaches *(Blatta orientalis, Blatella germanica, Periplaneta americana)* or beetles *(Blaps gigas, Tenebrio molitor).*

Female worms produce eggs that contain mature larvae (acanthors). The eggs are expelled with the host's faeces and once ingested by a cockroach or beetle they hatch in the arthropod's gut. The acanthors burrow through the intestinal wall to the haemocel, transform to the second stage larvae (acanthella), become encysted and develop to cystacanths. The development of the cystacanths induces behavioral alterations of the cockroaches (lower phototaxis, slower and prolonged movement) thus increasing the chance of being caught by a nocturnal animal (Moore, 1983; Wilson and Edwards, 1986). After ingestion of an infected arthropod host, the cystacanths are liberated. They then attach to the definitive host's intestine and become adults.

Geographical Distribution

M. moniliformis is a cosmopolitan.

Clinical Symptoms

Nonhuman primates: *M. moniliformis* is probably less pathogenic than the dangerous *Prosthenorchis* spp. and such infections are probably self healing like in other mammalian hosts.

Man: A four-months old boy suffered from episodes of diarrhea, undernutrition, underdevelopment, coughing, sweating and anemia with full recovery following loss of the worms (Moayedi et al., 1971).

Pathology

Nonhuman primates: In the tarsius the parasite possibly induced focal erosions of the mucosal membranes.

Man: No information on the kinds of lesions produced could be found.

References

1. Beck, J.W.: Report of a possible infection with the acanthocephalan *Moniliformis moniliformis* (syn. *M. dubius*). J. Parasitol. *45*, 510 (1959)
2. Brack, M. and C. Niemitz: The parasites of wild-caught tarsiers *(Tarsius bancanus)*; pp. 77–84 in "Biology of Tarsiers" (Niemitz, C., ed.). Gustav Fischer, Stuttgart (1984)
3. Cromptom, D.W.T. and D.L. Lee: The fine structure of the body wall of *Polymorphus minutus* (Goetze, 1782) (Acanthocephala). Parasitology *55*, 357–364 (1965)
4. Faust, E.C.; P.C. Beaver, and R.C. Jung: Animal Agents and Vectors of Human Disease. 4th ed., Lea and Febiger (1975)
5. Moayedi, B.; M. Izadi; M. Maleki, and E. Ghadirian: Human infection with *Moniliformis moniliformis* (Bremser, 1811) Travassos, 1915 (syn. *Moniliformis dubius*). Report of a case in Isfahan, Iran. Amer. J. trop. med. Hyg. *20*, 445–448 (1971)
6. Moore, J.: Altered behavior in cockroaches *(Periplaneta americana)* infected with an archiacanthocephalan, *Moniliformis moniliformis*. J. Parasitol. *69*, 1174–1176 (1983)
7. Noble, E.R. and G.A. Noble: Parasitology. The Biology of Animal Parasites. 4th ed., Lea and Febiger (1976)
8. Ruch, T.C.: Diseases of Laboratory Primates. Saunders, Philadelphia (1959)
9. Schmidt, G.D.: Acanthocephala of captive primates; pp. 144–156 in "Pathology of Simian Primates", Part II (Fiennes, R.N.T-W, ed.), Karger, Basel (1972)
10. Thatcher, V.E. and J.A. Porter, jr.: Some helminth parasites of Panamanian primates. Trans. Amer. Microscop. Soc. *87*, 186–196 (1968)
11. Wilson, K. and J. Edwards: The effects of parasitic infection on the behaviour of an intermediate host, the American cockroach, *Periplaneta americana*, infected with the acanthocephalan, *Moniliformis moniliformis*. Anim. Behav. *34*, 942–944 (1986)
12. Yamashita, J.: Ecological relationships between parasites and primates. I. Helminth parasites and primates. Primates *4*, 1–96 (1963)

10 Appendix

Appendix. *Y. enterocolitica* in Nonhuman Primates

Primate species	*Y. enterocolitica* serotype	Reference
Pan troglodytes	3	Szita et al. (1980)
Hylobates sp.	10	Szita et al. (1980)
Cercopithecus aethiops (2 ×)		McClure et al. (1971)
Cercopithecus cephus	3	Kageruka et al. (1972)
Cercopithecus ascanius	3	van Damme et al. (1978)
Cercopithecus petaurista	3	Poelma et al. (1977)
Erythrocebus patas (30 ×)	12	Otsuki et al. (1973)
Erythrocebus patas	5	
Erythrocebus patas	6	
Erythrocebus patas	14	
Erythrocebus patas		Skavlen et al. (1983)
Cercocebus torquatus		McClure et al. (1071)
Cercocebus aterrimus	3	van Damme et al. (1978)
Macaca nemestrina	3	Poelma et al. (1977)
Galago senegalensis	3	Poelma et al. (1977)
Galago sp.	5	Mair et al. (1970)
Varecia (Lemur) variegata rubra	2	Bresnahan et al. (1984)
Perodicticus potto	5; 5, 27	Poelma et al. (1977)
Perodicticus potto	3	Pattyn et al. (1978)
Aotus trivirgatus (29 ×)		Baggs et al. (1976)
Lagothrix lagotricha (2 ×)	15; 5, 27: a, b, c	Poelma et al. (1977)
Ateles paniscus chamek	3	Poelma et al. (1977)
Saguinus oedipus	9	Poelma et al. (1977)
Callithrix jacchus (2 ×)	5, 27	Poelma et al. (1977)
Callithrix spp. (6 ×)		Mollaret et al. (1970)

11 Subject Index

MIX
Papier aus verantwortungsvollen Quellen
Paper from responsible sources
FSC® C105338

If you have any concerns about our products,
you can contact us on
ProductSafety@springernature.com

In case Publisher is established outside the EU,
the EU authorized representative is:
Springer Nature Customer Service Center GmbH
Europaplatz 3, 69115 Heidelberg, Germany

Printed by Libri Plureos GmbH
in Hamburg, Germany